Stars Above, Earth Below

A Guide to Astronomy in the National Parks

Tyler Nordgren

Stars Above, Earth Below

A Guide to Astronomy in the National Parks

Dr Tyler Nordgren
University of Redlands Astronomer
Redlands
California
USA

SPRINGER–PRAXIS BOOKS IN POPULAR ASTRONOMY
SUBJECT *ADVISORY EDITOR*: John Mason, M.B.E., B.Sc., M.Sc., Ph.D.

ISBN 978-1-4419-1648-8 Springer Berlin Heidelberg New York

Springer is a part of Springer Science + Business Media (*springer.com*)

Library of Congress Control Number: 2009941872

Cover design: Jim Wilkie
Project copy editor: Dr John Mason
Typesetting: BookEns, Royston, Herts., UK

Printed in Germany on acid-free paper

Contents

	Preface	xxiii
	Acknowledgments	xxv
	How to use this book	xxvii
1	**Come see the Milky Way** *Our place in the Galaxy* (Yosemite, Big Bend, and all National Parks and Monuments)	1
2	**Black hole Sun** *Solar eclipses, black holes, and gravity* (Grand Teton and Chaco Culture National Historical Park)	35
3	**On the shores of the cosmic sea** *Tides on Earth, moons, stars, and galaxies* (Acadia and all coastal National Parks)	69
4	**Worlds of fire and ice** *Volcanoes in the Solar System* (Yellowstone and all volcanic National Parks)	105
5	**Red rock planet** *Mars* (Arches and all National Parks of the Four Corners Region)	147
6	**Glaciers and Goldilocks: a tale of three planets** *Climate change on Venus, Earth and Mars* (Glacier and all glacial National Parks)	187
7	**Autumn Moon** *The Moon and meteor showers* (Great Smoky Mountains, Acadia, and all eastern National Parks)	227
8	**Our cosmic connection** *Stars and archaeoastronomy* (Chaco Culture National Historical Park)	271

9 Worlds without number 317
The search for extrasolar planets
(Rocky Mountain, and all National Parks of the mountain west)

10 Far away and long ago: the Universe before you 357
Cosmology
(Grand Canyon National Park)

11 Starry sky national park 397
Light pollution and our night sky heritage
(Bryce Canyon, Natural Bridges, and all National Parks and Monuments)

Index 437

Preface

In the spring of 2005 I went camping in Yosemite National Park. That night, after a day of amazing hiking, I took part in a time-honored tradition of the parks and went to see the evening ranger program. As luck would have it, it was an astronomy talk and what I saw that night would lead me, over the next four years, to work with Park Rangers from Denali in Alaska to Acadia in Maine, and Big Bend on the Rio Grande to Glacier on the Canadian border.

I'm an astronomer – I love astronomy – and as I sat in the outdoor amphitheater that night I looked around and saw that I was far from alone. There must have been 60 or 70 people there in the audience with me and every one of them was absolutely mesmerized by the ranger talking about what we could see in the dark skies above America's national parks. In fact, we learned he was part of a small group of Park Rangers actively engaged in measuring just how dark those skies were so that parks like the one we were in could help preserve the views at night that were such an integral part of 'sleeping under the stars.'

Sitting there in the audience, I realized that the starry oases of the national parks were a natural outlet for professional astronomers, like myself, to talk to the public about our latest discoveries. For virtually everyone in the audience that night the Milky Way is something no longer visible from home. A truly dark sky with our Galaxy arching overhead is as strange and exotic as the glaciers, waterfalls, and geysers that draw visitors from all over the world to America's parks. And everyone there was dying to know more about what they could see what they'd been missing. I knew that night that the parks were the place to answer their questions.

After the talk I introduced myself to 'Dark Ranger' Kevin Poe, who himself was visiting from Bryce Canyon National Park in the heart of Utah's truly dark sky country and began the process of learning more about this small group of dedicated rangers and astronomers, a group I am now honored to be a part of: The National Park Service's Night Sky Team.

With their help, I began a project in the summer of 2007 to visit a dozen national parks over the course of 14 months working with the public and Park Rangers to educate both about all the ways that the parks are a natural setting to learn more about our Universe. In visitor centers, park lodges, and campground amphitheaters I talked about how the stars we see overhead are our window into the larger Universe around us. But I also helped show people how astronomy doesn't end with the dawn. Beneath us is the Earth, a planet with formations and features similar to features and formations on other planets. While by night we can see Mars shining blood-red in a pristine sky, by day we can walk amongst red rock features that proclaim our connection between the Red Planet and ourselves. And astronomy isn't new, there've been astronomers on Earth, and in the parks, observing the Sun, Moon, and stars for as long as there have been people.

This book is the result of that year in the parks. It's written for anyone who simply wants to enjoy the night sky, know what they're looking at, and know what it means. Every chapter was written specifically about what you could see for yourself when visiting the parks. So this book is yours: a personal guide to the stars above, and the Earth below, in America's national parks.

Tyler Nordgren
February 2010

Acknowledgments

Four years of planning and working on the travels and writing that led to this book could not be done alone. For that reason there is a long (and most likely still incomplete) list of people I want to thank and without whom you would not be holding this book. Louis Friedman and The Planetary Society supplied invaluable assistance in getting me out to the public in the parks. To be supported by the society Carl Sagan co-founded to promote the public's involvement in space exploration is an enormous honor. I would also like to thank Evelyn and Bill Rowland whose gracious gift to the University of Redlands helped me reach out to the public with our mutual love of astronomy. To keep this list from becoming longer than the book let me briefly offer my thanks, in the approximate order of the chapters with which they helped: Dustin Leavitt, Nick Myrman, Yvonne Flack, Susan Stolovy, the Rangers of Big Bend National Park, Bill Huntly, Eanna Flanagan, the Rangers of Acadia National Park, Rick Greenberg, Jim Halfpenny, Terry Hurford, John Spencer, the Rangers of Yellowstone National Park, the staff of Lowell Observatory, Jim Bell, Jeff Moersch, Marjorie Chan, Sierra Coon, Nancy Holman, the Rangers of Arches National Park, Megan Chaisson, Dan Fagre, Don Banfield, Mark Wagner, Matt Graves, the Rangers of Glacier National Park, Jay Melosh, Nick Schneider, the Rangers of Great Smoky Mountains National Park, Kim Malville, Russ Bodnar, GB Cornucopia, the Rangers of Chaco Culture National Historical Park, Bill Cochran, Jeff VanCleve, Jeff Maugans, Steve and Irene Little, the Rangers of Rocky Mountain National Park, Danny Dale, Chuck Wahler, David Smith, Judy Hellmich-Bryan, the Rangers of Grand Canyon National Park, Story Musgrave, Kevin Poe, Corky Hayes, the Rangers of Natural Bridges National Monument, Chad Moore, Dan Duriscoe, Peter Lord, Richard Blake, Chris Luginbuhl, the Rangers of Bryce Canyon National Park, David Batch and Abrams Planetarium, Peter Lipscomb and Gene Nordgren, my dad. In addition, there are four people who were with me from the beginning, every word of the way: Elizabeth Dodd, Angie Richman, Amy Sayle, and Julie Rathbun; I couldn't have done it without you four.

Lastly, I would like to dedicate this book to the all the Rangers in America's Parks.

You are our ambassadors to the world.

How to use this book

This book is a personal guide, showing you the reader, the astronomical phenomena that you can see for yourself when visiting the U.S. National Parks. Each chapter ties a specific astronomical phenomenon to a particular National Park or type of park. While it was written with the thought that you would read this all the way through, I know that when I'm visiting someplace special I like to read something specific to that location. As a result, the chapters can be read in any order. For those of you who would like to skip around based upon what park you are at, or a specific astronomical object you'd like to see, the Table of Contents includes what parks are associated with each chapter and what topic of astronomy it includes.

Since the guiding principle for this book is what you can see, each chapter concludes with a 'See for yourself' section that shows you how to see the planets, stars, nebulae, moons, etc. that are described within that chapter. They are activities that almost always require nothing but your own eyes, maybe a pair of binoculars, and rarely a small telescope.

For those chapters where it helps to have a star map to locate these objects, two have been provided; each map is separated by six months so that no matter what time of year you may be out, you will have a decent chance of being able to use one of the maps. The sky maps progress in order of the calendar through the book, beginning and ending with February and August in chapters 1 and 10, so that the entire year is covered. If you do choose to read the book in that order, you can see the slow progression of the sky and what follows what.

Each star map shows the sky as it looks during the evening about an hour or two after sunset when the last of evening twilight has faded. To orient the map with the sky, hold the book over your head with the top of the page pointed north. You will notice that while the positions of 'East' and 'West' on the maps appear backwards when looking down on the book, when held over your head with the top of the page pointed north, they now accurately reflect the correct directions. Anything along the perimeter of the sky map will therefore be found along the correct horizon, while the center of the map represents what is directly overhead. Depending on exactly what time of night, or what time of year you are out, the stars you see may be a little to the east, or a little to the west of what is shown in the star map; in each case though, the patterns of the constellations are the same and once you find one, all the rest should snap into place.

List of Star Maps (by month)

January - Chapter 9, page 354
February - Chapter 1, page 31 and Chapter 10, page 392
March - Chapter 3, page 101
April - Chapter 4, page 143
May - Chapter 7, page 269
June - Chapter 8, page 313
July - Chapter 9, page 355
August - Chapter 1, page 30 and Chapter 10, page 393
September - Chapter 3, page 100
October - Chapter 4, page 142
November - Chapter 7, page 268
December - Chapter 8, page 312

A Few Useful Terms

Zenith - The point in the sky directly overhead.

Arcsecond - An arcsecond is an angle. When we look at the sky we appear to be looking at the inside of a giant celestial sphere. Distances between two objects (or the apparent size of an object) on this sphere is measured in angles. From the horizon to the zenith is 90 degrees. One degree is divided into 60 arcminutes, while one arcminute is divided into sixty arcseconds. For reference, the Moon is 30 arcminutes in diameter.

Star (or stellar) magnitude - Numbers are used to describe a star's brightness: the smaller the number, the brighter the star. A 1st magnitude star is 2.5 times brighter than a 2nd magnitude star. A 3rd magnitude star is 2.5 times brighter than a 4th magnitude star. The brightest stars in the sky are typically 1st magnitude. Planets that are even brighter have negative values (Venus can be -4th magnitude). The faintest stars typically visible with the naked eye are 6th to 7th magnitude.

1 Come see the Milky Way

The clearest way into the Universe is through a forest wilderness.
John Muir

Come see the Milky Way. Come to the mountains, come to the forests, come to the national parks and see for yourself our Galaxy's power to inspire. Its glistening band across a jet black sky is the surest sight that we are part of a vast and complex Universe awaiting our discovery. And while each year there are fewer places where we can still hope to see the Milky Way, tonight, at least, I know I will be in one of them.

Pebbles and grit crunch beneath my boots on the trail that carries me up into a clear blue sky over Yosemite National Park. Above me it's a deep aquamarine, almost the color of lapis. As I lower my eyes I run through a virtual color wheel of blues until at the horizon I reach pastel. Blue all the way down. Tonight is definitely going to be a good night to the see the Milky Way.

Figure 1.1 Clear blue summer skies along the Panorama Trail above Yosemite Valley and Half-Dome (T. Nordgren).

Located in the Sierra Nevada mountains of northern California, Yosemite is only a seven-hour drive from my home at the southern end of the state. Back there in Los Angeles the blue rarely makes it to the horizon. There is no color of lapis. What blue there is, maybe a light powder blue at best, has given way to white, then grey and finally brown long before reaching down to the mountains and palm trees. Every grain of dust, every particle of smog, and (for those with more humid weather than we Californians) every drop of water vapor the atmosphere holds, scatters and reflects the light from the Sun until the very stuff we breathe becomes an opaque curtain hanging between us and the real sky beyond. At night the Sun's light is replaced by artificial city lights that shine upwards, their light scattering off the air itself to bounce around amid those particles of haze and dust. Ultimately, enough of that light is redirected back down to us on Earth that the sky itself appears to glow (and in fact it does) drowning out the much more distant glow of the stars.

Will that be a problem in Yosemite tonight? While this pollution by artificial light respects no city limits, the nearest cities are still fairly remote. No, the largest effect on what I'll see tonight will be the clarity of the air itself. As a simple rule of thumb, look to the horizon; it's there that we always look through the greatest amount of air towards the Universe beyond. If it's blue by day, you will be able to see faint stars by night.

That's why the clean air of the mountains, which rejuvenates me during the day, is what brings me here to stay at night. When the Sun sets and the first pin-pricks of light break through the twilight glow, I know that there is going to be as little between me and the stars as our atmosphere will allow. That's a reason why the astronomers put their observatories on distant plateaus and mountain tops. Yes, tonight is going to be a good night.

This night, beside my tent, far from any city, I will get a chance to see the sky as very few see it now outside one of America's national parks. In the last few generations humanity has created an artificial daylight that blinds and blankets the stars in the same way the smog and haze of LA sometimes blow in to quietly engulf the beauty of the Grand Canyon in Arizona. On those days when the North Rim disappears, visitors to the South Rim know exactly what they are missing. Sadly, there are very few today who remember what the sky is supposed to look like at night, and those who do remember have grown used to the idea that it's just the way things are now. The U.S. National Park Service's mission to preserve our natural beauty on the ground has become the greatest means for its preservation overhead as well.

The Sun is setting. Soon it is gone and along the eastern horizon the dusky red Belt of Venus rises. In some quirk of nomenclature it has nothing to do with the planet Venus, but rather it's the pinkish light of sunset reflected back off the gasses in the sky. Below it and above the horizon is the dark gray-blue of the Earth's shadow cast on its own atmosphere. As the Sun sinks lower below the western horizon the belt rises higher in the east while its colors dim as the light of sunset fades. Slowly the sky darkens. Ultramarine changes to slate changes to black. The first stars appear overhead. Any 'stars' visible at this time close to the eastern or western horizons are almost always planets as only they are bright enough to be seen there while the sky is still light.

The stars overhead shine all day long with the same intensity. Our ability to see them depends only on their contrast with the surrounding sky. During the day they are lost in the Sun's glare and the blue light scattered by our atmosphere, but as the daylight diminishes they begin to come out as darkness falls. First, only the brightest stars can be seen against the background sky and so

Figure 1.2 The starry night sky over Yosemite Valley as seen from Glacier Point. The National Park Service Night Sky Team is a small group of park rangers and astronomers working to measure and monitor the darkness of the sky above the national parks. The lights of campgrounds and lodges in the valley floor light the landscape (D. Duriscoe/ NPS Night Sky Team).

only the well-known constellations are visible. But then the fainter stars appear, stars of third, fourth, and fifth magnitude.[1] Eventually the familiar constellations fill in with stars not seen on the simple charts sold in the park's gift shop. Now that the constellations have become busy I begin to see a faint fuzziness overhead in Cygnus, the Swan, a wisp of 'cloud' where before the sky was clear. But this 'cloud' doesn't move, instead it grows lengthwise from northeast to southwest. Low on the horizon above Sagittarius another wisp appears, growing upward towards Cygnus to join it in a single pale band. The luminous haze seems

[1] One way that stars can be classified is based upon how bright they are. The magnitude scale in use since the ancient Greeks identifies the brightest stars as having a magnitude of one. Stars two-and-a-half times fainter than these are of second magnitude, while those an additional two-and-a-half times fainter still are of magnitude three. Thus, the higher the magnitude the fainter the star. If this sounds backwards, think about which is easier: identifying the very brightest star in the sky and calling it first magnitude, or finding the very faintest star visible and calling it first. Modern astronomy has expanded this scale to include negative numbers and zero for the very brightest objects and decimal values for more precise measurement. With patience and training the human eye can see as faint as magnitude 7.7 or 7.8 under the very darkest conditions.

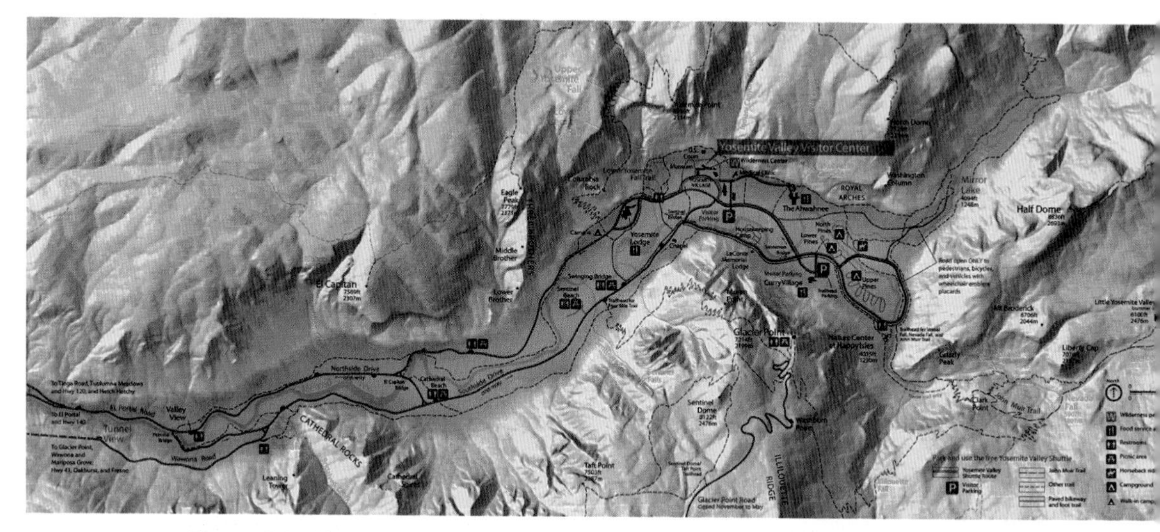

Figure 1.3 There are many places within the main valley of Yosemite National Park where you can stop and see the sky. Enquire at the Yosemite Valley Visitor Center to find out what evening astronomy programs are being offered (National Park Service).

brighter the darker the sky becomes. Twilight's gone at last and now the park's other marvel is on display. This is the Milky Way.

If like me, you're outside in a park after the stars come out, lie down on a picnic table bench or spread out in the grass. Let the sky fill your entire field of view. Typically by the time the Sun has been down for an hour-and-a-half your eyes and the sky cooperate and the landscape hidden by the light of Sun and civilization is fully revealed.

In the northern hemisphere, summer's the best time to be doing this. The Milky Way is at its highest and brightest then and temperatures are just right to let folks stay out late without discomfort. In summer, the southern horizon will be dominated by two giant constellations: Scorpius and Sagittarius. Those of us born in the late fall and winter recognize these as constellations having something to do with our horoscopes. As the Earth moves around the Sun, the constellation the Sun appears in front of (and thus the constellation that's not visible at night) slowly changes with the seasons. From November through early January it's Scorpius and Sagittarius that the Sun obliterates in the daylight's glare.[2] Six months earlier, however, the Sun is half way around the sky and so is down while these constellations are up. For an amateur stargazer they're wonderful constellations because they actually look like something recognizable. Unlike one of those other constellations where four faint stars in a 'Y' are

[2] From November 30th to December 17th, the Sun actually passes through a corner of the constellation of Ophiuchus located between Scorpius and Sagittarius. Strangely there are no Ophiuchan horoscopes.

Figure 1.4 The Milky Way in Scorpius and Sagittarius. The bright orange star to the right is Antares. Other prominent naked eye nebulae and clusters are shown (T. Nordgren).

supposed to be Cancer the Crab, Scorpius really looks like a scorpion. The brightest star in Scorpius, and thus the first star I see, is the dying and bloated, bright orange-red star Antares: the Rival of Mars, so named for its similarity in brightness and color to the Red Planet, Mars.

East of Scorpius is the giant teapot asterism at the heart of Sagittarius.[3] The teapot pours downwards onto the tail of Scorpius while magnificent clouds of celestial 'steam' billow between them before rising high into the sky. This 'steam' is the Milky Way and it's the reason I eagerly look for these constellations each summer. The ribbons of vapor rise northward from the teapot on the southern horizon all the way to the zenith where it forks in two and then fades away to the far northeastern horizon. If you're lucky enough to be in the southern half of the country the first puffs from the spout are the brightest part. As you travel north into Oregon or Pennsylvania, Sagittarius and Scorpius begin to disappear into the thick air of the southern horizon. The great mass of air through which we look scatters the faint starlight (and is also responsible for dimming the fierce glare of

[3] An asterism is a collection of stars made up from parts of a single constellation or from stars from different constellations. The Big Dipper, for instance, is an asterism of the far larger and less obvious constellation of Ursa Major, the Great Bear.

Figure 1.5 Native American pictograph showing the Milky Way as a celestial net of milkweed (T. Nordgren).

the Sun at sunset) and this part of the Milky Way loses much of its glory. In Hawaii, by contrast, this part of the sky is even higher and summer stargazers on the beach are rewarded with more than just a hint of its true magnificence.

Before there were artificial lights at night to draw away our attention and drown out the faintness of the evening sky, the Milky Way was a nightly occurrence with which everyone was familiar. There have been almost as many stories about what the Milky Way is as there have been civilizations on this planet to see it. E. C. Krupp of Griffith Observatory has documented many such legends and myths, including those of quite a few Native American tribes. One local tribe is the Luiseño people of the Los Angeles basin.

In a shallow canyon bottom, now surrounded by expensive suburban homes, there is still a cool creek flowing between green trees. Above this creek is a jumble of large boulders now isolated from the rest of the local geography. On the shadowed underside of one large boulder, where one is still afforded a view of the eastern sky, there is painted a curious netlike pattern. This pattern is reminiscent of a net woven from milkweed fibers that is featured in a Luiseño story related by Krupp in his compilation of star myths and legends *Beyond the Blue Horizon*:

> They stretched the net out upon the ground, put the Sun in the middle of it, and with magic chants and gestures bounced him into the heaven. At first the Sun went to the north, but everyone agreed that was not right. They put the Sun back down on the net and tried again. This time the Sun went south, but he came back again. So they made another attempt to launch him into the proper orbit. This time he went a little bit to the west and returned once more. Another snap of the net sent the Sun into the sky again, and this time they got him in the east where he belonged. With a little more effort, their songs put the Sun on a yearly course that never followed a straight line but carried him south and north in different seasons.

The interesting connection between this story and what we see above is that over the course of the year the noontime Sun really does move north and south in the sky. In summer months, especially in southern California, the Sun at noon is almost directly overhead while in the winter months the midday Sun is far to the south and shadows are longer. In addition, the slow path the Sun takes against the background constellations crosses the Milky Way from our perspective in two places. The first is in the vicinity of Sagittarius and occurs during the Sun's southernmost point in its motion, while six months later the Sun is at its northernmost point and crosses the Milky Way again near the constellation of Gemini. Between these dates, year after year, one can imagine the Sun bouncing back and forth from season to season between the Milky Way's net.

Given all the myriad explanations each culture has had for the Milky Way, the amazing fact remains that for all but the last 400 years of human history there was no evidence for any one explanation over any other. Was the Milky Way a net? Was it a seam in the fabric of the heavens? Was it literally the milk from the goddess Hera's breast (hence the name *milky* way)? While some ideas may have been more plausible than others, especially to European thinkers in the midst of

Figure 1.6 During summer weekends local amateur astronomy clubs set up telescopes on the heights of Glacier Point above the Yosemite Valley. They come each year to show visitors the beauty of a truly dark sky. On moonless nights the glories of the Milky Way are revealed stretching from horizon to horizon and arching high overhead (T. Nordgren).

the Renaissance, no evidence supported any one guess over any other. Imagine then the joy of discovery when for the first time since the very first human being looked up with wonder into a dark starry sky, Galileo Galilei pointed a telescope at the Milky Way and finally saw what it really was: the combined light of innumerable stars unknown and invisible to everyone's sight but his. These stars are simply so far away that to the naked eye they blur into one diaphanous glow like individual aspen trees melding into one continuous golden drapery on a distant hillside in fall.

Look up; let nothing impede your view of the sky. To my eye, the Milky Way's bright clouds appear to hang in front of the distant darkness as my brain interprets increasing faintness as increasing distance away from us. In my mind a two-dimensional sky becomes filled with three-dimensional depths. From high overhead down to the southern horizon the Milky Way's band is laid out like a strange and exotic mountain range seen from above. Bright regions become, to my mind's eye, mountains of pearly fluorescence that melt into dimmer plateaus and darker valley floors. Strange dark holes, streaks, and clouds, set among the stellar brilliance become silky fjords, bays, inlets and rivers of darkest jet between cliffs of pale light.[4]

What would it be like to walk among the spectral landscape of the Milky Way, to hike through Sagittarius and see these dark rivers and bright mountains of the imagination made manifest about me? Above the familiar teapot of Sagittarius there's a small bright concentrated cloud, a veritable Matterhorn in the mountain range of the night. This is the Lagoon Nebula. *Nebula* is from the Latin for mist or cloud. It is in fact a cloud of mostly hydrogen gas hanging between the stars, about 100 light-years across and 5,200 light-years away. A *light-year* is the distance light travels in a year. For comparison, the Earth is a little more than 8 light-minutes from the Sun, and the nearest star system to ours, alpha Centauri, is just over four light-years away. So now imagine a cloud of gasses much like a puffy white cloud of water vapor in a summer sky but

[4] You won't see these at first; your eyes will need to adjust. In the summer, when sunlight is intense and we have been outside in the bright light, your eyes can take several hours to become really dark adapted. As they do, look closely at the brightest part of the Milky Way in Sagittarius. Try looking at it out of the corner of your eye and see what pale knots and clumps jump out at you only to disappear if you look straight at them. This is a trick known to amateur astronomers called averted-vision. The optic nerve leaves for your brain from the center of your retina where you have very few cells sensitive to light. During the day when light is plentiful and your eye moves back and forth over brightly lit scenes your brain doesn't notice this spot. At night when light conditions are low and you try to fix your gaze intently on one small point, the details always seem to be just out of reach. One word of warning, a single glance at a bright car headlight will set your night vision back to square one. If you should need a light, use a dim red light such as a cellophane-covered flashlight or bicycle rear lamp. Keep it dim though; you'll be surprised how little light your dark-adapted eyes will actually need.

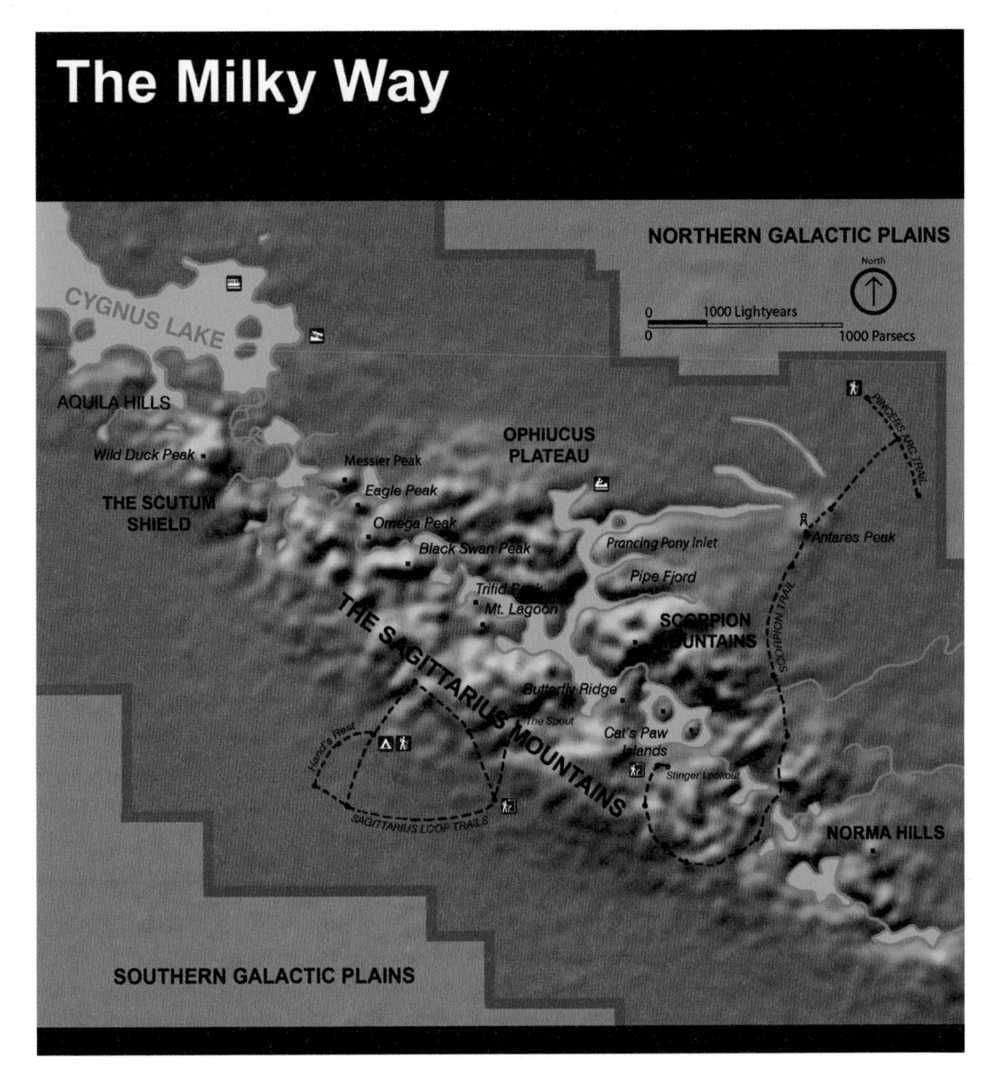

Figure 1.7 Visitor map of the summer Milky Way showing hills (bright gaseous nebulae and star clusters) with deep alpine lakes (dark dust clouds) lying along the plane of the Galaxy (T. Nordgren).

ballooning across space until stars like our Sun are no more than small birds drifting through its expanse. In these clouds stars are born out of the hydrogen gas and the ultraviolet light they give off (like that from the Sun which gives us sunburn) lights up the gas in a celestial 'neon light show'.

Most of the other tiny knots and clouds that I see with the naked eye in this part of the sky are in fact clusters of newborn stars, with hundreds packed into a space only a dozen light-years across. The night sky seen from a planet around a star in one of these clusters would be filled from horizon to horizon with stars as

Figure 1.8 The Lagoon Nebula (M8) as drawn through an amateur telescope. Notice the dark dust lanes and cluster of newborn stars visible through even a small telescope (T. Nordgren).

Figure 1.9 The Lagoon Nebula imaged by the U.S. Naval Observatory's 40-inch diameter telescope in Flagstaff, Arizona. The portion of the nebula seen here shows the inner portion of the nebula drawing (Stephen Levine, U.S. Naval Observatory).

bright as the very brightest we can see here on Earth. But we would likely be alone in seeing them. These stars and any planets they have would be no more than a couple hundred million years old, only one fiftieth the age of our own planet. On such a world it is unlikely that any life would yet have arisen at all, much less evolved to the point of stargazing.

As my eye treks the Milky Way north, I leave the heights of Sagittarius, cross low mountain passes and climb again as I enter the hills of Aquila, the Eagle, marked by the bright star Altair. Altair is one corner of the summer Triangle, an asterism of three nearly equally bright stars, each from a different constellation. Altair in Aquila, Deneb in Cygnus (the Swan or Northern Cross) north along the Milky Way and Vega in the constellation of Lyra, the Lyre (or Harp), just off the Milky Way to the west. While all three stars look virtually the same to the eye, each is revealed by modern astronomy to have a remarkable hidden beauty.

Altair is an average star like the Sun. It's a bit more massive, a little bit hotter, a little bit brighter, and a little less yellow. In one way though, it is quite different: Altair spins so fast that it bulges out at the equator as if it were in danger of flinging itself apart. Utterly hidden from sight before this century, astronomers have actually seen this rotational bloating with the latest generation of telescopes.

Vega is like Altair in that it too spins rapidly, but this time with its pole projected right at us so that we can't see that it is fatter at its equator than at its poles. But unlike Altair, Vega is a young star, less than 400 million years old and in the last decades of the 20th century it was one of the first stars around which astronomers observed a swirling disk of gas and dust. An entire solar system still in its infancy lies around that star in our sky.

Figure 1.10 The Summer Triangle as seen from the Valley View turnout within Yosemite Valley. Vega is at top, while Deneb rests to left above the granite cliff of El Capitain, and Altair is to the right above Bridal Veil Falls (T. Nordgren).

Figure 1.11 A model of Altair as it would appear relative to the Sun. Notice that Altair is darker and cooler around its bulging equator compared to its poles. For comparison, the Earth is about the size of one of the small sunspots visible near the center of the Sun's surface (Deane Peterson (SUNY Stonybrook) 2006, *Astrophysical Journal*, vol. 363, pp 1048, Reproduced by permission of the AAS).

Deneb's hidden beauty lies in the fact that to the eye it looks similar to Vega and Altair. It's this nearly common brightness of the three stars that has made the Summer Triangle a favorite of stargazers who regularly greet its appearance in the summer sky each year. A star's brightness depends not only on its own luminosity, but also on its distance from the Earth. Altair and Vega are nearby stars in our own little neighborhood. Light from Altair takes about 17 years to make the trip to our eyes each summer. Deneb, however, is close to 1,500 light-years away. The light you see tonight left the surface of that star during the age of King Arthur (or rather the historical period of battles between Britons and Anglo-Saxon invaders that would later give rise to the legend of Arthur). To appear as bright as Vega and Altair, Deneb must be 8,000 times more luminous than they are. Deneb is a supergiant star dwarfing Altair, Vega and our Sun just as a beach ball dwarfs cherries. Far more massive than the other two stars Deneb is consuming its available nuclear fuel so fast that, while still much younger than even Vega, it will die within the next few million years in a catastrophic explosion for a while more luminous than the rest of the stars in the Milky Way combined.

And these are just three of the strange and beautiful stars that make up the hundreds of billions of stars in our Galaxy, the Milky Way.

To walk further up the hills from Altair to Deneb is to walk the eastern edge of a great bay, the Cygnus Void, a celestial counterpart to San Francisco Bay off to the west of Yosemite. In Cygnus and Aquila the ghostly hills and valleys of the Milky Way fork and wrap around an emptiness where the western peninsula eventually breaks off and provides an outlet to the nighttime sea.

Where are the distant stars in this void?

What to our eyes appears as an absence of stars is in reality the shadow of an invisible presence. In addition to the luminous clouds like those in Sagittarius there also exist dark black clouds which give off no light of their own that can be

Figure 1.12 Dust clouds in the Cygnus Void from Deneb on the left to Sagittarius on the right. Antares is the bright orange star to the upper right, while Jupiter is the bright object at the bottom. The dark shadows to right are trees (T. Nordgren).

seen with the human eye. Rather we only see them because of their contrast with the surrounding glow of the Galaxy. These are the dust clouds that hang between the stars, so dense that no light from distant objects can be seen behind them. Those stars one does see in the heart of the void are only those stars that lie between it and our Sun. Where there are stars buried within these clouds, any planet around one would be completely shut off from the outside Universe and its night sky would be as dark and empty as a closed room at night. Tomorrow when I drive home to Los Angeles and all these stars disappear from my nighttime sky, I can't help but think I will know what it would be like to live on one of those depressing worlds.

North of Cygnus, the Milky Way grows quickly dim and patchy as I walk down rough slopes into broad rolling hills on my galactic hike. Just north of Deneb is the dark cavity of the Northern Coal Sack Nebula. Without the glorious bright heights of Sagittarius with which to give it contrast, it is only visible from the darkest locations. There are a few bright areas here but nothing as grand as the glories of the south. If it is fall and you stay up late you may notice that the Milky Way continues faintly across the northern sky and down past Orion. There it is nothing more than a low ridgeline that eventually fades back into the gloom of the southern horizon.

If the Earth were transparent you'd see that the Milky Way's band forms a ragged ring entirely around it. As the celestial sphere slowly rotates with new stars rising in the east and setting in the west, different parts of the ring are visible at different times of night. In addition, as the Earth goes around the Sun, different parts of the ring slip behind the Sun's obscuring light, and so every night we see a particular part of the band a little earlier than the night, week, or month before. Thus, as summer gives way to fall, the glories of the summer Milky Way slowly fade into the west after sunset. As it does so, the cold, stark, Milky Way in Orion, visible late at night in fall, gradually rises earlier each night until eventually it becomes the pale evening Milky Way we associate with winter. Not until summer comes again do we once more see the high majesty of Scorpius and Sagittarius in the evening sky.

I am not alone in my dream of hiking along the Milky Way's trail.[5] For many cultures, including many American Indian tribes, the Milky Way has always been a road, river, or bridge taken by those who've died. For the Navajo in the American southwest, the Milky Way is a path of corn pollen for the departed that brings balance between day and night, while for the Hopi and Pueblo Indians of the region, it as a ladder that leads from this world into the next. For the Chumash of the central California coast, near here, the Milky Way is a path taken by the souls on their way to Shimilaqsha, the Land of the Dead.

In the early 1920s, long after the Chumash culture was broken up and

5 The word 'way' itself has its origins in Middle English where it was used for 'road,' 'path,' or 'route.'

assimilated by western civilization, the ethnographer J.P. Harrington recorded oral narratives of those few remaining Chumash elders who could remember the stories of their youth. These tales have been compiled by Thomas C. Blackburn in his book *December's Child* and in Narratives 12 and 59 we hear a tale told by Maria Solares, an elderly Inezeño Chumash, of the path taken by the soul of a new bride, accidentally killed by her young husband in a moment of carelessness.

Heartbroken, the young man vowed to hide himself by where his bride was buried so that he could follow her wherever she went. When at last she arose, he followed her through the night to the land called Wit, where new souls are greeted by forever youthful widows.

> When dawn came he could see only her heels and what looked like mist, but when evening came again he began to see her more and more clearly as if she were alive. She told him to go back and said, "You have killed me, and now you are keeping me from my destination." The man replied, "It makes no difference, I will go even if I perish."

Passing through the Land of Widows, the two entered a deep ravine where the trail is cut up and worn by the passage of so many dead. Here in the land called 'Ayaya, two huge stones continually part and clash, crushing anyone that is caught between them. Here too, enormous ravens pluck out the people's eyes which the soul replaces with poppies that grow plentifully along the road, thus restoring their sight. Upon leaving the ravine the path crosses the home of *La Tonadora*, Scorpion Woman, who stings unfortunate souls with her tail if they stop to talk. After finally passing through all of these dangers, the young wife and husband came at last to the edge of a deep dark water with the entrance to Shimilaqsha on the other side.

> They reached the place and there was a long pole that kept rising and falling, touching the gate of Shimilaqsha and then rising until it brushed the sky. The wife said, "I will hurry and pass the gate. Be careful, there are two animals in the water that will try to frighten you." These animals emerge on each side of the pole and shout, and if a bad soul is passing it falls into the water and perishes but a good soul passes by safely....

First one, then the other crossed and entered Shimilaqsha. Here the road forked and the wife, who had died took one path, while the husband who hadn't was forced to take the other up to the Crystal House where Old Man lived. There in the Crystal House, Old Man challenged the young husband to stay awake for three nights with his wife without sleeping. If this he could do, then the two would be permitted to return alive to their village. For two nights he succeeded, but on the third night, no matter how much his wife shook him and pinched him the young man fell asleep. For his failure he was banished to return home alone where forever after he sadly told of what he had seen.

Like so much of Native American culture, we know very little of what this story may have meant to the Chumash. Their culture was already scattered and suppressed when Solares was born and those astrologer-priests with knowledge of the symbolism and tradition of these narratives were gone and unable to pass on the meaning behind them. One is then left, like a detective, to piece together hints and inferences, disparate pieces of information and, yes, guess and conjecture, to understand what is now lost. Putting these pieces together to understand the relationship between culture and the heavens is the goal of the interdisciplinary field of ethno-astronomy. In *Crystals in the Sky*, Travis Hudson and Ernest Underhay lay out the story of Maria Solares and find within it the intriguing possibility that in the story of the unhappy souls is a map of the Milky Way itself.

As the Earth revolves around the Sun, the stars overhead in the evening sky slowly change and from night to night gradually sink westward into the light of the setting Sun. In late December, as the Sun begins to move in front of the constellation of Sagittarius, the faint Milky Way towards Orion is rising and the forked band around the Cygnus Void arches down to the western horizon where one finds Altair glowing in the evening twilight. It is here over the western horizon that the Chumash believed the soul found Shimilaqsha.

Directly overhead at this time of year is the constellation of Cassiopeia which

Figure 1.13 The "Paths of the Dead" arch over Half-Dome and the Ahwahnee Meadow in Yosemite Valley. The lights of campers illuminate the valley walls as a meteor shoots overhead beneath the light of Jupiter (T. Nordgren).

sits along the Milky Way like an enormous letter 'W' and is the most conspicuous constellation along this portion of the Galaxy's band. As the gateway to the world above, this may be the first stop on the soul's journey: the Land of the Widows.

Just to the west, moving opposite to the direction I walked in my travelogue before, one comes to the region of the Milky Way north of Cygnus. The faint, scattered and chaotic nature of the Milky Way in this place would be wonderfully described by the treacherous boulders and ravens of 'Ayaya.

Passing through this region one comes to our constellation of Cygnus, the Swan. Here if one imagines the tail of the cross arcing up and west to the star Vega in the constellation Lyra we may in fact be gazing on the long ago Chumash constellation of Scorpion Woman. There she sits facing east to the approach of oncoming souls as she guards the Cygnus Void where long ago stargazers may have seen the Milky Way's diverging paths as a pole over dark and foreboding waters.

At their end is the star Altair, which the Chumash would see in winter hanging over the western horizon as the celestial marker for Shimilaqsha itself. Many stories like that told above include a forked trail or a 'Y' in the road for the souls of the dead. I can imagine a time long ago when on a dark night with the

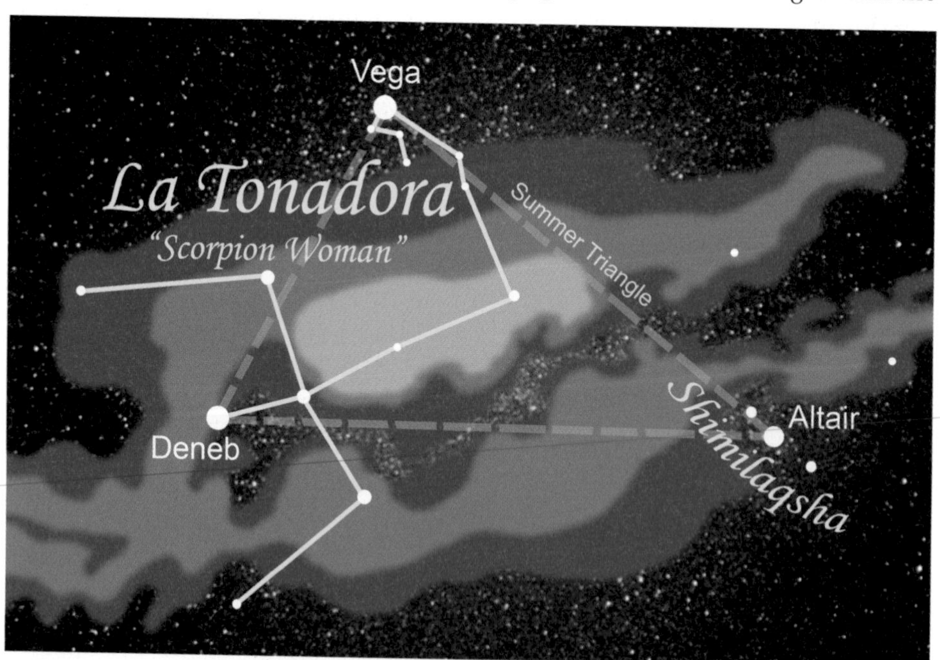

Figure 1.14 Chumash Indian legends speaks of a Scorpion Woman, *La Tonadora*, who guards the edge of a dark water that must be crossed to reach Shimilaqsha, the Chumash Land of the Dead. It is thought by some ethno-astronomers that these correspond to features along the Milky Way (shown in blue) and that today we may find them within the boundaries of the modern Summer Triangle (T. Nordgren after a diagram by T. Hudson and E. Underhay).

Milky Way high overhead, a young bride took the path to Altair while her husband was doomed to take the forked, lesser path, away from her.

Many other tribes in California, including the Miwok of Yosemite, had similar stories of husbands following their dead wives' spirits into the sky. Laying on my back among the grasses in the fields of the valley floor it isn't difficult at all to see the Milky Way as a spectral bridge joining together the enormous granite cliffs to north and south. What power this sight must have had to people who could look up and literally see the road taken by their ancestors, and thus presumably the path they themselves would one day take.

Whether these stories really were meant to be taken literally or simply as analogy or allegory is not clear. But what they all succeed in doing is to provide the listener with some frame of reference in which to make sense of a mysterious and often complex interaction between celestial objects and events, and doing so is a time honored tradition that continues to this day.

So the Milky Way's band is stars and clusters, glowing gasses and pitch-black dust clouds, rapidly rotating stars, and young stars with dusty disks where planets are waiting to happen. And all of this is arrayed in a circular band across the sky. How does this all fit together? Just what *is* the Milky Way?

The Milky Way is our Galaxy. Almost everything you see at night with your naked eye is in our Galaxy. The band of the Milky Way is simply the more distant parts of it grown faint and blurred together with great distance. Imagine the stars are the trees of an enormously large forest. All around me I see trees. Above me I see the widely scattered green tree tops but as my gaze falls lower to the horizon, perspective and distance cause the number of trees I see to increase. When at last my sight reaches the horizon I see a band all around me of trees merging together and overlapping with distance. What on the outside is simply a forest growing over a flat plateau, from the inside manifests itself differently. We are trapped inside our Galaxy, so our view of it is likewise distorted and incomplete.

If we could step outside it and look down upon it, what would it look like? William Herschel, one of the pre-eminent astronomers of the eighteenth century, was the first scientist to try just that in a systematic way.[6] When you look at the night sky, one of the first things you notice is that some stars are brighter than others. Why is this? One reason could be that the dimmer stars are just farther away. We see this every night with streetlights and car headlights. Our brains even do a bit of subconscious calculation because we have a gut feeling for how bright a car's headlight really is. If we know that, and we can see

[6] Prior to about 1925 the more-or-less typical view of the universe was that it was just large enough to hold our single galaxy. As a result, when astronomers like Herschel in the eighteenth and nineteenth centuries sought to measure the properties of the Milky Way, for them this was synonymous with measuring the properties of their universe. Not until 1925 when Edwin Hubble, for whom the space telescope is named, proved that the Andromeda 'Nebula' must lie outside our Milky Way, did the idea of galaxies as islands of stars scattered throughout a universal ocean of emptiness really catch on.

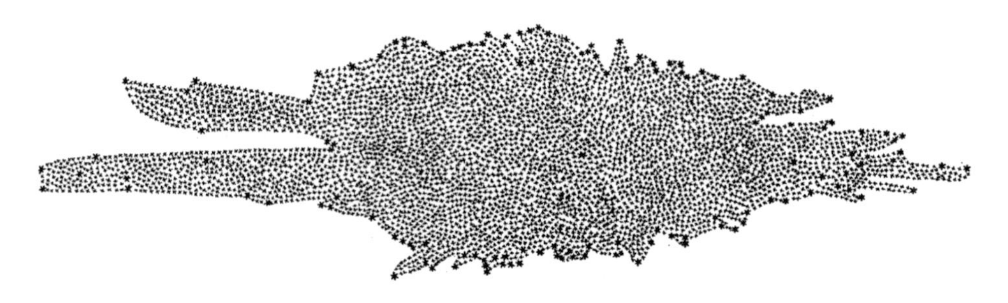

Figure 1.15 Herschel's map of the Milky Way for a paper entitled, "On the Construction of the Heavens," from the Proceedings of the Royal Astronomical Society, 1785. His map shows the position of the Sun represented by a large star near the center of the Galaxy (Dr Jeremy Burgess/Photo Researchers, Inc.).

just how dim the light looks, we have a decent feeling for how far away a car is and whether or not we have the time to cross a street. Herschel made the same calculation for stars. He assumed they all had the same intrinsic brightness as the winter star Sirius which is the brightest star in the sky. Based on this star he then calculated how many times farther away every other star in the sky was. If you know this distance from Earth, and you know where on the sky the star is, you can construct a three-dimensional map of the Galaxy.

Herschel's map looks like the flattened bug splatter on a car's windshield with our Sun near the center. You can see where this flatness comes from for yourself. The Milky Way looks like a band across the sky. If we lived inside a giant spherical ball of stars then there should be virtually the same number of stars in all directions and no reason for a band. And if we were off center of the ball, we'd see the majority of stars gathered over to one side of the night sky, and again, there would be no band. But we don't see this, we live in a disk and so like our hiker in the woods we see trees all around to the sides but very little above and below.

The reason the Sun is so near the center in Herschel's map is because in one very important way Herschel was completely wrong. While it is certainly true that some parts of the Milky Way are *brighter* than others, no matter which way Herschel looked in the band the *faintest* stars he could see were always equally faint. If the faintness of stars was only governed by how far away they were then this would imply that the stars extend equally far in all directions, and thus we must be near the center of the Galaxy.

Unfortunately, the space between the stars is filled with a vast invisible mist of gas and dust. Today we call this obscuring matter the interstellar medium. The farther away a star is, the more of this medium is present along the line of sight to the star and the more the star's light is scattered or absorbed before reaching our eyes here on Earth. Beyond a few thousand light-years there is so much interstellar absorption that stars quickly become invisible to the naked eye and eventually even large optical telescopes. Herschel's map, rather than being an accurate reflection of the distribution of stars in the Milky Way actually says more about the distribution of the obscuring dust and gas *within* our Galaxy.[7]

If, in addition to this nearly uniform obscuration, the Milky Way were also uniformly bright in all directions (where the *brightness* of the band is an indication of the *number* of stars along the line of sight) then you might still conclude the Earth and Sun were in the center of the Galaxy. But compare the bright, bold Milky Way of summer with the faint tenuous Milky Way visible in winter and you can tell for yourself that this is exactly not what we see. Our Sun and our Solar System lie somewhere in the galactic outskirts. If the biggest and brightest portion is in the direction of Sagittarius then perhaps that direction points the way to the galactic center. But if so, how far away is it?

The answer lies in the globular clusters. The biggest and brightest of these objects that is visible to we northern hemisphere observers is the Great Globular Cluster in Hercules (also called M13). Point even a small telescope at M13 and you will be rewarded with one of the most beautiful views in the northern sky: a shower of diamonds on a black velvet backdrop. Globular clusters are collections of between a hundred thousand and a million stars, packed tightly into a ball only a few light-years across. To be on a planet orbiting a star in one of these great accumulations of light would be to have a sky so ablaze with stars that even city dwellers would be unable to forget that there is a larger Universe around them.

There are hundreds of these globular clusters visible from Earth, the majority in the summer sky, with all but a dozen too faint to be seen with the naked eye. In the early twentieth century Harlow Shapley, a young astronomer at the Mt. Wilson Observatory in the mountains above Los Angeles, found a way to measure the distance to these objects. By mapping the positions of these clusters, just as Herschel first tried to map the Milky Way, Shapley found that they are gathered in a spherical halo wrapped around the rest of the Galaxy we see at night. The center of the halo shows us the center of our Galaxy. From the size and direction of that halo we now know that our eyes were not deceiving us. When you look towards Sagittarius, past all the stars, past the bright nebulae, and past the dark dusty clouds, you are looking towards the center of our Galaxy nearly 25,000 light-years away.

Figure 1.16 M13, the Great Globular Cluster in Hercules, contains several hundred thousand stars (N. Redfield, D. Slater, U.S. Naval Observatory).

[7] And actually it says even less than that because Herschel made one more mistake: in fact, all stars are not equally bright to begin with. But, in science as in real life, you have to start somewhere.

There, at the center of our Galaxy, are billions of old stars orbiting in random orbits like a swarm of bees around a hive. These random orbits create a great spherical bulge out of which appears a flat disk of stars in which you find all the types of things we have just seen in our summer sky. Between the stars in the disk there is the thin diffuse interstellar medium of hydrogen gas and dust, with the occasional thick dark cloud containing carbon dust grains and organic molecules such as carbon monoxide, ammonia, methanol and even water. Together, the diffuse and dense components of the interstellar medium are a little like the thin layer of lint and scattered dust-bunnies found under a bed. In addition to the dark clouds are the bright clouds of hydrogen gas lit up by the light of hot young stars forming within. As stars like Deneb grow old and die, they are replenished by these new stars that are born at a rate of about a half dozen stars each year, every year since the Galaxy formed around 13.5 billion years ago.

Over the last century, astronomers have continued to search for the structure of our Galaxy. Different investigations have looked for this structure in many different ways. Some measured the distances to young stars and the luminous

Figure 1.17 The hidden center of the Milky Way photographed using the Spitzer Infrared Space Telescope. The exact center is the bright area in the middle of the photograph. The disk of the Milky Way can be seen in the bright horizontal band. Distant stars hidden by intervening dust are shown in blue while dense star forming regions of gas and dust are shown in red. The area shown spans 890 light-years from side to side (NASA/JPL-Caltech/Susan Stolovy (SSC-Caltech)).

clouds where they form; others measured the distances to older stars and stars that had died. Astronomers observed that these phenomena are not found at just any distance from us, but rather are clumped at certain distances. By connecting these dots we see that are our Galaxy's disk isn't formless but rather has closely wrapped arms making many turns around the galactic center. From these data, astronomers can draw a picture for the first time, of what our Galaxy, our home, really looks like.

Continuing the work first seriously begun by Herschel, astronomers now survey distant stars using telescopes sensitive to certain wavelengths of infrared light. Unlike visible light, mid-infrared radiation passes easily through the intervening interstellar medium. Unfortunately, the same mid-infrared that passes through dust and gas is easily absorbed by water vapor in our atmosphere and so this distant starlight from the heart of our Galaxy is blocked in its last few miles of travel and never reaches the surface of the Earth.

Starting in 2003, astronomers have used the infrared Spitzer Space Telescope orbiting outside the Earth's atmosphere to peer through the interstellar dust and finally map the distant structure of our Galaxy. They do this by counting the number of stars in the Milky Way's band along lines of sight at different angles from the galactic center. Where the number of stars increases, they know they

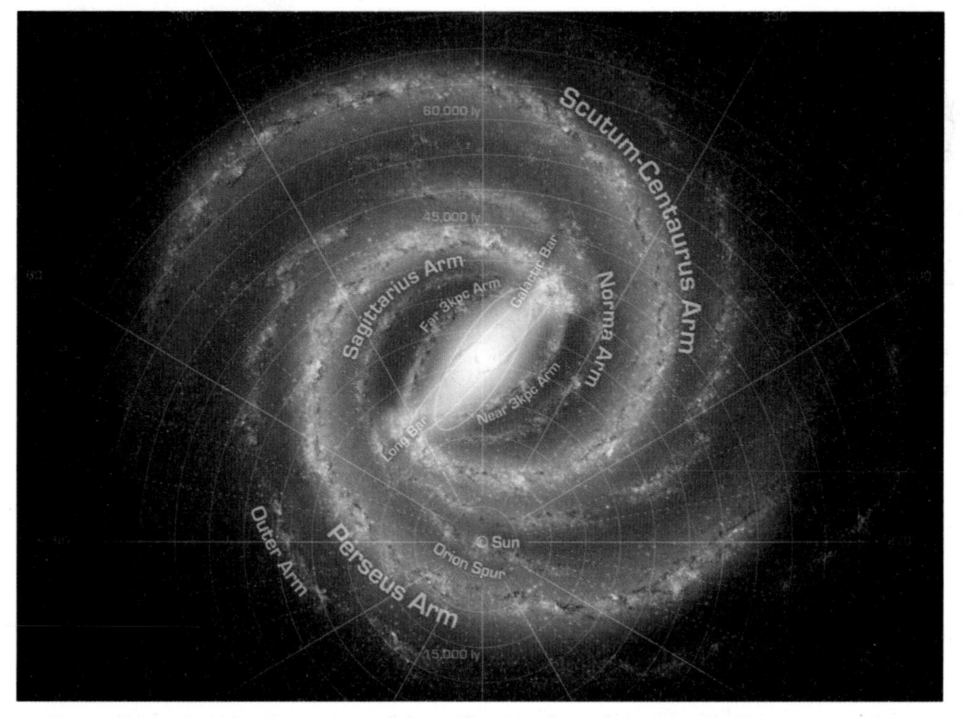

Figure 1.18 Artist's conception of the Milky Way based on Spitzer observations. The position of the Sun is shown as well as the names of and distances to the spiral arms (NASA/JPL-Caltech/R. Hurt (SSC-Caltech)).

are looking down the length of a spiral arm. Their new map shows we live in the outskirts of a vast pinwheel of two tightly wound arms that spiral off a central bar centered on the galactic bulge like a propeller trailing streamers. Prior to these observations no one knew how many arms our Galaxy even had. Now we know this giant whirlpool of stars, gas, and dust is nearly 100 thousand light-years in diameter and only about six thousand light-years thick. Shrink it down and our Galaxy's disk has the same dimensions as the quarter in your pocket.

Out here at the location of the Earth and Sun, half way from the galactic center, the average distance between stars is about seven light-years. In towards our Galaxy's bulge the average distance shrinks to only one. To get a sense of what these distances actually feel like, find a single grain of sand and place it in the palm of your hand: this is the Sun. Now imagine another grain of sand, just like our Sun, 20 miles (32 kilometers) away. This is the separation between stars in our part of the Galaxy. Now imagine every 20 miles another single grain of sand until you have over a hundred billion grains in a vast spinning disk stretching from here to the Moon. Close to the galactic center (half way to the Moon in our 'sand castle galaxy') the distance between sand grains shrinks to less than a mile with thousands crammed into the inner few miles of our model. This is the Milky Way.

Now look back at the single grain of sand sitting in your palm. On this same scale the Earth would sit on the tip of your index finger where it would be far too small to be seen with the naked eye. Light, the fastest thing in the Universe, would take a little over eight minutes to travel the length of your hand in this model. Everything we know about the Galaxy and the rest of the Universe travels to us at this snail's pace. There are around four hundred billion stars in our Galaxy, and there are believed to be around a hundred billion other galaxies in the Universe, and as far as we know from the information brought to us by starlight, ours is the only speck of sand around which life has arisen to contemplate this immensity.

Eventually the summer fades to fall and the folks who came to the mountains and forests head back to work and school. City lights take our minds away from the sky and winter's cold takes away any final thoughts of lingering outside under the stars. The bright starry shores of the summer Milky Way have long ago set behind the Sun when spring finally rolls around and a special opportunity begins to present itself. Since the Earth is a sphere, there should be some place on it, at a certain time of year, where to stand upright on the Earth would mean you were standing upright in the Galaxy.[8]

Think of the globes that used to sit on your grade school teacher's desk. As I looked at those globes, I distinctly remember wondering what it would be like if I

[8] Strictly speaking, since there is no 'up' or 'down' in space, what I mean is that you will be standing perpendicular to the disk of the Galaxy, parallel to its spin axis, with the entire disk arrayed around you.

Figure 1.19 The galaxy M109 in the constellation Virgo is considered an excellent candidate as the Milky Way's twin. Nearly a million light-years away it shows how alien astronomers in a distant galaxy might see the Milky Way (Swanson/Block /AOP/KPNO/NSF).

Figure 1.20 NGC 891 is also a galaxy believed to be much like our own. In our skies we see it edge on and so are treated to a view of its disk full of stars, dark dust clouds, and bright blue cluster of new forming stars (KPNO/WIYN/NSF/NOAO).

could stand at the North Pole and have the entire Earth laid out beneath me knowing I was standing on top of the world. I'd know I had accomplished this, that I had reached the North Pole, if I could look up and see Polaris, the North Star, directly above me. Polaris is an average looking star in our sky, over 400 light-years away in deep interstellar space, which only by chance is located directly over the north pole of the Earth.

In a similar way there is also a point in deep intergalactic space that is located directly above the north pole of our Galaxy. From our vantage point on Earth, this North Galactic Pole is located in the direction of the constellation of Coma Berenices. If we could go to that place on Earth where Coma Berenices passes directly overhead then we should see the Milky Way arrayed out around us, its band ringing the horizon, even if just faintly visible through the obscuring haze of our atmosphere.

This effect happens along the line of 27 degrees north latitude (similarly, the South Galactic Pole passes overhead for those at 27 degrees south latitude). In the northern hemisphere this passes through the southernmost extremity of the continental United States. The place where it does so, with the darkest skies and that's freely available to the public, just happens to be through Big Bend National Park in southwest Texas along the Rio Grande. The time of year to be there and see this galactic event is from February to April, when the weather is cool and the winds typically bring clear skies.

In early March I set out to visit Big Bend National Park and see this event that happens every year but for which I had never heard anyone describe seeing. The first thing I realize on my journey is that Big Bend is on the way to nowhere.

From the nearest dusty town it's still another hour of sun-baked driving just to reach the park boundary and from there another hour still to reach the turn-off

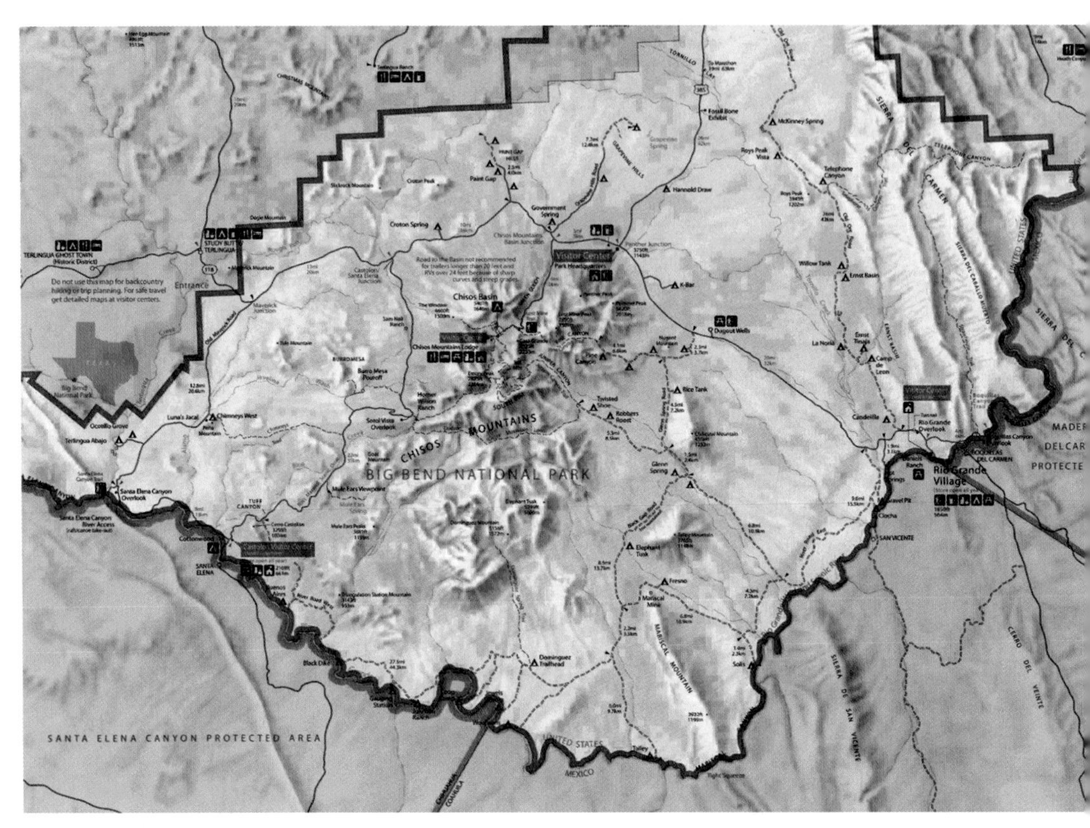

Figure 1.21 Portion of the National Park Service map for Big Bend National Park in western Texas. The backcountry campsites south of the Chisos Mountains have perhaps the darkest, least light polluted skies of any park in the continental United States (National Park Service).

for the place I'm heading. At every step of my travels into West Texas' Back-of-Beyond the roads have gotten smaller until now I find myself ten miles (16 km) down the narrow, gravel, backcountry road that is the final leg of this expedition. My knuckles are white from negotiating boulders while rocks bang against the underside of my truck and dry, prickly ocotillo branches rake the length of my Ford as the dirt road gets narrower still. But this drive is worth it because when the dust finally settles at the spot the park service has told me about, I have an uninterrupted view of the entire horizon farther away from any lights than I have ever been before in the continental U.S.

As the afternoon turns late the sky clears of clouds and I am conscious of the weight of the bright blue hemisphere hanging over my head. More importantly, the sky is once again a solid blue all the way to the ground. The clarity of the sky is a big issue here in Big Bend. Depending upon the direction of the wind, the view here can extend for what seems hundreds of miles or alternately shrink down to no more than a dirty few as smog from distant cities and power plants

Figure 1.22 The winter Milky Way silhouettes an ocotillo as seen from the Robbers Roost backcountry campsite in Big Bend National Park. The constellation of Orion and the faint pinkish glow of the Orion Nebula are visible to the upper right, just beneath the band of the Galaxy (T. Nordgren).

muddies up the sky. But the wind is out of the clean, cool north today so there should be very little dust and haze to obscure the stars as they rise above the horizon. I have chosen a night when there will be no Moon to interfere with the light of the faintest stars and in this place no city lights should be visible to my eyes. These are all important factors because for the event I'm here to see, the glories of the Milky Way will be found entirely at the horizon where there must be absolutely nothing to compete with the light of the stars.

Once the Sun sets and twilight fades I am not disappointed; overhead the sky is ablaze with thousands of luminous stars. From my backcountry camping spot there isn't even the light of another camper. I cannot see a single street light or house light. On the distant hills, even across the Rio Grande into Mexico, there isn't a single light of a farm or town. No remote radio towers flash red on the distant hillside, nor do I see even the faintest glow of a far away city over the dark horizon. To my surprise, there aren't even the lights of airplanes passing overhead. I suspect their absence is due to Big Bend's position in Texas where the border river loops southward, forcing any potential overhead flights to pass in and out of U.S. and Mexican airspace. For the first time in all my travels there is not a single source of artificial light to compete with the stars in the sky. The only sign I can see of the modern world is the faint steady light of a distant satellite reflecting sunlight off its surface as it passes by in orbit. Yet, as dark as it is, the landscape isn't black as all around me is a faint illumination by the light of the distant stars.

Sitting on the tailgate of my truck, cupping a steaming mug of coffee in my hands, I look to the west and see the light of the setting winter Milky Way, a pale sister to her showy sibling in summer. In places that are merely 'sort of' dark, one rarely sees this part of our Galaxy. I take another warm and comforting sip, then wrap my coat tighter. The clear sky possesses no clouds to trap in the day's heat and the night is already cold.

Midnight approaches and the winter Milky Way nears the horizon. I look up and see the spectacular sight of Coma Berenices near the zenith. What does this constellation look like? I can't really say. It's one of those constellations unfortunate enough to look like nothing; consult the star maps and it's just three stars of so-so brightness joined in a right-angle (not even a full triangle). But it's not the outline of these three stars that makes Coma Berenices noticeable, for even now I'm not sure which three stars I should be looking at.

No, what makes Berenices spectacular is that under the darkest of skies an enormous cluster of faint stars called the Coma Star Cluster (or, just as melodiously: Melotte 111) appears. Once every year or two, I find myself somewhere dark at just the right time and I am startled by its presence. What I am looking at is one of those clusters of newborn stars which we saw back in Sagittarius, but now nearby and much older. The center of this flock of stars is only 288 light-years away, and the stars I see are only 400 million years old. Trilobites were swimming the shallow seas of Earth when these stars were born. Under the influence of galactic tides these stars have begun to float away from one another and in a few tens of millions of years the cluster I see will have disappeared into the background stars. For now though, the North Galactic Pole lies just on the eastern outskirts of Melotte 111 (though of course vastly farther away): the power of its presence more in the idea than as an actual thing to be seen.

With time all the pieces move in unison and soon the winter Milky Way in Orion touches the western horizon as the point that is the North Galactic Pole reaches the zenith. But that's not where I'm looking now because to the

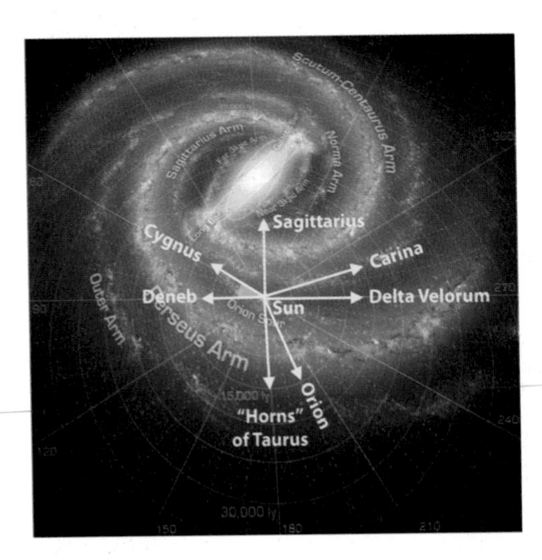

Figure 1.23 The direction to several prominent stars and constellations are shown superimposed on the artist's conception of the Milky Way. When looking towards Deneb we are pointing in the direction the Sun is traveling around the center of the Galaxy. The constellation of Cygnus that spans 30° along the disk of the galaxy is in the direction looking down the nearest spiral arm of our Galaxy. When looking towards the bright Milky Way in the direction of Sagittarius we are facing the galactic center, while the opposite direction towards the constellation of Taurus is the part of the Milky Way pointing directly away from our Galaxy's center. Around the figure are angles of galactic longitude where 0° is towards the galactic center and angles increase as you look along the band of the Milky Way counter-clockwise (T. Nordgren, (base map) NASA/JPL-Caltech/R. Hurt (SSC-Caltech)).

southeast I can begin to make out a faint glow backlighting the distant hills. At first it looks like great banks of steam rising off the distant Rio Grande until suddenly it occurs to me that they would have to be reaching thousands of feet into the dark night sky. All at once I see it for what it is: it's the first faint light of Sagittarius and Cygnus rising along the entire eastern horizon in unison. How often have I marveled at the beauty of a sunrise when before me now is not just one sunrise, but a billion? I get down off my truck and step out into the field of brush beside my camp. Before me I finally see what I have driven all this way to experience: the Milky Way is banding the horizon and standing here under a starry sky I know that I am standing upright in the Galaxy.

I look to the southeast towards Sagittarius and stare across the gap between spiral arms towards two distant bands, the nearby Scutum-Centaurus Arm and the much more distant (and slightly less euphonious) Three-kiloparsec Arm. The gas and dust that flow through each, blocks the starlight of what lies beyond. But to my eyes and to my imagination it doesn't matter. For behind both is the distant glow of the galactic bulge and its light of ten billion suns shines unimpeded as, in my mind's eye, its 25,000 years-old light illuminates my face. I raise my left arm straight out from my body and to the northeast I point to Cygnus where the Sun is now heading in its 250 million year voyage around the center of the Galaxy. The Milky Way is bright here because I am pointing straight

Figure 1.24 The Milky Way wraps around me as to right and left I can point to where our Sun has come from and towards where our Solar System is heading in its orbit around the galactic center before me (T. Nordgren).

down the Orion Spur: the local spiral arm that harbors our Sun. I raise my right arm and I point southwest in space to where the Sun, Earth and I are most recently leaving. This part of the Milky Way rarely makes it above the horizon for us astronomers in North America. Farther south in Hawaii and beyond this part of the galaxy towards the constellations of Vela and Carina is spectacularly bright because we are looking down a long bend of the Sagittarius Arm, the next closest arm to our own.

Behind me towards Orion, and the horns of Taurus the Bull, the Milky Way is faint because we are looking straight away from the great bulk of stars in our Galaxy. Here are the cold outer reaches of the Galaxy where observations reveal some strange form of dark matter makes everything revolve around our Galaxy faster than gravity says it should. Above me I look up out of the plane of the galaxy and stare into the vastness of intergalactic space. Invisible to my sight, but not to my imagination are clusters upon clusters of not just stars but entire galaxies. The Coma and Virgo Clusters of Galaxies are up there: hundreds of galaxies, each millions upon millions of light-years away. Beneath me through the invisible Earth at my feet are the South Galactic Pole and a million other galaxies farther away still.

I have come a long way to be here and see this, and a shiver runs down my spine. On this night, in this place, and for this moment, the Galaxy is laid out before me. My horizon extends a quarter million light-years in each direction, and I know exactly where I am in the Universe. I am standing straight and tall in my Galaxy, the Milky Way, my home.

See for yourself: our Galaxy

Binoculars

Point a simple pair of binoculars or a spotting 'scope at any of the objects described here and you will not be disappointed. Binoculars are a wonderfully simple and cheap way to observe the sky; you don't need an expensive or complicated telescope. What you do need, however, is a stable method of holding the binoculars. Use a camera tripod if you have one, or grip the binoculars in both hands while resting both elbows on a table, bench, or railing.

Scorpius and Antares

In summer look to the south or southeast 90 minutes after sunset for the constellation of Scorpius. Antares, the bright, orange star, at the heart of Scorpius is the first part of this constellation you will see. From Antares look upwards and to the west (to the right as you face south) and you see a crossbar of three almost equally bright stars. This is the head of the scorpion. From Antares down and to the east your eyes follow a chain of ten stars that curve toward the horizon and then hook back up to a set of stars in the shape of a stinger. The constellation is almost two full hand-widths side by side on the sky (Figure 1.4).

Sagittarius and the heart of the Milky Way

East of Scorpius (to the left as you look south) is the giant teapot at the heart of Sagittarius. It has a main body of four stars with a triangular lid formed by a fifth. Two additional stars on the left form a handle while another on the right forms a triangle for a spout. The Milky Way appears as steam rising northward out of the spout of the teapot. If you are in the southern part of the country or in a very dark location with no city lights to the south, then this region of the Milky Way will be the brightest part of the band and you will eventually see that the entire area between Sagittarius and Antares is aglow with the light of distant stars. Throughout this diffuse glow are also numerous dark patches caused by the dense clouds of dust blocking the light of the distant stars behind them.

The Lagoon Nebula (Messier 8 or M8)

In summer, look at the three stars that make up the spout of Sagittarius: two stars across and a third hanging down below. Imagine that third star swings up around the hinge formed by the top two stars of the spout. Where the third star comes to rest at its highest point there's a small bright concentrated cloud. This is the Lagoon Nebula. It is a cloud of mostly hydrogen gas hanging between the stars, about 100 light-years across. Many of the other tiny knots and clouds you see with the naked-eye in this part of the sky are in fact clusters of newborn stars, with hundreds packed into a space only a dozen light-years across (Figure 1.4).

In winter, the same type of object as the Lagoon can be seen in the Orion Nebula (M42) in the constellation of Orion. Find the line of three equally bright and equally spaced stars of Orion's belt low to the south. From the middle star of the belt, a chain of three dimmer stars hang down forming a sword. The middle star of the sword is fuzzier than the surrounding stars. The fuzziness is due to it being a cloud of interstellar hydrogen gas just like the Lagoon. Both are great objects to view through binoculars.

The Summer Triangle

Due east in summer is the Summer Triangle: three nearly equally bright stars, each from a different constellation. Altair is in the constellation of Aquila, the Eagle, and is the southernmost of the three lying just beneath the eastern edge of the Milky Way's band. It is closely flanked by two dimmer stars. Deneb in Cygnus (the Swan or Northern Cross) is the northern corner of the triangle and lies directly along the Milky Way at the northern edge of the great dark rift, or void, in the Milky Way's bright band. Halfway between the two and west of the Milky Way (which when looking east places it high overhead) is Vega in the constellation of Lyra, the Harp. These three stars are the brightest stars during the

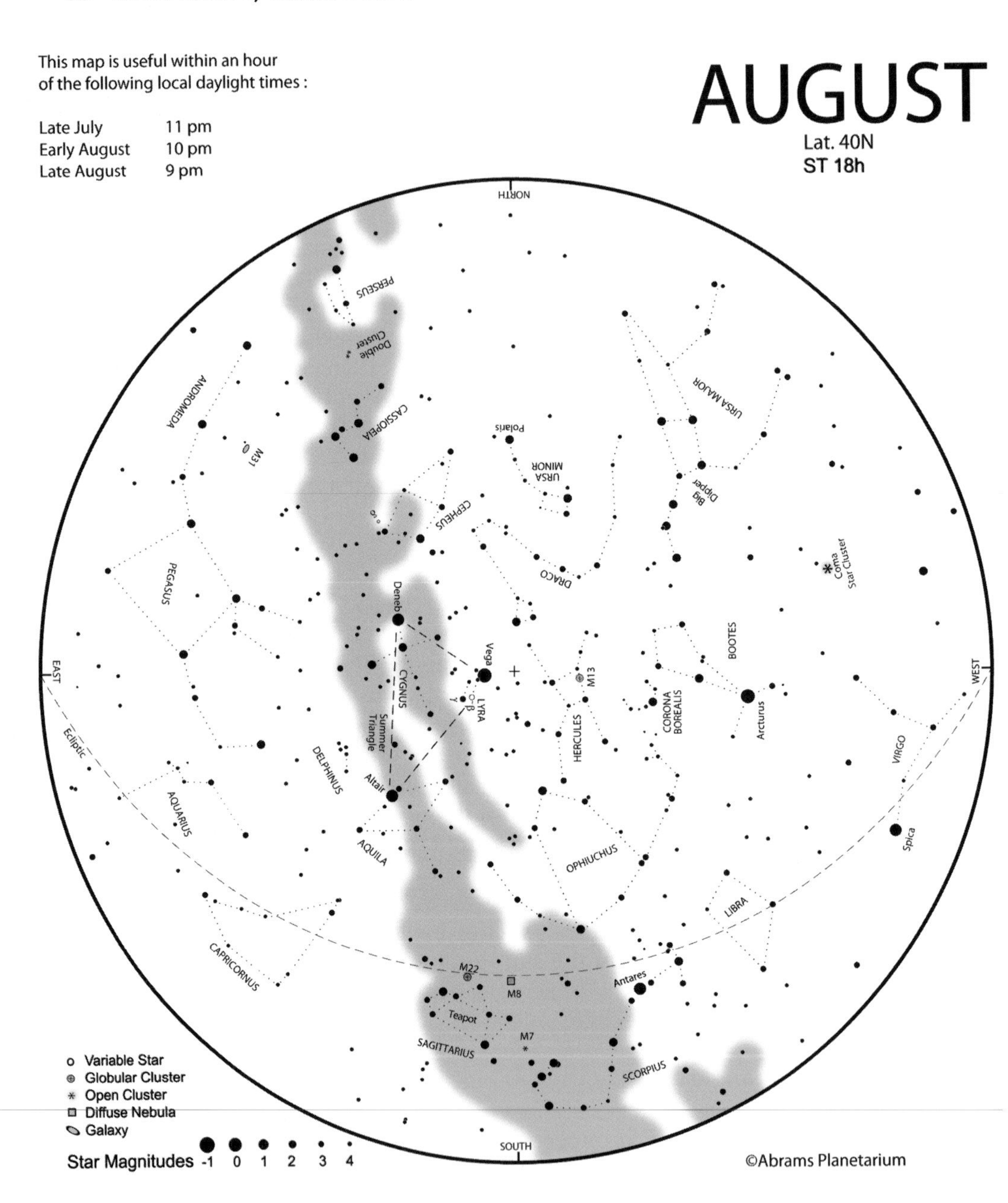

Hold the star map above your head with the top of the page pointing north. For those at mid-latitudes within the continental United States, the center of the map marked with a + will show the view directly overhead (the zenith) at the indicated times. For those farther north, objects on the map may appear a little farther south; for those farther south, objects will appear a little farther north.

This map is useful within an hour of the following local standard times :

Late January	10 pm
Early February	9 pm
Late February	8 pm
Early March	7 pm

FEBRUARY

Lat. 40

ST 6h

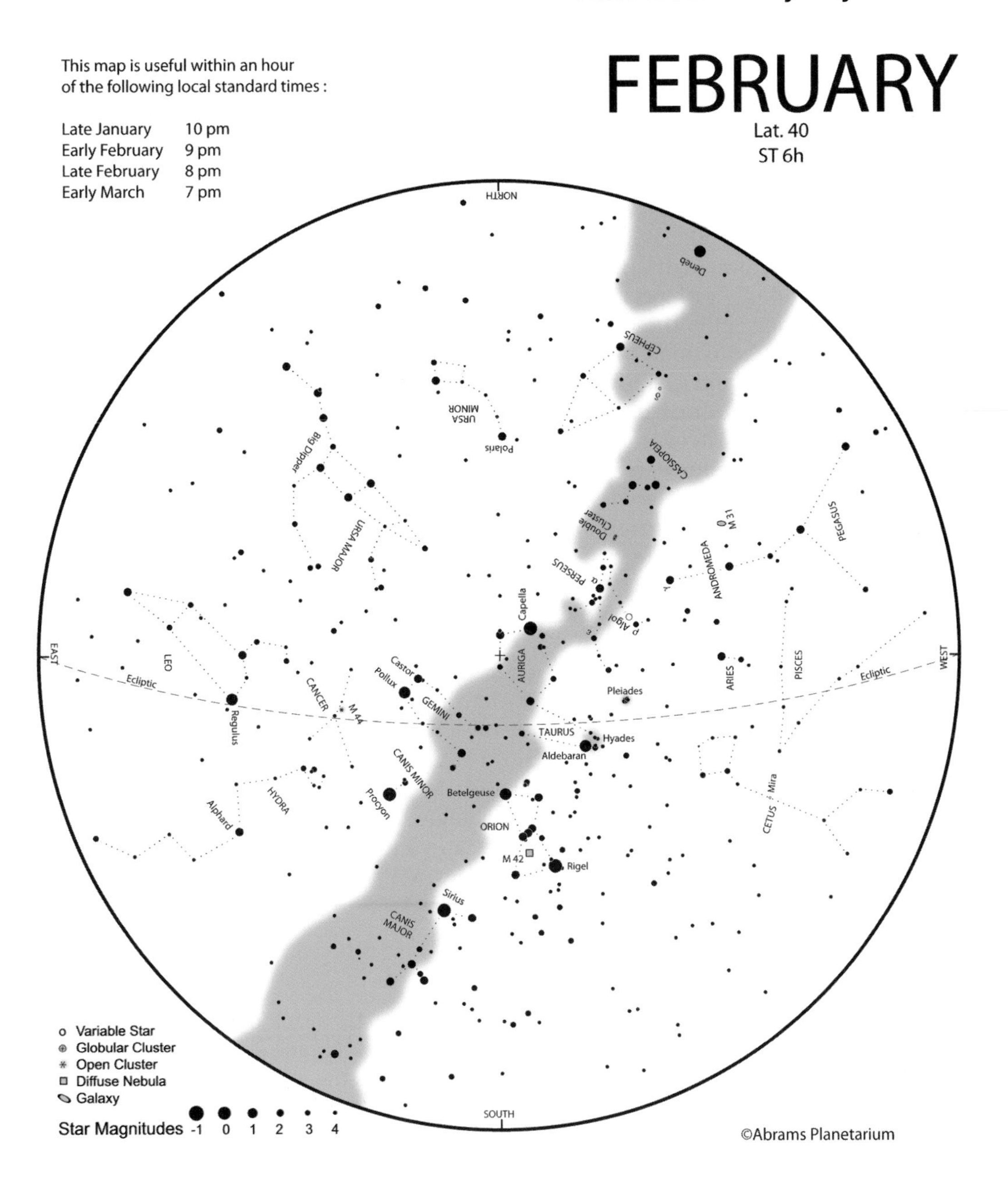

summer months and thus are some of the only stars visible from cities and bright towns. In fact the Summer Triangle only became of note during the last century as light pollution blotted out all other stars in urban skies except these three.

The Milky Way

In summer the Milky Way appears as a faint patchy band running generally north-south from Sagittarius up through Cygnus overhead and north to the constellation of Cassiopeia that appears as a large 'W' to the northeast. A large dark dust lane splits the band into two parallel parts for most of the distance between Sagittarius and Cygnus. The band of the Milky Way is at its brightest in both Sagittarius and Cygnus and gets noticeably dimmer towards Cassiopeia. As summer turns to winter the band of the Milky Way becomes more east-west and gets even fainter. By winter and early spring, the Milky Way runs through the constellations of Taurus, past the top of Orion and down through Gemini. These parts of the Milky Way are its faintest and often are only visible from very dark locations. Slowly pan a pair of binoculars along the band of the Milky Way to see the wonders hidden there.

The globular cluster in Hercules (Messier 13 or M13)

Look to the summer sky and see the Big Dipper to the north. The handle of the Dipper 'arcs' to the bright orange star Arcturus. From Arcturus, look back towards Vega in the summer triangle. Your line of sight crosses two constellations equally spaced along this line. One constellation is a crown of stars in the shape of a 'C': Corona Borealis. The other constellation is a rectangle that is the body of Hercules. Along the long side of the rectangle facing towards Arcturus there is, just barely visible under absolutely dark skies with no Moon, a faint fuzzy dot. This is the Great Globular Cluster in Hercules, M13. You can see M13 for yourself with binoculars, but only with steady support.

The North Galactic Pole

The North Galactic Pole is the point in intergalactic space directly above the north pole of the Milky Way Galaxy. To find it, look for the great empty region beneath the handle of the Big Dipper located towards the west in summer, or east in winter. If you are in a very dark location, you will see in this area a very large (almost a full hand-width wide) smattering of faint stars. This is the Coma Star Cluster (also called Melotte 111) lying 288 light-years away in the direction of the constellation of Coma Berenices. The North Galactic Pole is located on the eastern edge of this cluster.

Further reading

Cosmos by Carl Sagan (1983)
Random House, ISBN 0394715969

Beyond the Blue Horizon: Myths and Legends of the Sun, Moon, Stars, and Planets by E.C. Krupp (1992)
Oxford University Press, ISBN 0195078004

December's Child: A Book of Chumash Oral Narratives by Thomas C. Blackburn (1980)
University of California Press, ISBN 0520040880

Spitzer Space Telescope
http://www.spitzer.caltech.edu/

Abrams Planetarium Sky Calendar
http://www.pa.msu.edu/abrams/

National Park Service Night Sky Team
http://www.nature.nps.gov/air/lightscapes/

See the *Stars Above, Earth Below* page on Facebook for travel and astronomical updates.
http://www.facebook.com/

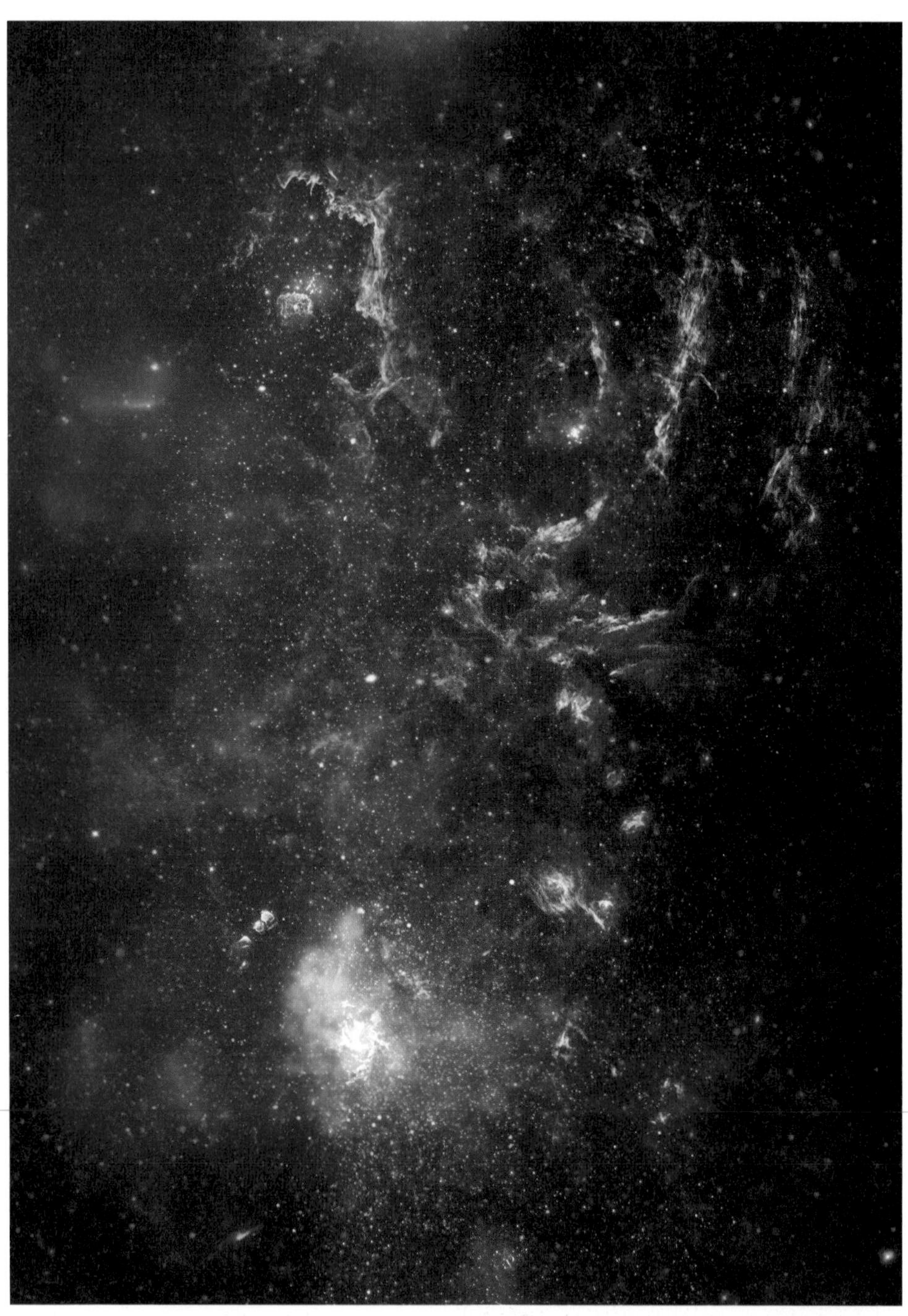

Center of our Milky Way Galaxy from NASA's three great space observatories: Hubble, Spitzer (Infrared) and Chandra (X-ray). Our central supermassive blackhole (Sgr A*) is the bright knot at the bottom (NASA, ESA, SSC, CXC, and STScI).

2 Black hole Sun

Usually it is a bit of a trick to keep your knowledge from blinding you. But during an eclipse it is easy. What you see is much more convincing than any wild-eye theory you may know.

Annie Dillard, *Total Eclipse*

A hush falls as the shadow sweeps over the crowd beside the road. My watch says 11:34 Mountain Daylight Time, shortly before noon, and drivers all over Grand Teton National Park in western Wyoming have pulled off onto the edges of roadways. The daylight dims and a strange twilight takes hold; what's been growing for an hour comes on like a sudden curtain's fall in the final seconds. When at last the Moon passes completely between Earth and Sun, the parting rays of sunlight stream down on me through valleys along the shadowed lunar limb. Beads of diamond light twinkle and shine where the Sun disappears; they are strung together on a ghostly ring encircling the silhouetted Moon.

Then it's gone.

Half way between south and east, horizon and zenith, a great gaping black hole stares down on me out of what's now an opalescent summer sky. For two minutes and 20 seconds I stand in the full shadow of the Moon. Pale, pearly streamers of the solar corona – the high, hot upper atmosphere of the Sun – fly like wind blown hair around the darkened disk. At all other times they are completely lost within the blinding glare of the solar surface. Venus, Mars and Jupiter appear amid winter constellations I haven't seen for nearly half a year. I have spent my entire life looking up at the sky and in all that time I have never seen anything else as awe inspiring as a total eclipse of the Sun.

That this event won't happen until August 21, 2017 makes it all the more

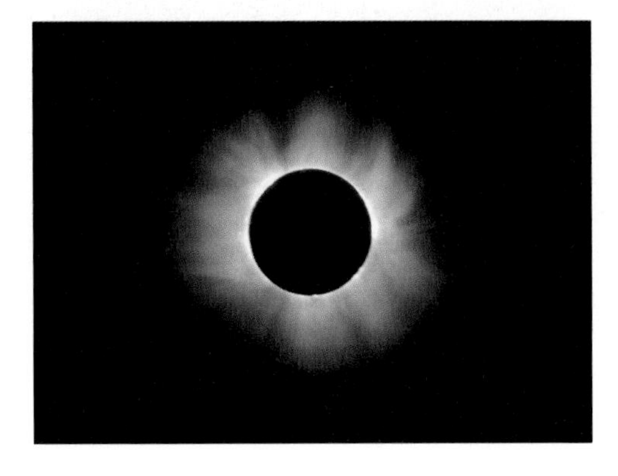

Figure 2.1 The August 11, 1999 total solar eclipse seen south of Budapest, Hungary. The faint solar atmosphere is only visible during those moments when the Moon's darkened disk completely blocks the blinding light of the Sun's disk. Bright red eruptions of hydrogen gas are visible around the edge of the lunar disk (T. Nordgren).

amazing. On that day, at that time, I know exactly where I need to be and where I need to look to see something that will not have been visible from the continental U.S. for 38 years. This is the predictive power of science.

Each and every month we can see for ourselves the celestial mechanics that lead to eclipses. The very word *month* reflects the orbital motion of the Moon around the Earth, going from west to east every 29.5 days as viewed relative to the Sun. While we see the Moon (and everything else in the sky) rise in the east and set in the west, this motion is really only due to the spinning of the Earth on its axis. To see the true motion of the heavens, step outside some evening when the Moon shows a beautiful crescent low in the west after sunset, and notice where it is relative to lampposts, mountains or buildings. Do the same thing the following night, at exactly the same time, and you and the Earth are once more pointing the same direction in space. But now the Moon is higher in the sky to the east (while also a little fatter as a bit more of its lit hemisphere is visible in the sky). The difference in position from night to night is about the distance between your thumb and little finger held outstretched at arm's-length and it reveals the distance the Moon actually moves through space in 24 hours.

Repeat this every night for two weeks and you slowly see the Moon march eastward in its orbit across the sky, night by night revealing more of its sunny surface. Two weeks after your first observation, the fully lit Moon rises in the east just as the Sun sets in the west and you have witnessed the actual orbital motion

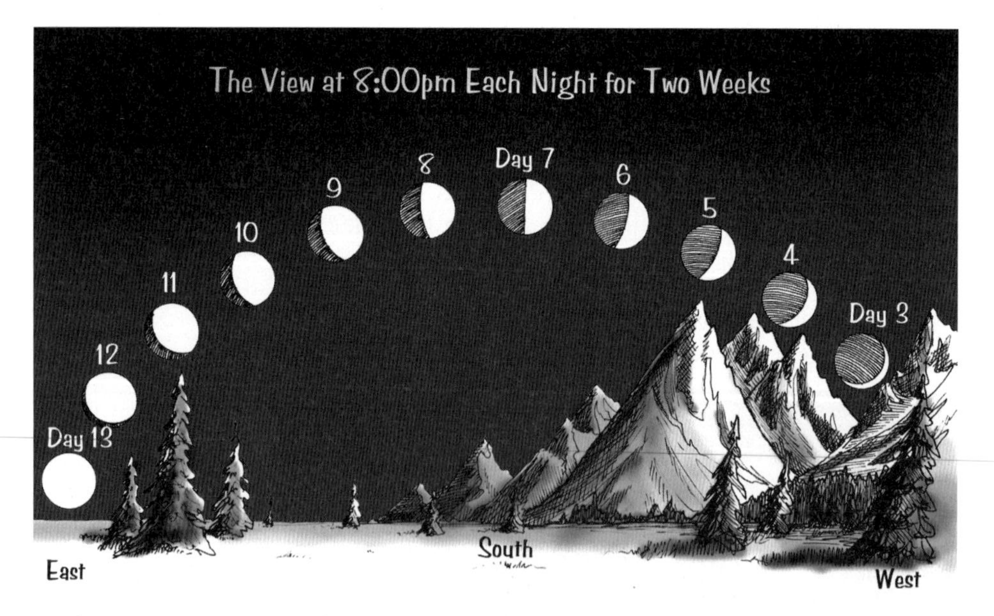

Figure 2.2 Go outside each night at the same time and you will see the Moon march slowly eastward against the landscape and background stars as it orbits the Earth. Each night a little bit more of the sunlit hemisphere is also visible from Earth making the Moon change phase from thin crescent to fat gibbous, and finally full over the course of two weeks (T. Nordgren).

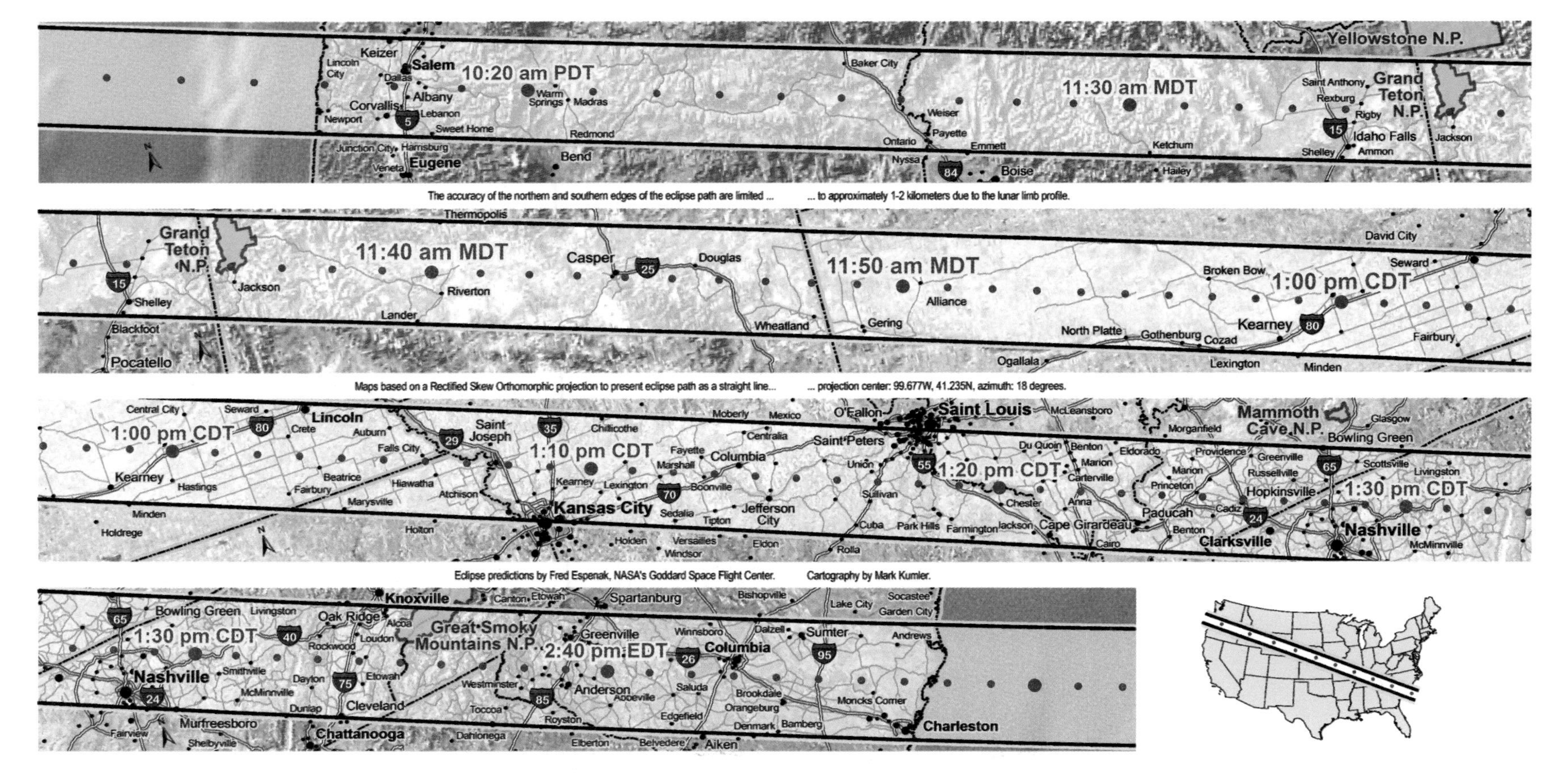

Figure 2.3 Path of the August 21, 2017 solar eclipse. To see the solar corona you must be between the black/yellow bands. The closer you are to the midpoint of the shadow (red dots) the longer totality lasts. Maximum totality lasts for 2 minutes starting on the Oregon coast, is 2 min, 20 sec in Grand Teton National Park and achieves a greatest duration of 2 min, 40 sec in Kentucky.

of the Moon half way around our planet. For the next two weeks the Moon continues its nightly motion rising later and later until nearly a month after you started, it rises in the east and sets in the west simultaneous with the Sun. At these times the Moon is invisible until a few days later when it appears as a thin crescent once more after sunset.

If the Moon went round the Earth in exactly the same plane as the Earth's orbit around the Sun, then we'd get eclipses every month as the Moon passed in front of the Sun. Sadly, the Moon's orbit is tilted by five and a half degrees with respect to the Earth's orbit. Most months if we could but see the Moon's completely shadowed dark side we would see it skim tantalizingly above or beneath the Sun from our vantage point here on Earth.

It is only at the moments when the Moon's orbit caries it from one side of the Earth's orbital plane to the other (the point of intersection is called the *line of nodes*) and does so exactly when the line of nodes points directly at the Sun, that we get eclipses. This happens roughly twice a year. In spring, for instance, a solar eclipse might occur somewhere on the Earth as the Moon passes through the line of nodes from north to south and the Sun, Moon, and Earth line up perfectly. As the Moon goes from above to below the Earth's orbital plane, it casts a dark circular shadow that moves across the Earth's daylight hemisphere. Depending on the geometry of the three celestial bodies, the narrow band of totality may be no more than 40 miles (65 km) wide from north to south. Only those observers within the thin band that the shadow traces out across the Earth's surface get to see the full glory of the solar corona. Roughly six months later, however, another solar eclipse happens on the other side of the Earth's orbit as the line of nodes once again points towards the Sun and this time the Moon passes through it from south to north.

In addition, two weeks before or after each of these magnificent alignments, lunar eclipses occur when the fully illuminated Moon passes through the line of nodes along the shadow of the Earth. Unlike total solar eclipses that are visible only to those within the narrow shadow band, lunar eclipses are much more democratic and visible to everyone on the Earth's unlit hemisphere where the Moon is above the horizon. This is what brings me to Grand Teton National Park today, August 27, 2007 – almost exactly ten years before the solar eclipse that will bring me back in 2017. For tonight will be a lunar eclipse visible to anyone in the western United States.

Perhaps you saw me. For the three days leading up to this eclipse I was the one with the compass and hand-made sextant making measurements at each of the scenic overlooks around the park. Just as I know exactly where I need to look to see the eclipsed Sun in the sky ten years from now, tonight I know exactly where I need to look to see the Moon pass through the different parts of the Earth's shadow. In such a beautiful location as Grand Teton National Park, I want the view to be in just the right spot where no part of the Moon's passage is blocked by mountain or tree. In the end I have decided to come back about 2:00am and set up my equipment at the Mountain View Turnout just north of the Jenny Lake Loop junction along the Teton Park Road.

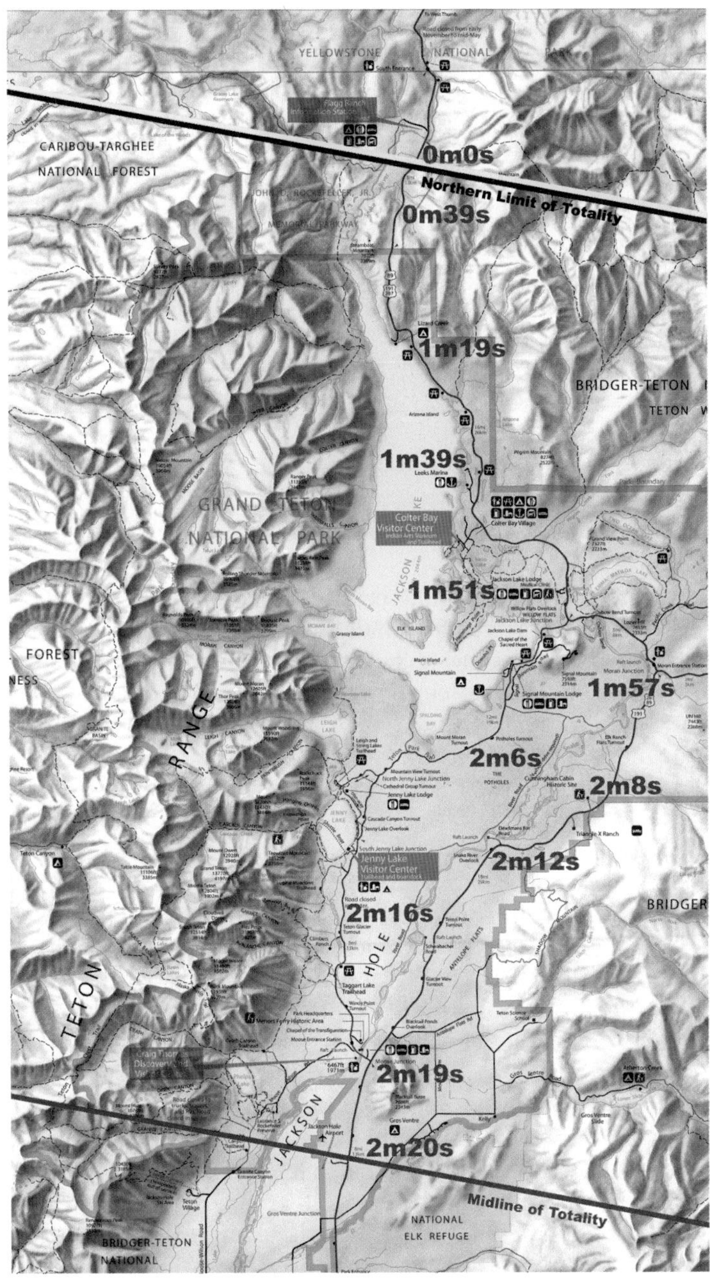

Figure 2.4 Detail of National Park Service map showing a close-up of Grand Teton National Park. The midpoint of the 2017 solar eclipse path of totality is shown in red with the northern limit, beyond which only a partial eclipse will be visible, shown in black/yellow. The duration of totality is given in minutes and seconds starting with the midline and progressing north through various park pullouts and lodges. Totality is markedly short near the northern end of the park. The midline passes directly through the Jackson Hole Airport (National Park Service, Eclipse Predictions by Fred Espenak, NASA Goddard Space Flight Center).

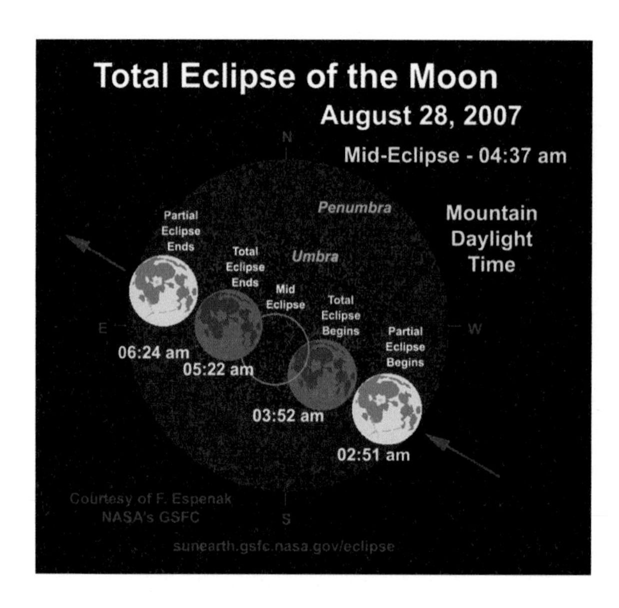

Figure 2.5 Schematic of the lunar eclipse of August 28, 2007. The penumbra is the region of the shadow where only part of the Sun's light is blocked. The umbra is the region of complete eclipse (Fred Espenak, NASA Goddard Space Flight Center).

Like its showier solar sibling, lunar eclipses are also a strange and awe inspiring sight. The Earth forever casts a dark shadow far out into space but like a raven at night, there is nothing to show that it's there. Only when the Moon casually passes into the shadow do we see the full extent of its presence. Slowly a faint darkening, the penumbra, begins to pass across the face of the Moon. Its shadow is so faint I wouldn't notice it if I wasn't paying attention. An hour after the first hint of the penumbra touches the eastern edge of the Moon (the leading edge as it travels eastward through space around the Earth), I see the first real dark brown bite taken out of the lunar limb. This is the umbra, the heart of the Earth's shadow, and over the next hour as it passes across the lunar disk it is increasingly obvious to all but the most casual observer that something is odd. Sure, we've all seen the Moon in its different phases each and every month, but this is different. The Moon shouldn't be this shape; it looks like a cookie with a bite taken out.

Ultimately, the entire disk plunges into the umbra at which point two amazing things happen at once. The first is that, where before there was hardly a star in the sky, suddenly a thousand stars burst forth and the Milky Way stretches the entire length of the celestial vault. Now you may be wondering why this is so amazing. After all, it's night, shouldn't there be stars, and especially so in a national park without the interference of city lights? While it's true, that the nation's parks are becoming the last remaining refuges of natural darkness where visitors can view a star-filled sky that's as strange and exotic as seeing glaciers or geysers, a full Moon can be as bright as an entire city. When the Moon is full, or even nearly so (as it has been for the last several nights before the eclipse), its reflected sunlight illuminates the night sky just as brightly as city lights do at home. However, at the instant that the Moon passes totally within the Earth's shadow, that reflected sunlight virtually disappears and suddenly the stars are revealed.

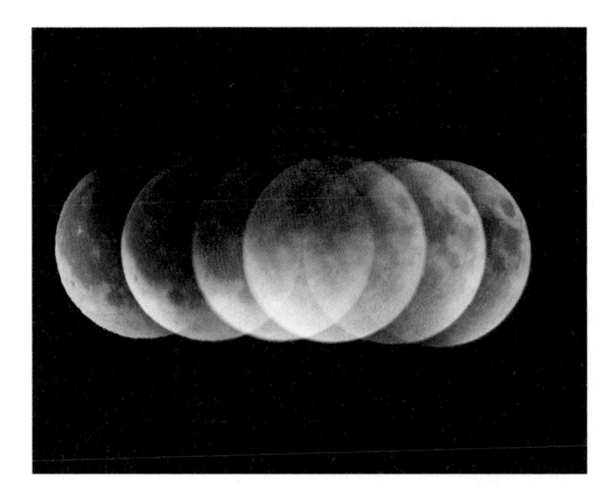

Figure 2.6 The Moon passing through the Earth's shadow in 2007. The multiple images show the actual motion of the Moon through the shadow and relative to the stationary background stars (one of which is visible in the upper left) (T. Nordgren).

I'd forgotten about this fact in all my efforts to get out here and get my cameras and telescope set up. Suddenly the summer Milky Way, all but invisible an hour ago, stretches clear across the sky and down into the jagged mountain tops of the Tetons. A meteor suddenly blazes overhead, and in the 90 minutes of totality's dark I see a half dozen left over shooting stars from the Perseid meteor shower that peaked two weeks earlier.

But even as stunning as this transformation may be, it's the second sight that leaves me speechless. For at the center of the star filled sky is an enormous blood red Moon, and for a few moments even an astronomer drops his scientific calm and is spellbound by its beauty.

During a total lunar eclipse the Moon turns shades of orange and rose, burnt umber and mahogany; it is unlike any other time in its monthly trip around the Earth. In my life I've surely seen a dozen lunar eclipses and each and every time the color is different. During the lunar eclipse of December 1992 I saw the Moon turn such a dark shade of red that it almost completely disappeared in the starry sky. To understand what causes this color, I imagine that I could witness these same events from the surface of the Moon.

In my imagination it's now two hours earlier, and I stand upon a blazingly bright lunar landscape vaulted by a coal black sky. The sunlight that falls around me is merciless in its light and heat; without an atmosphere, its brilliance shines more intensely than that seen from even the most desolate terrestrial desert. Had the Moon an atmosphere, I'd see waves of heat radiating off the rocks and along the crater rims that flank me on this smooth powdered plain. In the midst of the Moon's two week long period of daylight, it is 253° Fahrenheit (123° Celsius) at noon. All is quiet and motionless where there is no air to carry sound or blow dust across the distant mountain peaks.

Above me and beside the Sun, nearly completely lost within its glare, is the night-time Earth. If not for the overwhelming light of the Sun I could see those city lights that have rendered faint stars invisible back home. As the Moon's motion slowly brings the Earth in front of the Sun, I realize that while everyone back home is about to be treated to a total *lunar* eclipse, I alone on the Moon will witness a total *solar* eclipse.

For the first few minutes of the Earth's slow passage in front of the Sun, the light that falls around me gradually dims. It does so only faintly though, as the rest of the solar disk is still so bright. This then is the penumbra, that first faint wisp of shadow visible back in the Tetons. Slowly, the bright light dims as the Earth continues to cross the solar disk. Eventually the Sun disappears completely and suddenly darkness falls all around. This is what it means to stand in the umbra. I look around me on the lunar plain and can still see the distant mountains shining in the brilliant sun. The lunar eclipse visible back home is still not total, not until they too finally fall into the umbra and the entire hemisphere on which I stand is fully shadowed from the Sun.

But while the eclipse is still only partial for those on Earth, for me I am bathed in totality. The Earth appears four times larger in the sky than the Moon and Sun do on Earth. So while a total solar eclipse may last only a few minutes back there, for lunar observers it takes an hour and a half for the Sun to travel behind the Earth. While it does so, I am treated to a view no Earth-bound observer has ever seen. For while the Moon has no atmosphere, the Earth most definitely does. As I stand here on the Moon, fully within the shadow of the Earth, faint sunlight streams past the mother planet and lights up the ring of atmosphere I see separating the nighttime Earth from the day. All that gentle light passing from day to night in the dusty atmosphere of my home world forms a reddish ring in the darkness above. I drop my eyes to the landscape around me and I am bathed in this reddish light: it is the light of every single sunrise and sunset happening on Earth at this moment.

So, back on Earth, when you see that ruby red Moon overhead during the moment of totality, what you are seeing is the blended glow from all the colors of all the sunrises and all the sunsets shining up onto the Moon and reflected back to Earth. In that light is the rising Sun heralding every new workday in small towns and sprawling cities and every lazy Sun setting over crowded city streets and south Pacific seas.[1]

But even as beautiful as this lunar eclipse is, I long to return here in 2017 to see the awe inspiring spectacle of the total solar eclipse, 3,646 days hence. To be able to make such accurate predictions, to know where I need to be along the Moon's slender shadow's path at just the right time means we must really understand how the Universe works. We must really know how Moon, Earth and Sun move together through space, and know it so well that we can predict their paths perfectly over ten trips of the Earth around the Sun, and 123 and a half trips of the Moon around the Earth. And the proof for whether or not we truly

[1] The remarkably dark lunar eclipse I saw in 1992 was due to the eruption of Mt. Pinatubo in the Philippines the year before. So much volcanic dust and ash was sent into the atmosphere that brilliant red sunsets were seen all over the Earth thus also reducing how much sunlight reached the Moon during the total lunar eclipse.

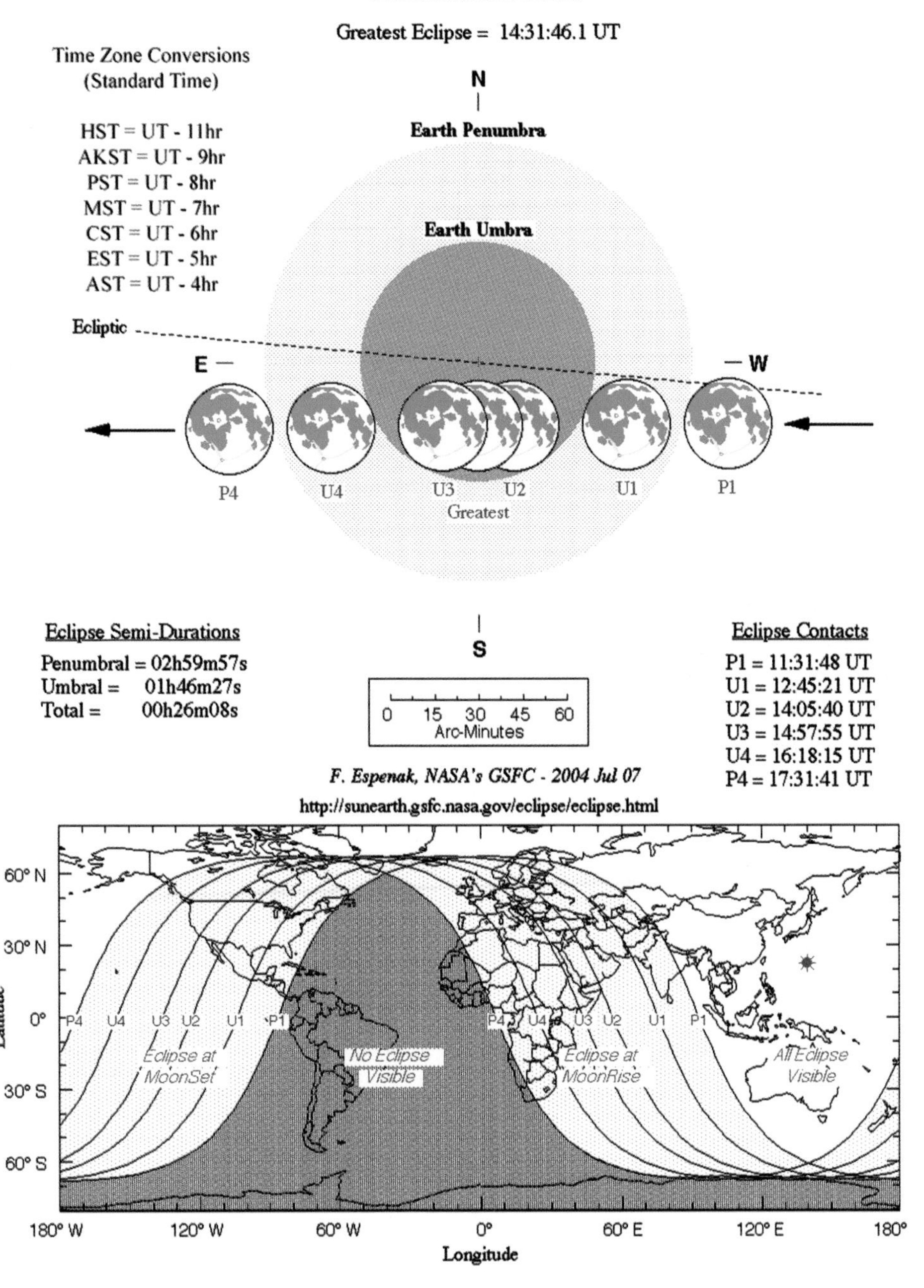

Figure 2.7 Observing information for the total lunar eclipse of Dec. 10, 2011. For North America the eclipse will occur early in the morning near moonset (Fred Espenak, NASA Goddard Space Flight Center: sunearth.gsfc.nasa.gov/eclipse/eclipse.html).

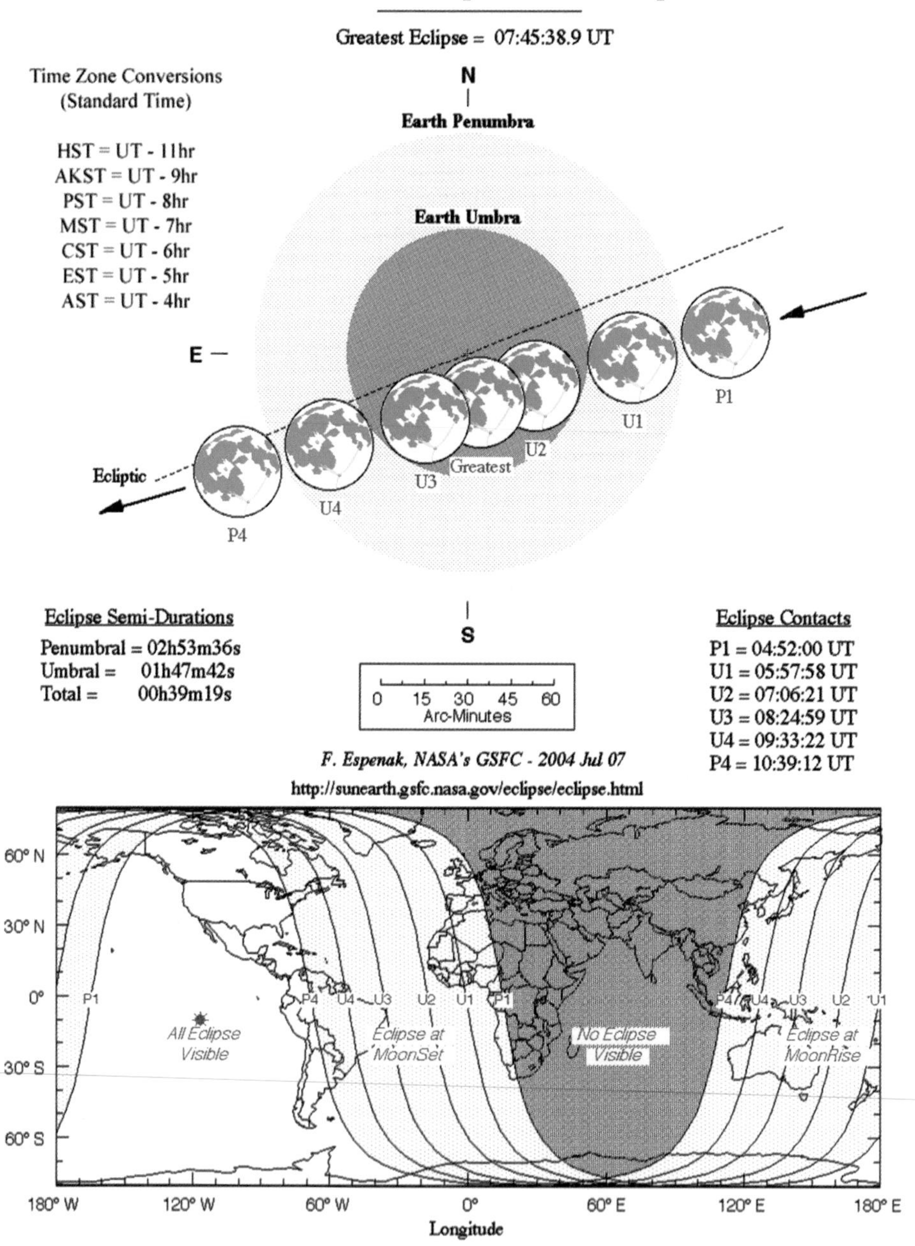

Figure 2.8 Observing information for the total lunar eclipse of April 15, 2014. For North America the eclipse will occur during the middle of the night (Fred Espenak, NASA Goddard Space Flight Center: sunearth.gsfc.nasa.gov/eclipse/eclipse.html).

understand this celestial mechanics is right there above me tonight just as it will be there for all to see on a sunny Monday morning in August, ten years from now.

It may sound strange, but I'll be in the right spot, at the right time, looking in the right direction because 400 years ago, a Lutheran astrologer named Johannes Kepler had an unshakable devotion to God. Five hundred years ago, in 1517, Martin Luther nailed his 95 theses to the Wittenberg Church door and started the split between Protestant and Catholic Churches. One of the founding tenets of Protestantism was that during the previous fifteen hundred years Christianity had become weighed down by successive layers of belief and practice that had nothing to do with real Christian faith. According to this thinking, the true Christian should be free to go back to the original holy texts, and without being forced to rely upon another's authority, understand the Creator's Divine plan without the accumulated millennia's worth of misleading or mistaken interpretation and translation.

Into this Protestant tradition Johannes Kepler was born in the afternoon of December 27, 1571, in the Imperial Free City of Weil der Stadt, Württemberg, in what is now Germany.[2] His father was a mercenary, uninterested in family life and when Kepler was still young, he went off to fight and die (presumably) for the king of Naples and was never heard from again. Young Johannes, with his brothers and sister, was raised by his mother, a harsh and intimidating woman. While the family was not rich, Kepler soon made a name for himself as being a brilliant student and he earned a spot in the scholarship system of the local duke which would pay for all his future education and eventually, he hoped, lead to a future in the Church. Kepler was fiercely devoted to his Protestant upbringing and education; the Protestant tradition that placed a paramount understanding of God through personal scholarship fit perfectly with his brilliant and inquisitive mind.

In Kepler's time, science, like religion, was subject to dogma. For nearly two millennia before Kepler, knowledge of the natural world had been seen through the prism of the Greek philosopher Aristotle. Aristotle had sought to apply order to the natural world and over time his writings had become the final word on the workings of Nature. But Aristotle's philosophy was rooted in the idea that the observed world could be understood simply by thinking about it.

In his view the world was made of four earthly elements: earth, water, air, and fire. The fifth element was the divine *aether* that composed the heavenly spheres and stars. Since the heavens obviously rise and set around the apparently stationary Earth, he concluded we must be the center of the Universe with each

[2] James A. Connor's book, *Kepler's Witch: An Astronomer's Discovery of Cosmic Order Amid Religious War, Political Intrigue, and the Heresy Trial of His Mother*, is an excellent biography and discussion of the various cultural influences that produced the first and greatest of mathematical astronomers.

element, from earth to *aether,* having its natural place in decreasing weight outward from this center. Because elements seek their natural place, components of earth held in the air aloft would always seek to fall back down when released. The more earth you hold aloft, the more it would seek to return to its origin. Thus is gravity explained.

Considering what one can see for oneself, this makes pretty good sense (be honest, have you ever seen an atom and have you ever seen or felt the planet zipping through space around the Sun?). But as reasonable as this theory of gravity may sound, it can be easily tested by dropping two objects of differing weights and seeing if they actually fall at different rates. While air resistance may keep the two weights from falling at exactly the same speed, the results are nowhere near what Aristotle's gravity would predict. But there is absolutely no evidence that it ever occurred to Aristotle or any of his adherents over the next one thousand, nine hundred years, to try such a test. In a conflict between the messy world and an elegant idea, especially an idea of Aristotle's, the idea won.

This is not the way science works today. For a scientist, and for the millions of people today who board airplanes, ride in cars, or take medicines and thus rely upon the products of science, the only thing that matters is if it works; our theories about Nature must actually match with what we see in reality.

In the mid-1200s a Dominican priest and scholar by the name of Thomas Aquinas combined Aristotle's writings with Christian texts to give the Church a widely accepted physical and philosophical support structure. Aristotle's ideas of the Universe now lent support to, and were in turn supported by examples in the Bible. As a result, by the time of Kepler and Galileo nearly four hundred years later, the principal way to decide the accuracy of a scientific theory was whether it agreed with Aristotle (which for all intents and purposes was now synonymous with saying it agreed with the Bible).

But the Protestant Reformation categorically rejected the idea of placing personal understanding under the yoke of accepted interpretation or translation. Since God created the Universe, Kepler saw the Universe as a wordless Testament reflecting God's will for the Cosmos. For Protestant Kepler, this meant the Universe itself must be studied in its original language – the language of mathematics. But the primary purpose of this study had to be understanding the Universe correctly, the way it really worked (even if this meant ignoring over two thousand years of Aristotle and the Church). To fail to do so would be to willingly place one's own wishes or beliefs above those of God's.

Kepler's Catholic contemporary, Galileo, also believed the Heavens were God's revealed plan, and thus should take precedence where conflicts with a literal interpretation of the Bible arose. It is this similarity between Galileo's views and the Protestant approach to interpreting Scripture that led directly to much of Galileo's eventual difficulty with the Catholic Church (which was busy at that time waging a spiritual war to crush Protestantism with the Counter-Reformation). For neither of these men was the quest to understand the Heavens ever a battle of science versus religion, as is often portrayed today. Both were devoutly religious; their single minded determination to reject the accepted dogma and authority of over two thousand years was simply an attempt to understand their God and His plan as closely as possible.

Kepler thought he had at last found this understanding in the orbits of the planets. In Kepler's time only seven planets were known: the Moon, Mercury, Venus, Sun, Mars, Jupiter and Saturn.[3] The widely accepted ordering of this Universe was Aristotle's Earth-centric model as interpreted mathematically by Ptolemy in the first century A.D. Over the next one and a half millennia, more and more precise observations of planetary motions led to layer upon layer of ever more complicated models of this Universe. The most important quality of Ptolemy's Universe was that the Earth was utterly motionless at its center. The Earth did not go around the Sun; it did not turn on its axis. The motion of the Sun across the sky was literally that: the motion of the Sun. This is actually the simplest answer to what we observe each and every day.

On average it takes 24 hours for the Sun to move once around the Earth, and so it must be traveling on a circle or sphere that carries it around the Earth with a period of 24 hours. Likewise the Moon also moves around the Earth, but in order to rise nearly one hour later every night it has to revolve around the Earth with a period of about 25 hours.

And what about the stars? Look at the stars each night at exactly the same time after the Sun goes down and you will notice that from night to night the stars slowly creep westward behind the setting Sun. In this way, the constellations we see in summer's evening sky (Sagittarius and Scorpius) slowly give way to Taurus, Orion, and Gemini in winter evenings. For the stars to do this, the celestial sphere that carries the stars must circle the Earth from east to west faster than the Sun. The period of the stars' orbit is therefore not 24 hours like the Sun, but rather 23 hours and 56 minutes.

And then there are the planets that wander against these stars, slowly moving through the constellations of the Zodiac. To do this, they must be on spheres that circle the Earth with yet another period. The rate with which each does so is again not 24 hours, but rather some slightly slower rate that lets each move slowly eastward through the Zodiac, each at its own peculiar rate.

To add final insult to injury, careful observation of the precise positions of the planets shows that their eastward motion isn't even constant. Sometimes a planet will slow, stop, reverse its direction awhile, then stop again and continue on as before (all of this happening while still circling the Earth once a day). To account for this *retrograde* motion that every planet but Venus and Mercury experiences, the planets were placed on small circular paths (called epicycles) where the center of the epicycle was actually the thing carried round the Earth

[3] At that time, a 'planet' was simply something that wandered through the background stars. The exact identities of the 'known planets' actually depended on whether or not you thought the Earth was the center of the Universe. If you did, then Earth wasn't a planet, but the Moon and Sun were. If you didn't, then the Sun was no longer a planet and the Earth was instead.

with a period a little less than 24 hours (but still more than 23 hours and 56 minutes).

Through this complicated series of epicycles and spheres, the Ptolemaic model of the Universe could be made to approximately match what was actually seen. But with every new improvement in observational precision, new layers of circles and cycles had to be added to keep the great whirling model in line with reality.

In 1543, nearly two thousand years after Aristotle's death and nearly thirty years before Kepler's birth, Nicholas Copernicus simplified the Universe. His model placed all of these bodies in perfectly circular orbits around the Sun, with only the Moon left circling the Earth. The nearly common period of 24 hours required for every other object in the Universe, was now simply the Earth turning once on its axis. All the slight variations in period were now easily due to the speed of the heavenly bodies and their distance from the Sun. While the new model often predicted the positions of the Moon and planets no better than the old system, it had the benefit of elegant simplicity.

With this newfound order, a person of a philosophical mind might then ask: Why are the planets at the distance they are found and why are there only six (now that the Moon was classified as something else)?

Kepler thought he found the answer in the five perfect Pythagorean solids, which he thought could be made to fit perfectly in the gaps between the planets.[4] With only five possible solids there could be only six possible planets and so for the first time in the history of human thought he believed he had found the hidden proof for God's blueprint for the Heavens. When Kepler published his *Mysterium Cosmographicum* in 1596 he became instantly famous as the man who had discerned God's framework for the Universe.

In keeping with his deep-held conviction in the Protestant ideal, Kepler spent the next decade seeking the observational evidence that would prove that his understanding of God's will was correct. In Tycho Brahe, the preeminent astronomical observer of the day, and the last great astronomer before the invention of the telescope, Kepler found the observations of the planetary positions that would yield the data he needed. But after years of laborious calculation using the planet Mars, Kepler was forced to the conclusion that his model of the Solar System and solids was utterly impossible.[5] The perfectly circular orbits he assumed for the Copernican model were completely inconsistent with what the observations for Mars actually showed.

[4] A perfect solid is a three dimensional object (a solid) where each face is exactly the same. A cube is a perfect solid composed of six equal squares. There are only five such solids, including the tetrahedron of four equal triangles, and the dodecahedron composed of 12 identical pentagons.

[5] These years also included titanic personality clashes with Tycho Brahe, Brahe's death, subsequent lawsuits from Brahe's heirs, and the persistent rumor that Kepler had murdered Tycho in order to steal his data and reputation for himself.

It must have been heartbreaking for Kepler. These were the ideas that made him famous. They were the window he thought he had opened into the mind of God: God the Geometer. And they and he were utterly wrong. How many of us in a similar circumstance would have swept the offending data under the rug, or done everything in our power to rationalize why it could be ignored or to shoehorn the observations into our cherished ideas? But Kepler didn't do any of these. Sticking strictly to what the observations actually showed, Johannes Kepler spent the next decade revising his most cherished ideas, and in the end discovering the three great laws of planetary motion that are still in use today. They are:

1. All planets orbit the Sun in an ellipse, where the Sun is located at one focus of the ellipse (a circle is just a very special case of an ellipse).
2. As planets orbit the Sun, a line joining them to the Sun sweeps out equal areas in equal periods of time (in other words, planets speed up when they are closer to the Sun and slow down when they are farther away).
3. The square of the period with which a planet orbits the Sun is proportional to the cube of the semi-major axis of the orbit (for a circle the semi-major axis is just the radius).

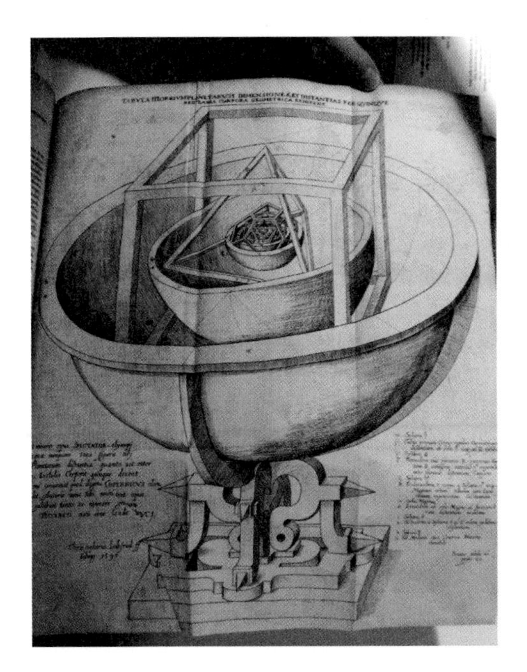

Figure 2.9 Kepler's model of the Solar System from his *Mysterium Cosmographicum* (Courtesy of the Division of Rare and Manuscript Collections, Cornell University Libraries).

In seeking to understand why the planets obeyed these laws, Kepler was the first to ascribe physical or earthly causes to heavenly motions. His three laws implied that some force must reside within the Sun that attracted the planets, speeding them up as they came closer while making distant planets move more slowly. The idea of a physical cause for why the heavens moved was revolutionary. From the time of Aristotle, the heavens were supposedly unique and separate from the elements and actions here on Earth. The role of the astronomical astrologer was, therefore to simply describe what the heavens *did.*[6] *Why* the planets did what they did was the provenance of God (or at the very least

[6] And thus be able to cast horoscopes for what this meant for your wealthy patron. What this meant for the planet was irrelevant.

separate from the impure and corruptible Earth). In looking for physical reasons for why the planets obeyed his planetary laws of motion (magnetism, perhaps?) Kepler stopped being an astrologer, and instead became the world's first astrophysicist.

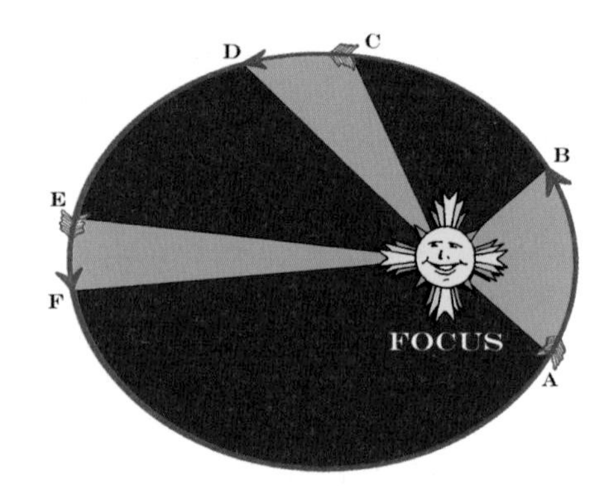

Figure 2.10 Kepler's first and second laws for planetary orbits. The Sun is shown at one focus of the ellipse. The path the planet follows over the course of a month (for instance) is shown by the three arrows. The area swept out by each arrow (shown in gray) is always equal. Thus the planet moves quickly when close to the Sun (perihelion) and slower when farther away (aphelion). The difference in distance between aphelion and perihelion is grossly exaggerated for planets in this image (T. Nordgren).

From Kepler's religious faith we see the birth of what would eventually become the scientific method: First (as Kepler saw it), it is vital to observe God's Universe. From your observations you create a model or hypothesis for how you believe the Universe works. But then, and here's the important thing, it must work. If your hypothesis fails to match what is actually seen, it's you that must change. It's you that are attempting to understand God; if your understanding is in error, learn from your error, and correct it. Do not fit false interpretations to God's observed work just because you think it should be the way you wish it. In time, it is only the hypotheses that survive this rigorous testing with observation and experimentation that survive to become laws, or what today we have come to call scientific theories.

And that's it. The scientific method we have today is not rooted in an anti-religious desire to prove there is no God, but rather it has its historical origins in a desire to understand God's Universe better. If Kepler and Galileo were alive and practicing science today, they would likely argue for the scientific theories of Evolution and the Big Bang, not because some perceive them as Godless, but rather because they are the best theories that describe what is really seen in Nature. Every 'religious' alternative fails to accurately reflect what is observed and so (according to the tradition of Kepler and Galileo) must therefore be perversions of the wordless testament that is the Natural world.

During his life Kepler made no secret that he would not bow to being told what he must believe. As the Lutheran Church sought to codify accepted tenets of belief for the new religion, Kepler fought for the right to accept and reject those parts as his understanding of the original texts warranted. He felt this was central to his faith as a Lutheran and was therefore central to the accurate understanding of the Universe and God. For this he was excommunicated by his own Church. But while Protestants rejected him, he could not reject Protestantism and so he was subsequently expelled from one home after another when

Catholic rulers demanded their subjects choose between Catholicism or exile. In the ensuing Thirty-Years War when Catholic went to war against Protestant, neither side trusted him and neither Church would accept him. During this time his wife died, his children died, and his mother was tried and convicted of witchcraft. Through all of these trials and torments he held true to his conviction in absolute honesty before Nature and God and as a result ushered in the modern age of science.

Look back to the stars. Computers using Kepler's three laws produce the numerical tables that tell me the Moon will eclipse the Sun in 2017 just as they told me the Earth would eclipse the Moon tonight. Astronomer Fred Espenak at NASA's Goddard Space Flight Center, maintains NASA's eclipse website where his calculations reveal the paths of solar eclipses for the next thousand years. When the Moon in its elliptical orbit passes between Earth and Sun at the far end of its orbit, the Moon appears too small to cover the entire disk of the Sun and a brilliant ring of fire, an annular eclipse is the result. When the geometry of the eclipse isn't perfect (or during totality where the majority of us are not fortunate enough to be within the central band swept by the shadow of the Moon) a partial eclipse is visible instead. For each of these types of eclipses, occurring all over the Earth, maps and times are already calculated for those willing to make the trip to see them.

But the calculations of Keplerian motion reveal not only future intersections of Sun and Moon, they reveal past ones as well. On the 11th of July, 1097 we now know the shadow of the Moon swept across the north Pacific Ocean and cut diagonally southeast across what would later be the western United States. During that early summer afternoon the shadow continued over the San Juan Basin in the Four Corners region of the southwestern U.S., and into what is today, Chaco Culture National Historical Park.

The arid canyon at the heart of the present park boundaries marks the ceremonial center of the Chacoan civilization, the ancestors of today's Puebloan peoples. In 1097 the vast Chacoan culture was at its peak and in the midst of nearly four hundred years of continuous occupation and ceremony amongst the dry sandstone walls of the canyon. From numerous apparent celestial alignments of rock carvings and the acre-sized buildings still found there, it is almost certain

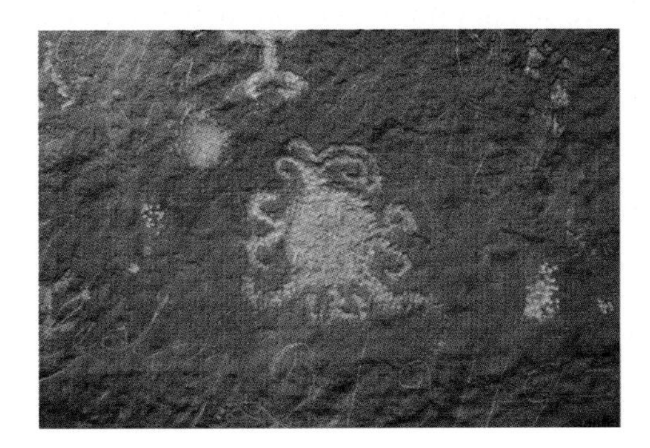

Figure 2.11 A total solar eclipse recorded in stone. Could this be the 1097 eclipse of the Sun with solar corona curling about the darkened disk? Venus is hypothesized to be the circle at upper left of the petroglyph. Many more current 'rock carvings' now surround the ancient observation (T. Nordgren).

that the heavens were an integral part of this culture living at the very edge of what the land could sustain.

On the outskirts of one of the park's ruins, on the east face of an enormous dun-colored boulder, a petroglyph was made unlike any other that can be found around the southwest. Someone carefully pecked out of the surface of the stone a circle with fanciful curls and loops extending outwards from all sides. To the upper left is another smaller circle whose form, depth and weathering suggest it was made roughly contemporaneous with the other. Kim Malville, an archaeoastronomer at Fort Lewis College in Colorado hypothesizes that these features could be the record of that solar eclipse, where the corona that is at all other times invisible is recorded in the strange waves and hoops radiating from the central disk. And the other, smaller circle?

Recall that during the moments of totality's darkness, when the Sun's glare no longer brightens the sky, the brightest stars and planets are briefly visible in a day-time sky. The residents of Chaco Canyon a thousand years ago would have seen the same thing, and may have accurately recorded not only the appearance of the Sun's corona, but the sudden emergence of Venus as well. From Kepler's laws we know the two circles sit in relation to one another exactly where you and I would have seen Venus at the time of totality, so many centuries ago.

Today the rock art of Piedra del Sol is surrounded by other 'artwork,' some much more recent and far less ceremonial in nature than the rest.[7] It is almost certain that the petroglyph's true meaning will never be known for sure. But through Kepler's work almost exactly five hundred years in its future, and the calculation of astronomers an additional four hundred years further forward in time, we can tell exactly what Chacoans saw in that long-ago afternoon sky. Whether they chose this rock to record this amazing apparition is almost beside the point.

Almost 50 years after Kepler's death, Sir Isaac Newton postulated a simple force of gravity at work between any two objects with mass. The greater the masses, the stronger the force, the farther apart the masses, the smaller the force becomes. This simple hypothesis of a fundamental physical force unified the everyday drop of an apple with the cosmic motion of the Moon around the Earth. Both follow the paths they do because of their attraction to the Earth. To see their similarity Newton imagined standing on a high mountain's edge. Drop any object with mass (an apple, say) and it will fall straight down to Earth as we expect. Now throw it with some slight velocity and while it still falls to Earth it does so a little farther away from the base of the cliff. Throw faster and each

[7] For this reason, as of 2008, the rock art of Piedra del Sol is not on display to the visiting public. While park rangers are looking to find ways to open this amazing panel to the public, they are careful not to do so until they are confident it won't succumb to the ravages of more modern rock art: graffiti.

Figure 2.12 Planetary orbits as devised by Newton. An object thrown with increasing velocity follows a curved path around the Earth until eventually an object thrown with sufficient speed loops back upon itself. Illustration from Newton's *A treatise of the system of the world*, 1728. A photo of this figure, from this exact copy of Newton's book was included on a golden disk placed aboard the Voyager 1 and 2 spacecraft now heading to the stars (Courtesy of the Division of Rare and Manuscript Collections, Cornell University Libraries).

successive toss sends the apple farther from the cliff. Gravity's attraction makes the apple forever fall towards the Earth while the speed of the apple simply determines how far around the Earth the apple flies before it eventually hits. Throw hard enough (perhaps with the aid of a rocket-powered cannon) and the apple flies entirely around the Earth, its path looping back on itself as it flies once more over the cliff where it started. Gravity's pull makes the apple fall forever around the planet, yet forever missing as it does so. The result is an Earth with an apple for a moon.

From Newton's law of gravity and the acceleration of bodies produced by this force, all three of Kepler's laws can be mathematically reproduced but now with the knowledge that it is gravity that connects the Heavens and Earth. From this discovery the mysterious quantity that related the square of the orbital period and the cube of the orbital distance in Kepler's third law is none other than the total mass of the objects. If you know the period of Earth's orbit around the Sun and how far from the Sun the Earth orbits, Kepler's third law reveals the total mass of the Sun (the Earth being microscopically tiny in comparison). Kepler's laws in this form then become models for everything a good scientific theory should do: they explain what is seen, predict the result of future events, and tie together a wide range of physical phenomenon in order to illuminate deeper truths about the world. While ancient astrologers may have been able to predict when eclipses would occur somewhere on Earth, today's astrophysicist can do that and predict the exact timing of eclipses for anywhere on the surface of Mars, as well.

Figure 2.13 A solar eclipse by the Martian moon Phobos as recorded from the surface of Mars. On the 45th Martian day, or sol, of its mission, the NASA Mars rover Opportunity photographed one of Mars' irregularly shaped moons pass in front of the Sun. Phobos is too small to ever completely block the solar disk when viewed from Mars. The eclipse progresses from left to right (NASA/JPL/Cornell).

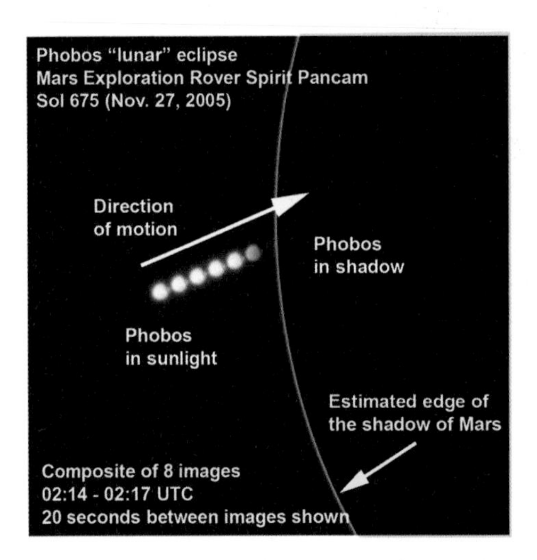

Figure 2.14 A lunar eclipse on Mars is observed by the NASA Mars rover Spirit. Here Phobos quickly fades from view as it enters the shadow of Mars. Crater Stickney is barely visible on the lower right limb of Phobos' disk (NASA/JPL/Cornell).

As a result, Kepler's third law (with Newton's addition) has become one of the most useful discoveries in astronomy in the last 400 years. While Kepler devised his law for planets in their orbits around the Sun, thanks to Newton we now know they hold equally true for any object that orbits any other. Point a pair of binoculars with a magnification of only 20× at Jupiter and anyone today can see the night-to-night motion of the four large moons Galileo discovered in January 1610. With Newton's refinement of Kepler's third law, knowledge of the period of the moons and the orbital distance of each tells us unambiguously that Jupiter is 318 times more massive than the Earth and the largest of all planets orbiting our Sun. Thanks to Newton and Kepler the heavens can be weighed.

Newton's example of throwing things from the tops of mountains leads to an intriguing result if followed far enough. The more massive the planet, the faster you have to throw or fly in order to go into orbit and escape the surface altogether. Jump up. Try it. No matter how strong a jumper you are, you will always come back down. The Earth has just too much mass to let you escape the surface without returning sooner rather than later. But the faster you jump, or the faster the rocket you use to jump for you, and the farther away you get before coming back to Earth. Alan Shepard, the first American in space, sat atop a

Mercury Redstone rocket that left the Earth with a speed of 5,134 miles per hour (8,214 km/h) and reached a height of 116 miles (186 km) before eventually falling back to Earth. In order to completely *escape* the Earth's pull and never return, he'd have needed to travel nearly five times faster.

The denser the planet (or star) the greater this escape velocity becomes. But what happens as the escape velocity becomes *really* big? Compress the mass of the Sun into a ball no larger than 4 miles (6 km) in diameter, a small town in size, and the force of Newtonian gravity at its surface will pull with such strength that not even light itself can exceed the escape velocity. The possibility of such "black holes" intrigued the eighteenth century Reverend John Michell, a trained scientist who also published papers on the strength of gravitational forces on Earth, and the likelihood that stars in the sky orbited one another under their own mutual gravity. Interestingly, it is the very fact that stars are attracted to and orbit each other according to Kepler's laws that allows us to test Michell's idea for what happens when a star's escape velocity approaches the speed of light. For how else is one to discover the massive unseen object from which even light is too weak to escape, if not by the gravitational influence it has on those objects that it encounters?

Deep in the core of the Milky Way there lurks a gravitational monster. Hidden from our eyes by 25,000 light-years of intervening dust, the center of our own Galaxy has always been a mysterious place. Since the invention of radio telescopes in the 1950s, astronomers have detected radio waves coming to us out of the hidden depths of the Galaxy beyond the constellation of Sagittarius. This radio source, called Sagittarius A* (pronounced Saj A-star) is believed to be at the very center of our Galaxy. For the last two decades Andrea Ghez, an astronomer at the University of California at Los Angeles, has been monitoring a cluster of young stars surrounding this radio source, using a new generation of giant infrared telescope able to peer through all the intervening dust and gas. Over the years, her patient and painstaking observations show the stars' orbital motion in the very inner one light-year of our Galaxy. Only because the stars move with

Figure 2.15 The center of our Milky Way recorded in infrared light using the Midcourse Space Experiment (MSX) satellite. Shown are distant dust clouds heated by faint starlight. The bright knot in the middle is the center of our Galaxy (MSX/IPAC/NASA).

Figure 2.16 Image of the center of the Milky Way made using the W.M. Keck Observatory. The position of the source of radio emissions that is believed to emanate from around a massive black hole called Sgr A* is labeled (W.M. Keck Observatory and the UCLA Galactic Center Group).

velocities of thousands of miles per second are we even able to see their changing positions from so far away.

A decade of observations reveal stars on great looping elliptical orbits that all have a single common gravitational focus consistent with the position of Sagittarius A*. Around this object, they, and by extension every dust cloud, cluster and star in our Galaxy, including our Sun, are all in perpetual Keplerian motion. From the periods and orbital sizes of these stars, Ghez calculates that the object about which they revolve must be 3.7 million times more massive than our Sun, all within a space smaller than our Solar System. Yet while her telescope reveals the light of the stars around it, it sees no sign of the light of almost four million suns at the single gravitational focus. In fact, it sees no light there at all.

However, an object this massive, in a region not much smaller than this, will have an escape velocity approaching the speed of light. For these reasons, astronomers are almost certain the object at the position of Sagittarius A* is a supermassive black hole. Black holes like this are thought to be found at the

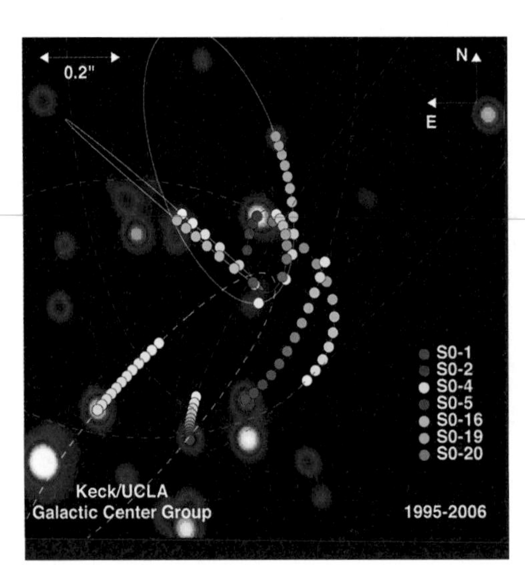

Figure 2.17 Stars at the center of our Galaxy are observed orbiting a common point at the position of Sgr A*. In the background is displayed the central portion of an image taken in 2004. While every star in this image has been seen to move over the past 15 years, only the seven labeled stars exhibit enough curvature of motion to construct their full orbits (shown as solid or dashed curves). Together they provide the best evidence yet for a supermassive black hole with 3.7 million times the mass of the Sun at their common focus (and where there is no corresponding object with the light of 3.7 million suns) (Andrea Ghez (UCLA) and the Galactic Center Research Group, data obtained with the W.M. Keck Telescopes).

centers of all large galaxies. Where they came from is one of the great areas of ongoing research. To understand the origin of these monsters may be to understand the origin of galaxies themselves.

Even our own Solar System has been no stranger to massive unseen objects. For all but the last two hundred years only the six planets of antiquity were known. In 1781, Sir William Herschel (who with his telescope first mapped the Milky Way's structure) discovered the slow motion of a new planet far beyond the orbit of Saturn. This new planet was named Uranus for the Greek god of the sky and father of Saturn. When moons were subsequently found orbiting Uranus, the application of Kepler's third law revealed it was 14 times more massive than the Earth. But subtle deviations in Uranus's position over the years implied the gravitational attraction of yet another, hitherto unseen planet. Neptune's discovery in September 1846, at the exact location where Uranus' orbital anomalies indicated, was a powerful testament to the predictive power of science in an age that had come to expect a perfectly clockwork heavens.[8]

But while astronomers of the 1800s applied Kepler's laws to the dark distant reaches of the Solar System, others sought to do the same in the fiery confines close to the Sun. Observations had long shown that Mercury too failed to perfectly follow Kepler's elliptical orbits. With each year's pass around the Sun, Mercury never retraced its orbit exactly. The point at which it passed closest to the Sun kept precessing or advancing with each passing orbit, so that instead of a single ellipse, Mercury instead traces out an enormous multi-petaled floral arrangement.[9] Such a pattern would result if another planet orbited even closer to the Sun than inner-most Mercury. Astronomers named this blazing unseen planet Vulcan and looked for its faint light near the Sun in the few moments of darkness during totality of a solar eclipse. But no matter how closely they looked, no planet was ever found. Something was wrong with gravity; the observations did not match the predictions.

The solution to this dilemma came in 1915 when a German-born ex-patent

[8] Pluto's discovery in 1930 was originally part of this progression as calculations showed an unknown planet's effects on the orbit of Neptune as well. However, with a mass less than a fifth that of our own Moon, there was no way Pluto could have produced the calculated affect on Neptune, a planet as massive as Uranus and 10,000 times more massive than Pluto. Eventually, the original orbital calculations were found to have been in error (there was no discrepancy) and Pluto's discovery in the correctly 'predicted' position was simply a fortuitous coincidence. Today, we now know Pluto to be one of the largest of a newly-discovered population of icy dwarf planets orbiting in a massive belt out beyond the orbit of Neptune.

[9] The amount of this advance is staggeringly small. Over a hundred years, the total error in position on the sky of where Mercury is, and where it should be due to all other known Newtonian effects, is no more than 43 arcseconds, (where one arcsecond is one 60th of an arcminute, which in turn is one 60th of a degree) or about one fiftieth the diameter of the full Moon.

clerk published a paper describing his General Theory of Relativity. In it, Albert Einstein presented mathematically a revolutionary new form of gravitation where, contrary to Newton, gravity was no longer a force between objects. Instead, gravity is a curvature in the fabric of a four dimensional space-time (three dimensions of space plus the single dimension of time) produced by the presence of a mass. Imagine a bowling ball placed on a trampoline. It warps the two-dimensional fabric of space that is the trampoline's surface into a pucker. Roll a baseball across the trampoline and the warp in the surface causes the ball to follow a curved path around the bowling ball. The path it follows is exactly what Newton's force of gravity predicts.

Similarly, if you've ever seen one of the plastic funnels on display in shopping malls or science centers, where a coin is rolled into the widely curved circular surface, you've seen how the coin rolls around and around the central depression. If you aim the coin just right, you can make it follow a fairly circular path around the central depression. If, however, you aim the coin closer to the central pucker, it will trace out an ellipse as it dips first down and then back out of the central well. On such a path as this, it picks up speed as it falls in close to the central hole only to slow down again as it rolls outward towards the farthest point in its orbit. If the curve of the funnel is built just right, then the path and speed of the rolling coin will follow Kepler's laws of orbital motion. Additionally, as the coin loses energy through friction with the plastic surface its elliptical orbit advances just as Mercury's does all the while spiraling downward. If there was no friction at all you could create a whole solar system of perpetually orbiting pennies and dimes.

Now if all General Relativity did was confirm the predictions of Newton, then there would be no reason to accept it over Newton's. After all, Newton's theory of gravity had worked pretty well for two hundred and fifty years. But in addition to explaining all the phenomena that we already knew, Einstein's theory made predictions for certain special cases markedly different from what Newton's law describes. According to Einstein, a planet orbiting deeply within the gravitational well of its star should experience a warp in both the fabric of space and the passage of time that constantly changes as its elliptical trajectory carries it alternately closer and farther away from the star. With each close pass to the star the orientation of its orbit will change, causing it to slowly precess around the star. This is exactly what is observed for Mercury.

That Einstein finally solved the problem of Mercury was quickly evident to astronomers, and an important observational support for his theory. After all, a good scientific theory needs to explain what is already seen. But Mercury's problems with Newton had been widely known for years. If scientists were going to accept Einstein's complicated new theory (it was widely stated that almost no one but Einstein could completely understand the complex mathematics and physical implications) a situation would have to be found where General Relativity predicted measurable results completely at odds with those predicted by Newton.

The scientific community didn't have to wait very long.

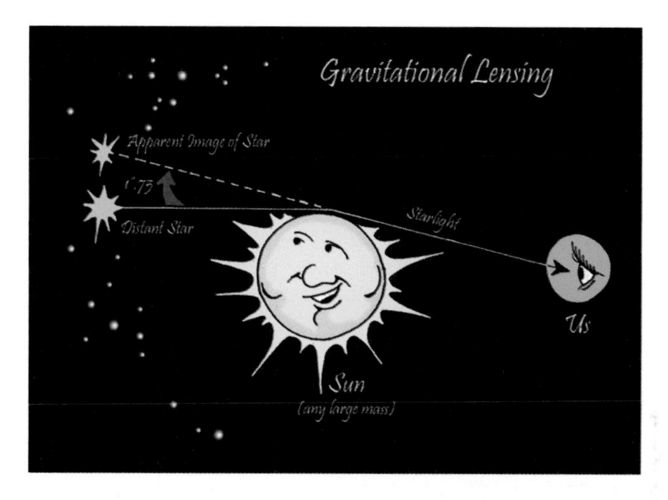

Figure 2.18 Diagram of gravitational lensing by starlight. Light from a distant star passes the limb of the Sun and is deflected towards the observer. The observer, however, only sees the direction from which the light reaches her eye. The apparent source of the starlight, therefore appears to be farther away from the Sun, than the true star's position (T. Nordgren).

On May 29, 1919, the shadow of the Moon would sweep across the South Atlantic Ocean and the stars near the Sun would become visible for the brief five minutes of totality. If gravity acted as Einstein's theory predicted, then rays of starlight passing by the limb of the Sun should have their direction of travel subtly changed as they passed through the gravitational warp of space-time around the Sun. The visible result would be that during the eclipse, stars appearing in photographs near the Sun's dark disk should be ever so slightly out of position, shifted away from the Sun, by an amount of 1.75 arcseconds when compared to a photograph of the same star field made when the Sun was not present. As small as this deflection might appear it was twice what Newton's force of gravity predicted.

To test Relativity's prediction the Royal Astronomical Society of Great Britain sent two expeditions by ship to locations along the path of totality (in the event that bad weather or mechanical problems could strike at just the wrong time during this rare event, it was always good to have a backup). One expedition went to Sobral in northern Brazil, while the other sailed to the island of Principe off the West African coast. Each would have to carry and assemble the delicate telescopes and photographic equipment necessary to photograph the Sun and the surrounding stars during totality's darkness. In addition, to verify the normal position of the stars without the Sun's influence, the expedition in Brazil would stay on for an additional two months so that the exact same star field could be photographed with the exact same equipment in the darkness of early dawn after the Sun had moved on.

Six months after the eclipse, Sir Frank Tyson, Astronomer Royal of the Royal Astronomical Society presented to a joint meeting of the Royal Society and Royal Astronomical Society the results of that May's expedition to demonstrate whether Einstein's predictions had come true:

> *The President.* I will call on the Astronomer Royal to give us a statement of the result of the Eclipse Expedition of May last.

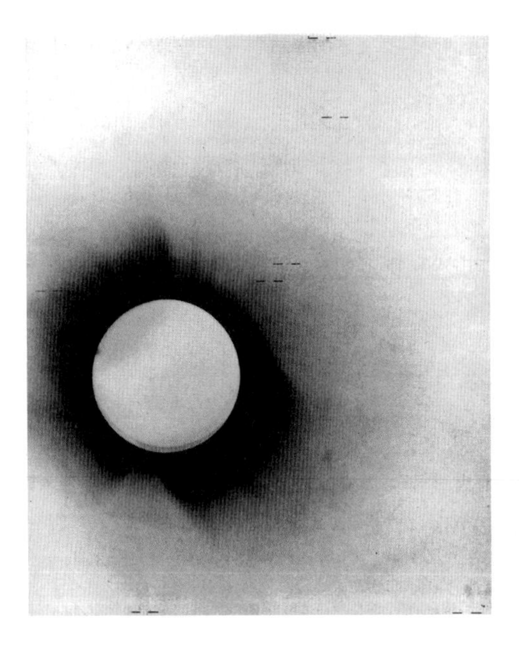

Figure 2.19 With this negative image of the totally eclipsed Sun (made from Principe Island in the Gulf of Guinea, on 29 May 1919) Arthur Eddington confirmed the prediction of Albert Einstein's General Theory of Relativity from the position of stars visible during the eclipse (marked by two horizontal lines) (Royal Astronomical Society/ Photo Researchers, Inc.).

The Astronomer Royal. The purpose of the expedition was to determine whether any displacement is caused to a ray of light by the gravitational field of the Sun, and, if so, the amount of the displacement. Einstein's theory predicted a displacement varying inversely as the distance of the Sun's ray from the Sun's centre, amounting to 1″.75 for a star seen just grazing the Sun. His theory or law of gravitation had already explained the movement of the perihelion of Mercury – long an outstanding problem for dynamical astronomy – and it was desirable to apply a further test to it. [*There then follows a lengthy description of the expedition and the experimental process.*]

After a careful study of the plates I am prepared to say there can be no doubt that they confirm Einstein's prediction. A very definite result has been obtained that light is deflected in accordance with Einstein's law of gravitation.

The President. I now call for discussion on this momentous communication. If the results obtained had been only that light was affected by gravitation, it would have been of the greatest importance.... But this result is not an isolated one; it is part of a whole continent of scientific ideas affecting the most fundamental concepts of physics.... This is the most important result obtained in connection with the theory of gravitation since Newton's day, and it is fitting that it should be announced at a meeting of the Society so closely connected with him.

The difference between the laws of gravitation of Einstein and Newton come only in special cases. The real interest of Einstein's

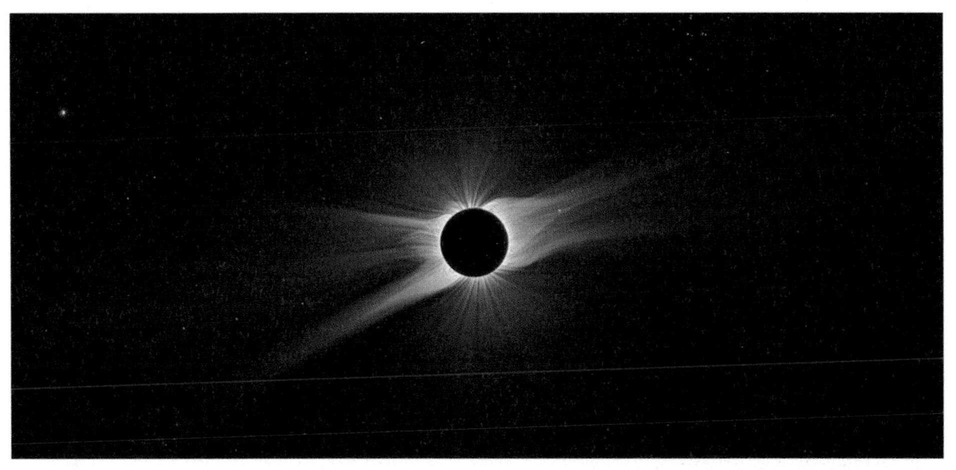

Figure 2.20 Total eclipse of the Sun as seen from Siberia in 2008. The solar corona extends for up to 20 solar diameters in this view that shows it against the background stars. The smattering of stars to the upper right is the Praesepe star cluster, also known as M44 or the Beehive cluster, in Cancer. The planet Mercury is visible to the upper left (Miloslav Druckmüller (Brno University of Technology), Peter Aniol and Vojtech Rusin).

> theory lies not so much in his results as in the method by which he gets them. If this theory is right, it makes us take an entirely new view of gravitation. If it is sustained that Einstein's reasoning holds good – and it has survived two very severe tests in connection with the perihelion of Mercury and the present eclipse – then it is the result of one of the highest achievements of human thought.

After 240 years, Kepler and Newton were supplanted by Einstein. As the president of the Royal Astronomical Society was careful to state, the earlier scientists weren't wrong, they were merely correct over a vast range of masses and motions that didn't take into account some very strange and exotic situations. Apply Einstein's theory of General Relativity to 'everyday' scenarios of moons, planets and stars at reasonable distances from one another and it simplifies to exactly the same relations giving exactly the same answers as Kepler and Newton. But approach the surfaces of stars where gravity's 'force' greatly warps the fabric of space-time and Newton utterly fails to account for all the wonderful effects to be found there.

If one could turn up the mass of the Sun, like turning up the volume on a radio dial (while keeping the size fixed) the warp that the Sun creates in the fabric of space-time would get steeper and steeper as it became increasingly deep. The light from distant stars that before would pass by and be deflected by just a few arcseconds, would instead be deflected by ever more degrees, until eventually like the coins in the plastic funnel, light itself would spiral down the central well never to escape. The surface inside which the walls of the well are too steep for even light to escape is called the event horizon and is not actually a physical surface, but merely the point inside of which all information is lost forever.

Here again we approach a black hole but now as an impenetrable chasm in the fabric of space-time. So while the Reverend Michell could hypothesize the presence of black holes in the 18th century, it is astrophysicists in this century who ultimately follow General Relativity down the gravity well to explore the warp of both space and time around a black hole's event horizon.

Meanwhile in the outer reaches of the Milky Way, one hundred and eighty degrees away from the supermassive black hole at our Galaxy's center, we continue to see the gravitational effects of even more unseen mass. Out at distances of 100,000 light-years from the galactic center, where Newton's laws should perfectly apply, we see stars and dust clouds orbiting our galaxy far faster than Kepler's laws say they should. In virtually every spiral galaxy we look at we see the same story. Instead of the distant edges of galaxies orbiting slower than the inner portions (in accordance with Kepler's third law) we see they orbit just as fast as stars and gas in the center. Just as with Uranus in our own Solar System, there must be hidden, unseen matter in the outskirts of the galaxies giving these stars an extra gravitational acceleration. Since this hidden mass emits no light of its own and is visible only by its gravitational influence we have come to call it *dark matter*.

In an ultimate marriage of Newton and Einstein, modern astronomers look to measure this hidden mass surrounding galaxies using a variation on the great solar eclipse experiment of 1919. As light from distant galaxies at the edge of the Universe passes by these nearby galaxies and clusters (where Newton and Kepler say this mass resides) the unseen mass warps the fabric of space-time deflecting the path the passing light takes. The gravitational warp acts as a lens bending the passing light: the greater the mass, the greater the lensing effect. From the warped image of distant galaxies we are able to precisely measure and map the distribution of dark matter in the Universe. The Universe is ultimately weighed by the weightlessness of light.

And then with a suddenness that startles me, sunlight once again falls down on me as the Moon emerges from the shadow of the Earth. On this night in 2007, in Grand Teton National Park, sunlight once more shines on the surface of the Moon as it begins to set into the nearby mountain's gap. There among the craggy mountain peaks it is followed down by the first reddish alpenglow of sunrise, the same reddish glow that just moments before illuminated the shadowed Moon. While we live only a brief time compared with the Earth, it is through the power of our science that we can see what has come before and predict what is yet to be.

And while it is not clear if or when the next great theory will arise to supplant Einstein, I do hope to someday see someone discover the nature of that mysterious dark matter. For science has shown that at every step where an extraordinary problem presents itself, in time there has always been an extraordinary answer. As Einstein once said, "The most incomprehensible thing about the Universe is that it is comprehensible." The ultimate proof of this statement, and science itself, is that I know exactly where I'll be ten years from now on August 21, 2017 at 11:34:48 am Mountain Daylight Time when the Sun is totally eclipsed by the Moon. Perhaps I'll see you there.

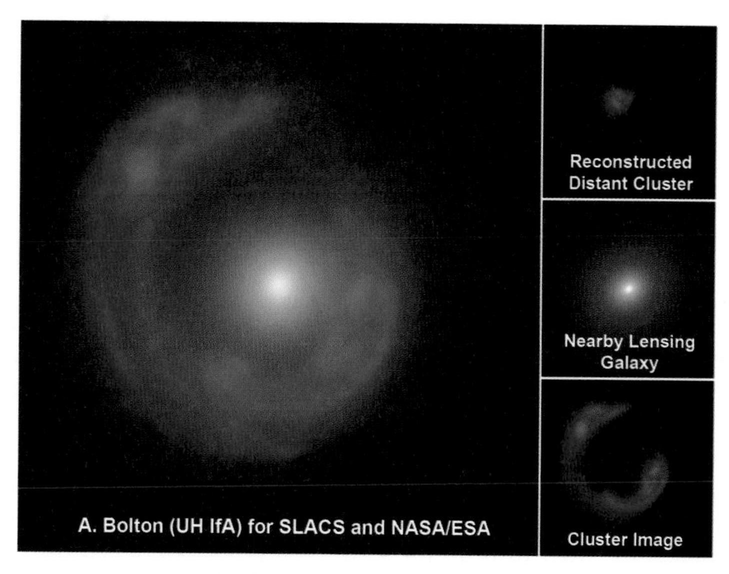

Figure 2.21 Gravitational lensing of a distant blue galaxy cluster by a nearby elliptical galaxy. The alignment between us, the nearby galaxy and distant cluster is so perfect that the image of the faint blue cluster forms an almost perfect ring. This is called an Einstein ring and can be used to measure the mass of dark matter in the nearby lensing galaxy. On the right are shown computer reconstructions of the distant cluster (top), lensing galaxy (middle), and the resulting image (bottom) (A. Bolton (University of Hawaii/IfA) for SLACS and NASA/ESA).

Figure 2.22 The eclipsed Moon of August 2007 sets within the gap of Cascade Canyon in Grand Teton National Park. The same alpenglow that lights the distant mountains is responsible for the reddish glow of the even more distant Moon. The image was made by photographing the eclipsed Moon every 10 minutes from the Mountain View Turnout along the Teton Park Road (T. Nordgren).

See for yourself: eclipses of the Sun, Moon and satellites

Viewing a solar eclipse (safely)

The following maps show the paths of total and annular solar eclipses. Annular eclipses occur when the Moon is too far away from the Earth to completely block the disk of the Sun. During this kind of eclipse the Sun looks like a ring of fire, but the stars and faint solar corona are not visible. The only time the Sun can be viewed safely with the naked eye is during a total eclipse, when the Moon completely covers the disk of the Sun. **It is never safe to look at a partial or annular eclipse, or the partial phases of a total solar eclipse, without taking the proper precautions**. Even when 99% of the Sun's surface is obscured during the partial phases of a solar eclipse, the remaining crescent Sun is still intense enough to cause irreparable eye damage. Specially designed solar eclipse glasses are commonly available for sale on the internet.

The safest way to observe a partial or annular eclipse is by projection. A small opening in a card is used to form an image of the Sun on a screen or sheet of paper. Multiple openings in a loosely woven straw hat, or between interlaced fingers, will also cast a pattern of solar images on a sheet of paper or a screen. Similarly, a tree with many broad overlapping leaves will create multiple tiny openings, each of which will project an image of the Sun forming hundreds of crescent-shaped images on the ground below.

Orbiting spacecraft

Eclipses of the Sun and Moon are rare. If only the Earth had more moons, then the predictive power of Kepler's laws would be a more common spectacle. In fact, the Earth is currently orbited by somewhere around 25,000 moons. All but one of which are small, metallic and artificial. We call them satellites and on any given night you can without much difficulty pick them out as they move across the background of stationary stars. Satellites normally betray themselves as a single, steady, white or possibly yellowish light that moves slowly and steadily against the background stars (airplanes always have more than one light – wingtip lights are red and green – while at least one light will blink). In as little as five to ten minutes satellites can move clear across the hemisphere of the sky. When you see this, what you are seeing is the light of the Sun shining off the surface of the spacecraft and reflected down to Earth. The larger the spacecraft the brighter it looks. Because of this, some tumbling cylindrical satellites, such as spent cylindrical rocket boosters, will brighten and fade or flash as different parts of their surface catch the Sun.

The brightest artificial satellite by far is the International Space Station (ISS) with acres of highly reflective solar panels. The ISS orbits the Earth every 90 minutes at an altitude of nearly 250 miles (400 km) above the Earth's surface. As it passes into the Earth's shadow (where it is eclipsed by the Earth) astronauts on board see the rosy red glow of sunset and those of us beneath see the yellow

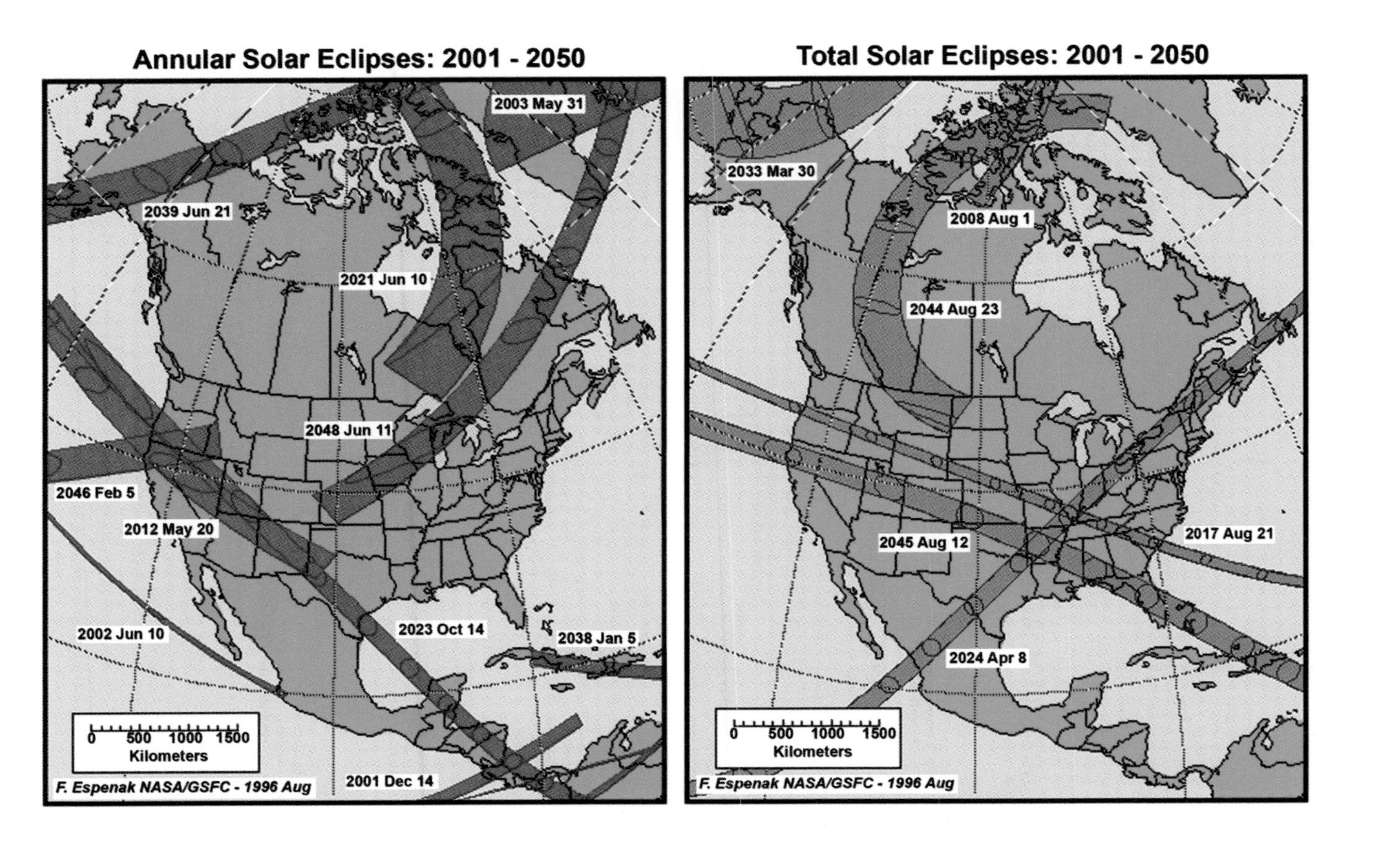

Map of annular and total solar eclipses visible from North America between 2001 and 2050 (Eclipse map courtesy of Fred Espenak, NASA Goddard Space Flight Center.

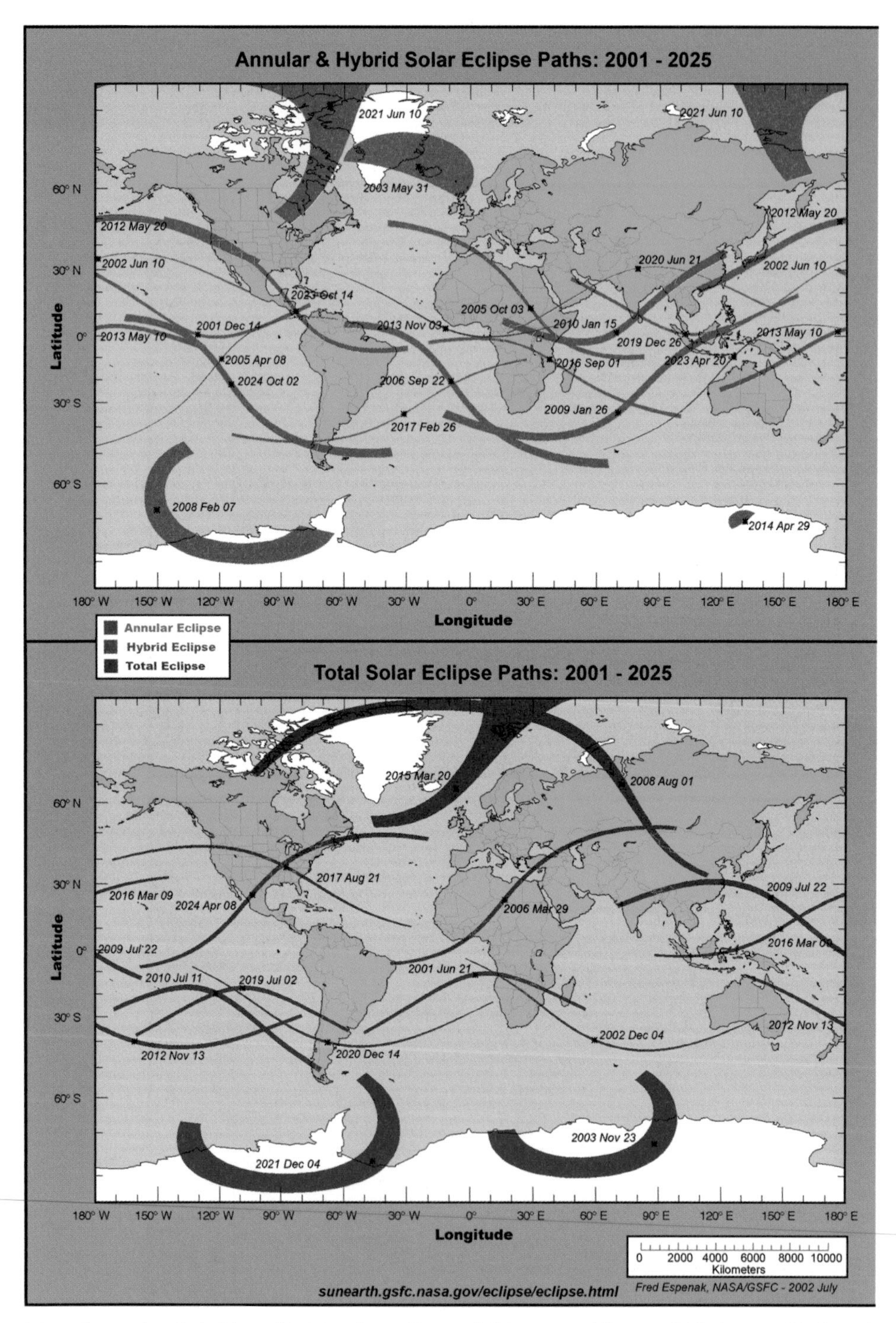

Map of annular, hybrid, and total solar eclipses visible around the world between 2001 and 2050. During a hybrid eclipse the Moon is too far from the western and eastern limbs of the Earth for a total eclipse. Viewers near the beginning and end of the eclipse track see an annular eclipse while those in the center of the shadow's path are on the part of the Earth closest to the Moon and so just barely see a total eclipse (Eclipse map courtesy of Fred Espenak, NASA Goddard Space Flight Center).

beacon turn red and dim as it silently disappears into darkness. While you may catch one of these events by chance, a number of websites predict these events for any date and location on Earth. The best site by far is Heavens-Above.com which will tell you when and where to look for everything from the ISS, to the Hubble Space Telescope to any number of satellites and tumbling abandoned booster rockets.

Table 2.1 Lunar eclipses (2010-2025)

Date	Eclipse Type	Eclipse Duration	Geographic Region of Visibility
2010 Jun 26	Partial	02h44m	e Asia, Australia, Pacific, w Americas
2010 Dec 21	Total	03h29m	
		01h13m	e Asia, Australia, Pacific, Americas, Europe
2011 Jun 15	Total	03h40m	
		01h41m	South America, Europe, Africa, Asia, Australia
2011 Dec 10	Total	03h33m	
		00h52m	Europe, e Africa, Asia, Australia, Pacific, North America
2012 Jun 04	Partial	02h08m	Asia, Australia, Pacific, Americas
2013 Apr 25	Partial	00h32m	Europe, Africa, Asia, Australia
2014 Apr 15	Total	03h35m	
		01h19m	Australia, Pacific, Americas
2014 Oct 08	Total	03h20m	
		01h00m	Asia, Australia, Pacific, Americas
2015 Apr 04	Total	03h30m	
		00h12m	Asia, Australia, Pacific, Americas
2015 Sep 28	Total	03h21m	
		01h13m	e Pacific, Americas, Europe, Africa, w Asia
2017 Aug 07	Partial	01h57m	Europe, Africa, Asia, Australia
2018 Jan 31	Total	03h23m	
		01h17m	Asia, Australia, Pacific, w North America
2018 Jul 27	Total	03h55m	
		01h44m	South America, Europe, Africa, Asia, Australia
2019 Jan 21	Total	03h17m	
		01h03m	Pacific, Americas, Europe, Africa
2019 Jul 16	Partial	02h59m	South America, Europe, Africa, Asia, Australia
2021 May 26	Total	03h08m	
		00h19m	e Asia, Australia, Pacific, Americas
2021 Nov 19	Partial	03h29m	Americas, n Europe, e Asia, Australia, Pacific

Date	Eclipse Type	Eclipse Duration	Geographic Region of Visibility
2022 May 16	Total	03h28m	
		01h26m	Americas, Europe, Africa
2022 Nov 08	Total	03h40m	
		01h26m	Asia, Australia, Pacific, Americas
2023 Oct 28	Partial	01h19m	e Americas, Europe, Africa, Asia, Australia
2024 Sep 18	Partial	01h05m	Americas, Europe, Africa
2025 Mar 14	Total	03h39m	
		01h06m	Pacific, Americas, w Europe, w Africa
2025 Sep 07	Total	03h30m	
		01h23m	Europe, Africa, Asia, Australia

Notes: Only lunar eclipses where the Moon passes into the Earth's umbra are shown. The time duration during which at least part of the Moon is within the umbra is shown, including (in **bold**) the length of time the Moon spends fully within the umbra if the eclipse is total. Credit: Eclipse table courtesy of Fred Espenak, NASA/Goddard Space Flight Center. For more information on solar and lunar eclipses, see Fred Espenak's Eclipse Web Site: sunearth.gsfc.nasa.gov/eclipse/eclipse.html

Further reading

Totality: Eclipses of the Sun by Mark Littman, Fred Espenak, & Ken Willcox (2008) Oxford University Press, ISBN 0199532095.

The Copernican Revolution: Planetary Astronomy in the Development of Western Thought by Thomas S. Kuhn (1957)
Harvard University Press, ISBN 0674171039.

Kepler's Witch: An Astronomer's Discovery of Cosmic Order Amid Religious War, Political Intrigue, and the Heresy Trial of His Mother by James A. Connor (2005) HarperOne, ISBN 0060750499.

Black Holes & Time Warps: Einstein's Outrageous Legacy by Kip S. Thorne (1994) W.W. Norton & Company, ISBN 0393312763.

NASA Eclipse Website maintained by Fred Espenak
http://eclipse.gsfc.nasa.gov/eclipse.html

Heavens Above, spacecraft observing predictions
http://www.heavens-above.com/

3 Tides on a cosmic shore

The surface of the Earth is the shore of the cosmic ocean.... Recently, we have waded a little out to sea, enough to dampen our toes or, at most, wet our ankles. The water seems inviting. The ocean calls. Some part of our being knows this is from where we came. We long to return.

Carl Sagan, *Cosmos*

The sea shares a kinship with space. Standing on the shore, staring out to sea, the limitlessness strains the senses. If you've ever been on a ship out of sight of land, the endless ocean makes even the largest ship shrink into insignificance. The same is vastly truer of space of course. Even here, with feet planted firmly on Earth, how many people feel small standing under a truly starry sky for the first time? How many let their imaginations run wild when they wonder what's out there; whether it's the dark depths of space or the equally dark ocean depths? In both cases, the impenetrable seems unknowable.

But stare long enough into either and their natural rhythms present themselves. Over a single night the stars rise in the east and set in the west. From night to night the Moon slowly moves from west to east. From year to year the planets slowly march from one constellation to another. Similarly on Earth, the crash of waves is hypnotic in its regularity: in and out, back and forth. From where I sit on a cobblestone beach the receding water sounds with the symphony of ten thousand tiny pebbles as each breaking wave rolls back out to sea. Stay for a while and you notice nuances.

Waves that once broke far down the cove now force me to move, lest my feet become wet with the rising water line. Eventually the entire cove is engulfed in

Figure 3.1 Acadia National Park is one of many national and state parks located along the country's coasts. As the nearly full Moon rises over the Atlantic Ocean, waves crash amid the cobblestones of Little Hunters Beach on Acadia's Mount Desert Island in southeast Maine (T. Nordgren).

the sea. But, if I stayed long enough, I'd see the cycle reverse until six hours later the sea level is a dozen feet lower and sea-foam and seaweed would rest once more on wet cobbles. These dramatic changes in the ocean's height are the tides and they are one of the most direct experiences with cosmic forces many of us will ever witness for ourselves.

Approximately every six hours we experience a new high or low tide. This pattern repeats itself, over and over, every day on shorelines all over the world. Here in Maine on the coast of Mount Desert Island and Acadia National Park the sea level regularly rises and drops by a dozen feet over the course of just six hours. Great and mysterious forces must be at work in order to make that much water simply go away. But where does it go, and why does it go there? From patterns come explanations. One such explanation comes from the Tsimshian people along the Pacific Coast of the U.S. and Canadian border:

> A long time ago, the old people say, the tide did not come in or go out.
>
> The ocean would stay very high up on the shore for a long time and the clams and the seaweed and the other good things to eat would be hidden under the deep water. The people were often hungry.
>
> "This is not the way it should be," said Raven. Then he put on his blanket of black feathers and flew along the coast, following the line of the tide. At last he came to the house of a very old woman who was the one who held the tide-line in her hand. As long as she held onto it the tide would stay high. Raven walked into the old woman's house. There she sat, the tide-line held firmly in her hand. Raven sat down across from her.
>
> "Ah," he said, "Those clams were good to eat."
>
> "What clams?" said the old woman.
>
> But raven did not answer her. Instead he patted his stomach and said, "Ah, it was so easy to pick them up that I have eaten as much as I can eat."
>
> "That can't be so," said the old woman, trying to look past Raven to see out her door, but Raven blocked the entrance. So she stood up and leaned past him to look out. Then Raven pushed her so that she fell through the door, and as she fell he threw dust into her eyes so that she was blinded. She let go of the tide-line then and the tide rushed out, leaving all kinds of clams and crabs and other good things to eat exposed.
>
> Raven went out and began to gather clams. He gathered as much as he could carry and ate until he could eat no more. All along the beach others were gathering the good food and thanking Raven for what he had done. Finally he came back to the place where the old woman still was. "Raven," she said, "I know it is you. Heal my eyes so I can see again."
>
> "I will heal you," Raven said, "but only if you promise to let go of

> the tide-line twice a day. The people cannot wait so long to gather food from the beaches."
>
> "I will do it," said the old woman. Then Raven washed out her eyes and she could see again. So it is that the tide comes in and goes out every day because Raven made the old woman let go of the tide-line.
>
> Tsimshian tale, How Raven Made the Tides

This narrative is from the book, *Native American Stories* told by Joseph Bruchac, a poet and storyteller of Abenaki ancestory (a people who today still live here along the Gulf of Maine). In it we see an awareness of the tide's patterns that has always been vital for anyone attempting to live along the sea.

Four hundred years ago along the Italian coastline, the astronomer Galileo became the first modern scientist to attribute the tides to, if not mystical causes, then at least cosmic ones. In the ocean's steady rise and fall in and out through the shipyards and canals of Venice, Galileo saw evidence for our planet's motion in space. For nearly two thousand years, the prevailing view in Europe had been that the Earth sat perfectly still and fixed while the entire Universe moved around us. Our language still clings to this commonsense notion that we are at rest. While millions of people each day enjoy the beauty of a sunset, even the most steely-eyed scientists I know never talk of ending the day with an "Earth-turning-solar-disappearance."

Now, we've all been told since we were young that the days are caused by the spin of the Earth, that a year is one full trip of our planet around the Sun. But what proof did our parents or teachers ever give us that we could see for ourselves? Galileo thought he had found this evidence with the tides. Many of the experiments Galileo did on what would later become the bedrock of physics involved dropping things, swinging things, and watching things roll downhill. In a swirling bucket of water, Galileo thought he found the reason for the tides. If the bucket is perfectly at rest he argued, then the water is too. The only way to

Figure 3.2 Even with foggy skies, the presence of tide pools left behind during low tide are evidence of astronomical forces. Entire ecosystems can be found here in these tiny, temporary islands of sea-life (T. Nordgren).

get the water to slosh back and forth was to set the bucket, and by extension the Earth, in motion. In the tides he saw the visible proof of Copernicus' idea that the Sun, not the Earth was the center of the Universe. So sure of this was he that this idea became the fourth and final chapter of his *Dialogue Concerning the Two Chief World Systems,* the book that eventually earned him almost 400 years of sanction by the Catholic Church.

Unfortunately, Galileo was wrong about the tides. Yes, they do have something to do with the rotation of the Earth; after all, the time between two sets of high or low tides is almost exactly 24 hours, which is how long it takes the Earth to turn once on its axis. So unless this is an amazing coincidence – and in nature there are very few amazing coincidences – then the rotation of the Earth must have something to do with the pattern of tides. What most of us have heard at some point, though, is that it is the Moon that is responsible. This we can partially see for ourselves. Notice that I was careful to say that the time between two sets of high tides is *almost* exactly 24 hours. Pick up a copy of local tide tables available in most seaside community shops and compare the times of high and low tides from day to day. Each day the tides occur about 50 minutes later than the previous day. If the tide tables don't also include moonrise and set times, watch the sky on those days when the clouds and fog are clear and you will eventually notice that the Moon also rises about 50 minutes later from day to day. Somehow these events must be related.

Forty-five years after Galileo's death, Sir Isaac Newton discovered the unifying force of gravity that governs all those things Galileo was interested in: why things fall, swing, roll, and orbit one another in space. In his theory of how gravity worked, Newton theorized that the strength of gravity's pull depends on only two things: (1) the product of the mass of the two things involved, and (2) the square of their distance from one another (i.e., their separation times itself). The second of these means that as the distance between a planet and a moon becomes two times greater, the force of gravity between them becomes four times weaker. If the distance decreases to one third of what it was, the force of gravity becomes nine times stronger. The distance separating things is therefore very important: small differences in distance can become large differences in gravity.

Tonight I'm walking along the sea-front in the town of Bar Harbor just outside the Acadia park boundary. From my position on the Atlantic Coast, I see the full Moon just begin to rise over the eastern horizon. Meanwhile, nearly a quarter of the way around the Earth to the east, colleagues of mine in Newton's England see the full Moon high overhead and thus are a little closer to it than I. Friends of mine in Hawaii, however, a quarter of the way around the Earth to the west, are on very nearly the opposite side of the Earth from the Moon and will not see it rise for nearly another six hours. I am therefore a little closer to it than they. All of this means that the parts of the Earth on which each of us is standing, myself, my friends and my colleagues, all feel a different gravitational pull from the Moon.

While rock is pretty solid and doesn't deform easily under what is really only a

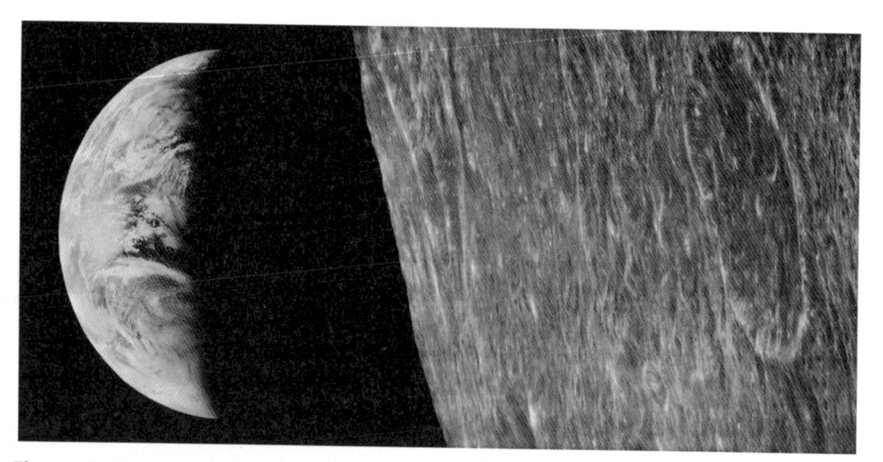

Figure 3.3 Different parts of the Earth are found at different distances from the Moon. Here, in the first image of the Earth ever taken from the Moon, Africa and the Mediterranean Ocean sit along the Earth's terminator and are the closest points on the planet to the Moon. This image was taken in 1966 by the Lunar Orbiter spacecraft and at the time was declared the Photo of the Century. In 2008 it was reprocessed to bring out the full beauty of the image (NASA/LOIRP (Lunar Orbiter Image Recovery Project)).

slightly different force, water is to say the least, fluid. The waters on the side of the Earth closest to the Moon are raised in a slight bulge outward from the rest of the Earth. Where I stand I see the effect of this lunar force as the waters are pulled away from my seashore to fill that bulge over by Europe and Africa. The result: I experience low tide while their tide is high.

Figure 3.4 A National Park Ranger demonstrates the cause of the tides using an assistant as the Moon and showing the two tidal bulges of water he raises on either side of the Earth (T. Nordgren).

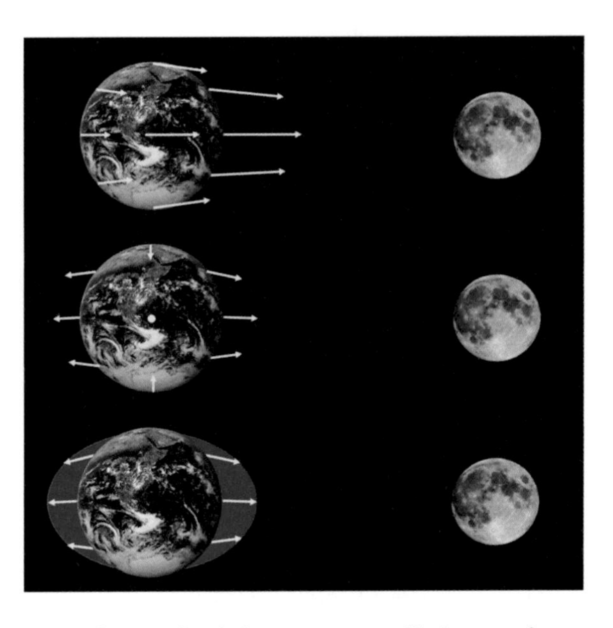

Figure 3.5 Top View: Different parts of the Earth at different distances from the Moon feel a different pull of gravity from the Moon (shown by the length of the arrows). Middle View: Because the different parts of the Earth feel a different gravitational pull *relative to the center* of the Earth, the *net* pull on each part, relative to the center of the planet, is as shown. Bottom View: Two bulges are therefore raised on the Earth. The closest is raised as the near portion is pulled away from the middle. The far portion is raised as the middle is pulled away from it (T. Nordgren).

What then of my friends in Hawaii? Are all of their waters pulled away from them so that they experience an even lower tide than I? If this were the case, and the Moon only pulled one bulge of water towards it, then as the Earth turned underneath that water we would experience high tide only once every day instead of twice a day as we do in reality. There must be two bulges, one on each side of the Earth, but why should the Moon cause water to bulge away from it? The answer is that just as my colleagues in England are closer to the Moon than I, and thus the water near their shores is pulled upwards away from the Earth, my friends in Hawaii are farther from the Moon than I and thus the Earth of which I am a part is pulled ever so slightly out from underneath the ocean where they are. Two bulges are formed, one because it is pulled with greater force than the rest of the Earth, the second because it is pulled less strongly than the Earth and is left behind.

As the Earth spins on its axis, the motion Galileo was interested in proving all those centuries ago, each place on Earth passes under each water bulge, and thus experiences two high tides a day. Since the Moon slowly revolves around the Earth every 27 days from west to east, every day the Moon appears to rise above each spot on Earth about 50 minutes later and the tidal bulges it induces come about 50 minutes later as they follow it around.

There are some special places along the Earth's shorelines where high and low tides are particularly extreme. As a college student, the last job I ever held before becoming an astronomer was as seasonal help in a fish cannery outside of Anchorage, Alaska. I worked twelve-hour shifts, seven days a week, and each day when the tedium overtook me, I'd gaze out the cannery door at the boat docks beyond. I distinctly remember the first time I noticed the hull of the ship outside replaced, six hours later, by no more than a view of its mast tied up to the pier.

Figure 3.6 In Bar Harbor, Maine, at the entrance to Acadia National Park, the sea level regularly changes by a dozen feet with the tides. Here are two images, from the same location, taken only six hours apart. Notice the height of the water and the white ship's hull relative to the end of the pier (T. Nordgren).

During the time in between, the tide had gone out and the ship's keel had come to rest a little over thirty feet lower than before. Some days it was left resting on the slick mud-flats altogether.

Here at the mouth of the Gulf of Maine the tides are also pretty extreme; a daily change of 12 feet in sea level (3.7 meters) is not uncommon. Farther up the Gulf in the Bay of Fundy National Park in New Brunswick, Canada, the change can be up to four times greater. Former Acadia Park Ranger Jim McKenna, now a professor at the Maine Maritime Academy, describes this effect, called *seiching*, this way: "Imagine the Atlantic Ocean is a giant bathtub. As the Moon passes over the tub it drags the tidal bulge with it. As the Moon passes over the western edge of the tub and the high water encounters land, high tide occurs and the bulge rebounds back east across the ocean. If the tub is just the right length so that the time it takes for the rebounding wave to reach the eastern shore is exactly the same as the time for the next bulge to come around again, the height of the wave is amplified."

This is exactly the same effect as pushing a child on a swing. If the rate at which you push a swinging child is exactly the same rate as which the child comes back to you, the amplitude of her swing increases. This is usually the goal.

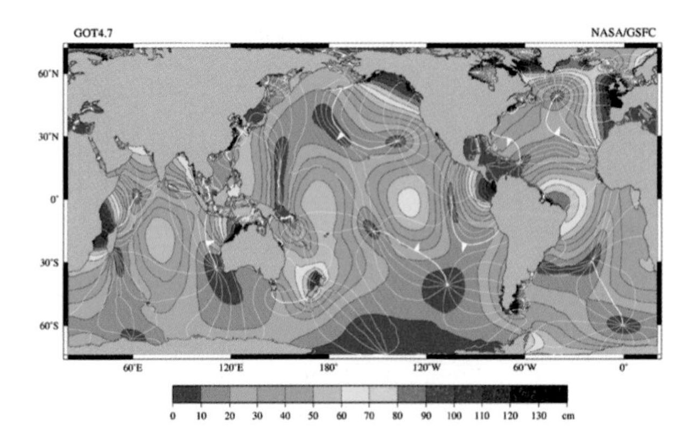

Figure 3.7 Map of the range of world tides. Red areas experience enormous tidal variations while blue areas experience very little (R. Ray, NASA Goddard Space Flight Center).

Push at the wrong rate and pretty soon chaos develops and the very irate child loses height. In regions of the world where the rate at which water moves in and out of a bay or inlet is very nearly equal to the rate at which the tidal bulge comes around, dramatic swings occur at the far end of the gulf (such as the Bay of Fundy at the far end of the Gulf of Maine).

The Sun also affects the tides. While the Sun is very much farther away from the Earth than the Moon, it is also 30 million times more massive than the Moon and so even it affects what we see in the pattern of coastal tides. Tonight is a wonderfully clear night, with just enough fog and clouds along the horizon to lend the sunset drama. Behind me, the Sun has set while before me the full Moon rises through the red glow of twilight. At this moment the tide is exceptionally low and before me a land-bridge to nearby Bar Island is revealed. Walking out onto the grey, gravelly sand bar where seaweed and crustaceans cover the

Figure 3.8 During low tides, Bar Island is joined to Bar Harbor by a wide causeway. Visitors can walk across to the "island" within an hour and a half of low tide. Six hours later the entire expanse is completely covered by the sea (T. Nordgren).

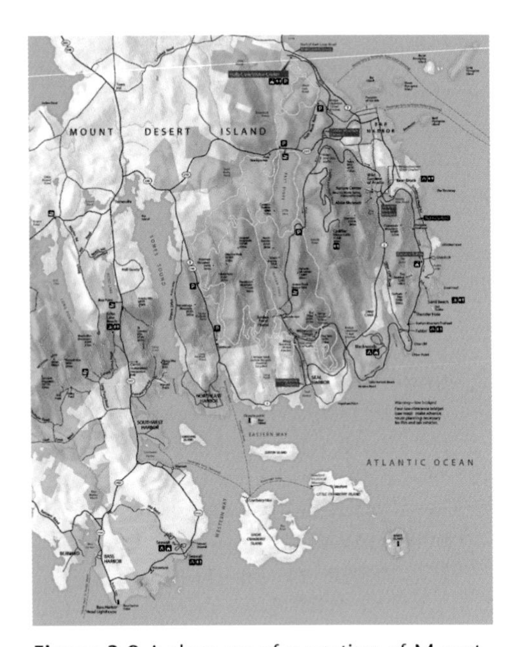

Figure 3.9 A close-up of a portion of Mount Desert Island that contains the majority of Acadia National Park. Bar Harbor with the periodically misnamed Bar Island is located in the northeast part of the island (upper left of the map) (National Park Service).

recently wet surface, I am awestruck to be standing on what, in six hours time, will once more be the sea floor.[1]

When the Sun, Earth, and Moon are all in a line, both of their tidal effects on the Earth are aligned as well, and low tides are lower, while high tides are even higher. This alignment happens near full Moon when the Moon is on the opposite side of the Earth to the Sun, and New Moon when the Moon is between the Earth and Sun. Check the tide tables that give the heights of the tides as well as the time they occur and you will see that about two days after each New and Full Moon the tides are exceptionally large.[2] These are called the *spring tides* (so named for images of the waves springing higher up the seashore) and this month in Bar Harbor, the height of the ocean will change by nearly 15 feet (4.5 m) over the course of just six hours. A week after full or new Moon, when the Sun and Moon are 90 degrees away from one another, their tidal pull is at cross-purposes and the difference between high and low tides is particularly small. These are called the *neap tides*.

Every seven and a half lunations (where a *lunation* is the 29 and a half days between two consecutive full Moons) an even more subtle phenomenon occurs. Because the Moon orbits the Earth in an elliptical orbit, sometimes the Moon is closer to the Earth (called *perigee*) than at any other time and the tidal forces are greater. When full or new Moon occurs while the Moon is at perigee, the combined effects of Sun, Moon, and orbital dynamics all conspire to produce spring tides even higher than those that occur every month. These are the *perigean* tides, sometimes called *astronomical* tides, but as you can see on any trip

1 The tides don't even have to be that extreme in order to open the passageway. Every day the tide drops low enough to open the connection to Bar Island. Just be sure to return within an hour and a half on either side of low tide, however, as the bar completely disappears during high tide.

2 The gap of a couple days occurs because the water bulge that the Moon and Sun pull around the Earth drags across the sea floor and so lags behind the particular phase of the Moon.

Figure 3.10 The full Moon rises over Egg Rock Light as seen from the Egg Rock overlook on the Acadia Park Loop Road. Because the Moon is forever tidally locked to rotate at exactly the same rate it circles the Earth, we always see exactly the same face. The dark markings that make up the Man in the Moon are actually smooth plains of dark volcanic rock that ancient observers once thought were seas (T. Nordgren).

to the coast, that's just being redundant. Over the course of just a couple weeks' vacation on any seashore you will witness for yourself the tug of war between astronomical forces, and there never needs to be a clear night to see it.

Ideally, the Earth's bulges should always stay pointed directly at the Moon with the Earth rotating underneath. But the Earth turns on its axis faster than the Moon orbits the Earth (the reason we see the Moon rise in the East and set in the West each day). As a result, friction between the seafloor and sea causes the bulges to be partially carried forward with the Earth's rotation relative to the position of the Moon. The Moon's gravity, however, continues to pull backwards on those bulges and ever so gradually the Earth's rotation is slowed as gravity, water, and rock slowly act against one another. The days are therefore getting longer by about 1.7 milliseconds per day per century.[3]

But these forces work both ways. Just as the Moon deforms the Earth, the Earth deforms the Moon. Since the Earth is much more massive than the Moon, tidal forces from

[3] We have direct evidence of this slowing from 2,000 year old Babylonian records of solar and lunar eclipses. The observations they recorded on clay tablets carefully note the dates and times these eclipses occurred (e.g., how soon after sunrise or moonrise). With modern understanding of the motion of Earth, Moon, and Sun we can calculate when these events would have occurred if the Earth's rotation was constant. To see the total solar eclipse of 15 April, 136 B.C., for instance, a clock set to the rotation of the Earth's spin must have lost a little over 3 hours over the last 2,000 years compared to an ideal clock keeping perfect time. Had it not, the solar eclipse would only have been partial as seen in Babylon, directly contradicting the view of the stars and planets a Babylonian astronomer reported at mid-day during totality. Incidentally, while the Earth is slowing down, the Moon is moving away. As the Moon's gravity pulls back on the Earth's tidal bulge, the bulge's gravity pulls forward on the Moon. This added bit of angular momentum causes the Moon to slowly spiral outward from the Earth. As the days get longer the Moon gets farther away.

the Earth long ago slowed the Moon's rotation so that its bulges (this time of solid rock) always stay pointed towards the Earth and the Man in the Moon never goes away.[4]

Many billions of years from now, a day will come when the Moon has slowed the Earth to the point that it turns at exactly the same rate that the Moon orbits; the two bodies will forever keep the same faces pointed towards one another. On that day anyone left on Earth will see the Moon cease to rise and set and only the weaker tides from the Sun will continue to move across the oceans.

There is nothing unique about tidal forces with respect to the Earth and Moon. As of 2009 there were 170 known moons orbiting planets in our Solar System, and nearly every one of them behaves in the same way as our Moon. Most of them were long ago tidally locked, always keeping the same face pointed towards their parent planet. Planetary scientists mapping the landscapes of these distant moons make reference, therefore, to a moon's sub-planetary point: the position on the surface of the moon on which, if you could but stand, the parent planet would forever hang directly overhead. For moons of the giant planets: Jupiter, Saturn, Uranus, and Neptune, their home planets are 10,000 to a million times more massive than their moons (as compared to only 100 times larger for the Earth and its Moon). As a result, the degree to which the Jovian planets deform their little moons is enormous.

Look at Jupiter through a pair of binoculars or a spotting scope with a magnifying power of no more than $20\times$ and you will see Jupiter's tiny disk just as Galileo did when he first pointed his telescope at the planet on the night of

Figure 3.11 A sketch of Jupiter and its moons as made through a replica of Galileo's telescope with which he first observed them. The magnification is only 25 times what the human eye can see, and is equivalent to what a pretty good pair of binoculars would show. To see these, both the telescope and the binoculars need to be steadied on a camera tripod (T. Nordgren).

[4] Many people suppose this means the Moon doesn't turn on its axis like the Earth. However, if the Moon didn't turn with respect to the stars, first we would see one side, and then two weeks later when the Moon had gone half way around the Earth we would see the other. In order to keep the same face turned towards us it needs to turn on its axis at exactly the same rate that it goes around us. Again, an apparent coincidence of that magnitude reveals the hidden beauty of the forces at work between Earth and Moon.

January 7, 1610. Steady the binoculars on a tripod or by resting them against the trunk of a tree and you will see the four tiny moons Galileo discovered arrayed in a line about the planet's disk. Since the common plane in which all four moons orbit is almost perfectly in line with us we always see them laid out like four tiny pearls on a string (although one or more may be obscured by Jupiter every so often).

Watch for only two nights and you will see the pattern of moons change as they make their way around Jupiter according to Kepler's laws of planetary motion. Innermost Io, about the same size as our Moon, takes 1.8 days to make one full orbit. Next out tiny Europa, only two-thirds as massive as our own Moon, takes 3.5 days to make the trip. The third moon is Ganymede, largest moon in the Solar System, circling Jupiter once every seven days, followed by Callisto at a little over 16 days. Watch these moons for a few months as Galileo did and you can pick out these periods for yourself.

Over the last four and a half billion years, tidal forces from giant Jupiter should long ago have pulled all four moons into perfectly circular orbits with each one forever locked with one face pointing inward towards the planet's disk. But, like rambunctious siblings, the gravitational yank and pull from each moon on another has kept Europa and Io, and to a lesser extent Ganymede, from ever settling into perfectly circular orbits. Europa's elliptical orbit therefore repeatedly carries it towards and away from Jupiter's enormous bulk. The tidal stretching that Jupiter induces therefore waxes and wanes repeatedly every three and a half days for the life of the moon.

The elliptical orbit also warps the moon in another unusual manner. Like our Moon, Europa keeps the same face pointed towards Jupiter by turning on its axis at the same rate that it orbits the planet. However, because Europa's orbit is elliptical, orbital dynamics embodied in Kepler's laws, say that the moon must speed up as it passes closest to Jupiter and slow down at the point that it's farthest away. Since the moon spins on its axis at a constant rate, this changing orbital speed means that some times the planet moves too fast to keep the same face pointed exactly at Jupiter while half an orbit later it moves too slowly. As a result, the face that always points towards Jupiter slowly swings back and forth ever so slightly relative to a line joining it to the planet.[5]

While the moon may swing, the position of the bulges does not. Jupiter's massive gravitational force keep the tidal stretching pointed directly along the line between moon and planet. For any one geographical spot on the small moon therefore, the bulges sweep back and forth across the sub- and anti-Jovian hemispheres.

[5] Because our Moon's orbit is also elliptical the same phenomenon happens here. Galileo first observed this effect, called *libration*, which allows us each month to see a little way around towards the 'farside' of the Moon. As a result, over the course of a month we can see almost 60% of the Moon's total surface area.

Figure 3.12 Jupiter, Io and Europa (top) taken from Voyager 1 as it neared the planet on February 5, 1979 (NASA Jet Propulsion Laboratory).

Figure 3.13 Europa rises over the limb of Jupiter. Notice the cracks crossing the face of the moon. Compare this image made by the New Horizons spacecraft in 2007, as it was passing Jupiter on its way to Pluto, to the similar image of the Earth rising over the limb of the Moon made by the Lunar Orbiter spacecraft in 1966 in Figure 3.3 (NASA/Johns Hopkins University Applied Physics Laboratory/Southwest Research Institute).

Both of these effects, the changing amount of the tidal stretching, and the changing position of the tidal stretching, cause the rock within the moon to slowly rub and scrape, back and forth. Deep inside its small body, every instance of friction between jagged surfaces generates a tiny flicker of warmth that keeps the interior of the moon perpetually aglow. Today, scientists think this heat has melted a planet-wide ocean of water on Europa, protected from the frigid vacuum of space above by a frozen icy shell.

Figure 3.14 The strangely smooth surface of Europa is revealed in this image from the Galileo spacecraft. Cracks overlap cracks without any trace of mountains or canyons and only one lone crater (compare this with our own Moon as seen in Figure 3.3). Three prominent cycloidal cracks cross through this image revealing the presence of a subsurface sea (at least at the time they were formed). Note the scale bar in the corner (NASA/JPL/University of Arizona).

That something strange was occurring at Europa was apparent from the very first images returned by the two Voyager spacecraft that passed by in 1979. Unlike our Moon there were no mountains, no giant impact basins, no maria, no canyons, and most important of all, virtually no craters. What there was, were innumerable cracks criss-crossing an improbably smooth surface, like a crystal ball that had been shattered with a hammer.

Cross-country ski across the frozen surface of Europa and from the summit of the highest icy plateau to the bottom of the lowest valley trough you will descend little more than a mile (no more than two kilometers). Along the way, in addition to crossing globe-girdling icy crevasses you will also see regions of vast broken ice-sheets where icebergs are forever locked in a frozen-solid sea. These features, first seen by Voyager, were imaged anew with greater detail when NASA came back for a longer look with the appropriately named Galileo spacecraft in the late 1990s.

For the last four hundred years, the bedrock requirement of science is that it's not enough to have a hypothesis that sounds good, but rather it is evidence that counts. For this reason, the Galileo spacecraft's mission at Europa was focused on finding evidence of this ice-capped ocean and clues as to how thick the obscuring layer of ice might be.

One of the first features visible to Voyager that were imaged in much greater detail by Galileo, were strange cycloidal ridges: great looping cracks that stretched for thousands of miles across the Europan landscape. In 1999, Rick Greenberg and collaborators at the University of Arizona Lunar and Planetary Laboratory published a paper in which they showed exactly how these strange fissures could have formed.

On Earth, we see our tidal ocean bulge by watching it rise and fall a dozen feet or so against the stationary coastline. On Europa there is no stable coastline as the entire surface rises and falls by nearly a hundred feet (30 m) over the course of a single three and half day orbit. But as the surface rises, stress on the icy shell causes cracks that slowly stretch across the surface at the same speed that you or I might walk.

From the perpetual swinging back and forth motion of the moon caused by its elliptical orbit, the direction that Jupiter's tidal stresses pull on the cracks slowly changes over the course of its half-weekly orbit. With each pass near Jupiter, a crack begins to grow, arc, and stop, only to start up again with each new pass around the planet. For the first time, a plausible explanation was shown for these beautifully strange features, but most important of all, they only worked if the ice being deformed is a thin shell on top of a liquid ocean. A solid icy moon will exhibit no such behavior.[6]

To be honest, however, all that the cycloidal cracks show is that the moon was liquid as recently as the last few million years. To understand what is under the ice right now, another piece of observational evidence was needed. Bob Pappalardo is a planetary geologist at NASA's Jet Propulsion Laboratory in California and since receiving his PhD in geology in 1994 he has been either at or near the gravitational center of one of the largest research groups working on Europa's mysterious interior. I was curious what he thought the single best evidence was for a liquid ocean on Europa - given that no one has ever seen beneath the icy surface. In neatly summarizing the last thirty years of theory and observation, he replied, "Theoretical modeling of tidal heating predicts that Europa could have a global ocean beneath the ice, and Europa's geology certainly supports this notion. But," he continued, "It is the magnetometer data from the Galileo spacecraft which is the best evidence we have that an ocean exists there today."

Like the Earth, Jupiter has a magnetic field. However, Jupiter's is 20,000 times

[6] Try it for yourself: squeeze an orange. An orange is a mostly liquid sphere with a thin solid shell on the surface. Now squeeze a similarly sized apple; an apple is solid all the way through. An orange will give and deform, the apple won't.

stronger and extends so far out into space that all four of its Galilean moons orbit inside its boundary. If you could look up into the sky and see Jupiter's magnetic field with the naked-eye, it would appear five times larger than the full Moon. If Europa were an inert lump of solid ice, a cross-country skier holding a compass on its surface would be guided outward into space towards Jupiter's North Pole, rather than anywhere specific on Europa.

The Galileo spacecraft carried a compass, of sorts, and as it swooped past the icy moon, Margaret Kivelson and collaborators at the University of California at Los Angeles found the direction of its 'needle' changed. Something within the moon alters the magnetic field of Jupiter in which it sits. The only way the moon could do that is if there is a substance inside Europa that carries its own electrical charge and is able move in response to the magnetic field it feels (and thus alter the net magnetic field around the moon). As explained by Pappalardo, "This indicates that there must be a conductor at a relatively shallow depth beneath the surface, and the only good candidate is a salty water ocean."

The question remains however, how far down beneath the ice do you have to go to finally reach the ocean? In answer to this question, there are currently two camps. One side claims a thin crust of no more than a few miles or kilometers, where cracks still form that stretch from the frozen surface all the way down to the liquid ocean beneath. The other side claims the evidence points to a thick crust around 15 miles (two dozen kilometers) or more, where the liquid sea is kept in utter isolation from the vacuum of space above.

Pappalardo favors a thick crust. "From the sizes and shapes of Europa's few impact craters, it appears that they formed in an ice shell that is about 20 km thick," says Pappalardo. "Most craters appear as if they formed in the solid ice, while the two largest show bullseye-like rings which suggest they have penetrated to liquid water."

For Greenberg, who proposes what he'd rather describe as a "permeable" model – where the crust is thin enough that the surface of the moon is in constant contact with the sea beneath – he sees global evidence for his hypotheses in the broad spectrum of cracks that cross the tiny moon's surface. As an example, he points to one of the most common features on Europa: strange double-ridged cracks.

> The most likely way these ridges occur is that tidal forces form cracks that open and close on a daily basis as a result of the variations in the stresses. When the cracks open, water is going to come up to the float line just like a crack in a northern lake in winter. When you go skating on such a lake you can find all these places where there are cracks and water has come up 90% of the way along the ice and then refreezes so you get these grooves that are just a few inches deep. So, a similar thing happens there on Europa and the water comes up, starts to freeze, but then a few hours later the stresses change and the walls close. As they do, they crunch up the fresh ice and squeeze it up as slush and broken ice onto the surface. Then a few hours later as the

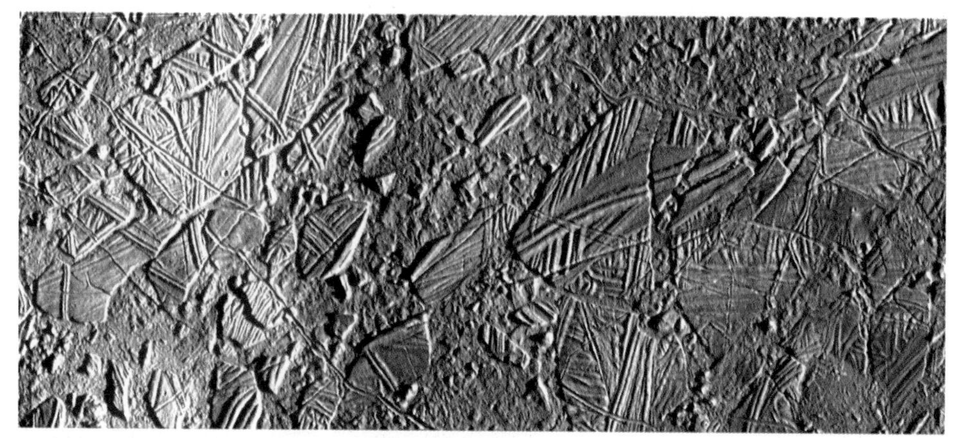

Figure 3.15 Icebergs locked in frozen seas are criss-crossed by double ridge cracks in this Galileo image of Europa's surface. The area covered measures about 45 by 20 miles (70 by 30 kilometers) (NASA/JPL/University of Arizona).

> moon comes around to the same spot in its orbit as before, the tidal stresses return and the crack opens up again and the slush and rubble is pulled to either side of the now open crack. Every day this happens and so eventually you build these double ridges. To have that happen though, you need to have the cracks open up all the way down to the liquid ocean.

In the end though, because the main high-bandwidth antenna by which Galileo was to have reported its science data back to Earth never fully deployed, the number of images that scientists could acquire was only a hundredth of what had initially been hoped. As detailed as the tidal models and calculations may be, the images upon which they are based and with which they must be compared, are painfully few in number. Where data are sparse, disagreements over theories arise. And in planetary science circles the debate between 'thin' camps and 'thick' is as heated as the moon's interior. There are scientists who hope the next NASA flagship mission to the outer planets is a Europa orbiter specifically designed to take off where Galileo ended. When asked what observation would definitively end the debate, Pappalardo replies,

> A spacecraft equipped with a radar sounder could definitely determine which model is correct. If the ice shell is thin, the radar signal will penetrate through the ice, reflect off the ocean, and be picked up again by the radar. If the ice shell is thick, the radar signal will fade slowly through the thicker and warmer ice, and reflections will occur that map out warm blobs of warmer ice that are predicted to be rising up through the thick ice shell.

But why is thick versus thin important? Watching how the gravity of Europa affected Galileo's trajectory as it passed overhead, astronomers determined the

Figure 3.16 Artist's drawing of the Galileo spacecraft with its umbrella-like high-gain antenna that never completely unfurled. All the images Galileo sent back to Earth, therefore, had to be done using the much smaller and far less powerful low-gain antenna shown here communicating with a probe Galileo dropped into Jupiter's atmosphere. Galileo arrived at Jupiter Dec 7, 1995 (NASA).

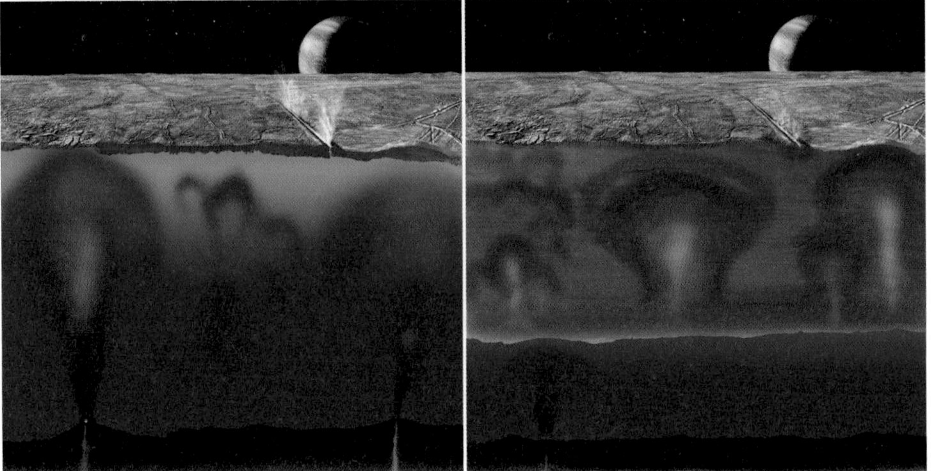

Figure 3.17 Artist's conception of two cross-section views of Europa. On the left is the thin, or porous, model where the surface of the ice is in direct contact with the liquid ocean beneath. Plumes of hot water from underwater 'black smokers' rise through the liquid sea. On the right is the thick crust model where cracks in the surface do not propagate all the way down to the liquid beneath. Here 'black smokers' heat the ice sending blobs of warmer ice rising through the cooler surroundings (Artwork by Michael Carroll, NASA/JPL).

total mass of the moon. From measuring the moon's size we know how dense the mass must be to fit within its volume. Because the average density is greater than the density of water we know the inner portion of the moon must be made of rock. Knowing the density of both rock and ice, if we can determine the thickness of the ice shell it will reveal the depth of the liquid ocean beneath. Using a conservative estimate for the thickness of the frozen ice shell (the 'thick' camp), the ocean on Europa must be at least 50 miles (75 km) deep. This is deeper than any ocean on Earth and the combined water on all of Europa is more than twice the waters in all the oceans of Earth combined.

Imagine what might be down there, forever hidden from view by any spacecraft passing overhead. Fifty years ago, biologists would have said nothing. All life on Earth known at that time needed sunlight to survive. If it didn't generate its own energy directly from sunlight, it ate something that did, or its food ate something that did (think about your own food). Out at Jupiter the Sun is weak. Its pale light is 25 times dimmer than here at Earth and what little light illuminates the icebergs on Europa surely doesn't penetrate dozens of miles through the ice into the sub-arctic seas. But all that changed in 1977 with the discovery of black smokers: towering submarine thermal vents fueling entire ecosystems in the permanent night of Earth's deep ocean floor.

From these undersea forests of strange stony shapes, dark water, rich in minerals, billows upward illuminated only in the camera lights of deep-sea submarines. This super-heated water, warmed by the volcanism of the Earth's interior pours into the cold ocean at temperatures greater than 660° F (350° C). As it mixes with the frigid waters over a mile beneath the surface of the sea, the minerals in the 'smoke' precipitate out building up giant chimneys over millions of years. They loom out of the darkness like a long-sunken city of steel factories left-over from an ancient unknown industrial civilization.

But rather than forming a submarine wasteland of coal tailings and slag, the volcanic heat and minerals are the basis of what before then was an entirely unknown and thriving ecosystem. In these sub-sea hot springs, life permanently cut-off from the energy of the Sun thrives off the internal volcanic heat of the Earth. And rather than just a simple evolutionary oddity, the life forms down there may reflect the very origin of all life up here. The chain of chemical reactions required for the green chlorophyll of plant life to convert sunlight to energy is actually quite complicated. By contrast, it's a chemically simpler process to extract energy directly from the Earth's heat. As a result, the first types of organisms to evolve on any world are most likely to be those able to get the most food, as quickly and as simply as possible. In the tube worms, snails, and clams of the sub-sea thermal vents we may see the shapes of life's first forms.

While many planetary scientists assume that the same tidal heating that keeps Europa's ocean liquid could also produce its own version of black smokers, Greenberg is skeptical. 'The problem with the black smoker story is there is no evidence for it on Europa.' After all, given the degree to which human beings have explored the Earth, it was not until deep-sea submersibles searched the ocean floors that we found evidence of black-smokers on Earth. For this reason, until submarines are sent to Europa, any hypotheses of Europan life based on their presence will be the worst kind of science: the kind with no evidence whatsoever.

But maybe there is another way for life; a way we have some hope of detecting? For Greenberg, the difference between 'thick' versus 'thin' is nothing less than the possibility of life itself. "The important thing about having the linkages between the ice surface and liquid sea is that it really makes it possible to

Figure 3.18 The summer Milky Way sets along Park Loop Road on a clear late-September evening. This view from just passed the Otter Cliff Lookout shows the beautiful starry skies over Acadia National Park. Views like this are possible from many national and state parks, far from city lights, seen here distantly backlighting the horizon (T. Nordgren).

have life on Europa. You have oxidants and other organic material on the surface, material brought by comets. You can only have organisms living in the ocean if you can get that oxygen, or organic material, down to the water." In Europa's tides, Greenberg sees the only possibility for discovering Europan life.

A stroll along any seashore, therefore, puts every one of us in direct contact with astronomical forces, and possibly even those forces responsible for life up there amongst the planets and the stars beyond. Two weeks after watching the full Moon rise from the town of Bar Harbor, I've gone about as far away from the town lights as I can get under a dark moonless night. Tonight, I'm walking alone along the quiet and blessedly dark Arcadia Park Loop Road that follows the

Figure 3.19 A panoramic map of the sky as seen from Otter Cliffs Lookout along the Acadia Park Loop Road. Members of the Island Astronomy Institute, working together with the National Park Service, routinely measure the brightness of the background sky (here shown in artificial color) in order to track the spread of artificial light pollution that blots out the distant light of stars. The brighter the background (brightness increases from black and purple all the way up through the rainbow, at left, to white) the fewer stars can be seen. In the top frame, a truly dark sky (before the invention of the electric light) would show a sky of purple and dark blue. The yellow over the horizon is from the local town of Bar Harbor. The light blue arc through the upper image is the natural light of the Milky Way. The image at bottom is a daytime panorama of the location for comparison. Zero degrees along the horizontal axis is due north, 90 degrees is due east (Island Astronomy Institute).

rugged and rocky coast around the island. At night the road travels under some of the most star-filled skies on the east coast of the United States. Coastlines are usually good places to see the night sky as there you are normally guaranteed that at least half the horizon will be free from the city lights that drown out the distant stars. This is especially true in Acadia National Park where Park Rangers consider the night sky above the island to be as precious and worth protecting as the rustic New England seascape with its rolling granite hills and bright fall colors. In cooperation with experts at the local non-profit Island Astronomy Institute, Acadia is one of the last preserves of starry-skies along the eastern seaboard. On any given summer night, stargazers gather here on the thousand foot heights of Cadillac Mountain or down on Sand Beach along the Park Loop Road for a ranger astronomy talk.[7] In the fall of 2008, residents of Bar Harbor even passed a city ordinance protecting this increasingly rare resource from the growing threat of light pollution.

So tonight, without a Moon to brighten the sky, a thousand stars coat the water's surface with light that has taken hundreds and thousands of years to reach me. One such star is omicron Ceti, or Mira, Latin for 'Wonderful.' Mira is

[7] Be sure to consult the local Visitor Center to see if this is happening on a night when you visit.

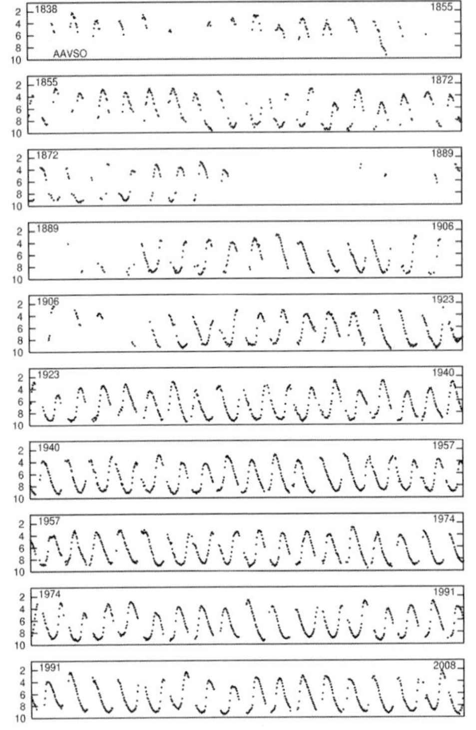

Figure 3.20 The changing, pulsating brightness of Mira as measured by dedicated amateur and professional astronomers for the last 150 years. Each individual dot represents the 10-day mean brightness of the star. The vertical axis for each row shows that at its brightest, Mira is about third magnitude (about as bright as Polaris, the North Star). At its dimmest Mira is tenth magnitude, invisible to the naked eye which has its limit around seventh magnitude (AAVSO).

part of the constellation of Cetus the Whale and appropriately enough for me this September night, it is just cresting the waves off to the southeast as midnight approaches (check the October star map in Chapter 4 to see what the sky looks like in September around midnight).

In 1596, Mira became the first star in history found to vary regularly in brightness. Over the course of 332 days, Mira slowly changes from one of the many moderately bright stars visible in the sky (noticeably red compared to surrounding stars) to utterly invisible and back again. It's a sun just like our own that a long time ago used up most of its nuclear fuel and has gradually swollen to a bloated red giant between three and four hundred times bigger than our own star today. When our own Sun goes through this process about five billion years from now it will swell up so large that it engulfs the Earth. Jupiter, five times farther away, will skim the Sun's tenuous outer layers.

As the interior of a star like Mira changes rapidly in response to its decaying condition, the surface spasms outward and back in regular cycles changing in diameter by a factor of almost 30% with each great gasp. Astronomers, professional and amateur alike, have been faithfully recording each and every resulting change in Mira's brightness for the last 168 years.

Soon after discovering Mira, astronomers surveying the heavens found a number of other stars that also changed their brightness in a regular way. One of the first after Mira was Algol, the Demon Star, 'Al Ghul' in Arabic. Algol is the head of the demon Medusa carried by Perseus to save the fair princess Andromeda from Cetus' attack.[8] Unlike Mira, Algol is nearly constant in brightness until suddenly it's not. Every two days and 21 hours, it visibly dims by

[8] In the fall, Perseus always rises before Cetus to protect us and Andromeda from the dangers of the deep.

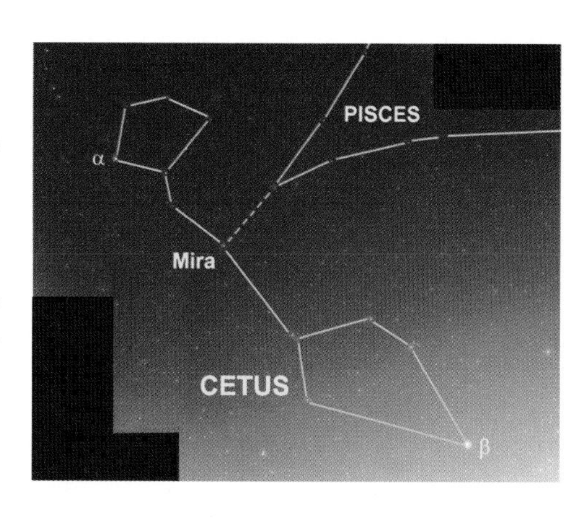

Figure 3.21 Finder chart for Mira (also called omicron Ceti) in the constellation of Cetus, the Whale. Mira brightens and dims over the course of 332 days. In this photo, the noticeably reddish (or at least noticeably not-blue) Mira is at its maximum brightness. At its faintest it will be invisible to the naked eye. To find Mira, use the star maps for the winter months to first find the constellation of Pisces, the Fishes. The star at the sharp bend in Pisces points directly to the position of Mira located between the circular 'tail' and spear-shaped 'head' of the whale. The table at the end of this chapter provides dates for when Mira is at its brightest and dimmest (T. Nordgren).

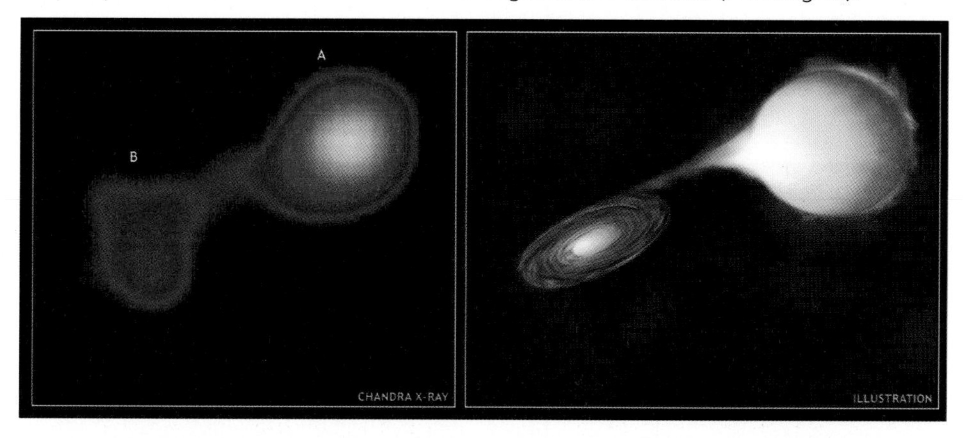

Figure 3.22 Chandra X-ray space telescope image of Mira (labeled A). Mira has long been known to have a companion (labeled B) and in this 2003 image we see for the very first time the gravitational attraction between stars pulling gaseous material from the atmosphere of Mira. This companion is slowly gaining mass from Mira while tides have possibly affected its shape. On the right is an artist's conception of the two stars (M. Karovska, NASA/CXC/SAO/M. Karovska et al.; Illustration: CXC/M.Weiss).

1.2 magnitudes (a little over 30%) for 10 hours. In 1783, English astronomers John Goodricke and Edward Pigott wrote of the intriguing possibility that this was the first discovery of a planet around another star. By their hypothesis, we Earth-bound observers just happened to be aligned so that once every orbit of the new planet we witnessed an eclipse of the distant star by its unseen companion. With Goodricke's subsequent discovery of more eclipsing binaries it was eventually realized that rather than planets, the new-found objects were instead stars too small, too dim, or too close to their primary star for us to see by the light of the primary's glare.

We now know the majority of stars in the sky come in multiples, with pairs, triplets, and pairs of pairs quite common. In this respect, our single Sun is in the

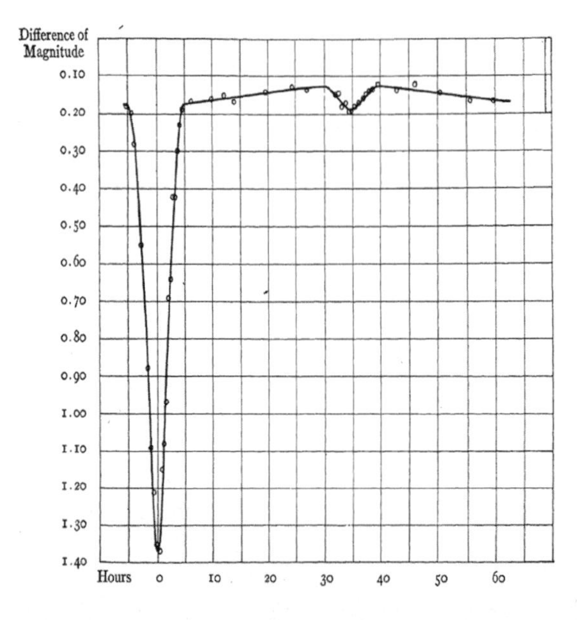

Figure 3.23 First light curve of the eclipsing binary, Algol, made using electrical detectors in 1910. The large dip in brightness occurs when the fainter companion star passes in front of the brighter primary star. The small, secondary dip occurs when the dimmer star passes around behind the brighter one. The rise in total brightness around the secondary minimum reflects that the face of the dimmer star tidally locked to forever point towards the brighter star, has been heated and so shines brighter than the side forever pointed away. A difference of one magnitude is a difference of 2.5 × in brightness (Stebbins, J., 1910, *Astrophysical Journal*, Vol. 32, pg. 185, Reproduced by permission of the AAS).

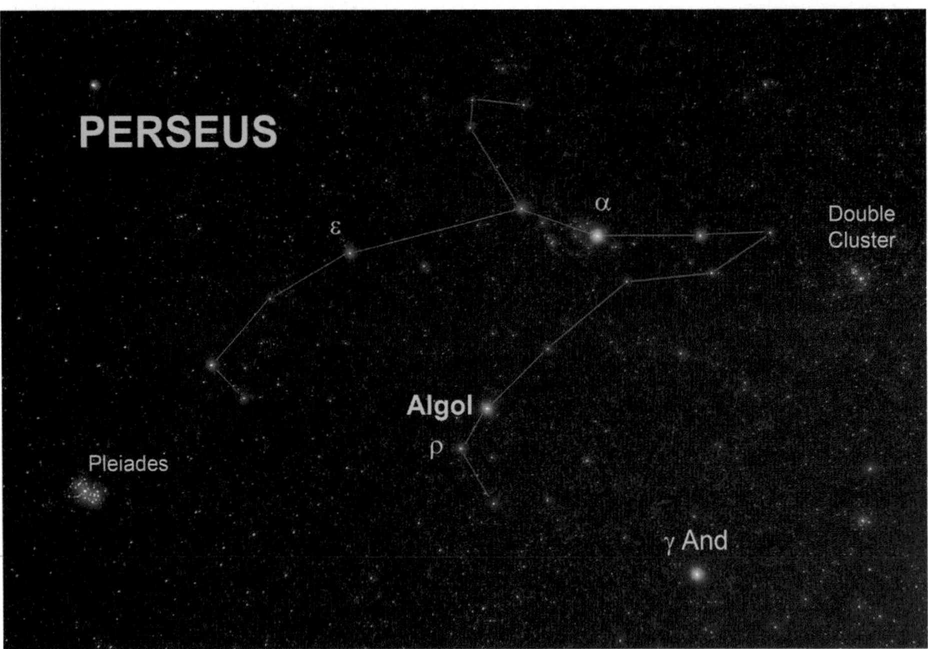

Figure 3.24 Finder chart for Algol, showing the constellation of Perseus and the nearby Pleiades cluster of stars. North is to the right. Compare Algol's brightness to the nearby star gamma Andromedae (γ And) and alpha Perseus (α) in order to detect if Algol is currently eclipsed. Normally, Algol will be nearly as bright as these stars. During the faintest 2 hours of its 10 hour eclipse it will appear slightly dimmer than nearby epsilon Persei (ε). This image shows it out of eclipse (T. Nordgren).

minority of all the stars we see at night. From these binaries, however, we learn about stars in ways no single star can possibly allow. For as these stars orbit one another we learn about their masses through application of Kepler's laws for orbital motion (the faster they orbit, the more massive they must be). Where stars eclipse one another we learn about their sizes and relative luminosities by seeing how much light is eclipsed as each passes in front of the other (the larger the unseen companion, the more light it blocks as it passes in front of the primary). The eclipsing binaries are our laboratories allowing us to measure and probe all the varieties of stars found in the sky and thereby determine how truly common or rare, our Sun and our Solar System might be.

One of those other eclipsing binaries Goodricke found is beta Lyrae in the constellation of Lyra the Harp. Tonight, Lyra is overhead and easily found as Vega, its brightest star, is also one of the brightest star in summer and fall. During September evenings this bright white star is near the zenith and just beginning its slow descent into the west. Vega is one third of the famed 'Summer Triangle' along with Altair and Deneb, and so is one of the few stars actually visible from cities. Thanks to Vega, beta Lyrae (also called Sheliak, Greek for harp) is relatively easy to find and tonight it's where my eye is finally drawn. It's an unassuming third magnitude star (about sixteen times fainter than Vega). But by its changing brightness, visible to the naked eye, we know that in reality it is two stars alternately passing one in front of the other with a cycle that repeats every 12.9 days.

Careful observation by dedicated observers shows something unusual about the combined light from these two stars and it's for this reason that my attention has finally settled here. Carefully compare the light curves from Algol and beta Lyrae. In general, the light from Algol is fairly constant except during those moments when one star passes in front of the other, but for beta Lyrae, the combined light from the stars is in constant flux.

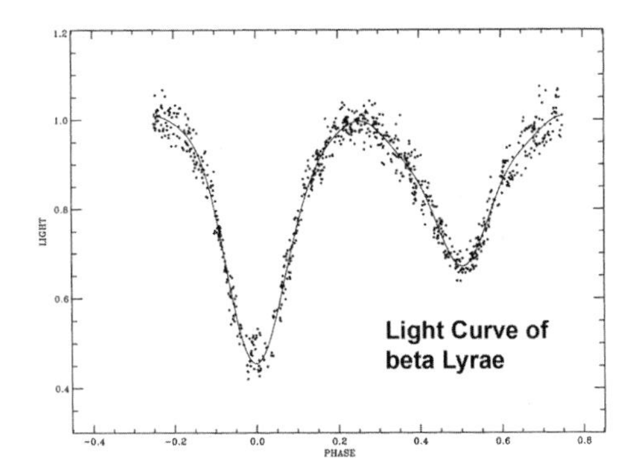

Figure 3.25 Light curve for the eclipsing binary beta Lyrae. The different size minima reflect the different brightness of the eclipsing stars. The continually changing brightness reflects the tidally distorted shapes of both stars; as they rotate, we see different amounts of surface area. Observations from multiple maxima and minima have been wrapped onto one single plot where 'Phase' represents the fraction of the time that has passed since the eclipse. The scatter in the observations at any single phase is a reflection of the gas and dust that surround the stars making the combined starlight change slightly from orbit to orbit (Van Hamme, W, Wilson, R.E., Guinan, E.F., 1995, *Astronomical Journal*, Vol. 110, pg 1350–1363, Reproduced by permission of the AAS).

Figure 3.26 Finder chart for beta (β) Lyrae, showing Vega and the Milky Way. North is up. Compare beta Lyrae to the nearby star gamma (γ) Lyrae in order to detect if beta Lyrae is currently eclipsed. Over its 13 day period, beta Lyrae will go from being nearly equal in brightness to gamma Lyrae, until at its dimmest it is the faintest of all the major stars shown linked together in the constellation. This image shows it out of eclipse (T. Nordgren).

In reality, the two stars in beta Lyrae orbit so closely that their surfaces nearly touch. As one star ages and slowly puffs up and swells like Mira, tidal forces from the other gradually pull it into a distended tear-drop shape. As it continues to swell tidal forces from the smaller, denser star begin ripping away its atmosphere until, with every passing orbit, the smaller star cannibalizes the larger. With time, the larger star loses mass and shrinks, while the smaller star, embedded in a thick disk of dust and gas, gains mass and grows.

The light we see from beta Lyrae is now almost entirely due to the light of the mass-losing star, its growing companion hidden completely inside its dark enclosing cocoon of accreting mass. We see the slow change in brightness between eclipses because as the mass-losing star turns to follow its companion around, we see the full range of its fat and bloated, tear-drop profile.

Since this process began a few thousand years ago, so much matter has changed places between the two stars that the roles have now reversed and what was once the smaller star is now the larger, estimated at nearly 12 times the mass of our Sun, while the star we see is now the smaller, and probably only twice as massive as the Sun.

Eventually as the mass in the disk slowly settles onto the enclosed star, tidal forces play one last role as they slow its rotation to exactly coincide with its orbit. Then like Europa around Jupiter and our Moon around us, it will keep one face pointed at its companion as they continue their slow spiral dance. Thanks to a new generation of telescopes we are able to see this gravitational tug of war first hand, and for the first time, see the gradual death of a star.[9]

But the tidal forces that destroy stars are also responsible for the most spectacular stellar births. Far beyond the stars I see with my naked eyes, entire galaxies of gas and stars with the combined mass of over a hundred billion suns gracefully glide through the darkness of space, each a hundred thousand light-years across. When two such leviathans approach on their Keplerian trajectories, tidal forces between the nearer reaches send the gas and dust between the stars spiraling

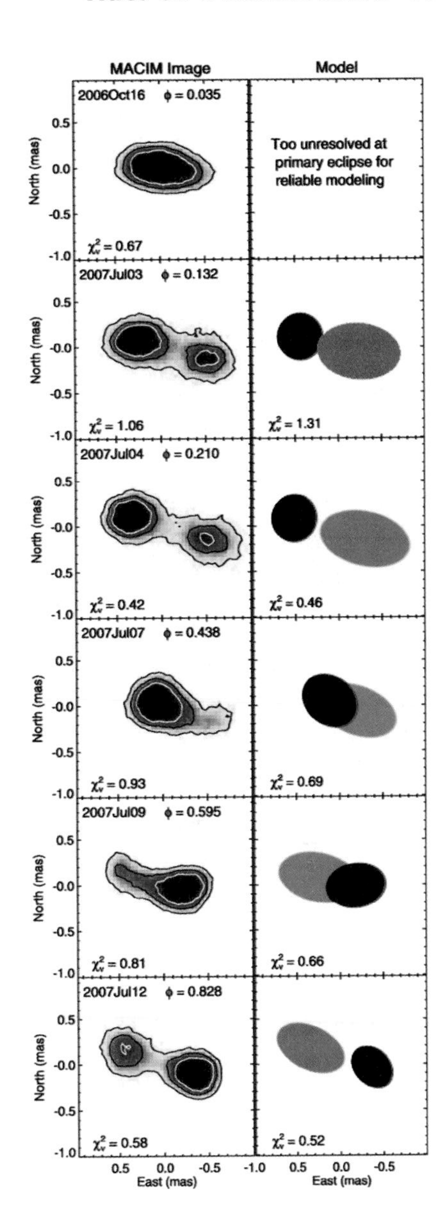

Figure 3.27 First image showing the tidal tug-of-war between the two stars of beta Lyrae. The left column shows the image of the orbiting stars made by the CHARA telescope on Mt. Wilson in southern California. Brightness increases from orange to black, with the black star representing the one that we see in the sky. The column on the right is a model of the two stars that best fits the data seen on the left. Notice how each star is best modeled as a tidally deformed ellipse rather than a perfect sphere. The typical separation of the two stars, as well as their size as viewed on the sky, is on the order of half a milliarcsecond (mas). A milliarcsecond is 1/3,600,000 of a degree. The Moon is only half a degree in size as seen from Earth and a child standing on the Moon would appear only 0.5 mas tall (Zhao, M. et al. 2008, *Astrophysical Journal Letters*, Vol. 684 pp. L95-L98, Reproduced by permission of the AAS).

[9] This new breed of telescopes called optical interferometers, combines the light from many widely separated mirrors to simulate the fine-detail possible with a single giant telescope. The CHARA interferometer on Mt. Wilson in southern California combines the light from four widely separated mirrors simulating the fine detail capable by a single mirror 1,075 ft (328 m) in diameter.

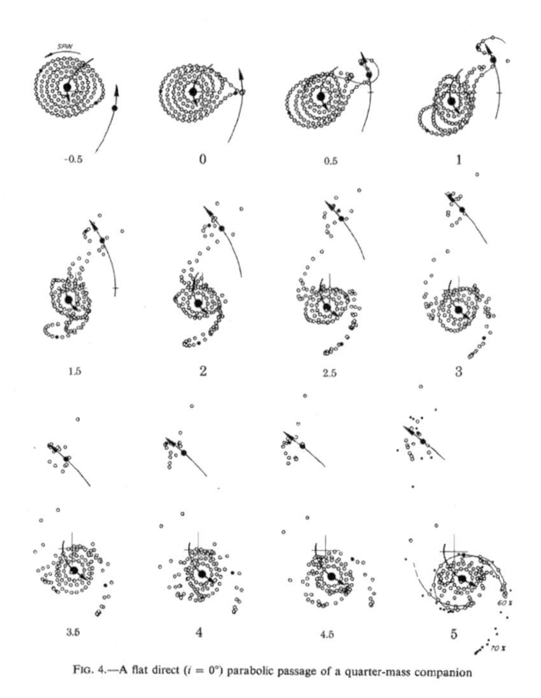

Figure 3.28 Computer model of tidal forces between galaxies during a collision in space. The primary galaxy is represented as particles on a series of concentric rings (this is the disk of the galaxy). The second galaxy is simply shown as a single large black dot representing its center of mass. Each time step in the collision (represented by the numbers beneath) is in units of 100 million years from the time of closest approach t = 0. Compare the model at t = 2 or 2.5 to the Hubble Space Telescope image of M51. As groundbreaking as these models were in 1972 when they were published in a landmark paper, today I can see these same interactions on the screensaver that came with my computer (Toomre, A. & Toomre, J., 1972, *Astrophysical Journal*, Vol. 178 pp. 623–666, Reproduced by permission of the AAS).

on new courses into the intergalactic space. The paths they follow depend on the rotation of the two galaxies as they approach one another. If one of the galaxies spins in the same direction as the path it takes around its neighbor (like an enormous gear locking and rolling around the teeth of a companion) gravity takes hold and draws a tidal bridge of gas between the two. A hydrogen stream ten thousand light-years long and a thousand light-years thick arcs through the void, tying the two massive galaxies together while gravity's tides send a counter tail spiraling outward in the opposite direction. Well over a hundred thousand light-years of new gas is thrown into the void in the galactic equivalent of the tidal bulges on Earth, stretched out from the force of gravitational interaction.

The collision of two galaxies like our own Milky Way is one of the most beautiful sights in extragalactic space. My favorite example is, appropriately enough for the sea, the Whirlpool Galaxy (M51) in the constellation of Canes Venatici near the end of the Big Dipper's handle. Here, not too long ago, a small galaxy passed too close to a grand spiral similar to our Milky Way. Today they are separated by 150,000 light-years as they continue around one another in their slow-motion passage. However, unlike waves upon a coastline whose splash is measured in seconds, the crashing of galaxies like the two in the Whirlpool is an event measured in hundreds of millions of years. While no single encounter could ever be witnessed in its entirety, from hundreds of examples at different stages of completion throughout the heavens, theoretical astrophysicists have modeled the process from beginning to end

Figure 3.29 Hubble Space Telescope image of the interacting galaxy pair comprising M51, the Whirlpool Galaxy, and a small galaxy (NASA, ESA, S. Beckwith (STScI), and the Hubble Heritage Team (STScI/AURA)).

In those tidal tails, gravitational forces bring together the gas and dust between the stars and spark the formation of new suns. Tidal interactions may be one of the most important mechanisms for creating new stars in galaxies. Where new stars form, our observations of the heavens show that the formation of new planets and new solar systems around them may not be too far behind. Where there are planets, there are surely moons, and the power of tides on those moons may even now be at work, making the conditions right for new life, in a new sea.

> I do not know what I may appear to the world, but to myself I seem to have been only a boy playing on the sea-shore, and diverting myself in now and then finding a smoother pebble or a prettier shell than ordinary, whilst the great ocean of truth lay all undiscovered before me.

Newton said these words with regards to his gravitational theories. I cannot help but think of them this night as I step down onto the still wet rocks, left behind by the vanishing tide. Above me is a clear starry sky filled with wonders of nature hidden from view by infinite dark depths. Beside me is an equally dark and no-less wondrous sea. I stoop to pick up a shell bright in reflected starlight set amongst the darker stone. All around me are fresh pools of water. The life that surrounds me here in Acadia may have first colonized the land from tide pools just such as these a billion years ago. In each one, I see the stars reflected while the mirrored image of Jupiter outshines them all. Amid all this beauty and power

Figure 3.30 Jupiter shines above the constellation of Scorpius and reflected in a tide pool along the coast of Mount Desert Island (T. Nordgren).

of sea and space together, I cannot help but look at that sparkle of planetary light and wonder about the possibility of other life, there in that other ocean, on that distant moon: perhaps impossible but for the power of the tides.

See for yourself: moons, planets, stars and galaxies

Jupiter and its moons

Find Jupiter in the sky this month and note its position relative to the stars. Wait a full year, so the Earth is back in the same spot in space and find Jupiter again. The difference in its position from one year to the next (shown on the sky maps) shows how far Jupiter moves through space over the course of 365 days. From this we find that Jupiter takes a little under 12 years to make one complete trip around the Sun and thus appear again where it started on the sky.

Through a pair of typical binoculars you can see the tiny disk of the planet and usually two or more of the Galilean moons. The moons will always appear in a straight line on either side of Jupiter. Innermost Io is never farther than 6 Jupiter (jovian) diameters away from the planet while outermost Callisto is never more than 13 diameters away. In many binoculars, the glare from Jupiter may hide the inner two moons unless they are at their maximum distance from the planet. If you have a pair of binoculars of at least 20× power (in a pair of 20×50 binoculars, the '20× ' refers to the magnification, while the '50' is the diameter of the aperture in millimeters) then you have at least as good a view as Galileo had 400 years ago. In addition, the larger the aperture, the more light is let in, and the more easily you will see faint objects in the sky.

Algol – the Demon Star

If you can see the Milky Way, look for Cassiopeia's big 'W' and then follow the Milky Way east to the next prominent grouping of stars. That's the constellation of Perseus. The brightest star in Perseus is alpha Persei, also called Mirfak. When not eclipsed, Algol, the second brightest star in Perseus (also called beta Persei) is nearly as bright as Mirfak and the nearby star, gamma Andromdeae. Use the image of Perseus in Figure 3.24 to find Algol in relation to these stars and compare their brightness. Every 2.87 days, Algol is eclipsed for 10 hours as its unseen companion moves in front of the brighter star. As the eclipse begins, Algol slowly dims for five hours. At its faintest, Algol dims to a level near that of epsilon Persei, less than half as bright as Mirfak. This minimum lasts for 20 minutes as the companion passes fully in front of the brighter star. As the dimmer companion moves on, Algol's brightness returns until five hours later it is once again as bright as before. In the end, a simple comparison of Algol to Mirfak or gamma Andromdeae will tell you if Algol is currently being eclipsed.

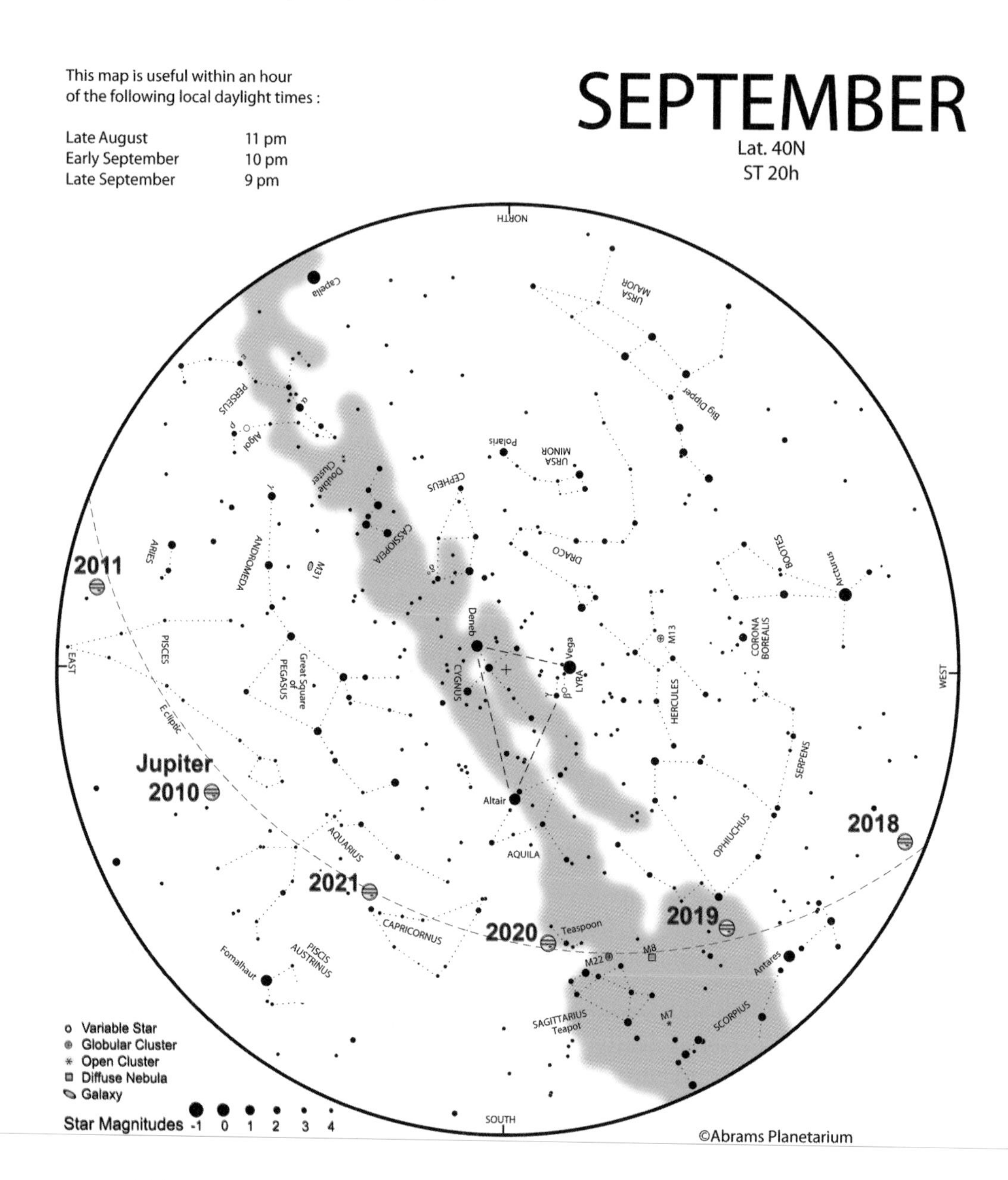

Hold the star map above your head with the top of the map pointed north. The center of the map is the sky straight overhead at the zenith. Jupiter's position is marked relative to the background stars for each September and March from 2010 until 2020 (almost one full Jovian year). To find Jupiter during another month, find its position between the two closest dates then consult one of the other monthly star maps to see if that constellation (and thus Jupiter) is above the horizon. While during any given month of any given year there may be other planets also visible, Jupiter is typically very bright and noticeably yellow.

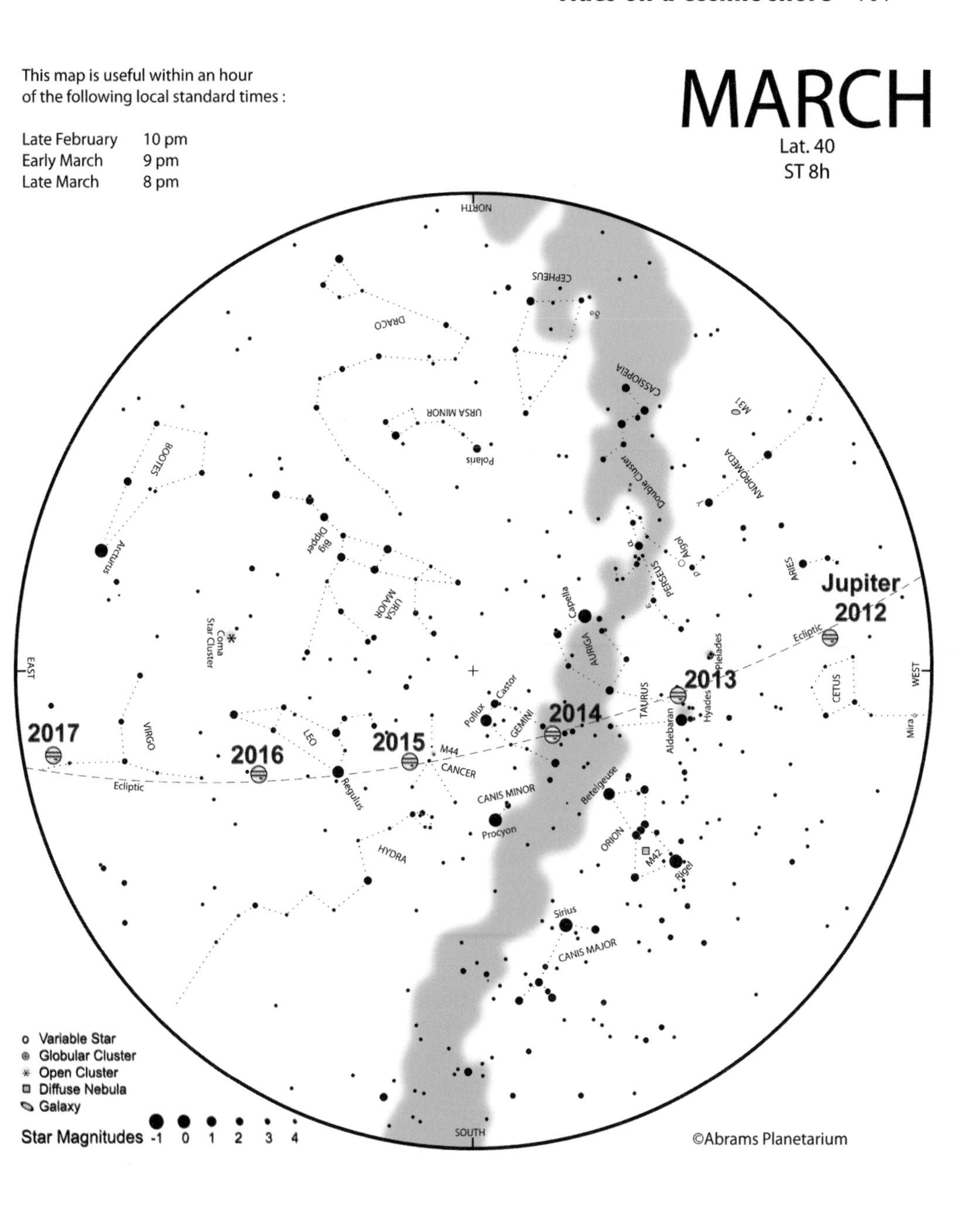
This map is useful within an hour
of the following local standard times :
Late February 10 pm
Early March 9 pm
Late March 8 pm
MARCH
Lat. 40
ST 8h
NORTH
SOUTH
EAST
WEST
CEPHEUS
DRACO
CASSIOPEIA
URSA MINOR
Polaris
M31
BOOTES
ANDROMEDA
Double Cluster
Algol
Arcturus
Big Dipper
PERSEUS
ARIES
URSA MAJOR
Capella
Jupiter
2012
Coma Star Cluster
AURIGA
Ecliptic
Pleiades
Castor
2013
Pollux
TAURUS
Hyades
CETUS
GEMINI
2014
2017
VIRGO
LEO
2015
Aldebaran
Mira
M44
2016
CANCER
Ecliptic
Regulus
CANIS MINOR
Betelgeuse
Procyon
ORION
HYDRA
M42
Rigel
Sirius
CANIS MAJOR
o Variable Star
Globular Cluster
Open Cluster
Diffuse Nebula
Galaxy
Star Magnitudes -1 0 1 2 3 4
©Abrams Planetarium

Beta Lyrae

The constellation of Lyra is best known for its brightest star Vega, which can be seen from even light polluted suburbs. Lyra the Harp is really two small geometric figures tied together. Vega is one corner of an equilateral triangle (all sides equal in length). From the southeast corner of this triangle, a parallelogram (two long sides and two short sides, joined at an angle) hangs down to the south. The two stars at the far end of the parallelogram are the brightest. Of these two, the one closest to Vega is beta Lyrae while the other is gamma Lyrae. At its brightest, beta Lyrae is very nearly the same brightness as gamma Lyrae. Over its 13-day period, beta Lyrae changes brightness continuously until at its faintest, when the brighter star is eclipsed by its fainter companion, beta Lyrae is as faint as the other stars shown in the constellation. See Figure 3.26 for a photograph of the constellation with the stars labeled.

The Moon

Point even a modest pair of binoculars at the Moon and you will see the craters, mountains, and smooth dark plains (the *maria*, Latin for 'seas') that first convinced Galileo that the Moon was a world like our own. Again, the best views are obtained when the binoculars are held steady on a tripod or with arms and elbows resting on a table.

Mira – the Wonderful Star

Use the accompanying sky maps to find Mira, also called omicron Ceti, in the constellation of Cetus the Whale. While Cetus is only just rising in the evening hours in September, use Figure 3.21 and the October sky map in Chapter 4, to see its full shape in relation to the surrounding constellations. If Mira is near maximum (roughly a month or so on either side of the dates below) it should not be hard to spot. See if you can detect its red color relative to the other nearby stars. Using the sky chart shown here, see if you can also spot Mira's changing brightness over the months relative to these nearby stars. Near its minimum, Mira won't be visible at all. The Association of Amateur Variable Star Observers (AAVSO) runs an extensive website with information on what variable stars are visible and what they are currently doing.

Table 3.1 The brightness of Mira

Date of Maximum	Date of Minimum
19 Nov 2009	2 Jul 2010
17 Oct 2010	30 May 2011
14 Sep 2011	26 Apr 2012
11 Aug 2012	24 Mar 2013
9 Jul 2013	19 Feb 2014
6 Jun 2014	17 Jan 2015
4 May 2015	15 Dec 2015
31 Mar 2016	11 Nov 2016
26 Feb 2017	9 Oct 2017
24 Jan 2018	6 Sep 2018
22 Dec 2018	4 Aug 2019
19 Nov 2019	1 Jul 2020

M51 – the Whirlpool Galaxy

M51 will not be visible to the naked eye, nor will it be noticeable in a pair of binoculars. However, with a small telescope, the centers of the galaxies can be seen from dark locations. A medium sized telescope of 8-inches (20-cm) in diameter will begin to show the spiral tidal arms. You can see where the Whirlpool Galaxy is by looking towards the handle of the Big Dipper. From the very last star in the handle, drop down at a right angle to the other stars in the handle. Imagine this star is the top of a small equilateral triangle where the corner on the bottom right is at the position of a nearly equally bright star there. The bottom left corner of the triangle falls on the position of M51. While there is nothing there for the eye to see, you will know that's where it is, 37 million light-years away. If looking for this in a telescope, finding it can take a bit of searching, don't give up. Take the image of M51 in this book out with you. Shine the light from a red flashlight on the image and notice how some of the detail in the galaxy is lost. This is what you are looking for through your telescope.

Further reading

Native American Stories by Joseph Bruchac (1991) Fulcrum Publishing, ISBN 1555910947

Beyond the Moon: A Conversational, Common Sense Guide to Understanding the Tides by James Greig Mccully (2006) World Scientific Publishing Company, ISBN 9812566449

Unmasking Europa: The Search for Life on Jupiter's Ocean Moon by Richard Greenberg (2008) Springer, ISBN 0387479368

David Levy's Guide to Variable Stars by David H. Levy (2006) Cambridge University Press,
ISBN 0521608600

Online tide tables for all over the United States.
http://www.tidesonline.com/

Sky and Telescope guide to the Top 12 Naked Eye Variable Stars
http://www.skyandtelescope.com/observing/objects/projects/3304276.html

Website for the American Association of Variable Star Observers (AAVSO)
http://www.aavso.org/

Galaxy Interactions page by astrophysicist John Dubinski
http://www.galaxydynamics.org/

4 Worlds of fire and ice

The noise made by the bubbling lava is not great, heard as we heard it from our lofty perch.... The smell of sulphur is strong, but not unpleasant to a sinner.

Mark Twain, *Letters from Hawaii*

Growing up in Portland, Oregon, I have always loved the mountains. The Cascade Range runs the length of the western horizon there and on clear days (a rarity in the Pacific Northwest) it's dominated by two giant mountains: Mt. Hood and Mt. St. Helens. Mt. Hood is the biggest and closest, but I loved Mt. St. Helens as a kid, because its perfect cone matched my ideas of what a mountain should be. So when small earthquakes began to rock the mountain in early1980 I started drawing daily pictures of the peak in my school notebook. On the first day, I remember something dark stained the summit. Within a week a small crater had formed and was eventually joined by another. On clear days from my house, I could see steam rising into the sky.

Then on the morning of May 18, a dark and cloudy day, my father woke me at dawn and drove me to the summit of a nearby hill with our little JC Penney telescope. He didn't tell me where we were going or why, but when we got to the small park at the top, he didn't need to say a thing. To the northwest, right where Mt. St. Helens should have been, a dark jumble of clouds towered into a sky the color of lead. Everything was shades of grey; only textures and motion distinguished the heavy clouds from the rolling, boiling, slow-motion plume. As far as I remember we spent hours looking through the telescope, spellbound by what was unfolding before us. It was both beautiful and terrifying.

That night it snowed great ghostly flakes of volcanic ash that I shoveled the next day from our driveway. When the skies finally cleared a few days later, the perfect cone was gone and the mountain looked like a crater on the Moon. The entire north face had vaporized and its insides hollowed out. It was eerie; it's not every day you get to witness the planet change before your eyes.

When most people think of volcanoes something very specific comes to mind. They usually think of something like my childhood Mt. St. Helens, or Mt. Fuji in Japan. They picture an enormous, steep-sided cone with a black cloud coming out the top and probably molten lava pouring down the side. Volcanoes of that shape are stratovolcanoes. Their thick, blocky, viscous lava forms a steep-sided cone as it slowly builds up around the vent. But this isn't the lava most people have in mind. For most people, lava brings to mind great glowing rivers of fire like that found in Hawai'i. But that lava isn't the result of stratovolcanos. Rather

that kind of lava is thin, runny, low viscosity magma that only builds up slowly like layer upon layer of flowing wax from a candle. This type of lava only forms big broad shield volcanoes.

So, while everyone thinks they know exactly what we mean by the word *volcano*, what they actually know is a conglomeration of many aspects of what a volcano could be. So now consider Yellowstone National Park in northwestern Wyoming. Ask any Park Ranger and they will tell you nearly every aspect of the park, its geology, its geography, its biology, chemistry, and history is "all about the volcano." But where is the volcano? There is no giant cone. There aren't even any mountains. There are no rivers of lava. There is no gaping crater. The list goes on and on, but in reality they are all here. They're just in forms that no one really notices. Except, of course, everyone notices the giant steaming pools and geysers.

I'm visiting Yellowstone during one of its snowiest winters in years.

Figure 4.1 Three views of Mount St. Helens. Top: As seen from Bear Cove, Spirit Lake in 1973 before the eruption (U.S. Forest Service). Middle: Aerial view during the May 18, 1980 eruption that sent volcanic ash, steam, water, and debris to a height of 60,000 feet (18 km). The mountain lost 1,300 feet (400 m) of altitude and about 2/3-cubic mile (3-cubic kilometers) of material (Austin Post). Bottom: Mount St. Helens reflected in Spirit Lake, two years after the May 18 eruption (U.S. Geological Survey).

Figure 4.2 Mauna Loa in Hawai'i Volcanoes National Park is the largest shield volcano on the planet. It towers 2.5 miles (4 km) above sea level, 30,000 feet (9 km) above the sea floor beneath (T. Nordgren).

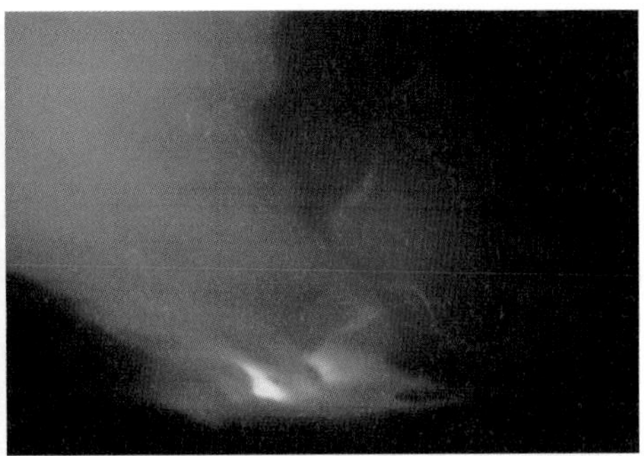

Figure 4.3 Lava from Hawai'i Volcanoes National Park pours into the sea along the Chain of Craters Road (T. Nordgren).

Figure 4.4 Castle Geyser bubbles and steams along the Firehole River in the Upper Geyser Basin of Yellowstone National Park. Beehive geyser erupts in the distance (T. Nordgren).

At Old Faithful it's been snowing for weeks and the world is a study in shimmering white. Mini-glaciers slowly advance down the roof-lines of every building, and icy stalactites hang low from every edge. Snow is packed over a foot deep on the boardwalks, and it's unspeakably delightful to snowshoe or ski in isolation where in summer there would be thousands straining for photos. The other amazing thing about this time of year is that it makes obvious that something unusual is going on beneath your feet. When everything is covered in three feet of snow except for the strangely warm spot of land next to the giant reeking pool of furiously bubbling water you know something odd is occurring.

At its heart, this is what volcanism is: it's the internal heat of a planet finding some way to escape to space. Big rocky planets like the Earth contain vast quantities of radioactive elements deep within their interiors. As these elements slowly decay over the millennia, the energy they release heats the interior of the planet, creating molten pockets of rock. Five billion years after the Earth formed, thanks to uranium, we still have a nice toasty planetary interior looking to find all sorts of creative ways to radiate that heat to space.

Again, when people think of volcanoes, they imagine that the lava which pours out of a great gaping hole reaches all the way down to the molten core of

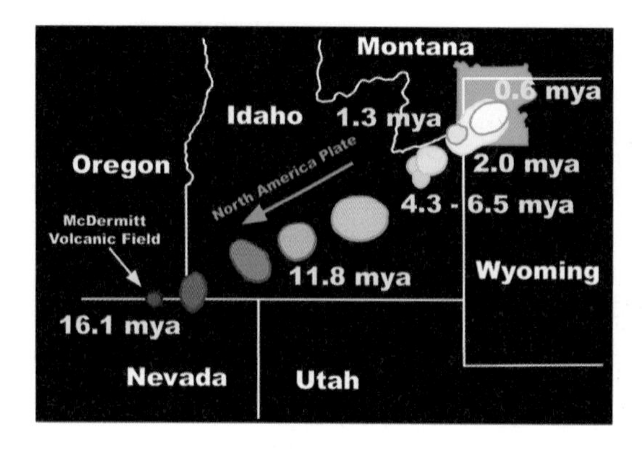

Figure 4.5 As the North American continental plate moves to the southwest (blue arrow) over the Yellowstone hotspot periodic eruptions leave a trail of volcanic features across the northwestern United States. Eruption dates are shown in millions of years ago (mya). The current boundary of Yellowstone National Park is shown in green (National Park Service).

the Earth. In reality, the molten rock volcanoes tap is usually found quite close to the planet's crust and comes from nowhere near the core. From seismic records, we know the hot spot that currently sits underneath Yellowstone National Park begins about 125 miles (200 kilometers) beneath the Earth's surface. From there it extends upward to about 50 miles (80 km) under our feet where it piles up beneath the North American plate that slowly carries our continent towards the southwest.

The molten plume blossoms outward under this plate, forming a mushroom cap thought to be 300 miles (500 km) wide. The heat of this magma mushroom melts great globs of the overlying rock which slowly rise through the surrounding material like a gargantuan Lava-Lamp (which is as accurate an analog as the name implies). All this molten 'goo' pools in a magma chamber located only five to eight miles (about 10 km) beneath our feet. Eons of earthquakes caused by this hot magma shifting around in the chamber have repeatedly broken and cracked the surrounding bedrock. Ground water that falls as rain or snow on this high plateau in the Rocky Mountains eventually gets channeled through this network of heated fissures resulting in a region-wide hot water plumbing system. Where this plumbing reaches the surface, pools or geysers form.[1]

In the 1980s and '90s a team of researchers including geologist and planetary scientist Susan Kieffer and astronomer Jim Westphal studied Old Faithful Geyser to determine the exact mechanism by which the geysers work. In 1992, they lowered an ice-covered miniature video camera into Old Faithful's vent to test different models of geyser activity. What they found surprised them. In a 1997 *Science News* interview, Kieffer explained, "I was assuming a simple geometry for

[1] The southwest motion of the North American continent over this hot spot is why it looks from the surface like the volcanoes in this area of the United States have been slowly moving to the northeast, starting first under Nevada, passing through Idaho, and now crossing the northwest corner of Wyoming. Had the National Park Service existed 16 million years ago, Yellowstone National Park would have been founded along the Oregon and Nevada border.

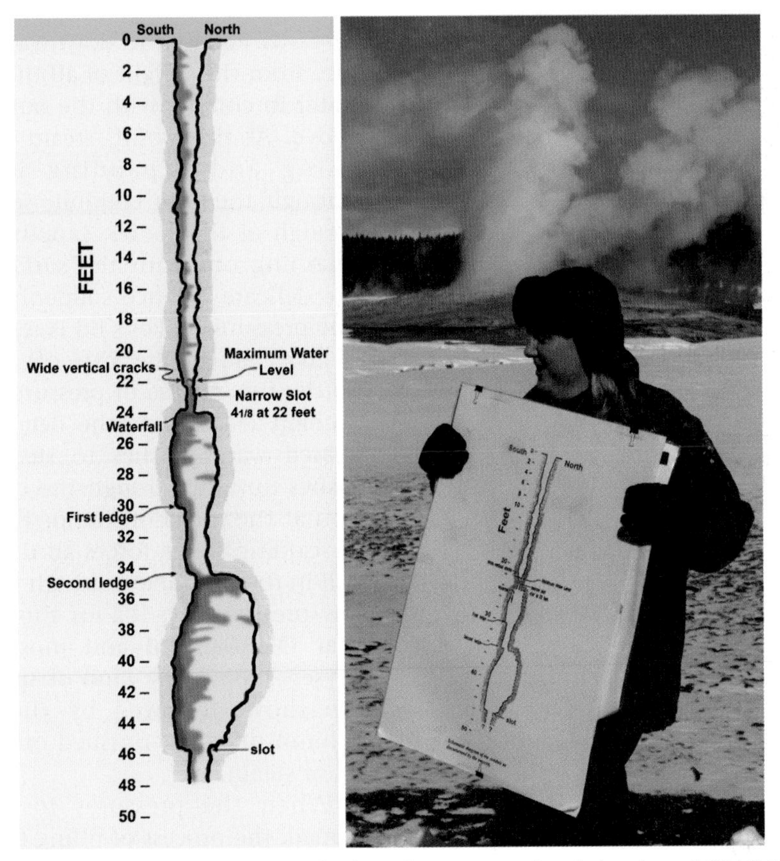

Figure 4.6 A National Park Ranger displays the cross sectional drawing of Old Faithful geyser that was revealed by lowering an ice-covered camera into the vent. Notice the tiny constriction in the tube at 23 feet (7 meters) beneath the surface that helps create the geyser (T. Nordgren, based on diagram by S. Kieffer and J. Westphal).

the fissure and expected a complicated physical explanation for the geyser's behavior, the video camera showed us it was just the reverse."

What they found is that Old Faithful, and presumably most geysers like it, is a rather complex series of fissures and channels where a main vertical tube is repeatedly filled by hot water pouring out of numerous intersecting cracks in the geyser wall. The most important feature for creating the geyser, however, is a narrow constriction at the top through which the rising hot waters must pass to reach the surface. This constriction, as little as a few inches wide, creates a bottle neck preventing the hot gas-filled water from moving freely up to the surface. The weight of the water in and above the constriction puts enormous pressure on the hotter water in the main channel below.

In exactly the same mechanism as a pressure cooker in a kitchen, the deep pressurized water rapidly becomes superheated far above the boiling point of water. Eventually, the water beneath becomes so hot that the pressure pushing

Figure 4.7 Hot water and steam vent into the air during an eruption of Old Faithful geyser. The hot water that rains back to Earth runs off in streams that melt the surrounding snow (T. Nordgren).

upwards equals the downward pressure from the weight of all the cooler water in contact with the outside air above. At this point, steam bubbles rising and expanding upward through the thin conduit, send just enough of the cooler, capping water splashing out onto the surface that the delicate balance suddenly shifts. The pressure cooker's lid is removed.

Like taking the cap off a soda bottle, this release of pressure begins a chain reaction as the deep superheated water flashes to steam and rushes upward through the constriction at the speed of sound. The blast of scalding steam forces all the water still in the tube above it high into the low pressure sky. It's for this reason that Old Faithful and most other geysers have some kind of splashing 'pre-show' followed by the main column of water, finished off with a jet of steam.

When the pressures inside are released and the water column is finally emptied, the process of filling the tube with ground water begins anew until the conditions for eruption are reached once more. This is why geysers like Old Faithful are faithfully periodic. Most importantly, watch any eruption and note that you are seeing one of the final energetic steps in a saga that started deep within the Earth. Geysers are a nuclear powered way the planet cools itself.

How does the process of planetary cooling play out on other planets? For smaller bodies like the Moon, their size works for them in two important ways. First, because of their small size, they simply contain less stuff, and hence have far fewer radioactive heaters, compared to larger bodies like the Earth. Without these elements, very little heat is ever generated in the hearts of asteroids and moons. Secondly, because they are small, they have relatively little volume per surface area exposed to space. What internal heat they do generate is rapidly radiated away to space. Look at the Moon and its lack of any recent geologic activity (other than being pummeled by space debris) is obvious. Since our Moon is one of the larger moons in the Solar System, its lack of internal heat and exciting geological surface processes should be about all you would find on other moons out there around the other planets.

That was the expectation; the problem was that prior to the Space Age, no one really knew. Given that a planet's surface is a window to its interior, consider

Figure 4.8 Family portrait of the Solar System. From top to bottom the major planets are Mercury, Venus, Earth (with its Moon), Mars, Jupiter, Saturn, Uranus and Neptune. Neither the sizes nor distances separating them are to scale (NASA/JPL).

how little we knew about our sibling planets prior to the 1960s. Mercury, closest to the Sun, is small, no more than 50% larger than our Moon. Because it is so close to the Sun in space, we never see it very far away from the Sun in the sky. When the Sun goes down (or shortly before it comes up), Mercury is right there on the horizon, dulled and distorted by the Earth's thick atmosphere. In telescopes large and small, Mercury is a fairly featureless disk that simply goes through phases like the Moon as it goes around the Sun.

Nearly equidistant between the Earth and Mercury is the planet Venus, a near twin to our Earth in mass and size. As Venus orbits the Sun we see the planet go through phases too. When it comes out from behind the Sun we see its fully lit side and it appears as a small circular disk. Slowly, however, as it travels around the Sun and begins to overtake the Earth, we see more and more of its unlit night side. To our eyes it goes from full to only half-lit or first quarter phase. Eventually, as it passes almost exactly between us and the Sun, we are presented with nearly the entire night side of Venus, and it appears as a large thin crescent through even a decent pair of binoculars. At these times Venus is closer to us than any other planet in the Solar System.[2]

Galileo saw these phases 400 years ago in a telescope no more powerful than a typical pair of binoculars, and it helped change our view of the Universe from an Earth-centered one to one centered on the Sun. But those phases are all anyone ever saw on Venus and pretty quickly astronomers realized that Venus was covered in perpetual clouds. What was under the clouds? No one knew.

The next closest planet is the one that since the late 1800s has been the most widely mapped: Mars. Prior to the space age, the prevailing view was of a vast, globe spanning network of ... something. While at the dawn of space flight no one still expected canals built by a dead or dying civilization, the nature of the fuzzy, dark features was a total mystery. Perhaps they were a lowly form of plant life with no chlorophyll (for there was no sign of chlorophyll in spectra taken of the dark regions). Perhaps they were volcanic sand alternately blown about or uncovered by the raging winds on the surface. Who knew?

These four planets (including the Earth) constitute the inner Solar System and all are located within an area that extends only one and a half times farther out from the Sun than the Earth. Beyond the orbit of Mars is a whole host of small rocky asteroids orbiting in a belt that serves as the dividing line between the inner and outer parts of the Solar System.

Farther out, the planets become huge with comparably large distances between them. Jupiter at its closest is 5 times farther from the Sun than Earth, while outermost Neptune is 30 times more distant than we. In the inner system, the Earth and Venus are the big kids on the block, with Mars and Mercury being 10 and 50 times less massive than us respectively. But in the outer system, mighty Jupiter, 300 times more massive than the Earth, dwarfs all of us puny Sun-huggers completely.

[2] See the 'See for yourself' section of Chapter 6 for a diagram of Venus' phases during its orbit.

I remember a book I had as a child showing a painting of Jupiter's surface where molten lava poured over dark cliffs to fill the giant lava sea that was the Great Red Spot. This spot has been visible since about the time of Galileo, and along with numerous dark bands, (and a few shorter-lived spots) is about all you can see through telescopes. But point even a small telescope at the planet and there in the field of view are four large moons. While each of these Galilean satellites is about the size of our Moon, before modern observatories they were no more than featureless dots.

Beyond Jupiter, the next closest planet and perhaps the most beautiful object visible in binoculars or telescope, is the ringed planet Saturn. Farther out there is blue-green Uranus (nearly 20 times farther from the Sun than the Earth) followed by its near twin Neptune. Each is a fuzzy little disk with virtually no visible features, but both shepherding their own retinue of moons. Out at Neptune the Sun's heat is 900 times less intense than what we experience on our own planet and the conditions out there must be very different than what we experience here in the inner Solar System.

Lastly, there is the small icy body of Pluto, even smaller than Mercury, in an orbit that carries it back and forth across Neptune's path. Since the 1990s, modern observatories have discovered that Pluto is simply the first discovered (and until the discovery of Eris in 2003, the largest) of a new class of icy 'asteroids' orbiting in a belt just inside and beyond the orbit of Neptune. These icy worlds, out in the cold outer reaches of the Solar System, comprise the Kuiper Belt, but only a few short decades ago they were completely unknown to Earthly astronomers.

And there in only a thousand words is just about everything that was known about the geography (and most of the geology) of eight planets and a dozen or so attendant moons. It's for this reason that prior to the invention of spacecraft, planetary astronomy was largely an unglamorous pursuit. But once you had spacecraft and could go there, the possibilities for discovery were enormous.

Less than a hundred years before spacecraft would start to map the unknown places in the Solar System, European and American explorers were still mapping the last remaining unknown places on Earth: the great interiors of Africa and America (including many of the places that would become our nation's national parks) as well as both north and south poles. James P. Ronda in his foreword to Donald Jackson's book *Thomas Jefferson and the Rocky Mountains: Exploring the West from Monticello,* writes that explorers of the eighteenth and nineteenth centuries, such as "Cook, Vancouver, La Pérouse, and ... Alejandro Malaspina saw themselves as part of a worldwide scientific enterprise." Their goal, along with the scientists and artist who accompanied them, was nothing short of the mapping of new lands and oceans, and the cataloging and categorization of the peoples, plants, minerals, and animals found there. "Enlightenment explorers," according to Ronda, "sought useful knowledge more than adventure and escape."

With few changes of wording, these goals are the same that NASA scientists pursue today: mapping the surfaces of new planets or moons, identifying their compositions and the geological processes at work, and perhaps most exciting of all, searching for evidence of life. Even the names they choose for their spacecraft speak to their history: they are Explorers, Pioneers, Surveyors, and Mariners. Outward from the Earth, we've sent Rangers, Voyagers, and Pathfinders searching for New Horizons.[3]

One of the most famous of the nineteenth century scientific expeditions to the American West was the Hayden Expedition that explored the Yellowstone plateau in 1871. For many years, adventurers and trappers known collectively as 'mountain men' had told tall tales of the mountainous West. The most famous mountain man of all, Jim Bridger seemed no exception when he told stories of the Yellowstone basin's mountain of glass (now called Obsidian Cliff), "the place where Hell bubbled up," (the geysers and hot pools that steam and reek of sulfur) and described standing trees petrified by volcanic sediments as "peetrified trees a-growing, with peetrified birds on 'em a-singing peetrified songs."

Ferdinand Vandeveer Hayden, geologist of the U.S. Geological and Geographical Survey, would lead the first scientific expedition to the area. Once there, Hayden was charged by Congress to investigate the geology, mineralogy, zoology, botany, and agricultural resources of the region. For this purpose he was given $40,000, 34 men, and seven wagons which set out for the territory in June, 1871. In *Windows into the Earth*, a popular account of the geology of Yellowstone and Grand Teton National Parks, Robert B. Smith and Lee J. Siegel describe how by the end of that summer Hayden had managed to piece together a surprisingly complete understanding of the Yellowstone country. In his report to Congress, Hayden wrote:

> This basin has been called by some travelers the vast crater of some ancient volcano. It is probable that during the Pliocene period [~3 million years ago] the entire [Yellowstone] country ... was the scene of as great volcanic activity as that of any portion of the globe. It might be called one vast crater, made up of thousands of smaller volcanic vents and fissures out of which the fluid interior of the earth, fragments of rocks, and volcanic dust were poured in unlimited quantities.

What Hayden was describing was the Yellowstone Caldera. This is a vast and subtle feature measuring nearly 50 miles (80 km) across the central portion of Yellowstone National Park. While today the major manifestation of the volcanic hot spot is the steaming pools and bubbling geysers, this was not always the case.

[3] Explorers were the first American satellites in Earth orbit, while the Pioneers went to Mercury, Venus, Jupiter and Saturn. The Surveyors and Rangers explored the Moon, while Mariners went to Venus and Mars. Pathfinder landed the first rover on Mars, while the Voyagers went to Jupiter, Saturn, Uranus, and Neptune. In 2007 New Horizons flew by Jupiter on its way to Pluto, which it will pass in 2015.

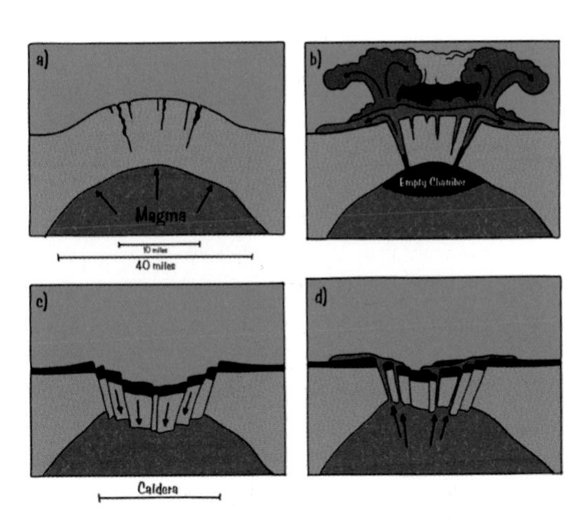

Figure 4.9 The steps in the formation of Yellowstone's volcanic caldera: a) A magma chamber swells up, raising a bulge and cracks in the surface above; b) Magma erupts to the surface through fissures that break down to the subterranean chamber; c) The surface collapses above the partially emptied magma chamber, forming a caldera; d) New lava flowing out of cracks in the caldera floor spread across the surface, filling much of the caldera (NPS/T. Nordgren).

Like geysers, the pressure that builds in the buried magma chamber lasts only so long before something blows. About every 650,000 years, the Yellowstone hot spot erupts in what scientists have come to call a super volcano. During the last episode 600,000 years ago, 240 cubic miles of rock and ash (nearly a thousand cubic kilometers) erupted through vents throughout what is now the park, partially emptying the magma chamber beneath. Without the support of this magma, the ground above collapsed, dropping like a freight elevator into the molten rock below, creating the caldera.

From this cracked and broken landscape, subsequent lava flows billowed upward and out like thick bread dough, covering much of the caldera floor. So much of the caldera was filled that today it takes a sharp eye to see the signs that are still there. You can see these ancient lava flows for yourself when you explore the Upper Geyser Basin and Old Faithful areas. Look around at the surrounding flat-topped hills with smoothly rounded, sloping sides. These are the now solid lava flows that welled up out of the cracked Earth after the caldera's collapse.

The entire southwestern quadrant of the park is a series of three overlapping calderas from the most recent eruptions. The youngest caldera is the most obvious, and is exactly what Hayden saw with his geologist's eye when he described the landscape 140 years ago from atop Mt. Washburn near the caldera's northern rim.

But as any visitor to Yellowstone knows, words alone cannot describe the strangeness and beauty of this place. According to Peter Hassrick in his book *Drawn to Yellowstone: Artists in America's First National Park*, Hayden made it a matter of course to hire artists as a way of enhancing his scientific observations and reports. "He understood full well the popular appeal and the scientific value of photographs and paintings. For him, 'collecting photographs and pictures of the West was similar ... to collecting fossils and natural history specimens.' "[4]

[4] In a direct testament to Hayden's vision, the Hubble Space Telescope owes its success to its hundreds of stunning photographs just as much as its many scientific discoveries.

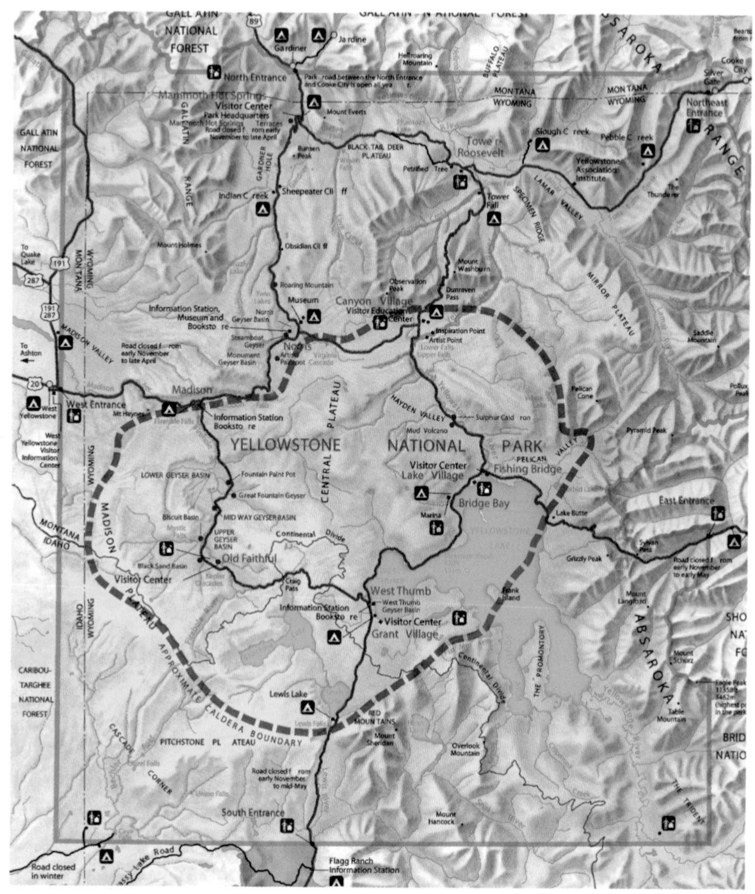

Figure 4.10 The most recent eruption of the Yellowstone hotspot, 600,000 years ago, created the Yellowstone Caldera (dashed purple outline) that Ferdinand Hayden first recognized in 1871 yet is often overlooked by many current visitors to the park (National Park Service).

Hayden included in the expedition Thomas Moran, a young Philadelphia painter eager to make a name for himself in the wide open wilderness out West, and William Henry Jackson, a pioneering practitioner of the new medium of photography. Jackson and Moran would work side by side in Yellowstone; what one would portray in colored paint the other would capture on black and white plates. The two became fast friends. From Moran's diary (with its eclectic spelling and punctuation) we can see the magnificent geology and exhilarating adventures they experienced together that summer as they explored the Yellowstone country:

> ...passed over the debris of a great land slide. where the whole face of the Mountain had fallen down at some time, laying bare a great cliff some 500 feet high. The view of the lake, as we approached it, was

Figure 4.11 Thomas Moran's watercolor painting of Castle Geyser in the Upper Geyser Basin, visited during the 1871 Hayden Survey to the Yellowstone territory. Compare this with the photograph from nearly the same position made by William Henry Jackson in Figure 4.12 (as well as a more recent photograph made from a position to the right of that shown here, in Figure 4.4) (NPS/Thomas Moran).

Figure 4.12 William Henry Jackson photograph of Castle Geyser taken during the 1871 Hayden Survey. For scale, notice the person climbing the right side of the sinter cone (not allowed today) (U.S. Geological and Geographical Survey of the Territories (Hayden Survey)).

very beautiful. It is a small pool formed by the widening of stream at this point, it is not more than half a mile in any direction. The Mountains surrounding it are about 11,000 feet high & about 3000 ft. above the level of the lake having snow still upon them.... For the first time in my life I slept out in the open air. During the night it rained a little but not enough to wet us to any extent. Got up early enough in the morning to get our Breakfast, & commence photographing as soon as the sun rose. The outlet of the lake is through an immense gorge in the Mountains bordered with great cliffs & peaks of Limestone some of them isolated & forming splendid foreground Material for pictures. Sketched but little but worked hard with the photographer selecting points to be taken.... Felt used up about 12 o'clock & started back to the camping ground where we prepared our dinner & rested an hour. Jackson got 13 negatives during the day which considering the difficulties quite a feat I think.... The view from The Mountains south east of our Camp & on the road to the lake looking toward the Yellowstone Country glorious, & I do not expect to see any finer general view of the Rocky Mountains.

Excerpt from Thomas Moran's diary, 1871

Together Jackson and Moran recorded for all time and for everyone to see the volcano made manifest. For an eastern public raised on landscapes of gently rolling hills and broad river valleys, the two artists presented an alien landscape of gigantic hot spring terraces, bizarre multicolored pools, geysers spouting from strange towers of sculpted stone, and the kaleidoscopic beauty of the Yellowstone's Grand Canyon. Commenting on an 1873 exhibition of landscape painters that included Moran's Yellowstone scenes, an art critic for the *Penn Monthly* wrote, "For myself, I prefer Thomas Moran's rich Yellowstone country.... The schools of science and impression in art seem to be growing farther and farther apart, and it is curious to see an observer like Mr. Moran turn aside to give us a couple of blots, fantastic 'suggestions,' he calls them, full of weirdness and dream."

Figure 4.13 Thomas Moran's watercolor painting entitled 'Tower Falls and Sulphur Mountain' based on his experiences during the Hayden Expedition of 1871 (NPS/Thomas Moran).

In the months after their return, Jackson's and Moran's artwork was circulated through the halls of Congress. U.S. Army Corps of Engineers Captain Hiram M. Chittenden wrote that their paintings and photographs "did a work which no other agency could do and doubtless convinced everyone who saw them that the regions where such wonders existed should be preserved to the people forever." Seven months after the expedition's return, the United States Congress made Yellowstone the nation's (and the world's) first national park: a national park for volcanism.

Ninety-one years later, the landscapes being photographed for the first time were on other worlds. Starting with Mariner 2's flyby of Venus in 1962, the current generation of planetary scientists has had the unique opportunity to be the first human beings to transform the faint fuzzy disks in the telescopes into actual worlds to be mapped and explored. While Hayden and the scientific explorers of the Enlightenment may have been the first white men to see their respective parts of the globe, today we know that with few exceptions, human beings had been living in all of these places for thousands of years. Indeed, many of these scientific expeditions relied upon the local peoples to show them the sights they would later 'discover.' But today, with each new spacecraft mission to visit new planets, moons, comets, and asteroids, we really are the first human beings to see and explore New Worlds.

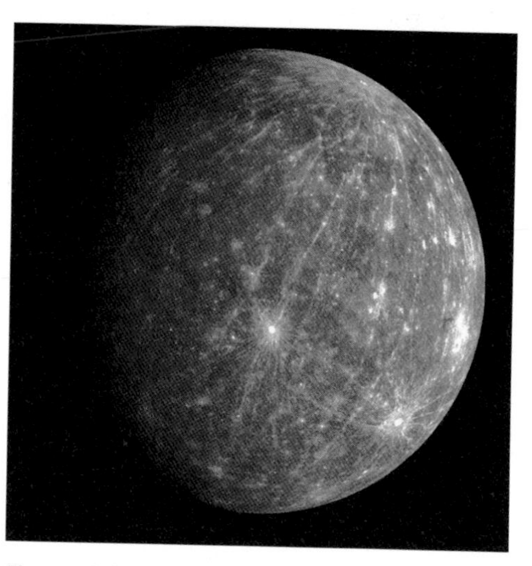

Figure 4.14 View of the planet Mercury taken by the MESSENGER spacecraft. Bright material ejected from multiple impact craters spread around much of the visible hemisphere. The part of the planet to the east of the central crater had never been seen before the 2008 flyby (NASA/Johns Hopkins University Applied Physics Laboratory/Carnegie Institution of Washington).

In 1973, the U.S. launched Mariner 10 which found Mercury to be a dark, dead, heavily cratered world like our Moon.[5] Thanks to Soviet and U.S. spacecraft in the 1970s, '80s and '90s, we now know that beneath the clouds of Venus the planet is a geologically active one covered with gaping rifts and meandering lava flows. An almost complete lack of visible impact craters means that like the Earth, some form of erosion must take place on the Venusian surface that wipes away these scars. But, on a planet where a runaway greenhouse

[5] At least it found one hemisphere to be Moon-like. On each of Mariner 10's three flybys it flew past Mercury when the exact same hemisphere was sunlit, placing the opposite hemisphere in absolute darkness. Only in 2008 with the arrival of a second spacecraft named MESSENGER was the other hemisphere finally seen. MESSENGER revealed a Moon-like surface with some subtle differences that will be the subject of continued study as the spacecraft eventually goes into orbit around the planet.

effect keeps the surface at a constant 900° F (480° C) with an atmosphere that is thick and sluggish and no water exists, the only forces to do so are the all-encompassing heat and volcanoes of globe-spanning proportions.

In the 1970s, Mariner 9 and Vikings 1 and 2 revealed a Mars with a canyon that could hold Arizona's Grand Canyon in an insignificant tributary and a gargantuan shield volcano whose summit touches the edge of space. And while there are ancient signs of flowing water everywhere, today there is not a liquid drop to be found. Whatever activity took place there seems to have happened so long ago that the only movement today is the gentle brush of sand blowing on the wind.

By the time Voyager 1 reached Jupiter in 1979, all the giant planets were known to be great gaseous worlds with no solid surfaces. The last remaining worlds to map and explore in our Solar System would therefore be their moons. The majority of these worlds are small icy rocks, no larger than our own Moon. Would they be nothing more than cold, grey, cratered worlds locked in perpetual deep-freeze by the feeble light of a dim, distant Sun? Or would there be surprises in store?

Torrence Johnson is now a senior scientist at the Jet Propulsion Laboratory in Pasadena California. Thirty years ago, in the spring of 1979, he was a member of the Voyager camera team. "Before Voyager, we knew from telescopes, from the color and spectra of the Galilean satellites, that they were going to be a little different from the Moon; we knew in some ways they were going to be weird." Three of the Galilean satellites, Europa, Ganymede and Callisto, were known from their high reflectivity and spectra to be made mostly of water ice. It was expected that ice slowly flowing across their surfaces had probably obscured many of the older craters, but in general they'd be pretty featureless, flat places with little activity to shape the landscape.

Io, on the other hand, the innermost large moon of Jupiter (and only slightly smaller than our own Moon) was particularly odd. It was noticeably red in views from Earth, and sulfur had recently been detected in its spectrum of reflected light. The prevailing view was that Io would turn out to have an old battered surface coated in sulfur compounds, rendering it a reddish version of our own Moon.

"When we saw the first light and dark markings on Io's surface they were interpreted as the lit and shadowed sides of topographic features. There were those who wanted to issue press releases saying we were seeing great circular impact features like those on the Moon. But we waited, because we knew that as we got closer the view would only get better. It was a very exciting time," Johnson vividly recalled.

Then, just three days before Voyager 1's closest approach to Jupiter, the March 2nd issue of the journal *Science* hit the scientists' mailboxes. It featured a paper by Stanton Peale at the University of California showing that radioactive elements are not the only way to heat a moon. Peale claimed that the combination of Io's orbit and Jupiter's mass could be surprisingly effective in heating Io's interior, and thus the moon might not be as cold and dead as our own.

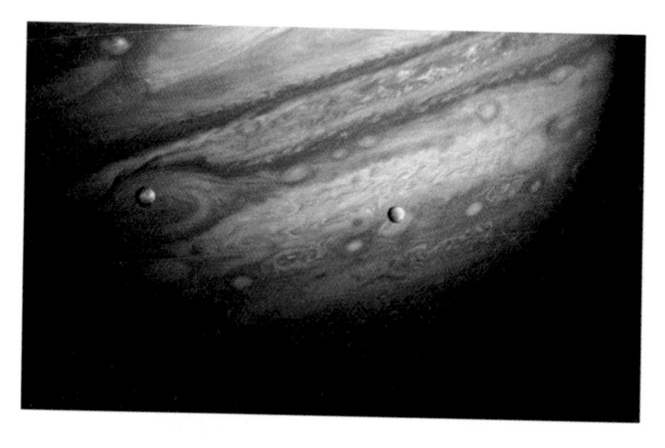

Figure 4.15 Voyager 1 images Io and Europa passing in front of Jupiter's cloud tops. Io is the orange one on the left passing in front of the Great Red Spot (NASA/JPL).

Here's how Peale calculated the process works: Jupiter's three large inner satellites (Io, Europa, and Ganymede) are locked into mutual orbital resonances. For every trip Europa makes around Jupiter, Io makes exactly two (and for every two that Europa makes, Ganymede completes one). The result is that for every four times Io travels around Jupiter, both Ganymede and Europa are back in their original positions relative to the inner moon and their combined gravitational pull keeps Io's path around Jupiter from becoming perfectly circular. Io's elliptical orbit means its distance from Jupiter constantly changes.

But the tidal forces from massive Jupiter that stretch the tiny moon (raising two bulges pointing towards and away from Jupiter's bulk) are highly dependent upon the distance between the two. With changing distances come changing tidal forces. As Io circles Jupiter, the gravitational forces continuously vary, deforming the poor moon over the course of its 1.8 day orbital period. These fluctuating forces stretch and squeeze Io's interior, generating friction that generates heat that melts rock into magma. Peale calculated that the more Io's interior melts, the easier it is for tidal forces to deform the crust, which in turn

Figure 4.16 This view of Io shows two simultaneously erupting volcanoes. One is visible on the limb, (at lower right) where volcanic material rises more than 150 miles (260 kilometers) above Io's surface. The second is visible along the terminator (the line between day and night) where the volcanic plume catches the light of the rising Sun. With this image Linda A. Morabito, a JPL engineer, discovered the first active volcano beyond the Earth. It was taken by Voyager 1 on March 8, 1979, looking back 2.6 million miles (4.5 million kilometers) at Io, three days after its closest encounter (NASA/JPL).

generates even more heat (melting even more rock) in a process he called "runaway tidal heating." The final line of the opening to Peale's paper said it all, "Consequences of a largely molten interior may be evident in pictures of Io's surface returned by Voyager 1."

"As a result of Peale's paper," Johnston recalled, "we knew to keep an eye out for the possibility of recent volcanism. But 'recent' probably meant within the last million years." Then Johnson chuckled. "As Voyager got closer, the views got clearer, until finally it became obvious. There wasn't just recent volcanism on Io; it was going on there right now."

Imagine being the very first human beings to see the rocketing plume of Old Faithful or the rainbow colors surrounding Grand Prismatic Spring. After traveling through green forests and grey mountains, you are brought face to face with yellow and orange canyons, aquamarine pools, and boiling red mudpots. Far from the dead worlds the lack of geological activity on the Moon

Figure 4.17 Like nearly every one of the images returned by Voyagers 1 and 2, this montage of Jupiter and its moons is an iconic image for me from my childhood. From left to right the moons (first visible to Galileo and still visible to anyone with a decent pair of binoculars) are Io, Ganymede, Europa and Callisto (NASA/JPL).

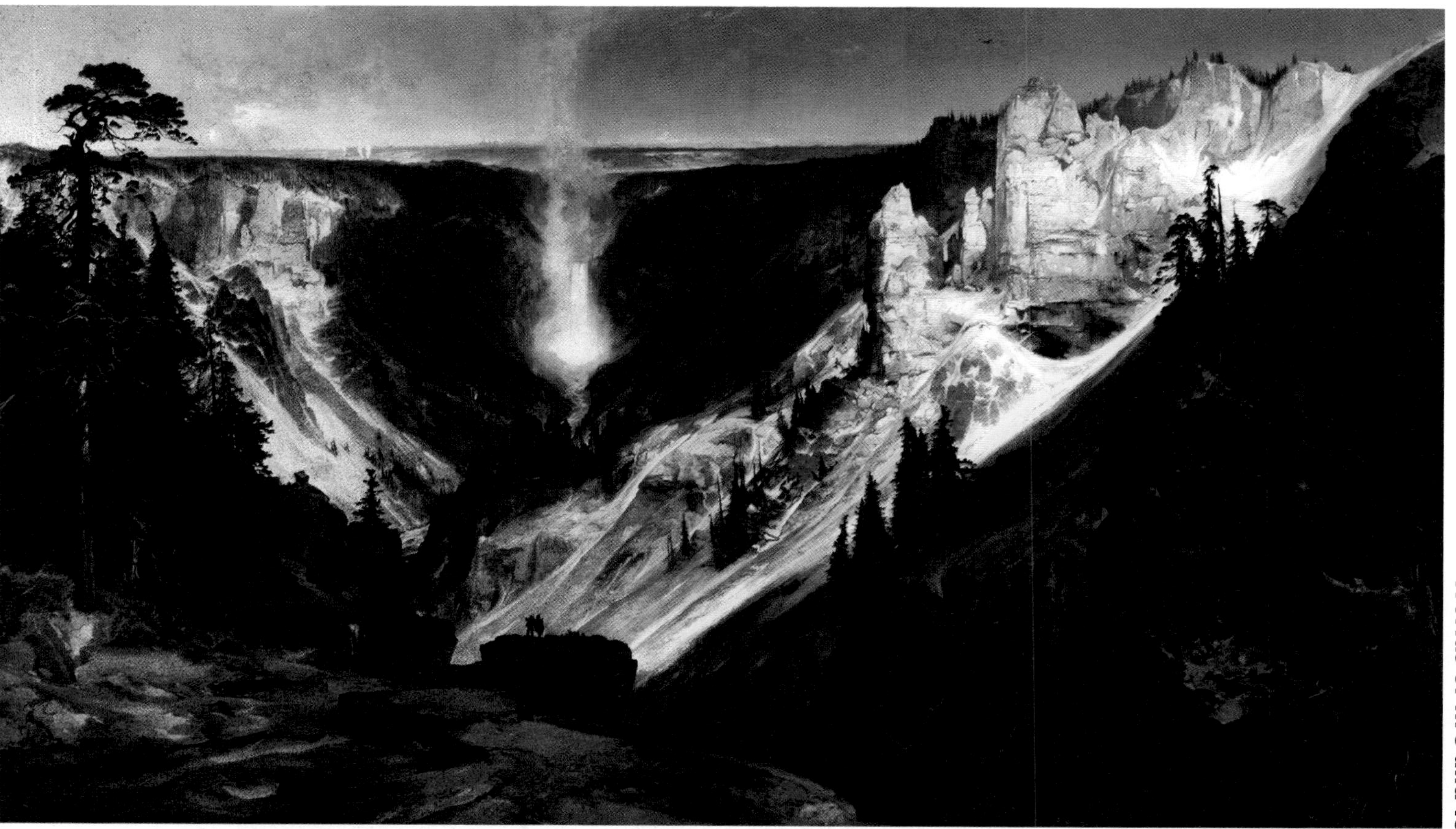

Figure 4.18 Thomas Moran's great masterwork, 'Grand Cañon of the Yellowstone, 1872,' oil on canvas, 84 × 144 in (213.4 × 365.8 cm) (United States Department of the Interior).

would lead us to expect, or even the moderate levels of activity that, "recent" volcanism might imply, Io was vibrantly alive in Technicolor brilliancy with enormous volcanic plumes erupting skyward above molten lava flows and lava lakes larger than anything on Earth. "During that mission," Johnson mused, "the moons went from being these little featureless spots in a telescope to real worlds in our photographs. Io was no longer just the reddish one, it was the world with *volcanoes*, while Europa became the one with *cracks*, and Callisto was the one with *craters*."

In 1872 New Yorkers crowded Clinton Hall to see Moran's great western masterpiece, *The Grand Cañon of the Yellowstone*. In 1979 I breathlessly waited along with everyone else to see the latest photos from Jupiter on my TV's nightly news. While Moran's kaleidoscopic painting became an iconic representation of the wonder and majesty of western expansion and exploration, Voyager 1's

Figure 4.19 This color image shows two volcanic plumes on Io. The bright plume on the limb fountains 86 miles (140 kilometers) into space. At bottom center a second circular plume erupts directly towards the camera from the volcano Prometheus. It forms where a dark black lava flow curls out of a volcanic vent and hits the cold sulfur dioxide frost on the surface. Every spacecraft sent to Jupiter, with just the right view of Io, has seen this volcano erupting, revealing that it may have been continuously active for more than 18 years (NASA/JPL/University of Arizona).

photos of Io with its volcanic plumes rising over a kaleidoscopic limb have become iconic images of humanity's outwards expansion and exploration of the Solar System. Every planetary scientist I know who worked on that mission, from senior researchers to undergraduate students, has a favorite photograph they still remember.

Look at a photograph of Io. Every single black speck and dark circle is a hot spot like the Yellowstone hot spot; every one is a volcano. The tidal heating that produces those volcanoes long ago boiled away all the water and ice that is still found on Jupiter's other moons. What's left is a hellish wasteland of sulfur compounds in neon yellows, oranges, reds, and whites. For anyone who's ever visited one of Yellowstone's major geyser basins, this description isn't too far wrong. Take a deep breath. Any future astronaut daring enough to brave Io's sulfur fields will be hard pressed to completely filter out that smell. For me, it's the distinctive smell of volcanism.

Yet as similar as Io's volcanic features may seem, in their number and scale they create a landscape utterly alien outside of nightmares. Yellowstone's most recent and visible caldera is less than 50 miles (80 km) across at its longest; this is no more than a typical, mediocre sized volcano on Io. By comparison, Io's largest volcano, and indeed the largest active volcano in the entire Solar System, is Loki Patera,[6] measuring almost 125 miles (200 km) across.

If, when visiting Yellowstone, you should make the drive from Tower Falls to Canyon Village, stop for a moment like Ferdinand Hayden did, and take in the view from Dunraven Pass. From here on the shoulders of Mt. Washburn, you can

Figure 4.20 An active volcanic eruption on Jupiter's moon Io was captured in this image taken on February 22, 2000 by NASA's Galileo spacecraft. This image (a composite of visible and infrared images) shows Tvashtar Catena, a chain of giant volcanic calderas. The orange and yellow ribbon at left is a cooling lava flow that is more than 37 miles (60 kilometers) long. The two small bright spots are sites where molten rock is exposed to the surface at the ends of lava flows. This picture is about 155 miles (250 km) across (NASA/JPL/University of Arizona).

[6] A *patera* looks like a caldera but we have no idea if the formation mechanism is the same. Just to be safe, we use this word which comes to us from the name for ancient Greek and Roman saucer-shaped vessels or dishes.

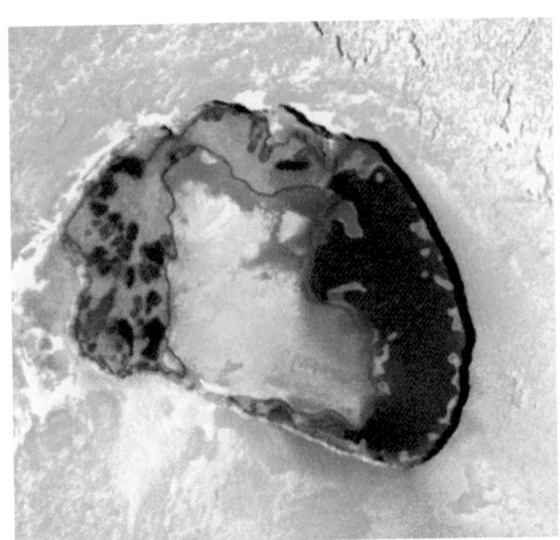

Figure 4.21 Different sulfur compounds on Io give rise to a riot of colors in Tupan Patera. Named after a Brazilian thunder god, the volcanic depression, possibly a caldera like Yellowstone's, measures 47 miles (75 km) across and is surrounded by cliffs 3,000 feet (900 meters) tall (NASA/JPL/University of Arizona).

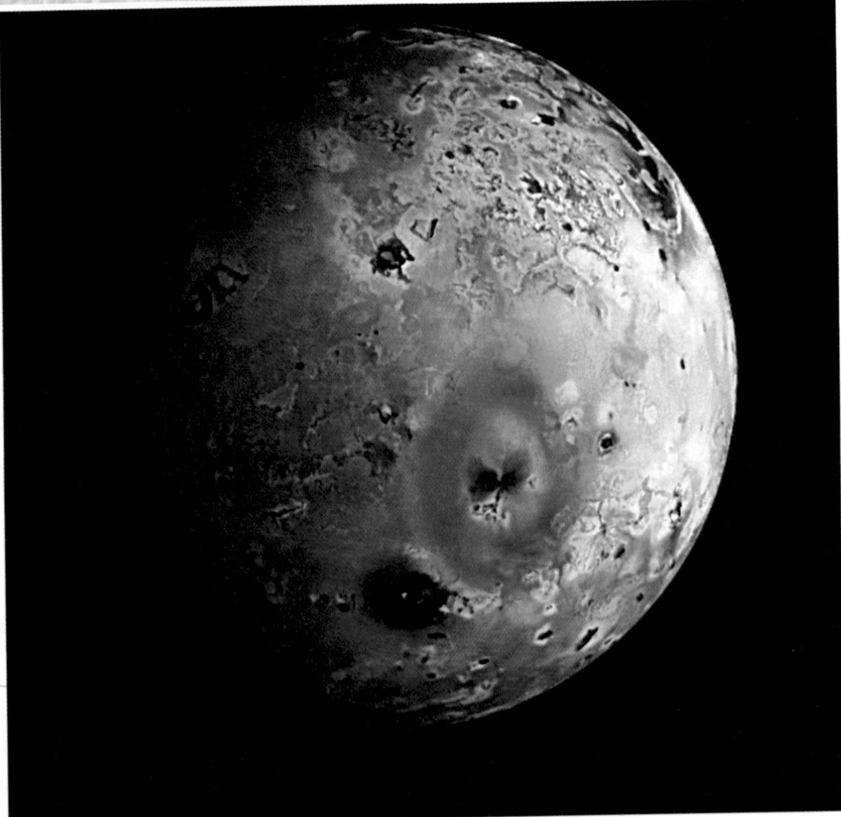

Figure 4.22 The images used to create this enhanced color composite of Io were acquired by NASA's Galileo spacecraft during its seventh orbit of Jupiter. The active volcano, Loki Patera, is the large, black, horseshoe-shaped lake near the boundary between night and day near the upper left. The big reddish-orange ring in the lower right is formed by material deposited from the eruption of Pele, Io's largest volcanic plume (NASA/JPL/University of Arizona).

see clear across the short-axis of the Yellowstone Caldera to its far wall in the distant hills on the horizon. If you could similarly stand on the rim of Loki Patera, the molten sea before you would extend four times farther and would be hidden from view as it passed over the distant horizon. Just as the very ground within Yellowstone National Park radiates 40 times as much heat as the average for the entire United States, so too does every square inch of tiny Io glow with 40 times the volcanic heat of Earth.

Sadly, Voyagers 1 and 2 were only passing through the Jupiter system on their way to the other outer planets. In their wake they left many questions: how many volcanoes were there on Io, how active were they, what were they made of, were they volcanoes like the Earth's or something more exotic like volcanoes of pure sulfur? Twenty years later NASA would revisit the Jovian worlds, but this time with an orbiter called Galileo that could follow up on the guesses and hypotheses of the earlier missions. Models could be made and predictions tested: science could be done.

Since the Voyager flybys, and continuing up through the Galileo mission, a tiny handful of planetary scientists have been watching Io – and Loki in particular – from distant hilltops here on Earth. What they lacked in quality of view through the Earth's turbulent atmosphere they made up for in quantity of observations. While spacecraft may get stunningly better images, their missions offer only a single snapshot in time of what's happening on a moon or planet's surface. Using telescopes located in both Wyoming (on the summit of Mt. Jelm a half hour's drive southwest of Laramie) and on Mauna Kea, a dormant volcano in Hawaii, these scientists have been watching Io's long-term activity for nearly a quarter century using infrared cameras sensitive to the faint light given off by the distant heat of Io's volcanoes.

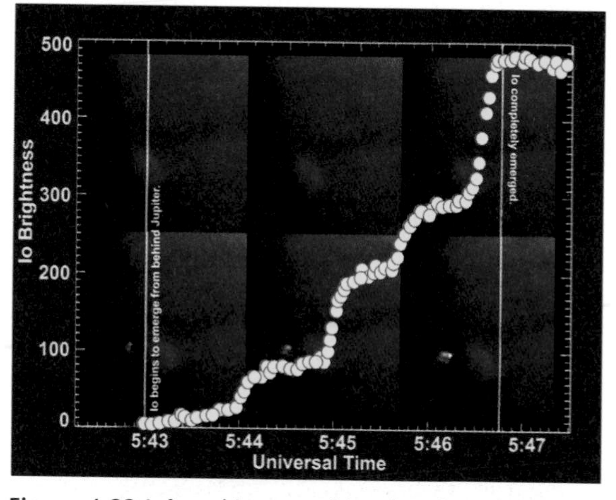

Figure 4.23 Infrared images of Io taken from NASA's Infrared Telescope Facility on the summit of Mauna Kea volcano on the Big Island of Hawai'i. Images were taken every couple seconds as Io (orange) appears from behind the disk of Jupiter (blue). The total infrared brightness of Io for each image is shown by the yellow circles. As each volcano emerges from behind the planet, the brightness jumps by an amount that depends on the size of the eruption. Six eruptions are shown (two appear simultaneously in the lower left panel). The brightest eruption is always Loki (J. Spencer/Lowell Observatory).

Once every 1.8 days Io passes behind Jupiter as viewed from the Earth. When the orientation between Earth, Jupiter and Sun is just right, the reemerging moon is briefly in Jupiter's shadow where no sunlight at all falls upon its surface. At these times the only light coming from Io is the infrared glow of its smoldering volcanoes. While Io itself may appear as nothing more than an indistinct and blurry blob, these planetary scientists are able to precisely pinpoint any eruption's location and magnitude by watching the eclipsed moon slowly appear from behind the planet's disk.

They accomplish this by looking at the total infrared power reaching their telescope as the moon slowly appears. If there are active volcanoes, then as each one appears, the infrared flux suddenly increases: the larger the eruption, the higher the jump. By timing when the jump occurs, they identify where on the moon (along a line north and south) the outburst occurs, and thus, which volcano it must be.

Julie Rathbun is the most recent planetary scientist monitoring Io's volcanoes. She first joined the project as a postdoctoral researcher at Lowell Observatory, set amid the volcanic cinder cones of Northern Arizona. To monitor the long term activity of Io's volcanoes, and Loki in particular, Rathbun uses observations of Io spanning all the way back to those begun in 1987 by Bob Howell, a geologist at the University of Wyoming.

From Voyager photos, Loki appears to be a molten lava lake with a central island. What exactly goes on under the molten surface is unclear but after nearly 20 years of observing its periods of eruptions and quiescence, Rathbun made a startling discovery. Loki's outbursts were periodic like a geyser's. Every 540 days Loki would flare up, putting out as much as 15% of the volcanic moon's total heat flow. This would last for about 230 days and then shut off. Looking back at the 20-year-old Voyager 1 images, Rathbun and her colleagues could faintly make out within the black lake a darker region near one shore. Was this newer lava, still dark and fresh compared to the older, weathered rock around it? Four months later when Voyager 2 passed by, the region had increased in size and grown around the island.

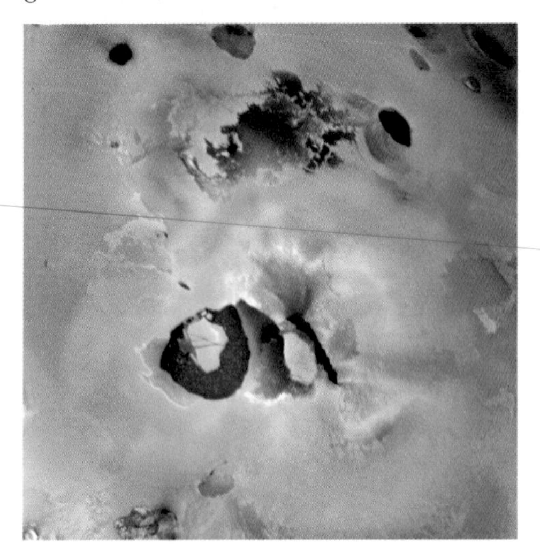

Figure 4.24 Loki is like nothing seen elsewhere on Io. The lava lake is the U-shaped dark area about 125 miles (200 km) across. In this specially processed image, darker, possibly molten lava is visible to the extreme left, while possible "icebergs" are imbedded within the hardened crust of the remainder of the lake. The dark lava of the lake is over 360° F (200° C) hotter than that of the surrounding area (NASA/JPL).

Whether on Earth or Io, molten lava is vastly hotter than the air or vacuum of space above. Once erupted onto the landscape, the surface of the molten lava cools and hardens into an ever thickening crust. We see a process like this every winter on Earth as northern lakes freeze over on cold days. The longer it's cold, the thicker the ice crust becomes. But water is different from rock. Ice is less dense than liquid water and so floats on a lake (or in your glass). Solid rock, by contrast, is denser than liquid, and so eventually sinks. Rathbun calculated that as Loki's lava crust thickens, the weight becomes so great that giant slabs should break apart and sink into the bright, hot, lower density lava beneath it. With the exposure of hot molten lava, observers on Earth see the beginning of a Loki 'eruption.' As adjacent slabs sink, the hole in the crust widens to eventually become a cascading overturn that sweeps across the surface of the small lava sea.

As each new mile of liquid lava is exposed, the eruption Rathbun sees at her telescope continues. When the last slab sinks and the exposed lava has once more begun to cool, harden and dim, the outburst ends. By the reflected light of Jupiter's disk, the lake sits silent and still, the crust growing thicker as it cools, until once again it becomes too heavy to float and the process begins anew.

Mark Twain, visiting the Kilauea caldera on the big Island of Hawai'i in 1866, wrote of seeing just such a phenomenon within what is today's Hawai'i Volcanoes National Park:

> For a mile and a half in front of us and a half a mile on either side, the floor of the abyss was magnificently illuminated.... The greater part of the vast floor of the desert under us was as black as ink, and apparently smooth and level; but over a mile square of it was ringed and streaked and striped with a thousand branching streams of liquid and gorgeously brilliant fire!... Occasionally the molten lava flowing under the superincumbent crust broke through split a dazzling streak, from five hundred to a thousand feet long, like a sudden flash of lightning, and then acre after acre of the cold lava parted into fragments, turned up edgewise like cakes of ice when a great river breaks up, plunged downward and were swallowed in the crimson cauldron. Then the wide expanse of the "thaw" maintained a ruddy glow for a while, but shortly cooled and became black and level again.
>
> Mark Twain, *Letters from Hawaii*, June 3, 1866

Nearly 130 years after Mr. Twain's observations, the Galileo spacecraft finally reached Jupiter orbit and over the next eight years passed by Io six times. Galileo carried infrared instruments specifically designed to measure the temperatures of Io's volcanic features. During the Io encounters these cameras discovered hot lava flows pouring from volcanic vents and enormous fountains of fire near Io's North Pole. In addition, the plumes observed by Voyager are now suspected to be due to molten basalt (like that found on Earth) pouring out of fissures onto a surface covered in a sulfur dioxide snow. Where the 1,800° F (980° C) rock comes in contact with the -240° F (-151° C) surface, the frost flashes to a sulfurous steam

Figure 4.25 Lava lake and fire fountain during the August 1963 eruption of Kilauea Volcano within Hawai'i Volcanoes National Park. For a sense of scale, notice the trees within the caldera in the upper left illuminated by the blazing fountain. Colorized version of photo by T. Miyasaki (U.S. Geological Survey).

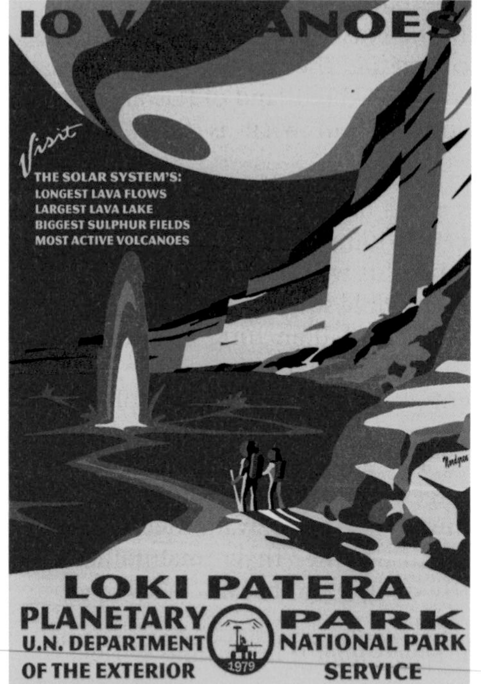

Figure 4.26 Design for a poster advertising the many attractions of Loki Patera for future tourists to the Io volcanoes (T. Nordgren).

that creates the enormous plumes we see arcing out into space. In this way, the particles in the plumes slowly settle over the entire surface of the moon, covering every square inch in the sulfurous stuff. Add up all the sulfur and rock erupted by all the volcanoes, and every spot on Io is buried to a depth of half an inch (1 cm) every year.[7]

For Loki, these infrared instruments showed that between two separate flybys warm temperatures swept around the patera in just the same manner and at just the same rate as predicted by the groundbased observations. By modeling all the data from three spacecraft and 20 years of observations, Rathbun and her colleagues were able to not only explain what the volcano was, but predict its next eruption.

Unfortunately, after publication of what's become known as the Rathbun Model, subsequent ground-based Loki observations

[7] Because sulfur rapidly settles on any exposed surface on Io, this is why fresh new lava in Loki's lake should be blacker than the older, cooler lava.

Figure 4.27 The Cassini spacecraft captures this image of Saturn and its rings upon its arrival in orbit in 2005. Light filtering through the rings colors Saturn's cloud-tops blue (NASA/JPL/Space Science Institute).

showed that what for two decades had been regular as clockwork suddenly changed completely. Sighs Rathbun, "I guess that's what we get for naming a volcano after a trickster god." But then Rathbun says, "Whenever I find myself thinking the model is completely wrong, I just look back at that Voyager image of Loki where you can actually see the different lava rafts floating in the lake, and I know that we'll figure out this new behavior eventually."

And so her observations continue.

Over a year and a half after passing by Jupiter and its moons, Voyagers 1 and 2 encountered Saturn.[8] Like Jupiter, Saturn is another mini-solar system with a giant gaseous planet surrounded by a horde of small attendant moons. And of course there are the rings. The rings themselves are billions of snowflakes, snowballs, and icebergs orbiting above the cloud tops in the finely ordered precision of Kepler's laws. More than 156,000 miles (250,000 km) in diameter, the rings are no more than 33 ft (10 m) thick in places. If Saturn and its rings were shrunk down to a diameter of 60 miles (100 km), just wide enough to fit within Yellowstone's boundary, the rings would be no thicker than the folded up park map you get when you visit.

Out there in the cold fringes of the outer Solar System, the rings aren't the only things made out of ice. The moons themselves are ice worlds: great balls of rock and water ice locked in a perpetual deep freeze. When the Voyagers flew by, they saw moons like our own, except every ridge, rift and crater is formed, not out of rock, but rather out of water, frozen as hard as granite. And, at long last, there were moons covered in craters, pummeled by eons of debris passing through Saturn's gravitational domain. Tiny Mimas was hit with an impactor so

[8] Voyager 1 passed by Saturn in November 1980, while Voyager 2 flew by in August 1981. Between watching these two encounters on TV, I saw the eruption of Mt. St. Helens.

Figure 4.28 The orange, smoggy atmosphere of Saturn's moon, Titan, is contrasted by the small bright body of another moon, Enceladus. On Titan, yet undefined processes are supplying the atmosphere with methane and other chemicals that are broken down by sunlight. These chemicals are creating the thick yellow-orange haze that is spread through the atmosphere and, over geologic time, coats the surface (NASA/JPL/Space Science Institute).

large that the resulting crater dominates an entire hemisphere.[9] There was also the two-faced Iapetus, with one side as bright as new fallen snow, and one hemisphere as dark as blackest coal.

But the largest moon of all was also the most interesting. Titan, as befitting its name, is larger than even the planet Mercury, with a thick and impenetrable atmosphere unlike any other moon in the Solar System. Were it not in orbit around Saturn, Titan would be a planet in its own right. When the Cassini spacecraft was sent to orbit Saturn 28 years after the Voyagers passed by, one of its primary missions was to divine the secrets of Titan, to finally look beneath the globe spanning clouds.

It is in these moments that one realizes what a privileged time it is to be a scientist. Here would be a new world, revealed to human eyes for the very first time. And unlike the explorers of the last couple centuries, we wouldn't be merely reporting back on the discovery of landscapes and perfectly happy people in no mood to be "discovered." In sailing beneath Titan's clouds and gazing on pristine panoramas we really would be the first people to see the Grand Canyon or dip our toes in the Pacific Ocean. With the exception of far off Pluto and the frozen ice-balls with which it shares territory in the Kuiper Belt beyond Neptune, Titan's would be the very last landscape totally new to human eyes in our Solar System.

In addition to cameras with which we'd attempt to peer through the murky clouds, Cassini also carried a radar transmitter and receiver designed to clearly map the surface of the moon. It works by bouncing radio waves off the surface at an angle as the spacecraft passes by: no radar return and we know a region is smooth and the radio waves have simply bounced away out to space; a bright

[9] And for a generation raised on Star Wars movies, we at last knew the Death Star was real!

radar reflection and we know the region is rough and jagged and the radio waves are scattering in all directions, some of them right back at the spacecraft.

Cassini's orbit around Saturn was devised so that it would make numerous passes by the cloudy moon and on each pass a swath across the surface is 'imaged,' slowly building up a map, one tantalizing strip at a time. But while Cassini is an orbiter, on board Cassini was a lander that would be dropped from on high as the spacecraft made its first approach. The Huygens lander, a European contribution to the international mission, would dive through the

Figure 4.29 The European Space Agency's Huygens probe captured these images of the surface of Titan as it dropped by parachute through the moon's thick atmosphere. Each row shows the view north, south, east and west, at different altitudes above Titan's surface: 100 miles, 20 miles, 5 miles, 1 mile, and 0.2 miles (160 km, 32 km, 8 km, 0.32 km). Bright, mountainous highlands contrast with dark, flat, lowlands into which a number of dry 'river channels' appear to flow (ESA/NASA/JPL/University of Arizona).

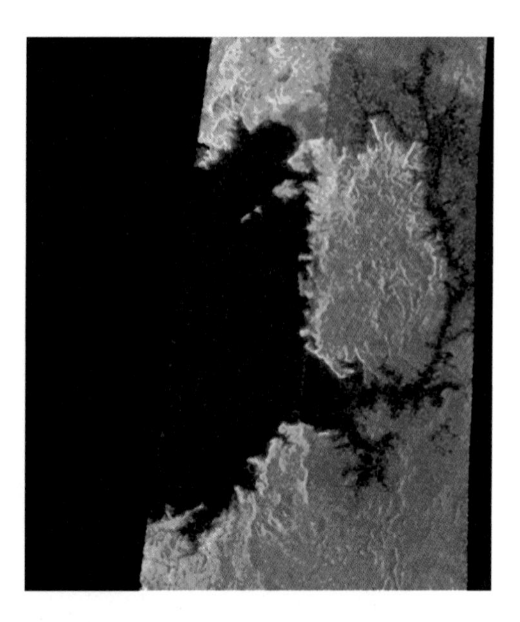

Figure 4.30 Cassini's radar instrument views a swath of Titan's north polar region. Through Titan's thick atmosphere we see, for the first time, features that look exactly like channels, islands, and bays typical of terrestrial coastlines. The liquid, most likely a combination of methane and ethane, appears very dark to the radar instrument and is shown in false color meant to emphasize the difference between the radar-dark lakes and rough and jagged, highlands. Some of the observed channels flowing into these lakes have well-developed tributary systems and drain many thousands of square kilometers of the surrounding terrain. These lakes are likely connected and may form part of a larger sea (NASA/JPL).

atmosphere, suspended by parachute, acquiring images in visible light as it saw what you or I would see if we ever came in for a landing on the frozen surface.

Because the moon *is* frozen. With an average surface temperature of –290° F (–180° C) the world is in a permanent deep freeze, but a special deep freeze near the triple-point of methane, one of the main constituents of its atmosphere. We live in a special place on Earth near the triple-point of water, where water can exist in all three of its phases: solid, liquid and gas. This morning I stood beside Morning Glory Pool in the Upper Geyser basin, and saw steam blow against the blue surface of the scalding waters set against the bright white of the surrounding winter snow. On Titan, the temperature may allow methane (the primary component of natural gas) to play the part that water does here and exist in all three of its forms.

From orbit, Cassini's radar shows great regions of the poles covered in something dark and perfectly smooth. These dark area's well-defined borders are evocative of liquid shorelines into which are what appear to be perfectly normal river channels cut through rough mountains. For all the landscape's apparent normalcy though, to peer down through Titan's cloudy veil is to see a frigid, alien Earth where methane rain falls on jagged hills of solid water, where methane streams grow to cut canyons through icy ground, and where methane rivers wash into arctic hydrocarbon seas. And everywhere is the sign of geological activity: coastlines and river channels, dune fields and seashores, weather and erosion. On no other world in our Solar System have we found something so like home, and yet so utterly alien.

As Cassini swept on from its first encounter with Titan, it passed by the tiny moon Enceladus. Enceladus is small enough to fit within the borders of Wyoming (about the same size as Great Britain), small enough that all internal

Figure 4.31 Design for a poster advertising the attractions of the Titan seashore for future tourists to this exotic moon (T. Nordgren).

heat should long ago have escaped and all geological activity slowed to a stop. But even as far back as Voyager 2's flyby in 1981, images of the moon showed a fascinating place where something special had happened, or might still be happening. Enceladus is the brightest moon in the Solar System; its surface is as white as newly fallen snow.

Here in the Old Faithful area, it's been snowing every day for weeks now, and snowshoeing across the drifts in the middle of the night when the stars are out, I wonder what it would be like to trek across the snows of Enceladus. Would I hear the fresh white powder crunch through the soles of my boots? I let my imagination take me to that distant moon and gaze in wonder at this new world around me.

As is true on Earth, there are relatively few craters here on Enceladus; what few craters are found depend on what part of the moon

Figure 4.32 Saturn's moon Enceladus is the brightest moon in the Solar System, as bright as new fallen snow. Its surface is a bizarre mix of craters and vast regions of nothing but ridges and wrinkles indicating that some geological process must be rearranging its surface. The bottom third of the visible disk is centered on the moon's south pole which is crossed by multiple bright blue crevasses (called Tiger Stripes). Infrared observations by Cassini show the south pole is warmer than the rest of the moon (NASA/JPL/Space Science Institute).

you are on. Like our Moon does to the Earth, Enceladus always keeps the same face pointed towards Saturn. On the leading hemisphere (the part of the moon that faces forward in its direction of motion around Saturn, where the planet appears to forever rest on the eastern horizon) there are almost no craters at all. While snow does cover the surface, there is no way that the thin layer could hide the types of impact scars typically found on our own Moon. No, something has erased the craters; some internal activity has erased what should be here. This means something deep down is still active, or was until recently. I see its presence, not just in the lack of craters, but in the cracks and ridges that criss-cross the countryside. Images from orbit show the entire South Polar Region is covered in wrinkles.

Ringing the pole itself is an almost continuous transition zone of ridges and valleys meeting at strange V-shaped intersections that look as if some giant child poked and pulled its finger through a tub of soft-serve ice cream. The entire pole is separated from the rest of the moon by this strange network of features, and to stand upon the final scarp overlooking the polar basin is to behold a view as strange and beautiful as that that Hayden saw from the edge of the Yellowstone Caldera.

Before me begins the first of the enormous parallel cracks called 'Tiger Stripes' seen in images from Cassini's third pass by the icy moon, the first to get a view of its southern pole. Each rift is about a mile and a quarter (2 km) across flanked by ridges running over 300 feet (100 m) above the surrounding plains. Within the bounds of these troughs the floor sinks away into darkness five times deeper, shadowed from the light of Saturn now resting upon the northeastern horizon behind me. Cassini's infrared cameras, similar to those carried by the Galileo spacecraft, passed over this rift system in 2005 and detected the first signs of heat radiating up from out of their depths.

A gravitational resonance with neighboring Dione, a moon fifteen times larger, keeps Enceladus in a modestly eccentric orbit around Saturn, just like Europa causes Io to do around Jupiter. The changing tidal forces from Saturn once again cause friction deep within the moon that generates the south polar heat. Why there's no comparable warming at the North Pole is still a mystery.

Figure 4.33 The Sun's rays backlight water geyser's erupting from Enceladus' south pole (NASA/JPL/Space Science Institute).

My presence here coincides with Enceladus' closest passage with Saturn. The minutes pass, and the geometry between rift and Saturn slowly changes until, just as I grow impatient, I see the first spout become a torrent as a jet of water vapor rockets from deep within one of the Tiger Stripes in the distance. First one, then others form. A dozen geysers soon touch the sky as tidal forces from Saturn pull apart different sections of the pole-crossing cracks, exposing to space unseen reservoirs of liquid water below.

Cassini first saw the plumes in 2005 when during one particular flyby it looked back to catch a photo of the lunar night-side. Exposed to the Sun was a razor thin crescent of daylight and beautiful back-lit streamers radiating from the moon's South Pole. The prevailing hypothesis is that the internal heat warms the ice creating giant reservoirs of liquid water underneath the polar ice cap. When the orientation with Saturn is just right, the planet's tidal forces open the cracks that cross the lunar surface. At these moments the high pressure underground waters erupt high into the surrounding sky, just like in Yellowstone National Park, 800 million miles (1.3 billion kilometers) away.

When planetary scientists first realized what these were, they called them Cold Faithful after the most famous geyser on Earth. The total output of water from these alien geysers is strikingly similar to the water output of their namesake back on Earth. About 8,000 gallons (30,000 liters) of water are shot into the sky during each outburst, but because of the lower gravity here on Enceladus, the geyser plumes seem to converge at a point high overhead, six hundred miles (1000 km) out into space. No need to crowd on the boardwalk to get a decent picture of that.

Figure 4.34 Design for a poster introducing future tourists to the attractions of icy Enceladus including the spectacular Cold Faithful geysers (T. Nordgren).

But part of the reason Old Faithful is famous, and the reason it has the name that it does, is because the mechanism that makes the geyser also makes it periodic. Old Faithful erupts nearly every 90 minutes once its plumbing system has refilled and repressurized after the previous eruption. According to Terry Hurford of the Goddard Space Flight Center in California, as Enceladus orbits Saturn the orientation of the crevices slowly changes with respect to the tidal forces from Saturn. At closest approach in its elliptical orbit, the gravitational forces from Saturn cause the cracks to open and the geysers to erupt. As Enceladus continues on in its orbit, however, the orientation and distance from Saturn changes and the forces that first

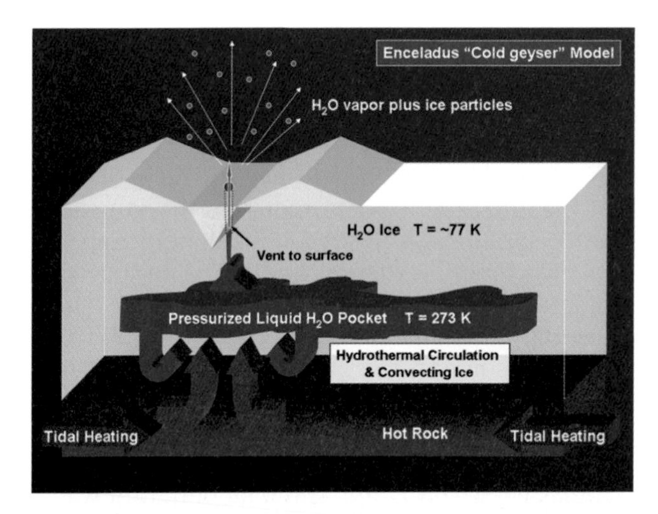

Figure 4.35 Model of how tidal heating warms Enceladus' interior, producing the liquid water that erupts from the south polar Tiger Stripes to produce the observed geysers (NASA/JPL/Space Science Institute).

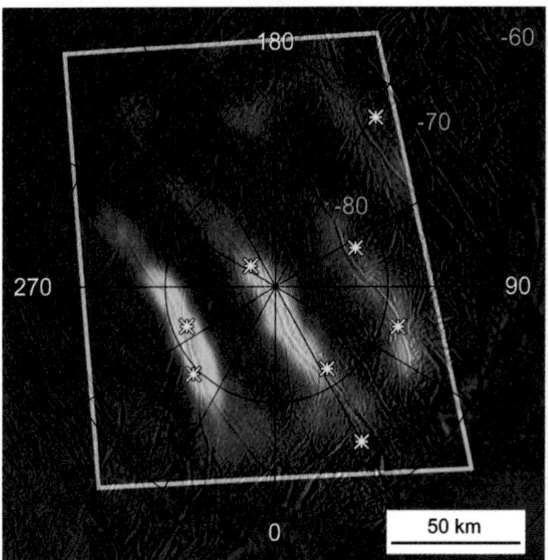

Figure 4.36 Heat radiates from the entire length of the 95 mile-long (150 km) 'Tiger Stripe' fractures stretching across the south pole of Saturn's ice moon Enceladus. The source of the geysers (shown by stars) tend to be located in the warmest parts of the fractures. The brightest fracture, Damascus Sulcus in the lower left, is at a temperature of at least –135° F (–93° C) compared with the surrounding polar region at temperatures below –330° F (–200° C) (NASA/JPL/GSFC/SwRI/SSI).

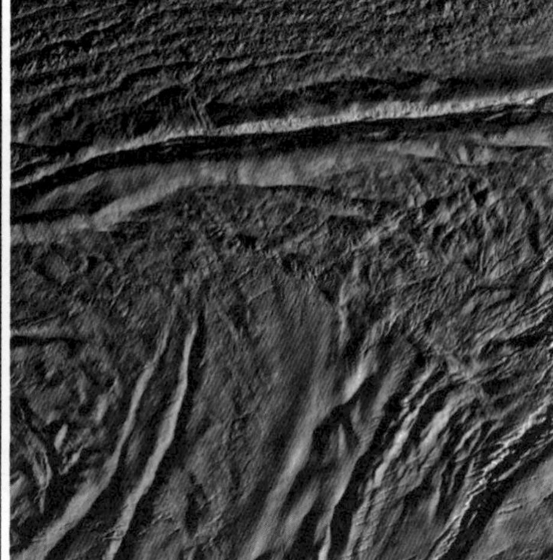

Figure 4.37 This image was taken during Cassini's very close flyby of Enceladus at a height of 2,947 miles (4,742 km) above the surface. Damascus Sulcus, crossing the upper part of the image, is the warmest of the Tiger Stripes and the source of two of the observed polar geysers. The smallest features visible within the Damascas crevasse are only a couple hundred feet across (NASA/JPL/Space Science Institute).

pulled the crevices open now begin to push them closed. One by one the geysers stop. But every 1.37 days as Enceladus circles round Saturn, the forces begin to build again and if observations confirm Hurford's calculations, Cold Faithful truly is.

It's to experience for myself what it might be like to see the eruption of Enceladus' geysers that I've fought my body's natural desire to stay warm, and gone out into a Wyoming winter to see Old Faithful erupt by starlight. Weather patterns in February make this one of the coldest places in the continental United States and on a clear night without the warming blanket of clouds it gets even colder. For whatever reason, right now before dawn is the one time the stars seem to come out and so four hours before twilight begins I'm out in the cold night air that makes my eyeballs freeze. I'm wearing a parka lent to me by an arctic researcher (it has a neon orange 'Polar Bear Alert' tag on the front) and it comes in handy as I snowshoe my way over to the quite empty Old Faithful viewing area. Starlight (and my head lamp) illuminates the snow-covered path as I attempt to avoid the bison in the area. I am told their eyes glow green in the light but suspect if I get close enough to tell I'm already too close. By the time I make it to the geyser I happily note Saturn in the starry sky above me, just as if I really was on the sub-Saturnian side of that tiny far away moon.

Figure 4.38 Old Faithful erupts by starlight in Yellowstone National Park. The hot water flashes to snow as it comes in contact with the cold night air. The bright white 'star' seen within the plume at center is the planet Saturn. Within that single, featureless dot is the planet, its rings, and moons, including Titan and Enceladus (T. Nordgren).

At 4:40 am under a moonless sky, I hear Old Faithful erupt just as I arrive. I plant my camera, snap my photos, and feel something begin to brush against my face. The super-heated water erupting from the hole in the icy ground is freezing in the air above me and raining down as gentle snow. I look up at Saturn. On Enceladus where the force of gravity is so much lower, most of this flash frozen water doesn't fall back to ground, but rather continues upward to go eventually into orbit around the planet. There it becomes Saturn's E ring, catching the light of the distant Sun to become one of the most beautiful sights in the entire Solar System. How

Figure 4.39 Lit by reflected light from Saturn, Enceladus appears to hover above the gleaming rings, its geysers spraying a continuous hail of tiny ice grains. Snowflakes that escape the tiny moon's gravity go into orbit around Saturn, creating its E-ring in which Enceladus resides (NASA/JPL/Space Science Institute).

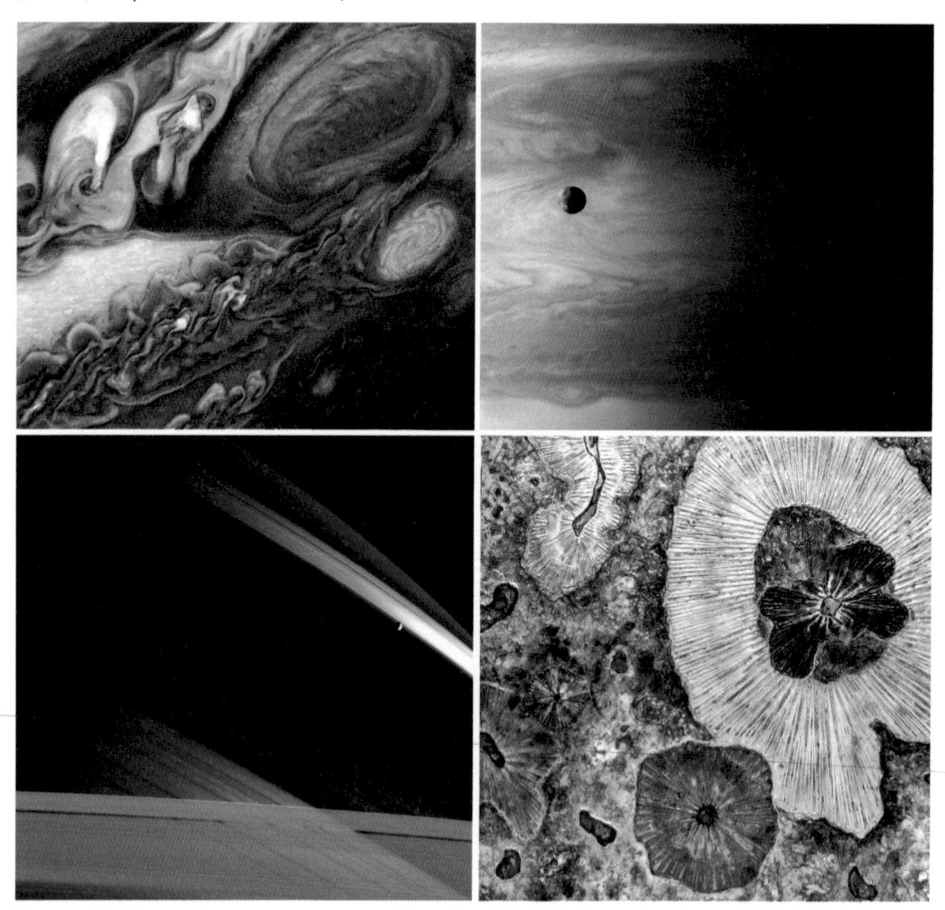

Figure 4.40 Art and astronomy are combined in the scientific exploration of the Solar System. Upper left: Jupiter's turbulent cloud tops from the Voyager 1 spacecraft (NASA/JPL); Upper right: Io hangs before the clouds of Jupiter (NASA/JPL/Space Science Institute); Lower left: Cassini arrives at Saturn and catches the tiny moon, Mimas, against the shadow of Saturn's rings (NASA/JPL/ Space Science Institute); and Lower right: Coming full circle, artist Monica Petty Aiello creates interpretations of planetary forms revealed by NASA spacecraft, in this case Pele's volcanism on the surface of Io (compare with Figure 4.22) (Monica Petty Aiello).

strange that something so alien should be the result of the earthly phenomenon before me.

As I see Saturn disappear into the geyser's plume, I reflect back on the explorers who've come before me here: from the first Native Americans to find this place to Jim Bridger and Ferdinand Hayden, from the artists who captured its likeness, such as Thomas Moran and William Henry Jackson, to the scientists who continue their tradition through the electronic eyes of Voyager, Galileo and Cassini. Exploration, art, and science are what make us human. Through them we learn about ourselves, our planet, and the Universe in which we all belong, and they are the passions that stir my soul here where the heart of my planet finally touches the sky.

See for yourself: the Solar System

Model Solar System

To put into perspective how large the Solar System is, here is a simple model you can make where both the sizes of the planets and the distances between them are to scale. The entire model, from Sun to distant Pluto in the Kuiper Belt, is 600 yards (or meters) long. This can easily fit within a large meadow, your neighborhood, or along the trail from Old Faithful to Castle Geyser in Yellowstone National Park.

The following table lists the distance of each planet from the Sun in yards/meters, the size of each planet in inches and millimeters, and a typical item of about the right size. Note: One yard or meter works out to about one large step. There is no need to be picky about getting this exact.

Table 4.1 Scale model of the Solar System

Planet	Distance from Sun	Size	Object
Sun	0 yards	5 inches (130 mm)	Grapefruit/large stone
Mercury	6	0.02 (0.5)	Poppy seed/sand grain
Venus	10	0.04 (1.0)	Candy sprinkle/tiny pebble
Earth	15	0.04 (1.0)	Candy sprinkle/tiny pebble
Mars	25	0.02 (0.5)	Poppy seed/sand grain
Asteroid Belt	30-50 yards	–	Dust
Jupiter	80	0.5 (13.0)	Candy M&M/small stone
Saturn	140	0.4 (10.0)	Candy M&M/small stone
Uranus	290	0.2 (5.0)	Peppercorn/large pebble
Neptune	450	0.2 (5.0)	Peppercorn/large pebble
Pluto/Kuiper Belt	600-?	0.01 (0.2)	Poppy seed/sand grain

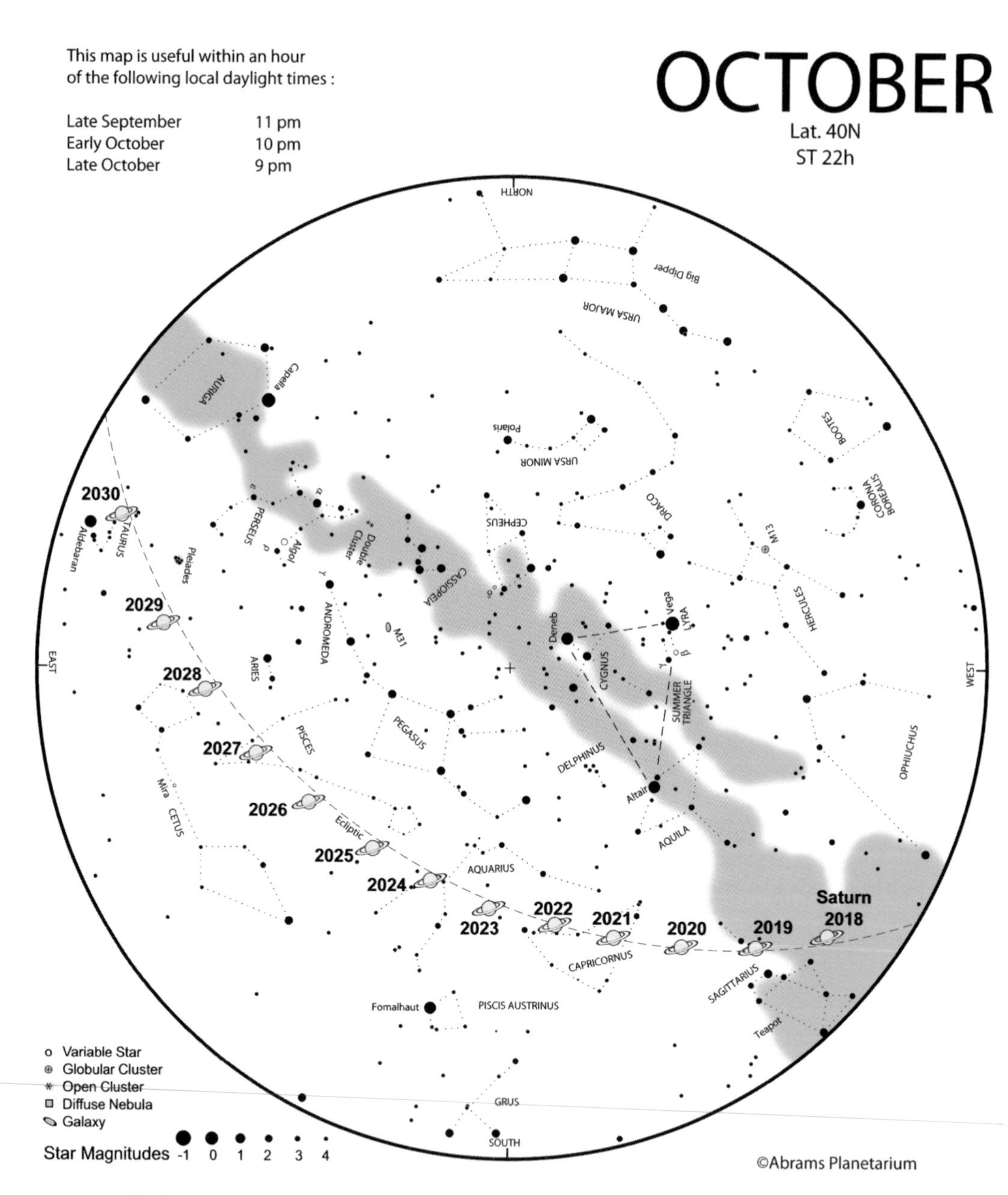

Hold the star map above your head with the top of the map pointed north. The center of the map is the sky straight overhead at the zenith. Saturn's position is marked relative to the background stars for each October and April between 2010 and 2038 (nearly a full Saturnian year). To find Saturn during another month, find its position between the two closest dates then consult one of the other monthly star maps to see if that constellation (and thus Saturn) is above the horizon. While during any given month of any given year there may be other planets also visible, Saturn is typically quite bright and noticeably

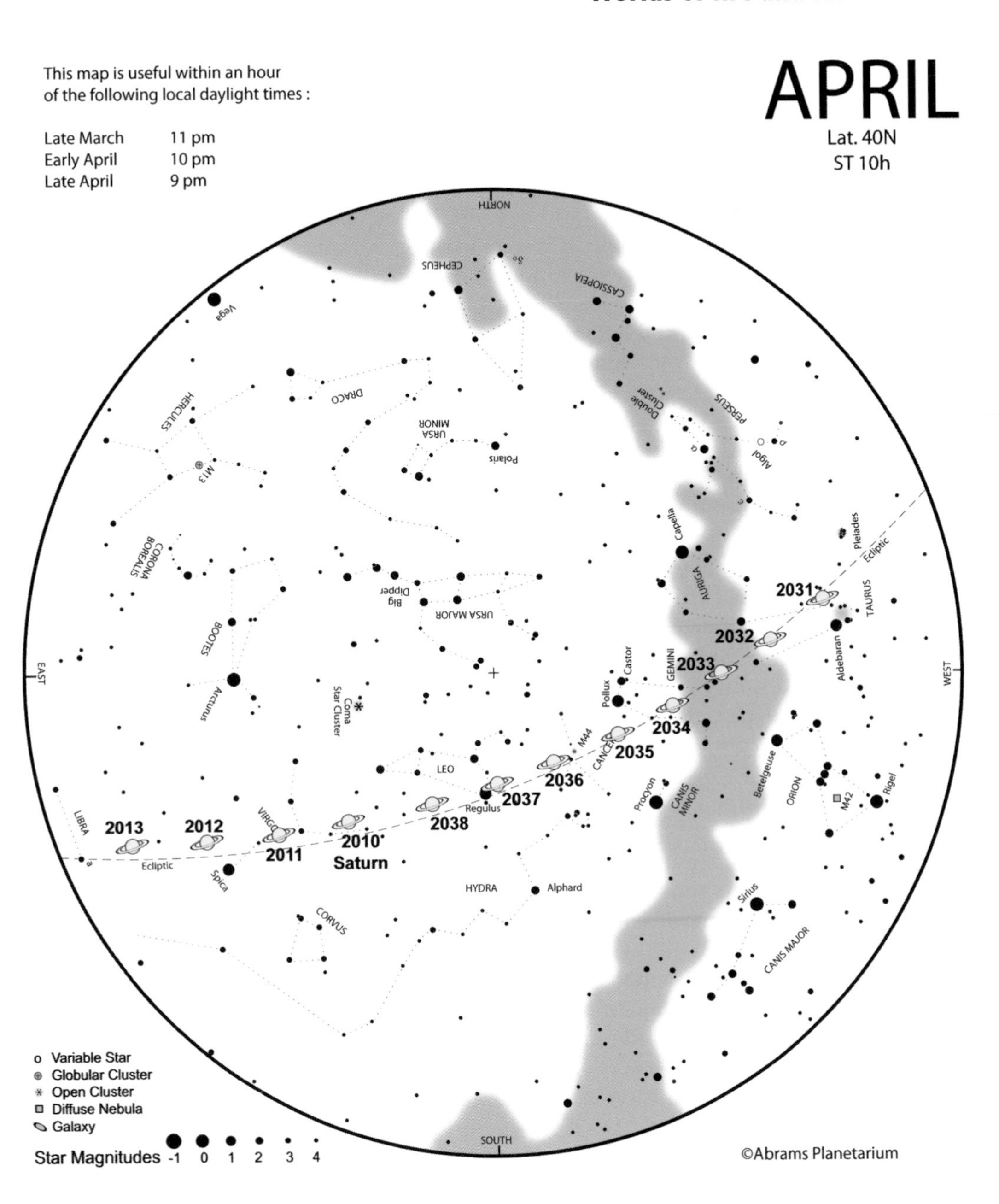

yellow (though not as bright or yellow as Jupiter). Note that between 2014 and 2017, Saturn is moving through the constellations of Scorpius, Ophiuchus, and Sagittarius and thus not visible on either of these maps. During those years, make a note of how far in the sky Saturn typically moves each year along the ecliptic and then check the summer sky maps that show these constellations the best.

Find a large stone in a central location or place a large grapefruit someplace you can see it (note that in the national parks you should not pick up and move rocks around). In walking from one planet to the next, be aware that in reality the planets are never all lined up as we are doing in our model. At any one time there are almost always planets on opposite sides of the Sun.

Once you get out to Jupiter and beyond, notice just how far these planets are away from the Sun. This is why it is so cold out here, and why nearly every moon is composed largely of ice. The planets in the inner Solar System, by comparison, are tightly bunched around the warmth of the Sun.

At Saturn: when you see its rings in a pair of binoculars, you are seeing light that has traveled the 140 yards/meters from the Sun, reflected off the ice crystals in the rings, and then traveled all the way back to observers on Earth, 135 yards/meters away.

When you get to Pluto and the edge of the known Solar System, look back towards the Earth. It takes light, the fastest thing in the Universe, 5.5 hours to get here from the Sun. If you walked the same distance in 15 minutes, you would be walking 22 times faster than the speed of light on this scale. The New Horizons spacecraft on its way to Pluto is the fastest machine ever built by humanity and it will take nine years to travel this same distance.

On the same scale as this model Solar System, the nearest star like our Sun is 2,400 miles (3,800 km) away. Now, hold up your thumb. On the scale of this model, the distance between the Earth and Moon is no more than the width of your thumb. This is the farthest human beings have so far personally traveled out into space.

Observing Jupiter's moons through binoculars

Point binoculars at Jupiter with a magnification of 18× or 20× and you will see one or more of its moons. You need to keep the binoculars steady by using a simple camera tripod. If you do not have a tripod, try sitting in a way that you can rest both of your elbows on a table or other solid surface. Alternately, stand next to a tree trunk or pole and press the side of the binoculars against the vertical support as a way to stabilize them. Lastly, there are some brands of binoculars that now include internal image stabilization to counter the natural shaking of our hands. Makers of the Galileoscope sell a plastic telescope for $20 with the same magnification as Galileo's original.

Use the September and March sky maps (Chapter 3) to find Jupiter's position relative to the background constellations over the next 11 years. It will be one of the brightest, if not the brightest 'star' in the sky when you find it. Observe Jupiter from night to night and you will see one or more of its Galilean moons. Io, the innermost and fastest moving moon, will even show motion after only a few hours.

On the lines beneath, try recording the positions of the moons over several nights. Each row shows a single night, with the central circle representing Jupiter's disk (if you have a telescope, see if you can detect the cloud bands or

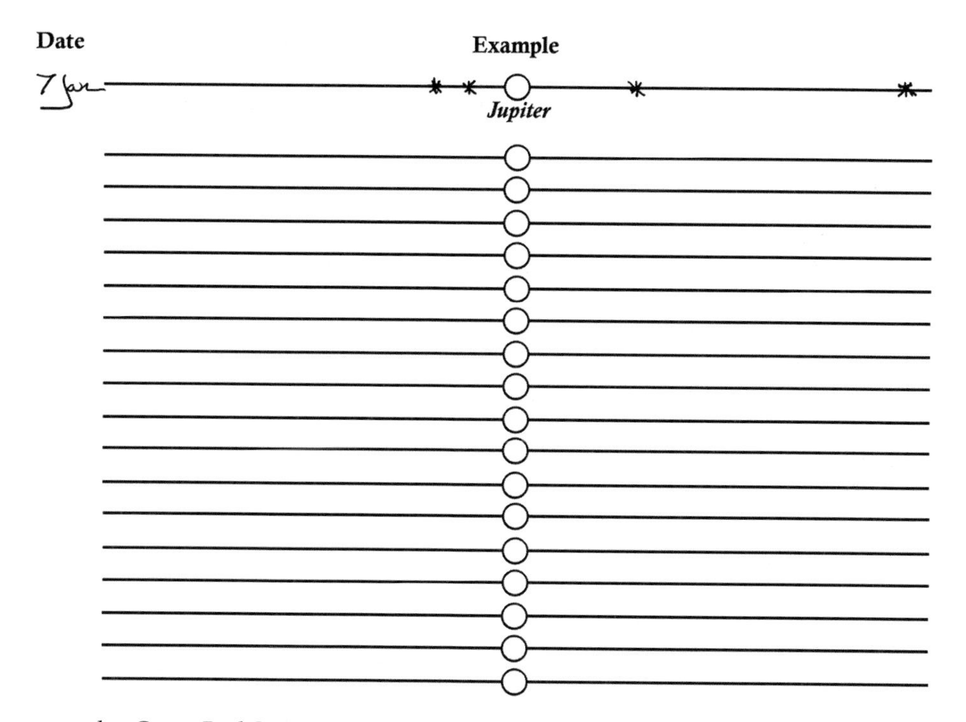

even the Great Red Spot on Jupiter). Use the size of Jupiter's disk in your view to gauge how far away each moon is. Callisto, the most distant moon, will never look farther away from the planet than 13 Jupiter disks (the ends of the horizontal line). In a little less than 17 days Callisto will make one complete orbit.

Observing Saturn through binoculars

Saturn, with its rings, is the most beautiful planet visible through a telescope or pair of binoculars. As shown in Figure 4.41, even with a magnification of just 20×, the rings are visible off to either side of Saturn's disk. Through higher power binoculars, or a small telescope, the rings are easily discernable as wrapping completely around the planet. In addition, you will also begin to see Saturn's moons, the brightest of which is Titan.

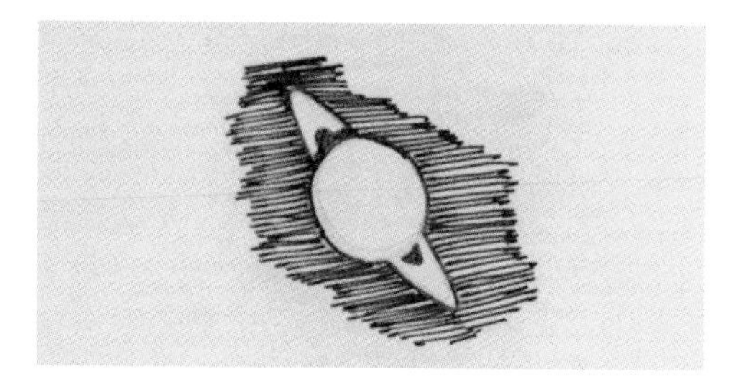

Figure 4.41 Drawing of Saturn through a 25× telescope. (T. Nordgren).

Further reading

Volcanic Worlds: Exploring the Solar System's Volcanoes ed. Rosaly M.C. Lopes and Tracy K.P. Gregg (2004)
Springer-Praxis, ISBN 3540004319

Super Volcano: The Ticking Time Bomb Beneath Yellowstone National Park by Greg Breining (2007)
Voyageur Press, ISBN 9780760329252

Windows into the Earth: The Geologic Story of Yellowstone and Grand Teton National Parks by Robert B. Smith and Lee J. Siegel (2000)
Oxford University Press, ISBN 0195105974
Drawn to Yellowstone: Artists in America's First National Park by Peter H. Hassrick (2002)
University of Washington Press, ISBN 0295981733

The Nine Planets Solar System Tour
http://www.nineplanets.org/

The NASA Photojournal of spacecraft imagery
http://photojournal.jpl.nasa.gov/index.html

Yellowstone's Old Faithful Webcam
http://www.nps.gov/archive/yell/oldfaithfulcam.htm

Yellowstone National Park Archive of Thomas Moran paintings
http://www.nps.gov/archive/yell/slidefile/history/moranandotherart/Page.htm

The Galileoscope, a high quality plastic replica of Galileo's telescope
https://www.galileoscope.org/

Monica Petty Aiello's Gallery of planetary surface-inspired artwork
http://www.tandmaiello.com/

5 Red rock planet

All this is the music of waters.

John Wesley Powell, 1895

Irrigation, and upon as vast a scale as possible, must be the all-engrossing Martian pursuit.

Percival Lowell, 1895

Follow the water.

NASA, official Mars exploration theme, 2003

Every two years and two months Mars glows red in a dark black sky. As the Earth overtakes and draws close to its slower cosmic kin, Mars grows brighter than any other star in the sky. Then, just as quickly, the process reverses; the Earth sweeps onward, and Mars fades away to become one more light in a cold starry sky. For those few months that we spend under a foreboding blood red star, it's no wonder the planet was named for the god of War.

But Mars is no harbinger of doom; it's simply a planet, the fourth planet from the Sun. It's smaller than the Earth, with a surface area a little more than a quarter that of the Earth's – or roughly equal to the dry land mass of our own world. Unlike all other heavenly bodies (save the Moon) Mars reveals a landscape to the telescopic observer. During each near encounter even a small telescope reveals a white polar cap of ice or snow. At these times, it is easy to see through telescopes that Mars is a reddish-ocher world with occasional markings of a darker blue-green. The movement of these markings across the disk reveals Mars has a day only 40 minutes longer than ours. Their movement also shows that Mars' axis of rotation is tilted only a couple of degrees greater than our own. Mars therefore goes through its seasons, just as we go through ours.

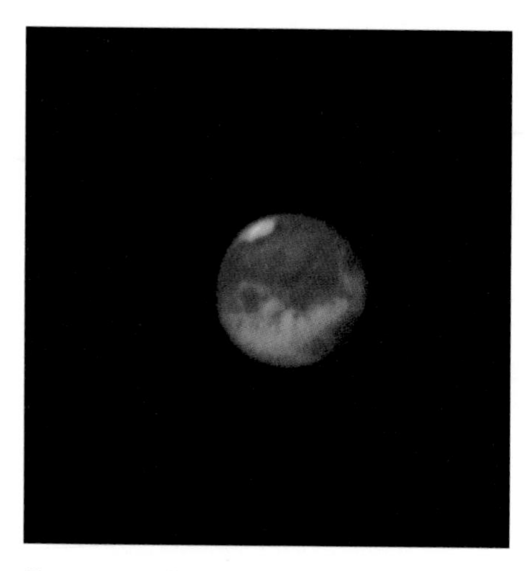

Figure 5.1 The planet Mars as seen through a small telescope in 2003 (T. Nordgren).

Through our telescopes we can see that Mars has an atmosphere. While it is thinner than ours, it is composed in large part of water vapor that keeps the Martian climate relatively mild though no trace of snow or ice is ever seen outside the polar caps. Observation of these caps and the blue-green regions show they change in size, darkness and color with the seasons. From winter to summer the polar cap shrinks. As it does so, a network of globe girdling lines becomes apparent and the equatorial blue-green regions grow in size and deepen in color. From all these observations we may logically conclude that Mars is a warm, dry, desert world where a globe spanning network of canals has been built to irrigate vast agricultural regions sustaining a civilization that, by its engineering skills, must be greatly superior to our own.

This is the state of our knowledge of Mars, as observed and deduced by the most prominent astronomer of today.

Provided that by "today" I mean 1895. A little over a hundred years later our view of Mars could hardly be more different.

In the spring of 1894 Percival Lowell, gentleman astronomer, and son of one of the rich and mighty 'Boston Brahmin' families of Massachusetts, built an observatory in Flagstaff, Arizona Territory, for the purpose of observing Mars. Prior to Lowell, it had been the practice for astronomers to build their observatories wherever they happened to already have their offices. The United States Naval Observatory, the observatory charged by Congress with observing stars and their positions in the aid of navigation[1] was located in the heart of Washington D.C. in a place called Foggy Bottom of all things.

Figure 5.2 Percival Lowell sits at the eyepiece of the 24-inch (61-cm) telescope with which he observed Mars (Lowell Observatory Archives).

Lowell was one of the first people to recognize that the quality of what can be seen through a telescope depends upon more than if it just happens to be a clear night. You can see this for yourself on a hot summer day as you look across a sun-

[1] The U.S. Naval Observatory, now the residence for the Vice President of the United States, was also where the moons of Mars, Phobos and Deimos, were discovered.

Figure 5.3 The 24-inch (61-cm) telescope at Lowell Observatory still offers views of the Red Planet from the top of Mars Hill in Flagstaff, Arizona (T. Nordgren).

baked parking lot. Waves of heat rise off the black surface, causing the horizon to ripple and wave in the turbulent air. Fine detail in the distant landscape is obscured in the distorting shimmer. We see the same effect every night in the sky as atmospheric turbulence and pockets of differing temperatures and densities cause the stars to twinkle.

Through a telescope, our atmosphere causes the view of stars, planets, and distant galaxies to dance and blur in the eyepiece or image. Wrote Lowell in his 1895 book *Mars*, "A large instrument in poor air will not begin to show what a smaller one in good air will. When this is recognized as it eventually will be, it will become the fashion to put up observatories where they can see rather than be seen."

Astronomers call the steadiness of the air the 'seeing.' Good seeing is a matter of the geography and the way in which the air flows over it. The American southwest with its dry conditions and high mountains has many locations of excellent seeing and Lowell's extensive testing of sites before settling on Flagstaff for his observatory was a harbinger of the great observatories like Palomar and Kitt Peak that would, in time, make the desert southwest their home.

As forward thinking as Lowell was, his observations and analysis of the Martian condition drew heavy criticism from other astronomers. And, as we now know, these critics were almost completely correct. The problem was the nature of observing Mars in the days before spacecraft could go there. As seen from the Earth, Mars is tiny and subtle details are far tinier still. But as bad as the situation may be, there *are* times when the view gets substantially better.

Mars appears at its largest – and brightest – when viewed through the telescope during times of *opposition*. Oppositions occur when the Earth overtakes Mars, and thus the two planets come closer than at any other time in their orbits. At these

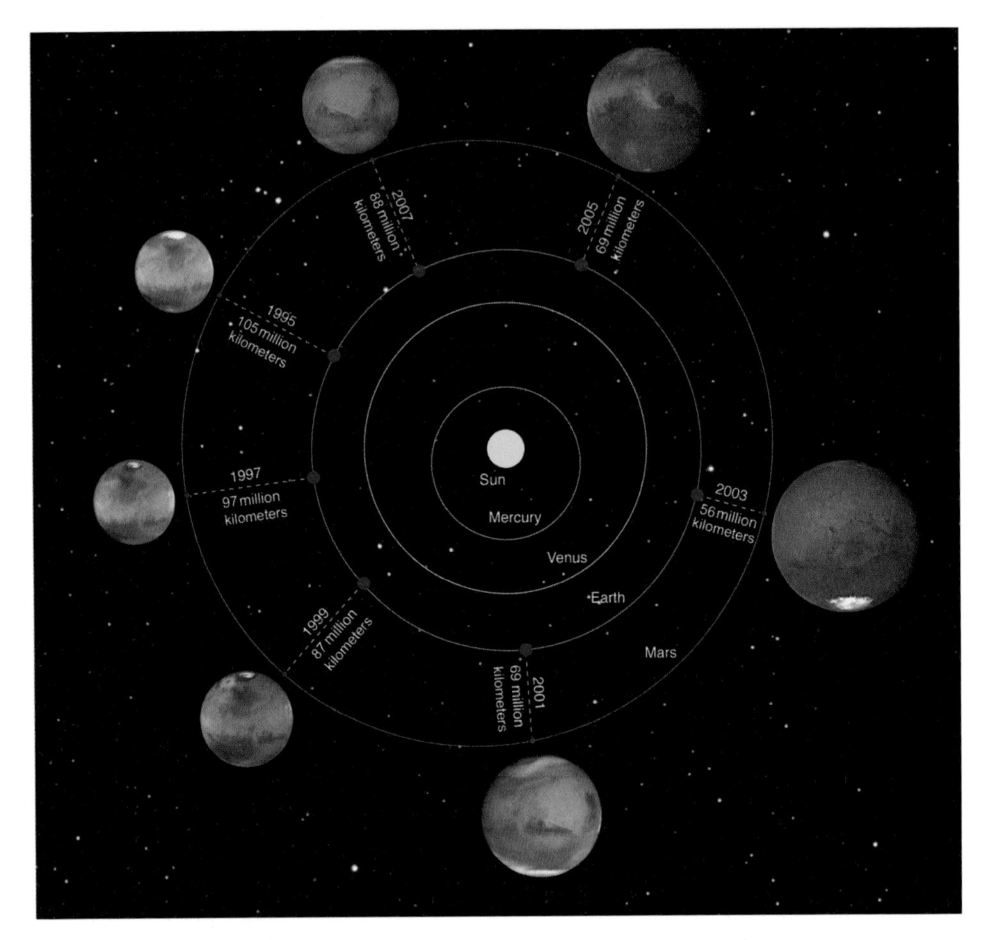

Figure 5.4 This illustration shows the positions of Earth and Mars at each opposition between 1995 and 2007. At opposition, the two planets are at their closest during which time Mars is on the *opposite* side of the Earth from the Sun. Because Mars' orbit is not circular, some oppositions are closer than others. These images of Mars show the planet's relative size as viewed by the Hubble Space Telescope in Earth orbit. Compare the different distances at opposition with that shown for future oppositions in the 'See for yourself' section (NASA/ESA/Z. Levay (STScI)).

moments, Mars, Earth, and Sun are all in a line and Mars appears *opposite* the Sun as viewed from the Earth.

Better still, because Mars orbits the Sun in an elliptical orbit, there are some oppositions where Mars is closer to the Earth than others. If Mars is at or near perihelion, (the closest part in its orbit to the Sun) during opposition, then the distance between Earth and Mars will be unusually small. This was the case for the opposition of 2003, when Mars was closer to the Earth than it had been at any other opposition in nearly 60,000 years. Yet, even then Mars appeared no more than 25 arcseconds wide. For comparison, the full Moon is only half a degree in diameter as viewed from the Earth, and so through even a telescope, the Martian disk is less than 1/60th the size of the full Moon.[2]

Typically the seeing renders anything smaller than one arcsecond as a featureless blur. But under moments of exceptional clarity, features two or three times smaller may be visible. So even at its best, only those features larger than 1/50th the size of Mars itself are really resolved into something other than a single blob. For comparison, this is equivalent to someone trying to read the date on an American quarter held by a friend more than two football fields away (about 200 meters).

At observatories today, I sit in a warm insulated control room. I direct the telescope operator to point the telescope at my target then command the computer to begin my photographic exposure. Shutters on a digital camera open to allow the faint light of stars or galaxies to slowly accumulate over many minutes or hours, building up detailed images of distant sights that can be measured and checked, quantified and calibrated at my leisure over the following hours, days or years.

Lowell had none of these. At those moments of exceptional seeing, where the image of Mars in the eyepiece suddenly crystallized into sharp relief, Lowell quickly sketched what he saw before memory and image faded and blurred. To add insult to injury, the opposition of 1894 occurred in winter. Clear night skies in Flagstaff, without the insulating effect of clouds, are exceedingly cold and in these conditions Lowell spent many hours in the open at his telescope's eyepiece. A dark winter's night became a long, cold collection of short, sharp sights, relegated to paper; the precision of the recording wholly dependent upon the artistic skill of the observer. Long after the night was over, the only record of what was seen was the artist's sketch, the image itself existing only in fading memory.

To make matters worse, the objects of most importance weren't the big obvious sights, visible to all, but rather those elements right at the limit of what could be perceived. A feature's importance to science therefore took on an inverse relation to its ability to be seen. What were faint, even invisible markings under normal conditions, under hard and difficult conditions became a network of connected lines, geometric shapes, and oval intersections, completely girdling the slowly spinning planet.

These were the canals of Mars, and astronomers of the day quickly became grouped into those who could see them and those who couldn't. Edward Emerson Barnard of Lick Observatory outside San Francisco, was renowned

[2] As a result of this opposition there is a Mars 'hoax' that makes its way around the internet every couple of years. Back in 2003, some knowledgeable astronomer wrote an email explaining that during the upcoming close-approach, Mars would look *through a telescope with 60x magnification,* just as big as the full Moon looks when *seen with the naked-eye.* Unfortunately, the letter long ago got garbled, and now the message states that this August (no year is included anymore) Mars will look as large as the full Moon. Trust me, it won't.

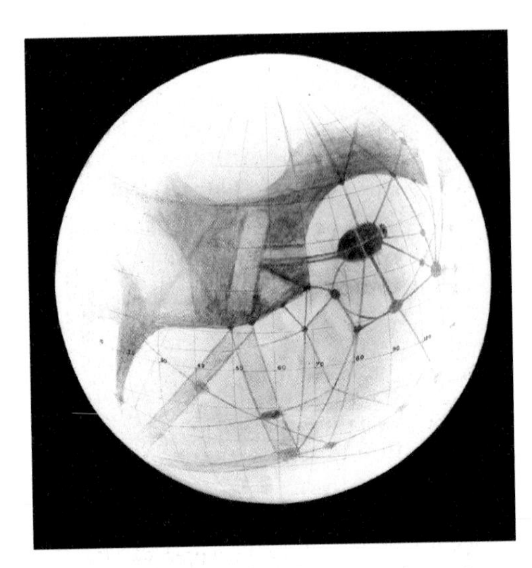

Figure 5.5 Composite drawing of Percival Lowell's observations of Mars. Solis Lacus (the Lake of the Sun) is the dark oval to the upper right from which many of the 'canals' radiate on their way around the planet (Lowell Observatory Archives).

amongst astronomers for the sensitivity of his eyesight, yet he argued vigorously that he could see no evidence whatsoever for these unnaturally sharp linear features. Lowell would argue just as vigorously that there was a difference between a sensitive eye and an acute one.

Despite these criticisms, Lowell would use his many observations of unnaturally precise features on another world (observations made under trying circumstances and at the extreme limit – if not slightly over the limit – of what was humanly perceptible) to draw logical inferences and make mathematical calculations that showed the most plausible explanation for what he was seeing was the one described in the opening paragraphs of this chapter.

Now it was not lost on Lowell that the Mars he saw in his mind's eye, if not through his telescope, bore a strong resemblance to the countryside beyond the borders of Flagstaff. The cold, high altitude summits of the San Francisco Peaks, visible from his observatory home, provided evidence to his mind for why life could easily survive the cold winters and low atmospheric pressure he calculated for Mars.

Outside the alpine forests of Flagstaff, the elevation drops two thousand feet and the landscape of northern Arizona and southern Utah is an arid plateau of red rocks, mesas and buttes where snow and rain rarely fall. In his 1906 book, *Mars and Its Canals*, Lowell described this desert's similarity to Mars' color seen through the telescope, "The pale salmon hue, which best reproduces in drawings the general tint of [Mars'] surface, is that which our own deserts wear. The Sahara has this look; still more it finds its counterpart in the far aspect of the Painted Desert of northern Arizona. To one standing on the summit of the San Francisco Peaks and gazing off from that isolated height upon this other isolation of aridity, the resemblance of its lambent saffron to the telescopic tints of the Martian globe is strikingly impressive."

For Lowell, this similarity was not by chance, but rather a reflection of the natural lifespan of the planets. From simple calculations Lowell showed that the smaller a planet, and thus the weaker its gravity (the force of gravity on Mars is only 38% of our own), the quicker its atmosphere should escape away to space and thus its seas evaporate and its surface turn to desert. Lowell then made the leap to advocating that this is the fate of all planets. Thus by understanding Mars

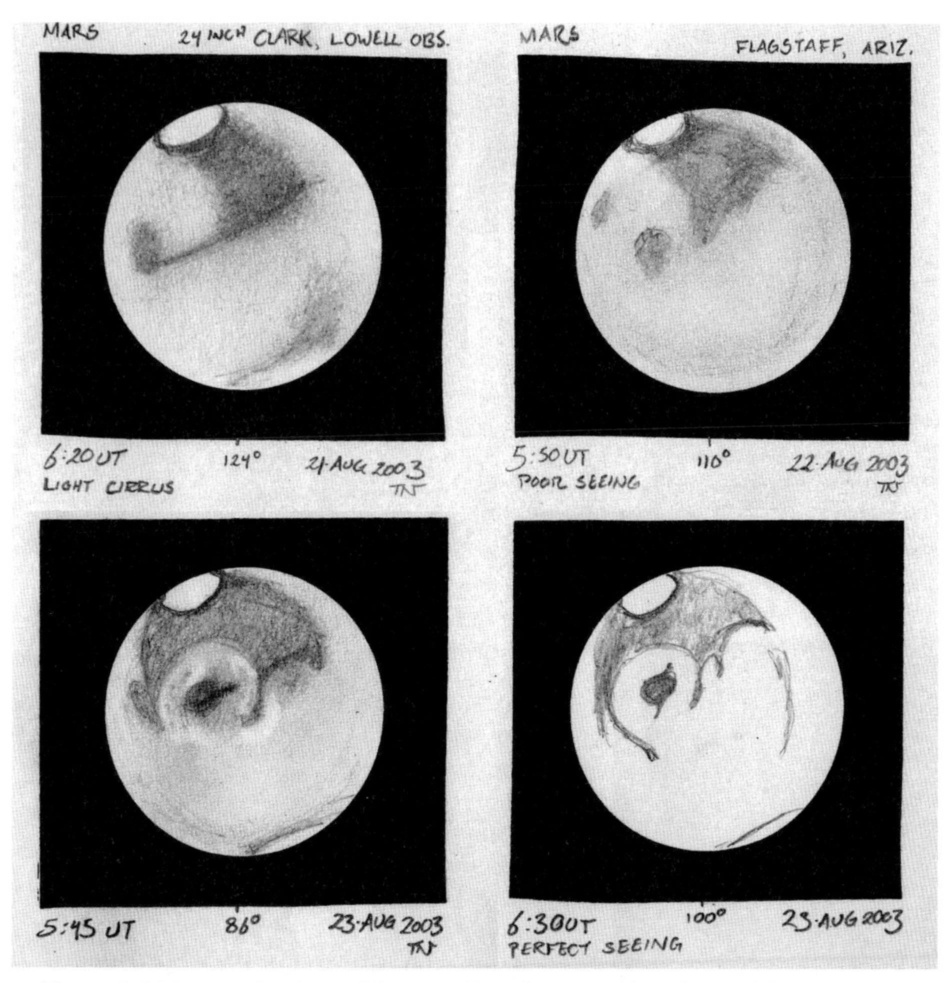

Figure 5.6 My own drawings of the same face of Mars made with Lowell's 24-inch (61-cm) telescope during the close opposition of 2003. Solis Lacus is visible on the left of each view. The different seeing on different nights clearly shows differences in visible detail. Nevertheless, I was still able to see a 'canal' for myself (T. Nordgren).

and the differing evolutionary states of the planets, he argued that we learn more of our own planetary fate.

Lowell saw evidence of this Martian "Death by Desert" in the Great Salt Lake of Utah with its surrounding dry salt wastes and ancient shorelines in the nearby hills. In his last book, *Mars As the Abode of Life*, published in 1908, Lowell presented further evidence of the growing desertification of the Earth in the form of ancient fossilized trees preserved in what is now Petrified Forest National Park, 100 miles (160 km) east of Flagstaff.

> Now these desert belts are widening. In the great desert of northern Arizona the traveler, threading his way across a sage-brush and cacti

Figure 5.7 The Little Painted Desert within the borders of the Navajo Nation northeast of Flagstaff (T. Nordgren).

> plain shut in by abrupt-sided shelves of land rising here and there some hundreds of feet higher, suddenly comes upon a petrified forest.... The land which once supported these forests is incompetent to do so now. Yet nothing has changed there since, except the decreasing water supply.... Proof of this is offered by the great pine oasis that caps the plateau of which these petrified forests form a part, and is kernelled by the San Francisco Peaks.... Two thousand feet upward the verdure-line has retreated since the former forests were.

All around him Lowell saw Mars, but a Mars the Earth was slowly becoming.

By the publication of Lowell's *Mars* in 1895, the nineteenth century had seen the greatest mass movement of humanity in recorded history.

'EARTH DOOMED TO BECOME A DESERT'

Water Will Be Dearer Than Gold, Says Prof. Lowell at College Club.

DRAWS CONCLUSIONS FROM MARS' CONDITION

Noted Astronomer Also Tells of Canals and Life on Neighbor Planet.

"It is the doom of the earth," said Prof. Percival Lowell yesterday afternoon in his lecture before the College Club on Commonwealth avenue, "to be covered with deserts like Mars, and the time will come when water in this world of ours will be far more precious than gold."

With the aid of stereopticon views, many of them photographic transparencies, Mr. Lowell sketched the fresh evidence which has been obtained at the Lowell observatory in Arizona of the existence of intelligent beings on the planet Mars.

"Schiaparelli," he said, "saw these lines in 1877, and his map of them promptly met with universal condemnation. But the world made a rash step in denying his observation of the canals, for time was perfectly sure to prove their existence. In 1877 only one man saw them, but we have now a dozen men who have not only seen them, but have seen them very much

Figure 5.8 Feature story from the *Boston Post*, December 8, 1907.

Lowell's relocation from Massachusetts to Arizona Territory was part of this headlong westward migration and eventual settlement of what was known as the Great American Desert.

For those of us born and raised in the Pacific Northwest or east of the Mississippi, Lowell's preoccupation with water and the growing desiccation of the planet may seem hard to understand. From my childhood in Oregon, my memories are of rain that once begun in October never ceased until the following summer. In my youth, soft, luscious moss grew on any surface that came in contact with the air, and the entire countryside was always green with life.

But the western American interior is an arid land. For those who live there, to *waste* water means to let even a single drop of water go unused. However, in the forty years previous to Lowell, it was the subject of conjecture that the very act of 'opening' the West with farms would bring the rains that would transform the bone-dry expanse into a new Garden of Eden. "Rain Follows the Plow" was the actual climatological claim[3] whereby Nebraska and Arizona would soon be as green as New York and Alabama. This bizarre conclusion was based upon nothing more than the unfortunate coincidence that the beginning of the westward expansion of the late 1870s and '80s was followed immediately by an extended period of above average rainfall that had never been seen before (or since). And, as fate would have it, the drought that inevitably followed was in full force when Lowell arrived in Arizona Territory to build his observatory.

But there'd been people living for thousands of years in the parched country the white men had just recently entered. John Wesley Powell – a retired Civil

Figure 5.9 The second Powell Expedition sets out from Green River, Wyoming, 22 May 1871. John Wesley Powell is standing on deck in the middle boat (Grand Canyon National Park Museum Collection).

[3] For a full account of American preoccupation and policy towards water in the West read Marc Reisner's *Cadillac Desert: The American West and its Disappearing Water.*

Figure 5.10 Around 800 years ago the pueblos that make up Wupatki National Monument were a flourishing meeting place on the Colorado Plateau. Today, it is an arid landscape as the Sun rises over the ruins of Wukoki Pueblo within the monument's boundary near Flagstaff, Arizona (T. Nordgren).

War Army Major, geologist, and university professor – set out from Green River, Wyoming in 1869 to explore the geology and people of the Colorado River canyon country (the last remaining blank spots on nineteenth century maps). Powell and his nine-man expedition were the first white men, and for all we know the first people at all, to navigate the length of the Colorado River canyons by boat. During his two historic expeditions through what would become a host of national parks and monuments (most famously Grand Canyon National Park) Powell mapped, measured and wrote about the spectacular landscape, all the while learning about the people who made this land their home. Powell was intensely curious; what were they like, what were their customs, how did they survive and prosper in a land of such magnificent desolation?

What Powell found, he spent the next thirty years popularizing to the country. The same 1895 that saw Percival Lowell publish *Mars* (the first of his wildly popular accounts of an ancient civilization coming to grips with the arid desert landscape of Mars) also saw Powell publish *Canyons of the Colorado*, his wildly popular account of his exploration of the Colorado River, the landscape, and people right outside Lowell's observatory doors.

"In centuries past," Powell wrote, "the San Francisco Plateau was the home of pueblo-building tribes, and the ruins of their habitations are widely scattered over this elevated region." Just a few short miles from Lowell's own observatory, Powell described cliff ruins of a "vanished" people that "built stairways to the waters below and to the hunting grounds above" in what is today Walnut Canyon National Monument. Farther south of Flagstaff, where the last great tributary of the Colorado drains much of present day Arizona and western New Mexico, Powell found:

> In the valley of the Gila and on its tributaries from the northeast are

> [the tribes of the] Pimas, Maricopas, and Papagos. They are skilled agriculturalists, cultivating the lands by irrigation. In the same region many ruined villages are found. The dwellings of these towns in the valley were built chiefly of grout, and the fragments of the ancient pueblos still remaining have stood through centuries of storm.... The people who occupied them cultivated the soil by irrigation, and their hydraulic works were on an extensive scale. They built canals scores of miles in length and built reservoirs to store water.

The American West in which Lowell made his observatory and observations was therefore awash in stories of ancient civilizations and an ongoing battle for water against the all consuming desert. Beyond the territory's borders, the late 1800s was also a time for enormous engineering marvels. Major canals were being built all over the world: they linked the major river systems of France; Germany linked the North Sea and Baltic; the Suez Canal created the first shortcut between Europe and the Indian Ocean; and the first attempt was made at a canal across Panama. All of these were major engineering undertakings, the talk of the world, and Lowell, as a gentleman scholar and traveler, would have had first hand-knowledge of their engineering and industrial power.

Whether Lowell fully realized it or not, the 'logical' conclusions he drew from his observations were certainly colored by the world in which he lived and worked. Desert planet, drought, ancient civilizations, engineering marvels, and canals; Earth or Mars? Who's to say which world he was really describing?

What is unambiguous is that thanks to Lowell, Martian canals were seared into the public consciousness. The Mars that the world saw, heard, and read about, was Lowell's Mars, not Barnard's. Martians were therefore everywhere. They were in the mass media from H.G. Wells' *War of the Worlds* novel published in 1898 to Orson Wells' 1938 radio broadcast of the same name. Thanks to Lowell, and the mass media he courted, for much of the early twentieth century, it was possible to look into the night sky at the ominous Red Planet and know, actually *know*, that you were seeing the abode of alien life. And that it was an *intelligent* life, gazing back down upon us, far older and more advanced than we. I can't imagine what that must have felt like, and I don't know an astronomer who wouldn't give anything to be able to feel that way today.

Figure 5.11 A favorite diner in Sedona, Arizona's red rock country pays tribute to flying saucers and Mars culture (T. Nordgren).

But by the late 1950s as eyepiece and pencil gave way to cameras and photographic plates, telescopic photos had failed to confirm any semblance of the artificially precise global canal system (although many Mars maps still showed some form of indistinct linear features). The changing nature of the 'blue-green' regions continued to support the

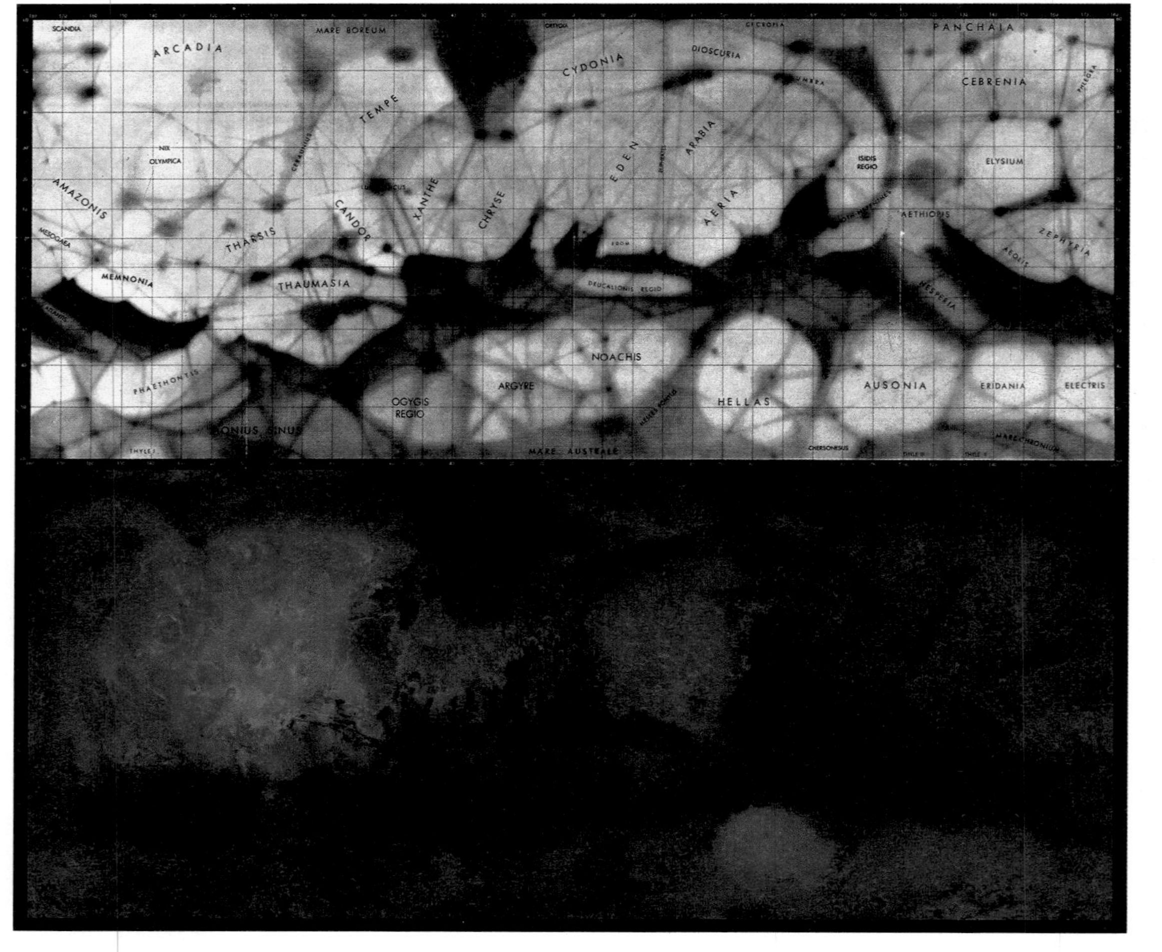

Figure 5.12 A Martian map published by the U.S. Air Force (based on Lowell Observatory drawings) in the late 1950s still shows some form of linear features. Even so, it compares well with a map made by the U.S. Geological Survey Astrogeology Division (located in Flagstaff, Arizona just down the road from Lowell Observatory) based on Viking Orbiter images in the late 1970s. Both maps are centered on Martian longitude 0 degrees and extend from Martian latitudes 60 degrees north to south. The Martian equator runs down the center of each map. Compare

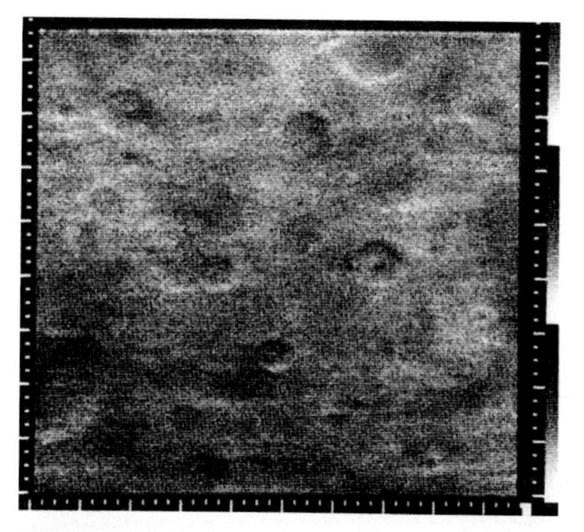

Figure 5.13 Mariner 4 image of Mars showing a heavily cratered surface like the Moon. The image was taken from 8,125 miles (13,000 km) and measures 158 by 140 miles (253 by 225 km) (NASA).

idea that at least some form of life, albeit probably not intelligent life, might be found there. In 1957, Walt Disney's weekly *Disneyland* TV show aired an episode in which Disney "edu-tained" American audiences about "Mars and Beyond.[4]" In his introduction, a touch of disappointment in his voice, Walt asks if upon our eventual exploration of Mars it is possible we will find "no more than a low form of vegetable life" as if that was the least we could hope for.

If only that were so. When the first pictures of Mars were sent back to Earth from the Mariner 4 spacecraft flying by in 1965 (the equivalent of children plastered to the car window gazing in awe at the roadside attractions with no time to stop for a souvenir) they showed a dry, dead, cratered world like the Moon. So much for Lowell.

But Mariner 4 (as well as the later Mariners 6 and 7) all flew past Mars's southern hemisphere. If the only things we knew about the Earth were gathered from a fleeting glimpse at our own planet's southern hemisphere, we might suspect our planet was covered by nothing but water.

When Mariner 9 went into orbit around the Red Planet four years later, it was for the express purpose of carrying out the first global reconnaissance. While its cameras certainly confirmed craters aplenty, the new views of the northern hemisphere revealed what appeared to be vast, dry riverbeds and canyon networks like those seen from a typical transcontinental flight over the American southwest. In addition, there in the window were volcanoes larger than any on Earth and great rift valleys as long as the United States. Together these features were stunning hallmarks of an active world, nothing at all like the Moon. While the planet might appear lifeless today,[5] orbital evidence showed a world that may not always have been so.

A hundred years after Lowell wrote *Mars*, I was an astronomer at the very same observatory. Even though I was working on star formation in dwarf irregular

[4] And Wernher Von Braun showed the plans he had for a fleet of atomic powered ion-propulsion saucers that would carry dozens of astronauts to explore Mars (all before the U.S. had succeeded in placing a single spacecraft into orbit).

[5] The 'blue-green' regions are merely areas of dark brown material periodically uncovered by seasonal sand storms (the color is an optical illusion where the eye sees in the darkness the complementary color, blue-green, to the ever present bright reddish-ocher surroundings).

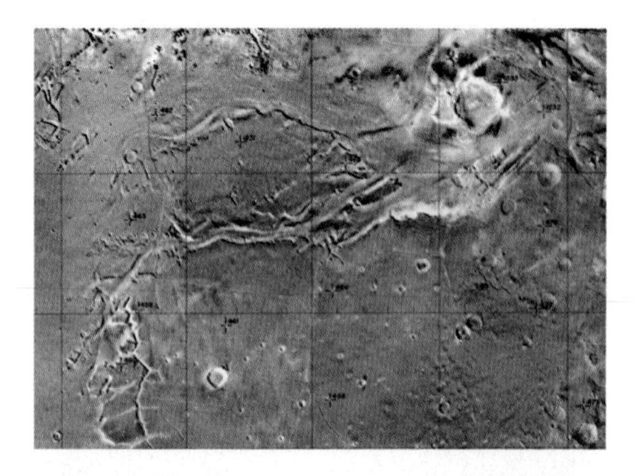

Figure 5.14 A mosaic of Mariner 9 images shows a world of difference compared to what Mariner 4 saw. While there are still craters, the part of Mars seen here has what appear to be dry river channels flowing from out of the cracks at the lower left. Notice the curved and wavy features along what looks like an ancient shore line along the top left portion of the mosaic (P. Thomas/NASA).

Figure 5.15 Mars mosaic made up of images captured by the Viking 1 Orbiter. The center of the Martian disk shows the 1,875 mile (3,000 km) long Valles Marineris canyon system. Along the western limb of the planet are three dark red volcanoes of the Tharsis Bulge (NASA/USGS).

galaxies, "Uncle Percy's" history was all around me. While working on a paper in which my colleague and I attempted to understand the complex structure of gas knots in a galaxy, I thought I saw signs of something odd. By linking together a number of the faint structures in an image, I thought I'd found evidence of a rare spiral arm. My colleague, however, was not so convinced. "This is Lowell," she said, "we're very careful here about trying to connect small, fuzzy blobs in a telescope."

During my weekends, I used to hike through the countryside of the Four Corners region, seeing the very places Lowell described a hundred years before. In the red rock country of southern Utah and out on the Hopi and Navajo Reservations I saw a land so red that on bright summer days sunlight reflecting off the mesas and buttes colored the undersides of the afternoon monsoon clouds a dusky rose and crimson. Set in a clear, dark blue sky, it seemed on days like this that I walked in a world of nothing but primary colors.

As of my time at Lowell, humanity had successfully landed only three spacecraft on the surface of Mars. In every camera frame was a landscape reminiscent of the dry, desert, red rock country of Utah and Arizona (just like Lowell described, minus the canals). But in 2003, during the unusually close opposition, NASA was planning to send a pair of rovers to Mars. Like Powell over a hundred years before, we would finally be able to get out and head to the horizon and explore. A former professor of mine at Cornell University, Steve

Squyres, was the lead scientist for the rovers and he asked me to help turn their color camera calibration targets into sundials as a way of reaching out to school children about the excitement of space exploration. So as I hiked through the landscape of Lowell and Powell, from the red canyons of Sedona, Arizona, to the red stone arches of Moab, Utah, I had Mars on my mind. I was curious to know how far the similarities with Mars actually extended. Looking around, just how much Mars could I see and had it always been this way?

Figure 5.16 Red rock and blue sky come together as a first quarter Moon peeks through a sandstone cleft in one of the iconic buttes within Monument Valley Navajo Tribal Park straddling the Utah/Arizona border (T. Nordgren).

Today is a hot sunny day in Arches National Park, in the heart of the red rock country outside Moab, Utah. My boots are powdered orange from the trail through the Park Avenue monoliths – towering fins of rust-colored sandstone that border a mostly dry wash on the edge of the park. In every direction I see red, layers and layers of it. Salmon-colored sand trickles off buff sandstone ledges bordering a pink and rosy trail beneath my feet while overhead stand gargantuan blocks of crimson stained by desert varnish in streams of maroon and purple.

Figure 5.17 An artist's illustration of one of the Mars Exploration Rovers. A camera on top of the forward mast returns IMAX quality images as seen from the height of a typical person. The 'Marsdial,' a sundial made from the camera's calibration target, sits on the back edge of the rear solar panel at left (NASA/JPL-Caltech/Cornell).

Figure 5.18 Sandstone sentinels like Courthouse Butte, at right, tower over the landscape in Arches National Park. How similar is this red rock country to what one would see standing on the surface of Mars? (T. Nordgren)

These awesome stone sentinels are an iconic vision of this country; Monument Valley Navajo Tribal Park, famous from innumerable western movies (and Road Runner cartoons) is only a couple hours' drive to the south. Even though no Martian lander or rover has yet photographed a towering landscape like what I see around me, it is not for its absence on the Red Planet. Rather, NASA scientists and engineers are conservative with their expensive spacecraft, and have taken great care to land them where they have precisely *because* they are devoid of towering buttes and breathtaking cliffs. But orbital imagery shows they are there if we could but drive far enough to visit them.

To experience Mars in person is a dream I have had since I was a boy. Long ago it became clear it would not happen for me. But to do so now, through the knowledge gained from the spacecraft that *have* gone there, all I need do is mentally manipulate the view before me and create, step by step, a Mars of my imagination.

First, take away the majority of the atmosphere and all the oxygen I breathe. As distressing as this might personally be, another more visible transformation suddenly takes place. Light from the Sun is composed of all the colors of the rainbow. On Earth, molecules in the air scatter the blue end of the spectrum in different directions. This blue light bounces from molecule to molecule before eventually finding its way towards the Earth and our eyes. From our perspective on the ground, we see this light reach us from every part of the sky, and thus the sky looks blue.[6] With virtually no atmosphere on Mars: good bye blue sky.

[6] The Sun's disk, on the other hand, now looks somewhat yellow since it still shines with all the colors of the rainbow but now with a little less blue than it did before. In addition, as the Sun sinks to the horizon and its light passes through more and more air and dust, so much blue is scattered away that the Sun takes on its familiar red of sunset.

Next, with less than one hundredth the atmospheric pressure of Earth (and correspondingly little of the greenhouse gasses that keep us warm) the surface temperature drops dramatically. Move the Sun 50% farther away, and it gets even worse. Only in summer near the equator do the temperatures on Mars top the freezing point of water. The humidity plummets, the atmosphere becomes bone dry and the puffy white clouds above me disappear. There are still clouds, but they are high thin clouds of tiny ice crystals.

The pressure and temperature over the rest of Mars means any water within a few meters of the surface is frozen solid, so there are no puddles, streams, or rivers. The tiny trickle of water at my feet from yesterday's spring rainstorm over Moab evaporates instantly as if it was never here, as indeed rainstorms on Mars may never have been.

With the liquid water, take away everything green. There's no plant or 'low form of vegetable life' for as far as the eye can see. This last one is a shock. Most places on Earth that are described as dry, arid, desolate, desert, usually have some kind of plant life, be it ever so drab or humble. Look around you and imagine your landscape, not just without any trees, eye-catching flowers, or bushes, but with nothing. Not a single scrap of grass, no matter how scraggly, not a weed, a lichen, not even a piece of dry broken twig. In fact, get rid of most of the dirt you see as, in addition to the minerals it contains, the soil beneath your feet is composed largely of decayed plant matter and moisture.

What then is my new Martian ground covered with? Rocks. Big rocks, small rocks, crushed rocks, pulverized rocks, powdered rocks, and dust. The last few billion years of sandstorms have blasted away at the stony landscape, weathering cliff and mountain down to pebble and grit. And without rain to wash away dust and cause it to clot, it works its way as a fine powder into every microscopic nook and cranny across the entire planet's surface.

Iron in the dust reacts with what little oxygen there is in the atmosphere (a byproduct of ultraviolet light from the Sun splitting apart the components of the thin carbon dioxide atmosphere) forming iron-oxide – rust – that turns the whole world red. Here, at last, is a change I don't need to make to my

Figure 5.19 Iron-oxide rust gives the Red Planet its name. Here, the Mars Exploration Rover Spirit climbs across hill tops in red rock country that extends as far as the eye can see. Parts of the American Southwest from the deserts of Arizona and Nevada to Death Valley National Park in California have been used to test Mars-bound spacecraft because of their strong resemblance to views like that shown here (NASA/JPL/Cornell).

surroundings. The red rocks of Mars are red for the same reason the red rock country is too. Mars is Moab made global.

In addition, without any moisture to hold and clump the Martian 'soil,' the loose red powder fills the air and is picked up by the winds. Great globe-girdling dust storms blow the iron-oxide particles high into the thin Martian atmosphere, turning the sky forever pink and ocher. The Technicolor palette of the southwest transforms Mars to variations on a russet theme. The grandeur of the southwest, however, is not just in its color but also in the gigantic forms that every year draw visitors from around the world. What are the processes that make these mesas, and buttes, arches and hoodoos? Are they the same processes at work on Mars? And if so, what changes must I make, to make them fit in my mental Martian scheme?

The blocky monoliths we see in Arches and all over the southwest are made of great slabs of sandstone. Each layer reveals a moment in the history of this place; each layer tells the story of alternating seas, and seashores, dune fields and alluvial fans. Reading the rocks, I see that seas have come and gone 29 times where I stand in this arid part of Utah. In that time, nearby mountain ranges have risen and fallen, and new ones have taken their place. Rivers, rain, wind and ice broke those now forgotten mountains down and washed their remains downhill to west and east covering this region after the seas dried up.

Each epoch of erosion left its sand, silt, salt, mud and pebbles. New erosion buries these and, over time, lithofies the sediments (cementing them with pressure and heat) to form solid rock like sandstone, siltstone, mudstone, and shale. The nature of the conditions under which the sediments came to rest determines many of the chemical and physical properties of the rock they became.

Entrada sandstone is the smooth dark red blocks that make up the fortress-like walls of the monumental towers around me. It was laid down about 160 million years ago as sand dunes near an in-land sea that was flooded only occasionally during storms or high spring tides.

Beneath it, farther back in time at 180 million years ago is the pale pinkish-ocher accumulation of multi-banded arcs that make up the Navajo sandstone. These are the fossilized remains of enormous sand dune fields: an ancient Sahara covering the southwest before the most recent seas appeared. Millennia of dry, shifting, sand dunes slowly drifting across the endless miles are recorded in the Navajo's sweeping, crisscrossing bands.

Between the Entrada and Navajo sandstones is a darker, lumpy, red-brown layer called the Carmel (or sometimes Dewey Bridge) formation that has its origin in wet, muddy, tidal flats along the eastern shore of the eventual inland sea. Together these three layers form a stack reflecting a 20 million year snippet of time here, but are themselves just a few of the many layers that make up the Arches area.

Erosion and time eventually buried these layer stacks, while tectonics (the stretching, pulling, cracking and shifting of a planet's crust) subsequently raised them far above sea level. Later, rain water and snow-melt, washing down to the sea, revealed the layers once more in newly formed cliff walls.

Figure 5.20 Blocky red sandstone at top, streaked with desert varnish, is an example of the 160 million year old Entrada Sandstone. Beneath it are visible several layers of the dark, clumpy Dewey Bridge (or Carmel) formation. Beneath these dark red layers, and visible in the foreground are the pale, pink, Navajo Sandstone formed from ancient Sahara-like sand dunes. These layers are visible throughout Arches National Park, including here at the main entrance (T. Nordgren).

Figure 5.21 Water pouring off the dark, red Entrada sandstone at top (and seeping through more porous rock beneath it) washes away softer layers of stone beneath. Eventually the overhanging Entrada collapses and the small canyon in which I stand grows just a little bit longer. This process is called sapping, and is at work all across the canyon country of the American southwest. Visible in the pool of water is the reflection of Delicate Arch above, shaped from the same forces of water and erosion around me (T. Nordgren).

Depending on how fractured, hard, or porous layers are, different layers erode away in varying amounts under the scouring action of water. Where harder rocks sit on top of softer ones, erosion washes away the softer rock beneath. The harder cap rock is undercut and eventually collapses in a process called sapping. Narrow canyons increase in length and width as the harder rock above falls inward after rushing water washes away its support. In time, canyon networks spread and broad plateaus crumble into disconnected mesas, gradually narrowing into lone buttes that slowly wear away to solitary pinnacles and hoodoos. Eventually, the last bit of cap rock falls down and the plain upon which this all once stood washes clean as the next layer down begins to erode.

Balanced Rock within Arches National Park is nearing the end of this process as a cap of smooth Entrada stone sits atop a conical tower of the crumbling

Figure 5.22 The hard cap of Entrada Sandstone will someday collapse as the softer Carmel formation making up the pedestal, slowly weathers away. Eventually, Balanced Rock will do so no more (T. Nordgren).

Carmel mud and siltstones that rests upon a broad flat base of Navajo sandstone. In time the Carmel pedestal will weather sufficiently far that the Balanced Rock will no longer do so, and off it will go. Once it does, the soft Carmel mound will be fully exposed to the elements and quickly weather away to nothing.

When seas more ancient than the Navajo sandstone's evaporated, they left behind thick layers of salty residue. The Great Salt Lake to the west is an example of what was once here as each sea receded. Over millions of years the salt from those seas was buried beneath the thick layers of sandstone above. Since then, tectonics has slowly pushed the salt westward into long north-south running masses that built up under this part of the state. As the weight of the overlying mass pushed down elsewhere, the salt had to go somewhere, and so welled upwards under Arches. The upwelling salt cracked the thick sandstone stacks above it. Rain and ground water seeped downward, dissolving the subterranean salt. As the salt dissolved, the landscape above buckled and drooped into the cavity that formed. Along each shoulder of what became Salt Valley, the overlying Entrada sandstone cracked again, forming great parallel fins as the central valley sagged.

The arches we see today are caused by sand weathering off these sandstone fins.[7] Carbon dioxide from the atmosphere, reacting with rainwater, forms a very weak carbonic acid loosening the minerals that hold the sand grains together. Sand accumulating around the base of the fins acts like a sponge trapping even more corrosive rainwater against the stone. Cavities appear. Water, running down the sides of the fins and into the cavities, widens the opening further.

[7] The arches of Devils Garden and Fiery Furnace form within the weathered fins running along much of the eastern edge of Salt Valley. The arches and windows of the Klondike Bluffs and Windows section of the park are forming within the sandstone fins along the west rim of the valley.

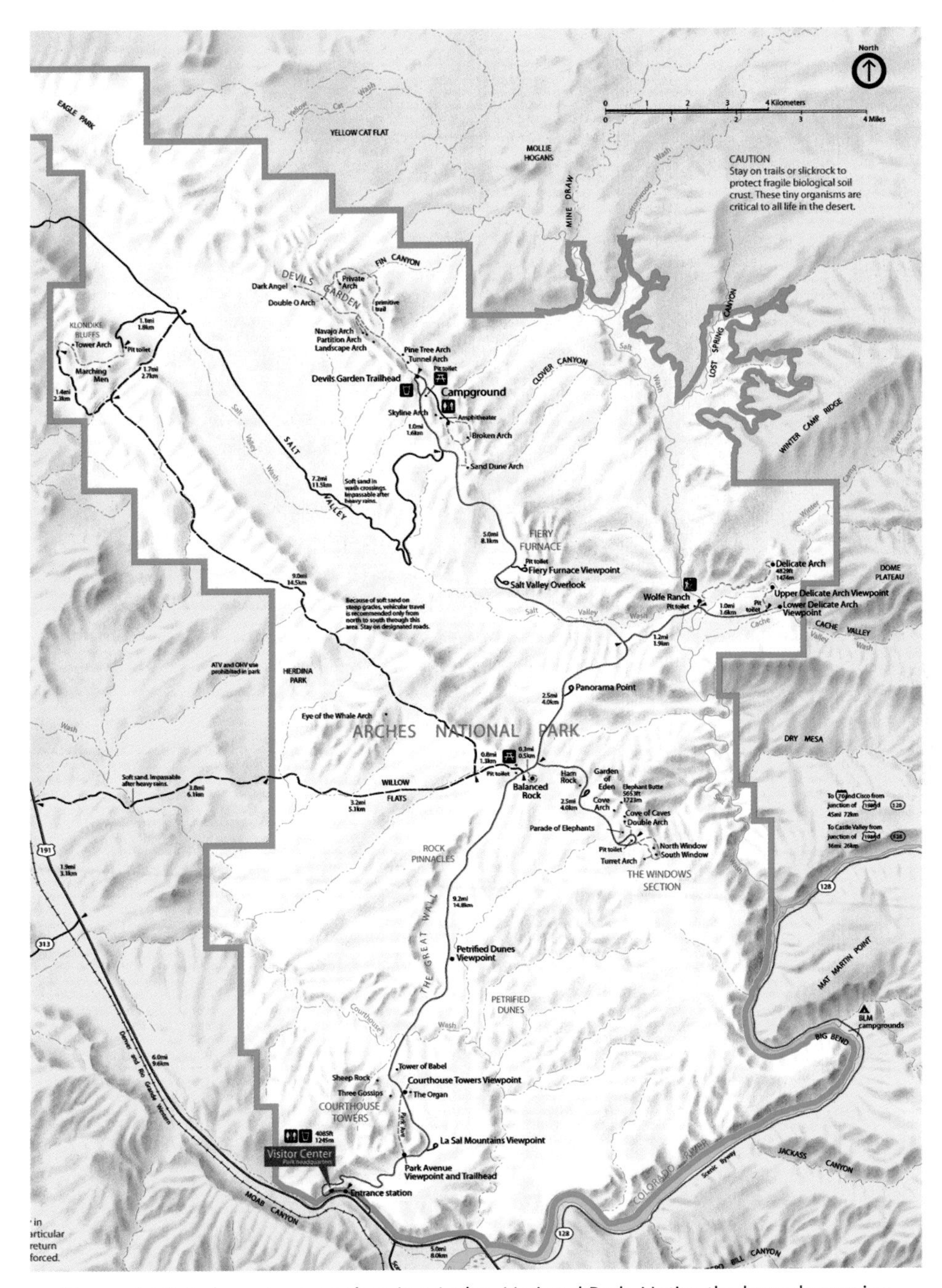

Figure 5.23 Park service map showing Arches National Park. Notice the long depression of Salt Valley coming in from the northwest. When ground water dissolved the buried salt, the overlying layers of sandstone buckled and collapsed. Rain water has subsequently eroded away arches, windows, and fins along the cracked valley shoulders in such places as Devils Garden and Fiery Furnace (National Park Service).

Figure 5.24 Gravity acting on the overhanging sandstone, causes rocks to fall into the cavity beneath, gradually widening the span of Double Arch. I stand in one of the openings, backlit by the light of the setting Moon (T. Nordgren).

Eventually the cavity breaks through the fin and you have a window. As windows widen, gravity dislodges the hanging rock, forming small arches that grow with time. Eventually, the rest of the fin may wear away completely, leaving something like Delicate Arch behind.

But these are only transitory phases in an arch's life. Wait long enough, and gravity, rain and sand will wear an arch away to nothing. There have been many more arches in the history of the park than there are arches today, and there will be many more in the future. We see each arch and window at a special time in its geologic life. The beauty and grandeur of the parks and monuments of the Four Corners region is a testament to our active planet's past. If the history of the southwest had been monotonous and unchanging, then there would be no differing layers, and the landscape we see would be far less varied and beautiful. Thus the majesty of the southwest is due to fluctuations in its past, and for much of that time, up to and including the present, it is a past heavily influenced by water.

Is there an Arches National Park or a Monument Valley on Mars?

We know that liquid water can't exist on the surface of Mars today. But ever since Mariner 9 it was apparent that water did play some part in Mars' past. Recent images from the Mars Reconnaissance Orbiter, around Mars since 2006, reveal landscapes of layered canyons, mesas, and buttes. From an altimeter aboard the European Mars Express, in orbit since 2003, we know that some of these desert buttes tower several thousands of feet (hundreds of meters) into the

Figure 5.25 A spring storm brings rain to Canyonlands National Park. As arid as this canyon country may be, water's results are everywhere. The large, branching canyon at center is just one example of what eventually happens when water acts on stone for long periods of time. A long time ago, this impressive canyon system probably had its origins as no more than a tiny wash like that in Figure 5.21 (T. Nordgren).

Figure 5.26 On the left are the buttes and canyons visible from Dead Horse State Park between Arches and Canyonlands National Park in central Utah. On the right is a section of Candor Chasma on Mars as viewed by the High Resolution Imaging Science Experiment (HiRISE) on board the Mars Reconnaissance Orbiter. Part of the Valles Mariners canyon system, it shows many buttes and canyons made from layers of apparently sedimentary rock like those seen at left on Earth (T. Nordgren (left), NASA/JPL/University of Arizona (right)).

Martian sky, taller than the cliff walls around us. So here, at last, these monoliths remain in my vision of Mars.

But why are they there, where did they come from?

Jeff Kargel in his book *Mars, A Warmer Wetter Planet* describes in great detail the evidence for water's role in the creation of the Martian landforms we see today. Three main types of landscape present in Martian orbital images cry out for liquid water as an explanation. The first are great rounded lobes or petals that appear around many impact craters across the planet. Unlike craters on the dry Moon, where perfectly straight rays of blown material spray outward from the impact site, on Mars a large number of craters look as if something large was dropped into wet mud. These lobate shapes look as if the force of impact threw up giant fountains of rock mixed with water (whether the water was already liquid or was ice liquefied by the heat of impact is not clear).

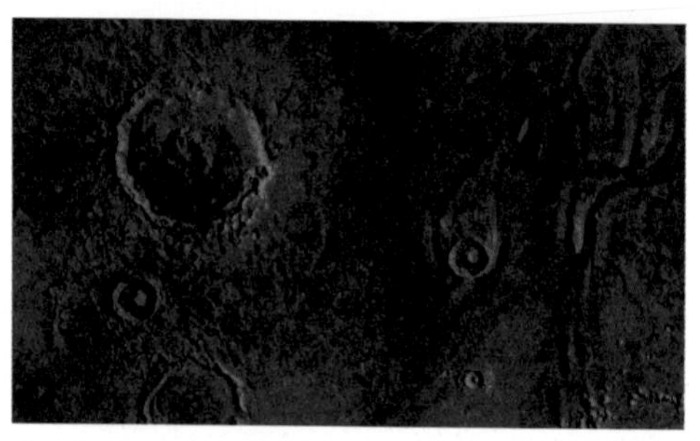

Figure 5.27 Viking Orbiter mosaic showing multiple clues that Mars once contained liquid water. On the left are three large craters that appear to have formed when objects hit a surface that was wet (or at least contained a lot of water that melted upon impact). Drop a rock into thick mud, and it would create the petal-shaped lobes around the hole, just like those seen around the craters. On the right are tear-shaped islands that look as if great amounts of water flowed across the surface, washing away the terrain around high craters and hills (NASA/USGS).

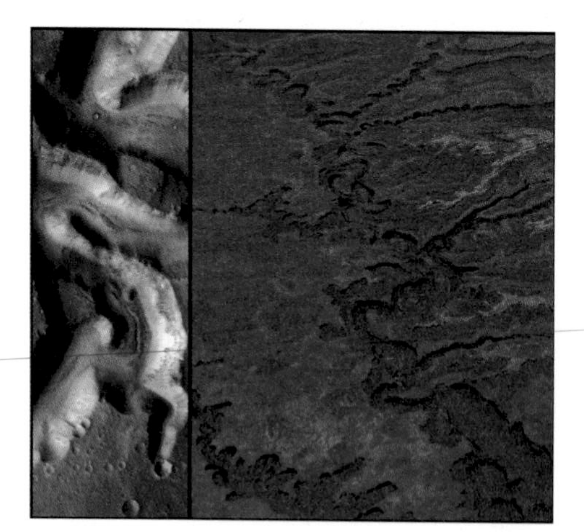

Figure 5.28 Valley networks on Mars and Earth. On the left, the HiRISE camera captures the meander of Nanedi Valles. On the right is a meandering canyon my camera captured on a recent flight from Los Angeles to Denver. Groundwater sapping (where running water undercuts overhanging layers that subsequently fall into the canyon, extending the canyon's length) is the leading theory for the Martian valley. Canyons that form this way on Earth typically have constant widths along their path, as well as rounded (amphitheater-shaped) heads. These are visible in both the Mars image on left (notice the rounded valley that ends at the crater) and the image from Earth on the right (as seen in the tributary valley in the bottom left of the scene). The rounded canyon heads are also visible from the ground in Figures 5.21, and 5.25. The meandering nature of Martian valleys suggests persistent or repeated flow just as they are required to form meanders in streams and valleys on Earth (NASA/JPL/Univ. of Arizona (left), T. Nordgren (right)).

The second type of feature is what looks to be enormous dry river beds where flowing water carved channels and created tear-drop shaped islands around craters and hills in its path. These features all seem to originate out of great circular basins of jumbled chaotic terrain as if some massive underground reservoir had been tapped and the overlying ground collapsed as the waters flowed out. Huge regions of the landscape appear to have been swept aside in massive floods, yet all on a planet where today liquid water cannot exist.

The third category of features are numerous examples of small branching channels that look as if waters draining from a plateau had slowly grown together to form deep river canyons on their way to some long gone lakes or seas. In nearly every respect they look like winding canyons in the American southwest where water has sapped away at overhanging rock, creating long, branching canyons, just like those Powell and his crew explored but on a planet far drier than any he ever saw.

All these features appear in hundreds of pictures sent back by decades of spacecraft. But pictures can be deceiving. Even when we have beautifully clear photographs of great branching canyon networks, or large fans of material flowing from out of high mountains down onto lowland plains, it is tempting to interpret these in the light of what we see around us and know so well. By this thinking, the canyon networks become the dried remains of water drainage systems like the Colorado River. Conical shapes become alluvial fans as water washes out of mountains, spreading rock and sand out onto flood plains. But interpret the picture wrong, and a chain of very reasonable, logical conclusions can lead to a modern picture of Mars as mistaken as a world of canals and ancient intelligences.

Jeff Moersch is a planetary scientist at the University of Tennessee in Knoxville. Raised in southern California, he earned a Masters degree in Geology at Arizona State University, in between a Bachelor's degree in Physics and PhD in Astronomy at Cornell University in New York. He has spent years tramping around the southwest studying the rocks, learning how to read the countryside, and one of his interests is the alluvial fans on Mars. He is a science team member for one of the cameras on board the Mars Odyssey spacecraft, in orbit around the Red Planet since 2001. The camera he uses is called THEMIS (the THermal EMission Imaging System) that photographs the planet in infrared light, recording the heat from the landscape as it warms and cools over the course of the Martian day and night.

Though the Sun has now set, I can still feel the rock radiating its daytime heat here in the Devils Garden Campground at Arches. From my campfire, I take out two baked potatoes, one very much larger than the other. While the big one stays hot for quite a while, the little one cools quickly as its smaller volume radiates heat through the greater proportion of surface area. Stand at a distance with an infrared camera and even if you couldn't see the difference in potatoes with your eyes you could easily determine which was which by how quickly each cooled.

That's exactly what Moersch and his colleagues hope to do with the potential alluvial fans on Mars. Because it's the size of the sand and rock at different parts

Figure 5.29 Three views of alluvial fans where water appears to have washed out of hills spreading rock and sediment in its path. At top is a HiRISE image showing what appears to be an alluvial fan pouring out of the high rugged wall of the aptly-named Mojave crater in the Xanthe Terra region of Mars. It looks remarkably similar to alluvial fans in the Mojave Desert of southeastern California (middle). Alluvial fans form as a result of heavy desert downpours, typically thundershowers. Because deserts are poorly vegetated, heavy and short-lived downpours create a great deal of erosion and mass movement of rock and debris onto nearby low lying regions. Channels in the Mojave Crater fan begin at the top of ridges, consistent with precipitation as the source of water, rather than groundwater which would form gullies beginning lower down. At bottom is the alluvial fan visited by Moersch and his students in Death Valley National Park (notice the road across the base) (NASA/JPL/Univ. of Arizona (top), T. Nordgren (middle and bottom)).

of the fan that reveal if flowing water truly created these features. Flash floods flow out of high hills as veritable rivers of moving earth, carrying everything from sand to boulders along in their wake. As the water reaches the valley floor it loses energy and slows. As it does so, it loses the ability to carry the largest boulders first, even as it still flows fast enough to carry the rest of its muddy content along. Eventually though, the slowing water drops ever smaller rocks, pebbles and stones, until finally, at the limit of the water's flow (a gentle wash no more than an inch or two in height) tiny sand grains gently come to rest. Measure the cooling pattern of possible Martian fans and you determine the size distribution of the surface material confirming the mechanism that formed them.

In March 2007 I joined Moersch and his postdoc, Chris Whisner, and graduate student, Craig Hardgrove, in a Mars expedition to Death Valley National Park in California. There we traipsed through the area around the Eureka sand dunes to film the ebbing heat of an actual alluvial fan on Earth and compare its infrared signature to those that were seen from Martian orbit.

At night the infrared camera whirred away hour by hour recording its movie of cooling sand (something only slightly more interesting than drying paint) while by day we huddled in the rented RV's air-conditioning to escape the scorching heat. At the end of the weekend, Moersch and his group packed up and flew back to Tennessee and I drove home to southern California. It will take time for his group to analyze their data and compare it to Mars in what will eventually become Hardgrove's PhD thesis. When they do, they will have gathered important 'ground-truth' for understanding what they observed from space.

These are the lessons we learn from Lowell.

Recalling that trip to the Eureka dunes, I cup my hands into the soil back here in Utah and let it fall through my fingers, feeling the different grain sizes. The sand grains I feel are made of quartz. Quartz crystals are the product of erosion wearing away at granite which originally formed as molten rock cooling deep inside the Earth under incredible pressure. On Mars there are only trace amounts of granite and thus hardly any quartz. Without much quartz the Martian sand is most likely made of pulverized volcanic basalt (the most common type of rock seen on Mars so far) and thus while there may be Martian sandstones, they will be unlike any sandstone I see around me. The chemical make-up of what's in my hand is a story of the water history of this planet, in this place.

To learn about the water history of Mars is the primary mission of the two rovers NASA sent to Mars in 2003. They were designed to explore two different places where orbital images said water may once have been. From the Martian surface, these rovers, Spirit and Opportunity, would examine the physical and chemical properties of the rocks to find that ground-truth Moersch was looking for in Death Valley. Only from the rocks themselves would we read, at long last, the history of Mars.

The two landing spots for the rovers, which were expected to last for 90 sols (a sol is one Martian day) and drive no more than maybe a kilometer, were chosen specifically for their likelihood to have once had water. Spirit would go to Gusev Crater, an enormous impact basin with a flat crater floor, 100 miles (160 km) across. From orbit, it has what appears to be a river canyon flowing into it, raising the possibility that Gusev may once have been a lake. Perhaps in its bottom would be the telltale signs of sedimentary rocks, sandstones and siltstones, or at the very least the chemical signs of rocks awash in water.

Opportunity was sent to the opposite side of the planet to explore a broad flat expanse called Meridiani Planum. From orbit, infrared observations detected the chemical signature of hematite, a dark grey iron mineral, that on Earth is usually formed in the presence of water (although sometimes through volcanism). Together, the two rovers were sent to places that scientists suspect should have once been awash in water – one on the basis of the shape of the landscape, the other on the basis of its chemical signature.

Figure 5.30 Ma'adim Vallis is the 375-mile-long (600-km) channel that appears to run upwards in this image draining into the 100-mile (160-km) wide-impact basin of Gusev Crater. If water ever did flow through this valley, the elevation of the surrounding landscape shows that it almost certainly should have pooled inside Gusev, perhaps forming a lake. The floor of Gusev Crater is the landing site for NASA's Spirit rover (NASA/JPL/USGS).

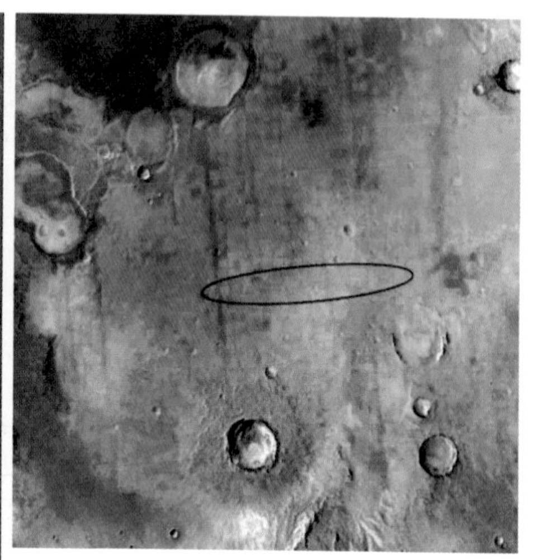

Figure 5.31 Colors show the presence of the mineral hematite on the flat Martian plains of Meridiani (shown in grey). The location and abundance of hematite was mapped by the thermal emission spectrometer on NASA's Mars Global Surveyor orbiter. Red and yellow indicates higher concentrations while green and blue areas denote lower levels. The black ellipse shows the area where NASA's Opportunity rover was sent (NASA/JPL/ASU).

The summer of 2003 was therefore an exciting time to be looking at Mars; NASA was sending these two spacecraft to Mars at precisely the time that Mars would be closer to the Earth than it had ever been before in recorded history. I flew out to Florida to see the launch of Opportunity, the second of the two rovers. Out there on the sands of Cocoa Beach outside Cape Canaveral, I joined the other members of the Rover team for late-night music and beer, as everyone gathered to blow off steam from preparing the rocket for flight.

My first night there, Squyres came up to me and gave me a big happy handshake. I had been a teaching assistant for him at Cornell, and had known him since he first submitted his rover proposal to NASA twelve years before.

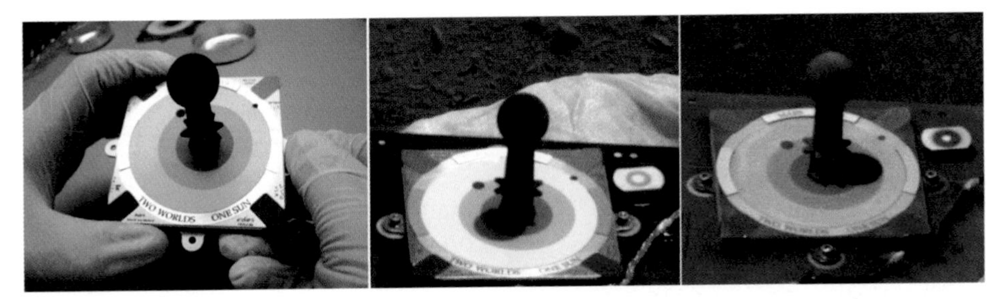

Figure 5.32 Three views of the camera calibration target ('Marsdial') sent to Mars onboard the Spirit and Opportunity rovers in 2003. At left is the dial at the Jet Propulsion Laboratory. Colored tabs and grey rings provide scientists with objects of known color when sitting on the surface of Mars (middle and right). The central post casts a shadow so that scientists can accurately adjust the color of their images to reflect the true color balance for objects in both direct sun and shadow. If you know the orientation of the rover, then the position of the shadow will tell you both the local time and season on Mars: hence, it's a sundial. On the surface of the sundial is the date we landed and the name of Mars in 16 different languages. A requirement of all sundials is a motto: ours is "Two Worlds, One Sun." The middle image shows the Spirit dial upon landing, while the right image shows the Opportunity dial after 5 miles (10 km) of travel (NASA/JPL-Caltech/Cornell).

From Florida's east coast, Mars was a brilliant red jewel in the sky rising over the Atlantic Ocean. Lifting his arm to point out over the sea at the planet now brighter than nearly any star in the sky, he said "That's where we're going." I could hear in his voice a career's worth of effort and planning to get to this point, and now it was all coming true for him. There before us was Mars-rise, reflected in water; and it was evidence of water we hoped to find when we got there.

Spirit landed first in January 2004. I remember standing on my doorstep watching Mars shine in a clear California sky. On NASA TV behind me, mission controllers at Caltech's Jet Propulsion Laboratory called out the entry and descent benchmarks: that Spirit had entered the atmosphere, that its chutes had opened and, finally, that it was on the surface and alive. The first image I remember seeing on the television that night was a tiny thumbnail sent back by Spirit of the sundial I helped work on. Whatever else I may do in life, a small part of me will now forever be on Mars.

In the weeks that followed, Spirit trundled about the hard-packed russet sand and scraped at the nearby rocks littering the surface. It analyzed their color and composition, photographed them in visible and infrared light, peered at them microscopically, and in the end determined that everything in its immediate area was volcanic basalt. There wasn't a sign of water anywhere to be found.

Sixty-five sols later, after methodically examining and photographing its surroundings, Spirit finally reached a nearby crater called Bonneville where the rover team hoped that the impact had exposed actual bedrock in the crater's floor. What they found was nothing but more sand dunes and basalt. For a rover only 15 sols away from the end of its warranty and nearly at the expected limit of miles driven, the only thing different in any direction were the distant hills on the

Figure 5.33 Spirit climbs into the Columbia Hills. In the distance the rim of Gusev Crater is barely visible. Our Marsdial casts a long shadow as Spirit climbs upward in the late afternoon hours before sunset (NASA/JPL/Cornell).

horizon, over a mile away (about two kilometers). Without any hope that they'd actually make it, mission planners sent Spirit driving for the horizon.

Meanwhile, on the other side of Mars from the rover that found itself on a plain inside a crater, Opportunity landed inside a crater on a plain. Opportunity's first images showed it to be inside a shallow crater only 65 ft (20 m) across and less than 6 feet (two meters) deep in the middle of a vast parking-lot of an empty plain. But there in the rim of the crater were the first bedrocks ever seen on Mars. Bedrock, just like the canyon walls of Utah, tells you the history of where you are. It tells you exactly what the conditions were under which it was formed and by looking at what sits on top of what, it tells you the order in which it happened and how long it lasted relative to what came before or after. Bedrocks are novels to be read, whereas the loose rocks that littered the surface of every other previous Martian landing site were merely random pages scattered about from unknown manuscripts.

Figure 5.34 This approximate true-color panorama, dubbed 'Lion King,' shows the Opportunity rover's landing platform set inside Eagle Crater and the surrounding plains of Meridiani Planum. Behind the lander is the outcrop of rock exposing the first bedrock seen anywhere on Mars (NASA/JPL/Cornell).

Figure 5.35 The exploration of Martian bedrock is shown in four panels. In the upper left is a wide angle view of the broken segments of the rock outcrop within Eagle Crater. The colored box around one rock in the outcrop, dubbed Stone Mountain, is enlarged in the upper right image. It shows the fine layers in the outcrop, plus the Martian ''blueberries'' littering the darker soil. The grey square in the center of Stone Mountain is enlarged at the bottom right. Clearly visible are the thin sedimentary-like layers and a Martian blueberry in the process of being weathered out of the rock. The image at lower left is a fish-eye view of Opportunity bringing its instruments in contact with the outcrop (NASA/JPL/Cornell /USGS).

From the very first images of the bedrock it was also clear that the outcrop was layered, finely layered. Two processes tend to lay down layers, volcanoes, and water. These are the same two processes that give rise to hematite on Earth and which process had been at work on the plains of Meridiani was exactly what Opportunity had come to determine.

Outside the crater, a flat nearly featureless racetrack of smooth sand stretched to the horizon. But the sand itself wasn't featureless. Everywhere the rover's cameras looked, inside and outside the crater, tiny grey spheres the size of small marbles littered the surface. The landscape was covered with millions of them, and soon the science team labeled them "blueberries" because of their color relative to the ever-present red of the Martian countryside. Their source appeared to be the layers in the rock wall, where the cameras could see signs of blueberries still imbedded. Over the millennia, the constant wind and sand appears to erode the layers, causing the relatively harder berries to pop out and roll across the surface.

Using a smaller version of the infrared camera that first detected hematite from orbit, the Opportunity team found that wherever there were blueberries there was the chemical signature of hematite, and where the blueberries were scarce, the hematite was not to be found. This relationship was confirmed when the rover team managed to get a spectrum of a small pile of berries, gathered in a depression, playfully called the "Berry Bowl." The tiny blue spheres were the hematite they'd been looking for.

Back on Earth, Marjorie Chan, a geologist at the University of Utah, saw pictures of the Martian blueberries and told herself, "I know what those are; there's been liquid water there." Chan and her collaborators had just written a scientific paper describing the origin of strange iron spheres, informally called Moqui Marbles, found littering the surfaces of a number of outcroppings of Navajo Sandstone around southern Utah. In these areas, the normally iron-red sandstone has been bleached nearly white of its iron. Where the sandstone is bleached, that's where you find the marbles, some still embedded in the rock layers. Because Navajo sandstone began as dry sand dunes, it is relatively porous

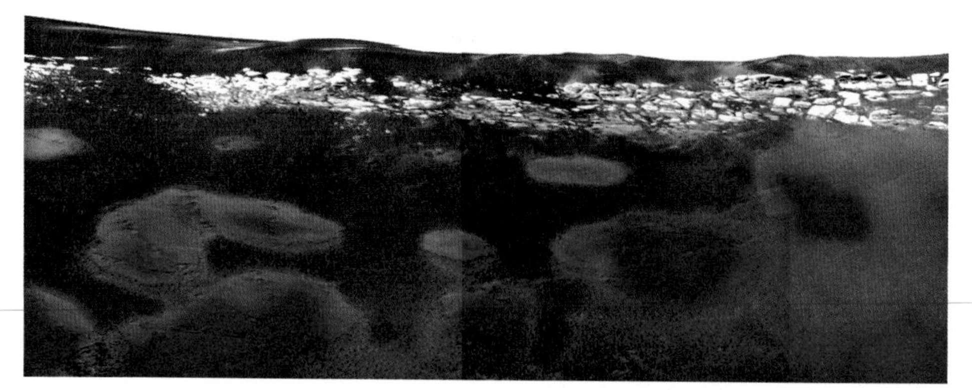

Figure 5.36 The area inside Eagle Crater showing the outcrop of Martian bedrock. In the distance are the plains of Meridiani Planum. The colors indicate location and abundance of the mineral hematite using Opportunity's onboard infrared imager. This is a smaller version of the instrument on board NASA's Mars Global Surveyor spacecraft that recorded the hematite abundance from orbit in Figure 5.31. Like that image, red indicates high hematite abundance, blue is low. The blue circles show where Opportunity's landing bags pressed the blueberries into the soil, hiding them from view. High hematite abundance is where the blueberries still cover the surface (NASA/JPL/ASU/Cornell).

compared to denser sedimentary layers that formed from wet sand on ancient seashores. Groundwater easily seeps through the gaps between the sand grains using the layers of the Navajo formation as a subterranean plumbing system in much of the area where the Navajo is found.

Chan explains the relationship between the sandstone and the marbles this way: the colors we see in red rock country occur because of iron-oxide that forms around individual grains of sand that make up the sandstone. As groundwater moves through the more porous Navajo sandstone, buried hydrocarbons (the oil and gas in the ground that makes this region so enticing for exploration companies) dissolve the iron and carry it along in the water. As the sandstone loses its iron, the sandstone bleaches white. When the iron-rich water encounters oxygen-rich groundwater of differing chemistry, the iron re-precipitates out of the solution. Slowly, the solid iron minerals bind together grains of sand, forming a well-cemented sphere called a concretion.

One such iron mineral that does this is hematite. With time, the hematite spheres grow until either the water movement stops, or the iron supply ceases. Eventually, when this area is raised by tectonic forces, the sandstone weathers away around the harder iron sphere. When it does, the marble drops out and rolls across the surface of Utah just like the blueberries on the planes of Meridiani. While the exact chemistry is a little different on Mars (no one expects to find oil reserves there) the result is the same.

Using Chan's paper as a guide, I went in search of these marbles for myself. Along the way, I crossed paths with John Wesley Powell and his expedition in the small, isolated town of Bullfrog, Utah. The town sits on the north shore of modern-day Lake Powell (a reservoir created in the 1960s when the Colorado River was dammed) but a hundred and forty years ago, when Powell and his crew came this way, it was the north rim of Glen Canyon. Today, a lonely ferry makes its way back and forth between the two sides of the lake in the middle of scenery still very much like what Powell saw when he climbed to the canyon's rim. From just north of Bullfrog, the Burr Trail, a dusty, desolate dirt road, climbs up and over the Waterpocket Fold within Capitol Reef National Park.

I'd been this way once before, and have still never encountered anyone else along this section of road in what is as close to the back of Beyond as my travels have ever taken me. All around is a rainbow of rock. When I get out of my truck to take a picture; I don't even bother pulling over, I just stop in the middle of the track. The hillsides are a pale yellow-white, and the ground beneath my boots is a soft, cake-like consistency made up of the eroded sands washed down into the valley where the road makes its way.

And then I see them. Partially buried and littering the surface all around me are what I thought were pebbles, but in reality are really dark grey iron spheres. Just like Opportunity that came looking for hematite only to find it in the first place it looked, the hematite marbles of which I came in search were in the very first place I stepped. They're no bigger than the tip of my fingers; some as smooth as marbles, others knobby like fossilized raspberries. Some are split open revealing sandy or hollow interiors, others are tiny barbells where two spheres

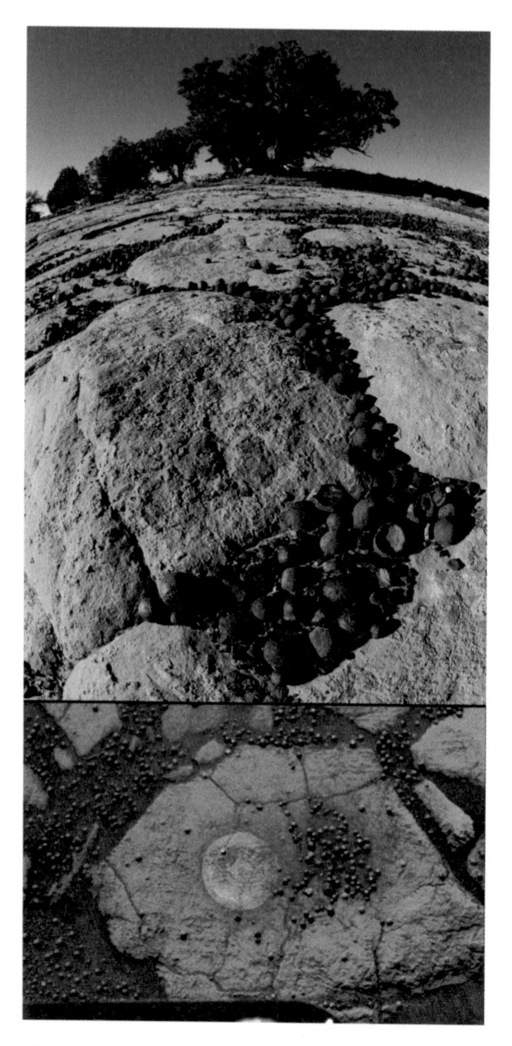

Figure 5.37 Hematite spheres litter the sandstone surface of Grand Staircase-Escalante National Monument (top) and sedimentary bedrock in Meridiani Planum, Mars (bottom). While the hematite spheres in this part of Utah can be a wide range of sizes with some up to an inch (~2 cm) in diameter, the Martian blueberries seen in the "Berry Bowl" are only a tenth of an inch (~2 mm) in diameter. The 1.8-inch (4.5-cm) circle imprint in the rock at bottom is from the Rock Abrasion Tool (RAT) onboard the Opportunity rover (part of which is visible at the bottom of the image) (T. Nordgren (top), NASA/JPL/Cornell (bottom)).

long ago grew together. They look like nothing else I've ever seen in Nature, and holding them in my hand I look around me and am thankful for the ingenuity of physics and chemistry.

Later that day, as I camp out in a broad valley within the Grand Staircase – Escalante National Monument, I am once more surrounded by marbles, this time, big ones and small ones rolling by the millions across a plain of weathered Navajo sandstone bleached as white as straw. Beside my campsite, are new marbles only now weathering out of the sandstone outcropping, just like those seen on Mars, on another plain, 50 million miles away. Today is my ground truth.

But as with Lowell a hundred years ago, even the best scientists can be fooled by what they want to find. This is a possibility that scientists struggle with whenever they discover something new or something long hoped for. While for Chan, her experiences on the ground in Utah meant that the Martian blueberries were evidence enough for her, for the members of the Mars Rover team, they were going to do everything in their power to question their own conclusions, until there were no more viable alternatives to what they were seeing. Over its first 26 sols on Mars, Opportunity would compile a virtually complete story of Meridiani as told in that single outcrop of rock a dozen steps away from where it landed. Numerous lines of independent experiment and imagery all flowed together into what emerged as a single unifying history of water.

First there is the fine rock layering not unlike anything you or I might find in the sandstone layers of the southwest. Within the layers are the hematite spheres, looking, in every respect, exactly as concretions should. In addition, the layering itself traces out tiny ripples in exactly the right shape and pattern as if the layers of which the rock were made were laid down by water waving and lapping across its surface.

The rock layers themselves are composed of salts, just as if shallow briny lakes or seas had washed across the surface and then dried on a warm Martian day. Different layers were composed of different types of salt and in different concentrations, just as one would expect in a pond where the more water that evaporated the more salt would be left behind in the liquid that remained. Look west to the Great Salt Lake and the process is at work on Earth today.

Chemical analysis by one of the rover instruments revealed that one of the salts making up the Martian outcrop is a mineral called jarosite. It's not a common mineral on Earth, but it's one that has actual water within its crystal structure. So not only does the presence of salt speak of water's once distant presence, but one of the types of salt present still contains traces of water itself.

Meanwhile, all over the surface of the bedrock layers are tiny indentations or tabular-shaped pits. To all appearances they look like places where crystals grew, pushed aside or replaced the material that was there, and then dissolved or weathered away. On Earth, gypsum salts form little crystals like this as minerals precipitate out of rock awash in water. In time, as conditions change, the crystals weather away and all that's left are their little empty molds. These features are called vugs.

Six months after Opportunity landed I spoke to Squyres at a meeting we were attending. I hadn't seen him since Opportunity's launch, and his team had just announced their conclusion that liquid water had once flowed across the surface of Mars. I asked him what that moment of discovery had been like and he told me a story he later recounted in his book *Roving Mars*, "Ever since we landed," he wrote, " I've been resisting the idea that there was water here at Meridiani. Maybe I just want to be very, very certain before declaring victory." But, after the layers, and blueberries, and ripples, and salts, and jarosite, and vugs, after all these separate lines of enquiry:

> It's the preponderance of the evidence.... It's been fascinating to watch the whole team as these clues have been revealed to us, sol by sol over the past few weeks, like some kind of weird Martian mystery novel. Each of us has come to this remarkable experience with our own background, with our own set of prejudices. And each of us has reacted to the mounting evidence for water here differently. Some people leapt joyfully off the cliff when we got our first hint of the outcrop's composition.... Others still aren't convinced yet.... But today did it for me. I simply can't see how you can make rock like this without a lot of water being involved.... There's no doubt in my mind now that some of these rocks were laid down, long ago, in liquid water.

Figure 5.38 My mental transformation of Moab to Mars is complete in this image of my imagination. After all we've come to learn about Mars and its history, perhaps there really is a landscape like this to be found there, different from, but still quite similar to what I see around me in the red rock country of Utah (T. Nordgren).

In the end, if Squyres and his team are right, Mars really once was warmer and wetter at some point in the past, with salt water lakes or seas rippling and retreating across the red Martian landscape just as it did where I stand here in Utah. I look around me at the landscape of the country I love so much and I wonder, as similar as these landscapes may be today, was there a time when the comparison was even closer? Was there a time when the Martian sky was blue? When the streams did run? When the grass was green and the Red Planet wasn't nearly so red? And even with all we now know, might we still be wrong about Mars?

For the great majority of human history Mars was probably viewed by shamans or holy men as a fearful god, or mystical force in the sky; Mars was Ares, the God of War. They were wrong. A thousand years ago, astrologers thought Mars was a perfect celestial light set among the perfect crystal spheres in the sky. Four hundred years ago, Galileo's telescope showed they were wrong.

Three hundred fifty years ago, the astronomer Christian Huygens discovered surface features on Mars and surmised there might be Martian astronomers looking back at us. Being an astronomer, it seems, was no proof against also being wrong. Two hundred years ago, William Herschel detected the first signs of an atmosphere on Mars and deduced it probably had a climate much like England's. He was very wrong. A hundred years ago Percival Lowell's Mars had canals and canal builders. As we know all too well, he too was wrong.

Fifty years ago, astronomers thought that Mars might yet be home to great swaths of simple lichens and grasses. Sadly, they were wrong. Forty years ago, planetary scientists thought Mars was a dead, dry, lifeless world that had always been so. It now looks like they too were wrong. Today we think Mars is a dry world that has not always been dry, a cold world that has not always been cold, and a lifeless world that may yet turn out to not always have been lifeless (or may not even be lifeless today).

Are we wrong? Only our children will know. The one thing science teaches us over the centuries is the one sure way to never be right is to never risk being wrong.

In 2011, NASA will launch a new generation of Mars rover, designed to operate, not just for many sols, but for many Martian years. It will carry a fleet of instruments and experiments to explore the chemical history of Mars' warmer, wetter past now that we know what to look for and where to look for it.

Figure 5.39 Some day we will make our own tracks across the Martian surface (T. Nordgren).

Like Percival Lowell, John Wesley Powell, and everyone else who has ever been drawn to the stark beauty of the American southwest, we are following the water where it leads us on this new planet, reading the history it reveals, and wondering what life may have been here before us. While today we climb mountains and cross distant horizons through robotic eyes, someday we will be there in person, our boots caked in the red dust of a whole new planet to discover.

As with Spirit and Opportunity before it, this new rover will carry its own sundial to help calibrate its color camera. Like all sundials, the one on this new rover, carries a motto. Its motto is our motto: To Mars. To explore.

See for yourself: Mars

Mars at opposition

Mars is at its brightest approximately every two years and two months as we approach opposition with the Red Planet. The following table identifies the dates, constellation, brightness, distance and size of Mars at each opposition for the next 25 years. Notice that some oppositions are better than others; when Mars is closer it is bigger and brighter. From the constellations listed here, use the monthly sky maps at the ends of other chapters to find Mars in the sky. At opposition, Mars will be one of the brightest, if not *the* brightest star in the sky. And it will be red.

Table 5.1 Oppositions of Mars (2003–2035)

Date	Constellation	Magnitude	Distance (millions of km)	Ang. Size (arc sec)
2003, Aug 28	Aquarius	-2.88	56	25
2005, Nov 7	Ares	-2.33	69	20
2007, Dec 24	Gemini	-1.64	88	16
2010, Jan 29	Cancer	-1.28	99	14
2012, Mar 3	Leo	-1.23	101	14
2014, Apr 8	Virgo	-1.48	92	15
2016, May 22	Scorpius	-2.06	75	19
2018, Jul 27	Capricornus	-2.78	58	24
2020, Oct 13	Pisces	-2.62	62	23
2022, Dec 8	Taurus	-1.87	81	17
2025, Jan 16	Gemini	-1.38	96	15
2027, Feb 19	Leo	-1.21	101	14
2029, Mar 25	Virgo	-1.34	97	14
2031, May 4	Libra	-1.80	83	17
2033, Jun 27	Sagittarius	-2.51	63	22
2035, Sep 15	Aquarius	-2.84	57	25

Martian surface

To the naked eye, Mars never appears as anything other than a red 'star.' Through a pair of binoculars with a magnification around 20×, the small red disk of the planet is just barely discernable. A small telescope about 6 inches (15 cm) in diameter, with a magnification of about 60×, will just begin to show surface markings within a month of opposition. The most noticeable feature is usually the white polar cap. Also likely to be seen are the bright circular Hellas impact basin (bright because its low-elevation bowl is usually filled with white clouds), and the dark Syrtis Major region. All of these features will be subtle amid the all encompassing bright orange of the planet's disk.

The best time to see detail on Mars is when Mars is high in the sky. When Mars is low on the horizon you will be looking through the greatest amount of the Earth's blurring atmosphere. Avoid setting up a telescope near black asphalt, or where you have to look directly over someone's house. These objects warm up during the day and radiate heat during the night causing the air above them to ripple and wave, ruining the seeing.

The following graphic is a simulated map of Mars showing the most common bright and dark features visible on the disk. On any given night, the portion of Mars that is visible will be a circle extending from pole to pole, centered on the equator and a particular line of longitude. A number of websites such as *Sky and Telescope*'s will tell you the central Martian line of longitude for any given date and time.

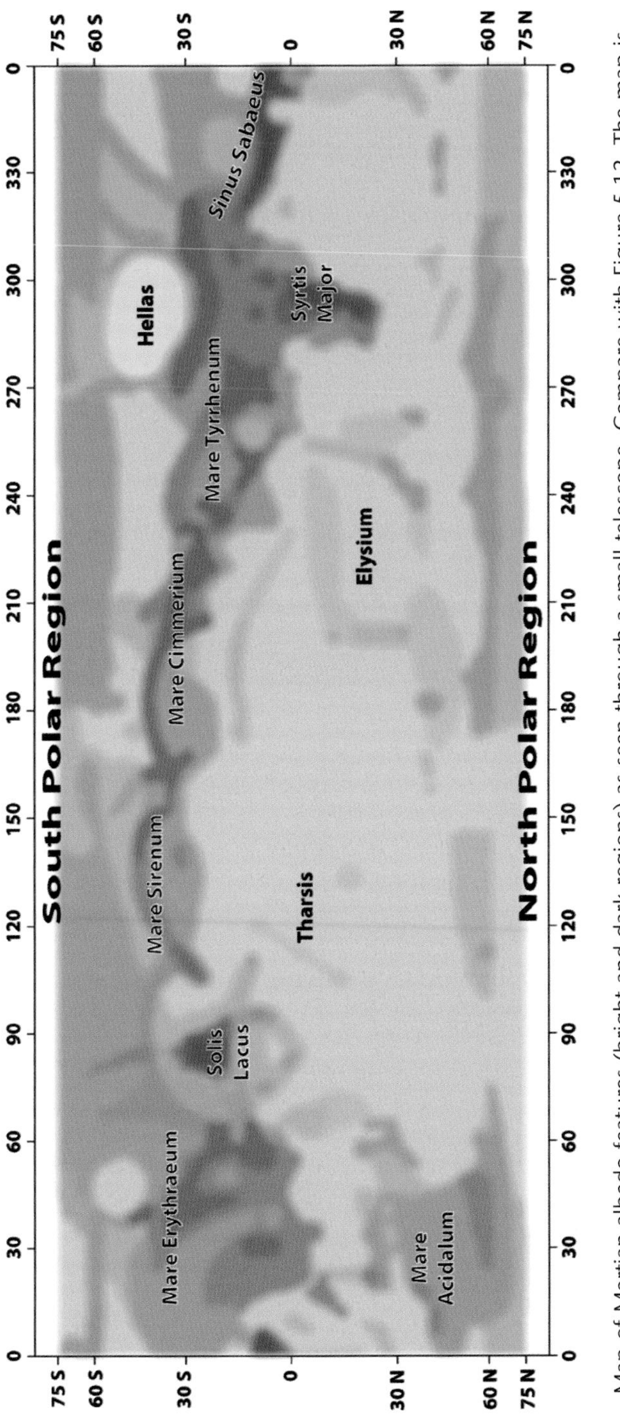

Map of Martian albedo features (bright and dark regions) as seen through a small telescope. Compare with Figure 5.12. The map is shown upside down (South at the top) and reversed (East to the left) to match the view an observer has looking through a refracting telescope like Lowell's. For observers looking through a reflecting telescope, East and West will be reversed (the view you see through the eyepiece will be a mirror image of the map). For observers using either of these telescopes with an eyepiece diagonal (a right-angle device on the end of the telescope that allows ease of viewing) the view will be flipped North and South (Roger Venable (Association of Lunar and Planetary Observers)).

Further reading

Roving Mars: Spirit, Opportunity, and the Exploration of the Red Planet by Steve Squyres (2006)
Hyperion, ISBN 1401308511

Postcards from Mars: The First Photographer on the Red Planet by Jim Bell (2006)
Dutton Adult, ISBN 0525949852

Mars As the Abode of Life by Percival Lowell (2008)
Kessinger Publishing LLC, ISBN 0548985251

The Exploration of the Colorado River and Its Canyons by John Wesley Powell and Wallace Stegner (2003)
Penguin Classics, ISBN 0142437522

Geology of Utah's Parks and Monuments ed. Paul B. Anderson, Thomas C. Chidsey Jr., and Douglas A. Sprinkel (2003)
Bryce Canyon Natural History Association and Utah Geological Association, ISBN 1882054105

NASA's Mars exploration homepage
http://marsprogram.jpl.nasa.gov/

Lowell Observatory's website
http://www.lowell.edu

The Association of Lunar and Planetary Observers
http://www.alpo-astronomy.org/

Sky and Telescope Mars interactive Mars observing resources
http://www.skyandtelescope.com/

6 Glaciers and Goldilocks: a tale of three planets

The wise man will be guided by the stars.

Benjamin Franklin

Glaciers speak to me of ancient forces. There is something primeval about them; they evoke images of cavemen and mastodons and a time before history. And time is crucial, because without time, great heaping swaths of time, there can be no glaciers and no glacial landscapes. To climb up a great glacial valley is to lose oneself beneath sheer mountain heights like battlements from an age of warrior giants that's faded into legend. On my path beneath these ancient ice-carved ramparts I am powerless with awe at what has gone before.

My trek to reach this spot began a long way down where the dusty path first crossed one of the primary signs of a glacier's past presence. The terminal moraine is a hill formed of rocky debris where the glacier's flowing river of ice dumped boulders, pebbles, stones and sand that it once carried on its surface. Acting like a frigid conveyor-belt, debris that accumulates on or within the glacier's ice is carried down the glacier's length until it's dumped where the river of ice ends. When glaciers advance, the rubble is flattened and ground down by the mass of the advancing ice; when glaciers retreat, the mound remains, marking its furthest advance. Wherever you see a moraine in glacier country, here you can be assured that a glacier lingered before retreating back up into the mountains from whence it came.

Beyond the terminal moraine, the shape of the valley itself bears witness to what was once here. The crushing weight of the glacier's motion flattened the valley floor just as my scuffing boot on the trail lies flat the dirt beneath my feet. From this floor, the bounding mountains curve rapidly upwards forming the tell-tale U-shaped valley profile that marks this as a former glacier's home. Along the solid rock walls, weathered gouges run parallel to the motion of the ice that once flowed here. Boulders the size of cars and houses plucked from the mountain heights further up were pushed along the cliff face, grinding back the high rock walls and leaving behind rough remnants of their work.

As I climb higher on the trail, the rock walls beside me reveal the layering of former seas and flood plains that were here long before the glaciers were. These are what the glacier cut through on its passage to the lowlands beyond and so we learn the history that preceded the ice when the climate was different. Tiny ripples and cross-bedded layers reveal small dunes, some only a few inches high: perhaps a shallow seashore once washed upon these stones?

A few miles more and I reach a natural mountain amphitheater: the glacial cirque. Here, in a rocky arena a million times larger than any constructed by humans, more snow fell in winter than ever melted in summer and so the icy mass accumulated with the years. Under daytime temperatures in spring, melting ice water seeped into cracks in the mountainside. Night came and the temperature plummeted. Water turned back to ice and, unlike nearly every other element, ice expands when frozen. Shots must have rung out like a pistol as rock cracked away from where it formed and the glacial cirque widened with time into the familiar shape of a bowl. Cirques are concert halls for the ringing songs of giants.

Over long decades, the combined weight of each new snowfall flattened the snows from previous years into downward sloping sheets. High pressures decrease the melting point of water (the reason skaters glide on a film of water created by the pressure of a skate's knife edge). Water films that formed deep down between the ice sheets acted as lubricants. Eventually the mass began to move; the ice began to flow. A glacier was born.

Before the awesome weight of a billion tons of ice, every irregularity in the valley below was subjected to the pressure of this inexorable force until eventually the landscape itself was ground down before it. This story was repeated in every other valley up and down this lonely range of mountains. Where two glaciers flowed side by side, nothing but a narrow fin of rock, called an arête, remained between them.[1]

While everything I see is the product of water, snow, ice and time, the only thing left here now is time. Under a red sky the rock continues to crumble, but the glaciers are gone. They disappeared three billion years ago. Snow no longer falls, and water no longer flows. The atmospheric pressure on Mars is now so low, that only the dry, silent rock is testament to the forces that shaped it.

I, of course, have never hiked a Martian glacial valley, and so have never seen these sights with my own eyes. Mars today is a cold, dry, barren, wind-swept desert, which would make even the most inhospitable wasteland on Earth look like the Garden of Eden. But over the last forty years, spacecraft orbiting the Red Planet photographed what look to be meandering river valleys, scoured flood plains, branching river deltas, and glacial U-shaped valleys. Rovers on the surface have found evidence of sediments laid down by rippling waters flowing across now-barren plains. Where Mars must once have been warm and wet, today it is not. Mars' climate has changed.

On Earth, climate change is all around us. My travelogue through a Martian Glacier Park is actually (with the exception of red skies and a size way too big) exactly what I and everyone else sees who make the 11-mile (17-km) round trip hike to Grinnell Glacier within Glacier National Park in northwestern Montana. It's a trip made by thousands of visitors every year, including myself, making

[1] When three or more glaciers erode a mountainside from different directions, a single towering, straight sided horn results. The Matterhorn in Switzerland is perhaps the most famous example on Earth.

Figure 6.1 Ophir Chasma is a tributary canyon of Mars' great Valles Marineris (a massive canyon as long as North America is wide). Down these dark-floored valleys, some planetary scientists see evidence of long vanished glaciers. Cirques, moraines, and U-shaped valleys have all been identified within the 3,125 ft (950 meter) deep chasm here imaged by the European Space Agency's (ESA) Mars Express High Resolution Stereo Camera (ESA/DLR/FU Berlin (G. Neukum)).

Figure 6.2 There is no question that enormous glaciers once carved these valleys within Glacier National Park in northwestern Montana. Here in the Many Glacier valley of the park, the Continental Divide runs along the tall thin arête that marks the jagged horizon bounded by the U-shaped valley walls overlooking the shores of Swiftcurrent Lake in the foreground (T. Nordgren).

Figure 6.3 A solar panel is visible 'out the window' as cameras onboard ESA's Rosetta spacecraft look down on the red Martian landscape. Passing by below is the Mawrth Vallis region of Mars, where the Mars Express spacecraft has detected evidence of hydrated clay minerals, signs that water may once have flowed across the now dusty surface (CIVA/Philae/ESA Rosetta).

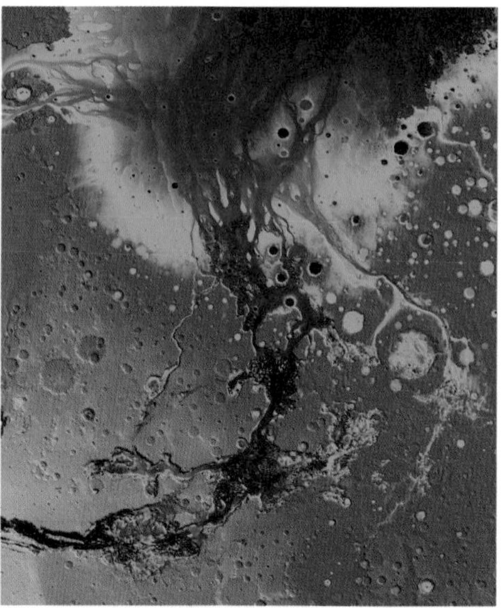

Figure 6.4 Topographic map of outflow channels on Mars. Beige areas show heavily cratered terrain through which lower elevation green, cyan, and blue channels cut. Their paths and shapes look exactly like those carved by rivers and flash floods on Earth (USGS/NASA).

Grinnell one of the most visited glaciers in North America. And each time I make the trek to its cirque, those glacial sights I see along its path never fail to inspire me.

All those arêtes, cirques, and horns that together form this narrow jagged range of peaks in the Rocky Mountains are why the local Blackfeet Indians call this the "Backbone of the World." But the glaciers that carved this rugged alpine silhouette were over half a mile (a full kilometer) thick, far bigger than any glacier that's existed in Glacier National Park for over 7,000 years and larger than any that exist in the continental United States today. The climate has changed.

Even the composition of the rock through which the glaciers once plowed reveals changes in Earth's ancient climate. High in the vertical cliff walls of the Grinnell glacial valley you can easily see alternating bands of red and green stone rising up the mountainside. The red-colored mudstone of the Grinnell Formation gets its color because of hematite, a red mineral that forms when iron in the sediments comes in contact with oxygen. These bands were once the muddy shore-lines of a shallow inland sea. Oxygen in the water and air reacted with the iron in the mud to form the brick-red deposits that time and pressure would later compress to stone. All along the Grinnell Glacier Trail red stone slabs reveal fossilized water ripples and cracks baked hard in mud under a hot noonday Sun a billion years long ago.

Figure 6.5 The glacial carved mountains of the northern Rocky Mountains are called the "Backbone of the World" by the Blackfeet Indians who still make this region their home (T. Nordgren).

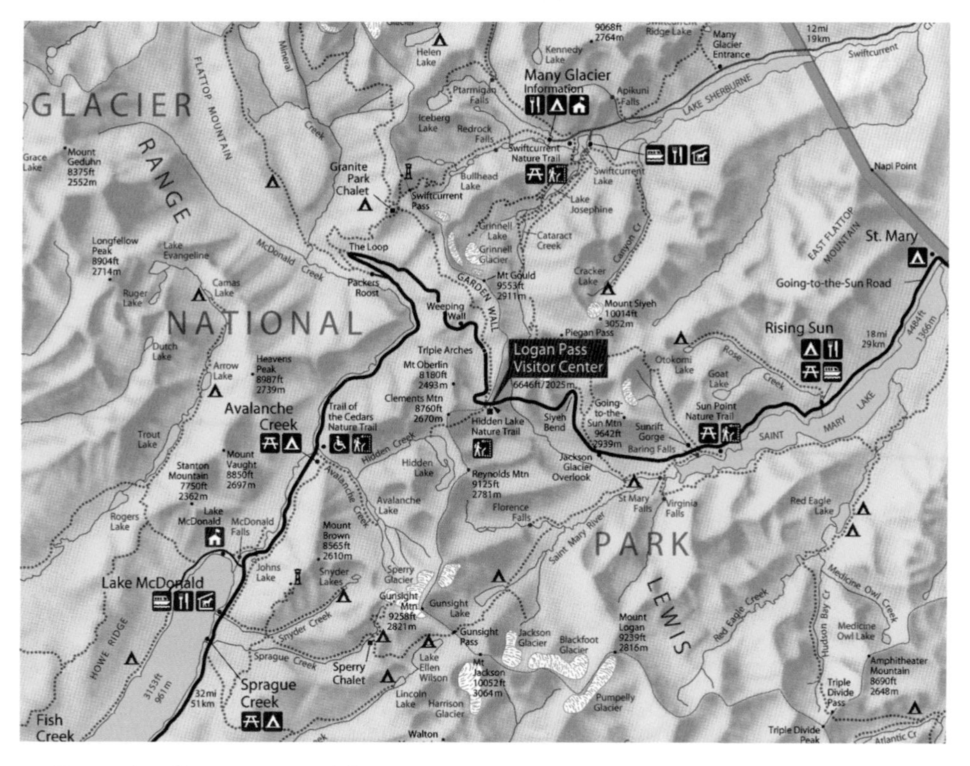

Figure 6.6 Close-up view of the map visitors receive upon entering Glacier National Park. The trail leading to Grinnell Glacier is shown by purple dots leaving from near the Many Glacier campground. Take a series of boats across Swiftcurrent Lake and Lake Josephine, and the hike to the glacier is much shorter. All of the images of Glacier National Park in this chapter were taken from the Grinnell Glacier Trail, Granite Park Chalet, or from overlooks along the Going-to-the-Sun Road (NPS).

Figure 6.7 Mud cracks (left) and ripples (right) are captured forever in the now hard stone that used to be a shallow, muddy seashore. Today these rocks are just one layer in a mountainside resting 5,800 feet (1,770 m) above sea level (T. Nordgren).

Figure 6.8 Every layer in these rock walls along the trail to Grinnell Glacier is a record of the conditions here a billion years ago. The sun, weather, wind and water that shaped this place, over each day that has ever dawned here, are visible from the trail that cuts through their record (T. Nordgren).

A hundred feet (30 meters) lower down the cliff wall, layers of green sedimentary stone show that a few million years before these shallow tidal mudflats existed, this exact same region was at the bottom of deep seas where there was little oxygen to react with iron in the thick muddy depths. Alternating layers of red and green attest to changing conditions that have sent ancient seas back and forth across this land, laying down and burying mudflats and seafloors and, with the passage of time, compressing it all into cold hard stone.

What's more, the conditions that made these rocks are not the same as the climate that created the glaciers that revealed them. And those conditions are not the same as the one that produced the glaciers we now see within the park, which in turn, is not the same as the climate that is inexorably causing their extinction.

It's a common misconception that the glaciers we see today in Glacier National Park are the remnants of the same ones that carved the dramatic landscape we see around us. During the last million years, the Earth has been undergoing a regular series of ice-ages. During the most recent ice age that ended only ten to twelve thousand years ago, great rivers of ice, thousands of feet thick, began to pile up and flow out of these northern Rocky Mountain valleys. Look at any topographic map of the park, especially the large three-dimensional models on display in Lake McDonald Lodge or the Apgar and St. Mary visitor centers, and you will see rounded cirques scattered throughout the mountains of the park as if some giant ice-cream scoop had been taken repeatedly to the landscape. The ancient glacier that carved the valley where Grinnell Glacier now huddles against the sun flowed 30 miles (50 kilometers) out onto the eastern Montana plains and left its claw and scrape on the surrounding mountains two thousand feet (600 m) above the valley floor.

After the hike to Grinnell Glacier, turn around and you see a chain of lakes, Grinnell, Josephine, Swiftcurrent, Sherburne, and distant Duck Lake on the

Figure 6.9 From the Grinnell Glacier cirque, looking east, a chain of glacier-fed lakes is visible in the U-shaped valley carved during the last great ice age over 10,000 years ago. Their extent marks the position of the giant ice sheets that once flowed out of these mountains carving the vertical cliff walls around me. When combined with a boat trip, the trail up to this spot begins at the end of Lake Josephine, the middle lake in this view (T. Nordgren).

horizon, each sitting in a successively lower section of the glacier-carved valley. Lateral moraines, where the glacier's force pushed aside debris and rubble, form the long low mountains hemming in Lake Sherburne. Each major lake and valley within the park to east and west of the Continental Divide was formed in just such a way.

Yet as massive as these rivers of ice were, maybe thirty to forty times thicker than today's Jackson Glacier (visible from a roadside pullout along the Going-to-the-Sun Road), they were utterly dwarfed in size by the ice sheets moving south from the Arctic. To the east, the Laurentide Ice Sheet towered 10,000 feet (3050 m) high and ran as far south as Colorado, Missouri, and Ohio. It carved out the many long narrow valleys of the Finger Lakes in western New York and gouged out the much more massive inland freshwater seas that, in time, would become the Great Lakes. To the west the Cordilleran Ice Sheet swept south into Washington and Oregon. When it melted, flood waters ran at a rate ten times greater than all the rivers of the world combined.

On Mars, the Argyre Basin is an ancient crater in the cold, thin air of the southern highlands. Six hundred miles (1000 km) across, its heavily eroded crater walls have long since broken up into a series of ragged mountains: the Charitum Montes. While no spacecraft has ever visited these 20,000-foot high (6,100 m) peaks, for the last forty years NASA has been photographing the landscape from orbit with a progression of ever higher resolution cameras. Today, modern explorers need not tramp through dark forests and cold alpine rivers to discover new canyons and mountain-peaks; instead they must wade through tens of thousands of images sitting on hard-drives in cluttered university offices around the world.

Figure 6.10 Today, the Bering Glacier is the largest in North America. It is 120 miles (190 km) long and is visible in southeast Alaska from airline flights between Seattle and Anchorage (on the right side of the plane as you head north). The Bering Glacier is a piedmont glacier, a glacier that begins in a valley but then flows out onto a plain. As such, it is an excellent model of the glaciers that once formed in the Many Glacier, St. Mary, and Two Medicine valleys of Glacier National Park before flowing out onto the plains of eastern Montana (T. Nordgren).

Jeff Kargel, is a planetary scientist and Professor of Hydrology and Water Resources at the University of Arizona. In 1989, he was a graduate student working on his PhD at the University of Arizona's Lunar and Planetary Lab where he was studying icy satellites in the outer Solar System. One day, while taking a break from his thesis, he casually pored over Viking Orbiter photographs of Mars. When his wanderings finally brought him to Argyre he noticed features unlike any he was expecting to find there. At the base of the Charitum Montes, on the smooth

Figure 6.11 The mountains of southern Argyre on Mars show signs of ancient glaciation. This Viking Orbiter data show meandering ripples on the plains to the upper left, thought to be glacial eskers. Notice from the mountain shadows, that the Sun is shining from the upper left, meaning that the ripples are raised above the surrounding plain. Sharp mountain ridges are thought to be glacial arêtes, with numerous amphitheater-like cirques present in their shadows. On the left, a raised rocky mass (perhaps a rock covered glacier) is visible flowing out of a valley and surrounding the mountain base (USGS/NASA).

plains of Argyre Planitia, Kargel saw a network of long sinuous ridges that looked just like eskers he'd seen left behind by the Laurentide ice sheet in Ohio.

An esker forms when melting water, dripping inside a glacier, forms a stream or river between the buried ground and the glacier above. As the river grows larger it melts the ice around it until eventually an aquatic subway tube snakes along the ground under an arching, icy roof. Dirt and rock, suspended within the melting ice, settles to the floor of the sub-glacial river, confined in place by the walls of ice on either side. Eventually the glacier disappears completely and when it does, what remains is the raised rock bed, confined for all those years by the now vanished ice tube through which the stream flowed. These are eskers, and these are exactly what appeared to be meandering across the valley floor heading out of the ragged mountains on the southern edge of Argyre.

Once Kargel made this connection and knew what to look for, circular valleys in the Charitum Montes became glacial cirques, thin mountain ridges became arêtes, and tall lonely pyramids were Martian Matterhorns. Far out on the Argyre Planitia were ridges that now looked like glacial moraines, while the valleys they flowed from had U-shaped cross sections as perfect as Grinnell or St. Mary Valley back on Earth. Soon a number of planetary scientists looking at other mountains and canyons saw glacial signs all over Mars. While individually, each of these features could be explained by other phenomena, Kargel's glacial hypothesis had the benefit of being one single explanation that described them all. That is a powerful trait in a scientific theory.

Today the atmospheric temperature and pressure on Mars are too low to support glaciers. Water ice sublimates (turns from a solid directly into a gas) and the only ice sheets still moving across the Martian surface may be rock glaciers: glaciers covered by layers of rubble, sand and boulders that protect the ice from sublimating away in the thin Martian atmosphere. Orbital spacecraft would eventually reveal many examples of what appeared to be dirt-covered ice-sheets.

Figure 6.12 NASA's Mars Reconnaissance Orbiter captures this image of what may be a rock glacier flowing out of an alcove out onto a valley floor, similar to piedmont glaciers (like Alaska's Bering Glacier) on Earth. Radar observations from MRO indicate that this region of Mars has a large reservoir of subsurface ice, supporting the possibility that this is an active Martian glacier (NASA/JPL/University of Arizona).

In 2005 the European Space Agency's Mars Express orbiter photographed, on the plains of Elysium, what looks like a dust-covered sea of broken ice-rafts just like those in polar seas on Earth. But are the coverings of dust and rock enough to protect the ice of this potential ocean (and the other glacier-like features). Or, had all the ice sublimated away a long time ago in the thin Martian air (if, in fact, it was ever there to begin with)?

Proof that there was still water ice just beneath the Martian surface came three years later, when NASA's Phoenix Mars Lander photographed bright white ice just beneath the surface dust of its landing site near the Martian North Pole. Ice like this may therefore be locked away in cold storage in many areas of present day Mars. But what altered Mars' atmosphere? What caused the climate to change so dramatically that once flowing rivers are now dry with sand dunes, and glaciers that once carved glacial valleys that dwarf Earth's own are now gone or trapped forever beneath the rusty red surface?

Whatever Mars may once have been like, however warm and wet, however Earthlike, whatever life may once have evolved there – or been just on the verge of doing so – all of it was most likely doomed from the very beginning. In the end, Mars most likely died because of its size; Mars is small.

The story begins with Mars' interior. Kargel and colleagues propose that while Mars has always had water, the cold temperatures there have always kept much of this water frozen. Ice has therefore been just one more building block of the

Figure 6.13 Mars' Elysium Planitia is a low elevation plane that may once have been an ocean when conditions on the Red Planet were different. This image, taken by the High Resolution Stereo Camera (HRSC) on board ESA's Mars Express spacecraft, shows what looks like dust covering an ancient sea of floating pack ice, similar to what we find in the Arctic Ocean on Earth (ESA/DLR/FU Berlin (G. Neukum)).

Figure 6.14 ''Holy Cow'' exclaimed mission scientists upon getting a look beneath their spacecraft. When NASA's Phoenix Mars Lander set down near the Martian North Pole its descent rockets (visible above) cleared away just enough dust and rock to expose the frozen water ice that may have once flowed across the surface of the planet (Marco Di Lorenzo, Kenneth Kremer, Phoenix Mission, NASA, JPL, University of Arizona, Max Planck Inst., Spaceflight).

Martian crust: a global permafrost like that found in northern countries here on Earth. But, with this Martian permafrost perpetually near the melting point, all it would take to be unleashed as water would be relatively small changes in the atmospheric temperature and pressure. Atmospheres are a reflection of planetary interiors. When planets are warm inside, the rocks, metals, liquids and gasses of which they're made slowly differentiate. Heavier metals ooze downwards towards the core, while lighter elements like rocks and ice rise upwards to the surface. Gasses, the lightest of all, escape to the surface through the explosive force of volcanism. These planetary burps are the origin of atmospheres. If the planet is large enough, its gravity holds these gasses and the atmosphere builds and thickens; too small, like the Moon or Mercury, and the gasses drift away to space.

The primary gasses that volcanoes release are carbon dioxide and water. Mars possesses the largest volcanoes in the Solar System, and while they appear to be extinct now, at some point in the distant past their eruptions must have been awe inspiring. Every mission to Mars' surface has found volcanic rocks littering the surface, and until NASA's Opportunity rover in 2004 found salts and sedimentary rocks, nothing else had ever been seen from the surface.

Olympus Mons on Mars is a single shield volcano 16 miles (25 km) high, three

Figure 6.15 Alaska's Mt. Redoubt erupts on March 30, 2009. Volcanic gasses, including water vapor and ash rocket 15,000 feet (4.5 km) into the air (Heather Bleick/Alaska Volcano Observatory/USGS).

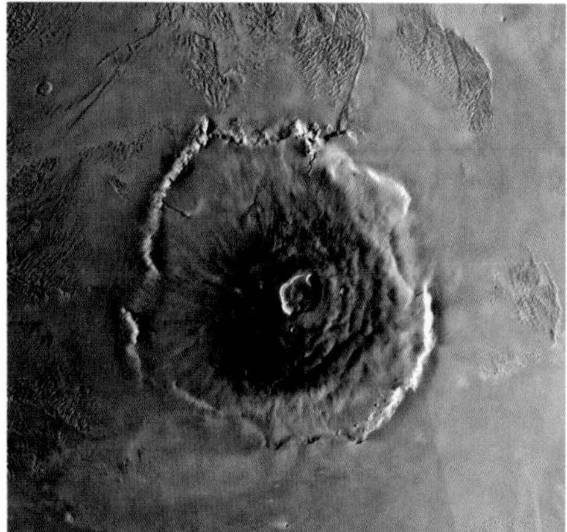

Figure 6.16 Viking Orbiter image mosaic of Olympus Mons on Mars. Rising 16 miles (25 km) above the surrounding plains, Olympus Mons is the tallest volcano in the Solar System, and is the size of the state of Arizona (or the entire Hawaiian island chain) in area. For comparison, Mt. Redoubt, a typical volcano on Earth, is only a little over 1.7 miles (2.7 km) tall. The darker area surrounding the central caldera in the image shows where the summit of Olympus Mons rises up beyond much of Mars' thin atmosphere (Jody Swann/Tammy Becker/Alfred McEwen/NASA/USGS).

times higher than Mt. Everest on Earth, and as wide as the state of Arizona. By comparison, the largest shield volcano on Earth is Mauna Loa in Hawai'i and it rises only 6 miles (9 km) above the submerged Pacific Ocean floor. As the largest volcano to have ever erupted in our Solar System, the subterranean heat Olympus Mons must have generated would easily have melted the Martian permafrost. Gargantuan floodwaters must surely have raced across the surface in tidal waves that scoured out flood channels still visible from orbit and left behind enormous regions of collapsed chaotic terrain. The carbon dioxide and water vapor released by these volcanoes would have entered the atmosphere and raised the pressure and global surface temperature with their thermal blanket of greenhouse protection.

To understand how the greenhouse effect does this, consider that whether light can pass easily through a gas depends on both the type of light and the type of gas. White light is made up of all the colors of the rainbow. However, there are more types of light than what we can see with our eyes. Beyond the red end of the spectrum there is infrared light. In the same way that very hot objects like the Sun or an electrified wire in a light bulb give off visible light, cooler objects like you and me, or even a black pavement on a summer day, give off light (or energy) in the infrared. This is why, for the vast majority of warm things we encounter in nature, we associate infrared energy with heat.

Certain gasses like the oxygen in our atmosphere let all the colors of sunlight's rainbow pass easily through. Other gasses, like carbon dioxide, are opaque to infrared. Visible light from the Sun passes downward through the carbon dioxide in our atmosphere. When the energy that sunlight brings in heats up the Earth, the infrared energy it emits cannot pass back out unobstructed.

While carbon dioxide makes up only 0.037% of the Earth's atmosphere, this plus the water vapor in our atmosphere (another greenhouse gas) is just enough to keep average surface temperatures above the freezing point of water; without this greenhouse effect, the world's oceans would freeze. When we stand on a mountain top we experience a little of what our planet's true global surface temperature would be like without the benefits of greenhouse gasses. With them we are a Goldilocks Planet: neither too hot, nor too cold for the life as we know it to have evolved and spread across its surface.

Figure 6.17 Granite Park Chalet and the mountain peaks of Glacier National Park are located at elevations where average temperatures are closer to what the Earth would normally have without the warming effects of greenhouse gasses. The Milky Way rises up from snow-flecked Mt. Cannon to arch over one of the last remaining back-country hiker's chalets within the park (T. Nordgren).

So while today Mars's climate has more in common with high altitude deserts on Earth – like the cold, dry, South American Atacama Plateau at an average elevation of 13,000 ft (4,000 m) – back when Olympus Mons was active, the conditions were radically different. Liquid water, freed from the icy confines of permafrost, could have run freely across the Martian surface. Over time, according to Kargel and others, the waters would gather by river and stream to form a Borealis Sea in the great northern lowlands. The now frozen Sea of Elysium may have been one component of this aquatic hemisphere. There the waters would be joined by rainfall from an atmosphere newly supplied with abundant water vapor and heat.

To the south, in the high-elevation southern hemisphere, snow would fall, blanketing the mountains and craters in the cold, thin, air. In time, the snows that gathered on the alpine plains and along shady canyon walls formed the glaciers whose work we see in Argyre and the tributary canyons of Valles Marineris.

Eventually, though, Mars' inside cooled. Rocky worlds like the Earth are kept warm inside through nuclear power. Radioactive elements, like uranium, are a natural component of the rocks out of which the planets formed. As these

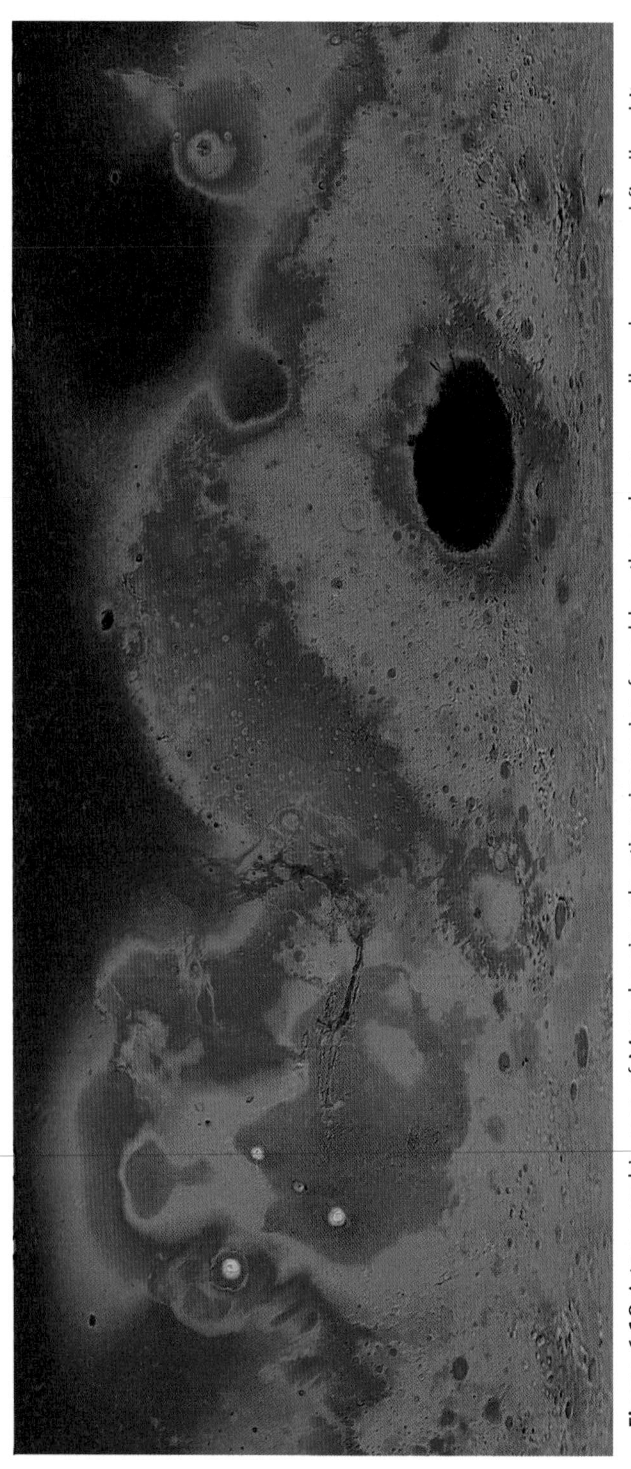

Figure 6.18 A topographic map of Mars showing elevations increasing from blue, through green, yellow, brown, and finally white. The entire northern hemisphere (top) is located at a lower elevation than the rest of the planet and many locations have been observed that may mark the shoreline of an ancient polar ocean. The ‘frozen sea’ of Elysium Planitia is the light blue channel on the east edge of the map between the southern highlands and the isolated ‘island’ of ancient volcanic peaks. The Tharsis bulge is the brown mass at left that is home to the largest volcanoes in the Solar System (Olympus Mons is the one to upper left). From out of Tharsis stretches the great deep blue gash of Vallis Marineris apparently emptying into the ancient Borealis Sea. Argyre is the light blue and green basin at bottom center, while the dark blur Hellas Basin (left of center) is the lowest point on the entire planet (USGS/NASA).

elements decay they give off heat keeping their planetary interiors warm. Small planets are doubly cursed. Not only do they have less mass (and thus less radioactive elements) to generate heat, what heat they do generate radiates away to space quicker. As Mars cooled, volcanism sputtered and slowed. Without new gasses replenishing its atmosphere, carbon dioxide carried out of the air by rain reacted with the rocks and sea water, forming carbonate sediments like limestone. The pressure dropped and the greenhouse effect faltered; temperatures plummeted and the sea froze over.

Perhaps this warming cycle happened many times, as volcanoes periodically belched out their remaining greenhouse gasses. In time, though, the cooling effects of the planet's small size won out. The volcanoes died, the surface cooled. The frozen northern ocean and southern glaciers gradually evaporated into nothingness. If you've ever seen 'dry-ice,' the solid ice of carbon dioxide, steam and smoke as it sublimates, then you have seen the fate of the Martian glaciers. Eventually the only water that remained on Mars was frozen solid, locked beneath the protective surface.

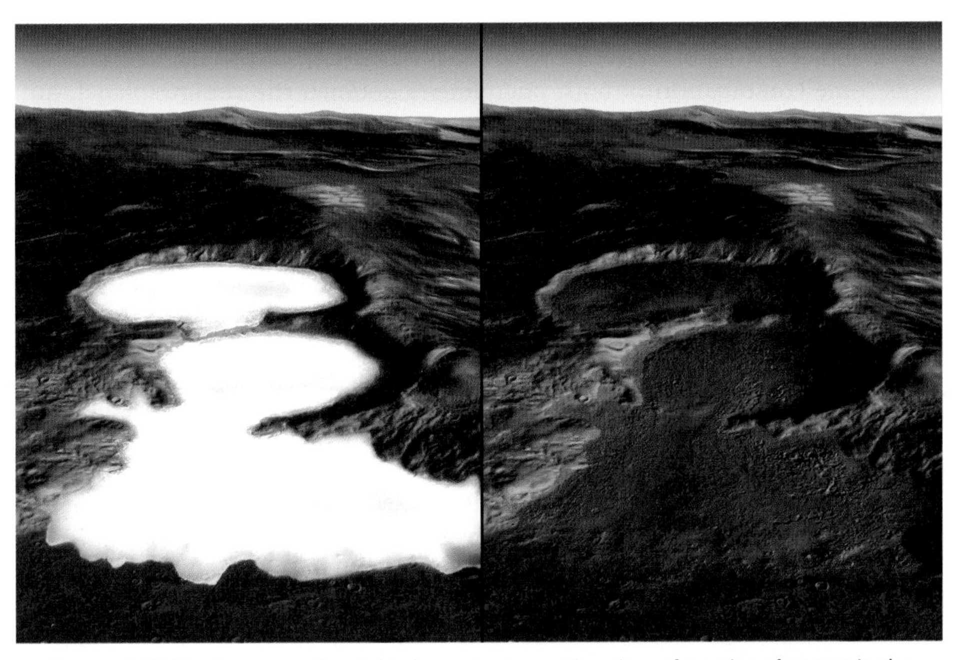

Figure 6.19 The image on the right shows a perspective view of a series of craters in the Hellas region of Mars as they appear today. Radar measurements from the Mars Reconnaissance Orbiter spacecraft show that buried water ice is 820 ft (250 m) thick in the upper crater and grows to as thick as 1,500 ft (450 m) at the bottom. Today this ice is protected from sublimation in the cold, thin Martian atmosphere by overlying rock and dust, but at some point in the past, the conditions were almost certainly different enough that the ice glaciers freely flowed across the surface (shown in an artist's conception on the left) (NASA/JPL-Caltech/UTA/UA/MSSS/ESA /DLR/JPL Solar System Visualization Project).

But this wasn't the only way Mars' small size doomed it from the beginning. Deep inside the Earth, our molten iron core's constant motion produces our planet's magnetic field. Mars too would once have had such a magnetic field fueled by its own molten iron center. But, once again, Mars cooled faster than we and as it did so its molten metallic core slowed and began to solidify. Mars' magnetic field weakened and failed as the core gradually solidified and stopped. Without its protective magnetic umbrella, Mars fell victim to the harsh weather of the solar winds. Protons and electrons erupting off the Sun stream outward past the planets at speeds of up to 250 miles per second (400 km/s) or more. Atom by atom, Mars' atmosphere was stripped away as these particles tore into its atmosphere, no longer deflected by the planet's former magnetic field.

The thick carbon dioxide and water vapor atmosphere that protected Mars' surface and kept the days and nights above the freezing point of water was ripped from the planet and with it the temperature and pressures at the surface plunged. Once more the Borealis Sea froze. The rivers dried. The Martian glaciers evaporated as no new snow accumulated and the atmosphere leaked away to a point where the rivers of ice slowly disappeared. Measured in the long history of planets, what was one moment solid as a rock, the next moment became as ephemeral as gas.

As some combination of both phenomena doomed the cooling planet, any Martian glaciologists three billion years ago would surely have taken careful notes of the great plumes of steam rising off the icy fields. They would have measured the positions of new moraines deposited as the glaciers' advance halted and the ice disappeared around rocks newly revealed imbedded in the glacial walls.

Once enough of Mars' upper atmosphere was lost to space, ultraviolet rays from the Sun could penetrate down into the thinning atmosphere and break apart what little water vapor was still present in the air. The light hydrogren atoms escaped to space. The newly freed oxygen chemically reacted with the dry rocks now littering the planet's surface. Iron in the surface became iron-oxide. Mars rusted. Three billion years ago, Mars' death turned the Red Planet red.

On the timescale of those hypothetical Martian glaciologists, the change almost certainly wasn't overnight and for a long time the evidence must surely have been doubted and debated. As the solar activity peaked and paused over an 11-year cycle there must have been times where it looked as if the changes were stopping, or perhaps even a wet warm summer and winter here and there gave hope that the trends had stopped and reversed. Perhaps a particularly cold year, followed by a few less cold would lead some to claim that the planet was now, thankfully, once again getting warmer. I imagine those mythical Martians, and wonder what they would have thought when the news finally hit home that their planet was dying. How long would it have taken them? What would they have done? Who would they have blamed?

Mars' glaciers dried up and evaporated. Ours are warming and melting. Though the mechanisms are different, Mars illustrates the power of climates to

Figure 6.20 Venus sets in the evening sky above Jackson Glacier within Glacier National Park (T. Nordgren).

change and to do so permanently. What then of the other planets? Is there any light they may shed on what is happening on our own planet today?

On a clear evening low to the west, a star brighter than any other is often visible long before the other stars have emerged from the light of sunset. And if not, wake before dawn and it will be there, low to the east, shining brighter than all other heavenly bodies besides the Moon. From our vantage point farther out in the Solar System, the planet Venus never wanders very far in the sky from the Sun that it orbits.[2] Approximately every 584 days, the second planet in our Solar System passes nearly between the Earth and Sun and changes from an evening star to a morning one.

The Romans associated Venus with the goddess of beauty, and its symbol is a stylized mirror, the symbol for Female. The reason Venus is so beautifully bright in a languid summer sky is a combination of several factors. First, it passes closer to us than any other heavenly body except the Moon. Second, it is twice as big as Mars, which is the next closest planet to us, and third, the surface is shrouded completely in clouds, making it one of the most highly reflective bodies in the Solar System. In fact, as Venus begins to draw close and overtake us in the evening sky, there is a period of several weeks where Venus is so bright that the planet is visible in broad daylight, all day long, high in a clear blue sky.

The question of what lay beneath the bright, all encompassing clouds has long been a source of mystery. Edgar Rice Burroughs, taking his cue from the best that planetary science had to offer in the early twentieth century, wrote of steamy jungles with dinosaurs and beautiful princesses adventuring across a cloud enshrouded planet. Today we know that shroud hides a surface under an atmospheric pressure 90 times that of Earth at sea level. Dive six tenths of a mile down into the ocean here on Earth and the pressure of nearly a kilometer of seawater above you will give you a good approximation of the crushing pressures weighing down on anyone merely standing on the surface of Venus. Composed almost entirely of carbon dioxide, its atmosphere's greenhouse properties hold the planet at a nearly constant surface temperature of 900° F (480° C, hotter than most ovens). Together the combined pressure and strength of Venus' carbon dioxide

[2] For example, you will never see Venus high overhead in the sky at midnight.

blanket mean the scorching temperatures never vary or cool from pole to pole, or from noon to midnight.

While prior to the space age no astronomers really held out hopes for beautiful Venusian princesses, the modern picture of Venus is not what most expected to eventually find there. Venus and Earth are twins; it is almost exactly the same size and mass as the Earth and therefore has the same density, implying the same composition. The outgassing that produced our planetary atmospheres should have produced the same gasses and produced them in the same quantities. So how did the Earth and Venus wind up so different?

Figure 6.21 Venus photographed by NASA's Pioneer-Venus spacecraft in ultraviolet light (to show detail in the clouds) (NASA).

On the Earth, the water that we've outgassed over the eons has condensed under the cool local conditions to cover three quarters of our planet as oceans. The carbon dioxide, in turn, dissolved in water where it undergoes chemical reactions to form carbonate rocks. The grey-colored limestone you see in layers of the Helena Formation throughout Glacier National Park are just such carbonate rocks.[3] An inventory of carbon-rich sediments on Earth reveals that we have 170,000 times more carbon dioxide locked up in our crust than is found in our atmosphere and that pound for pound our planet must therefore have outgassed just as much carbon dioxide as Venus. The limestone rocks you see are one of the reasons we never became Venus, and we wouldn't have them had we not had liquid water whose presence is all around us. It is thanks to our Goldilocks status, neither too hot nor too cold for liquid water to exist on our surface, that we have the conditions right for life today.

Perhaps then the only important difference between the Earth and Venus is that Venus is closer to the Sun. There but for the grace of orbital dynamics go we? If that is true, then what if some unforeseen and completely unimaginable force moved the Earth to the same position as Venus, only 72% as far from the Sun as we are now? Perhaps the same fate would befall us still? At such a distance the increased sunlight raises our planet's temperature by about 55° F (30° C), causing a global average of over 110° F (43° C). Imagine a world where a simple, average, ordinary day was as hot as Phoenix, Arizona at noon in summer. Lakes and

[3] While the limestone that makes up the Helena Formation is grey, the part of the limestone in contact with the air slowly turns tan as it weathers. High on the mountain-sides look for the tan layers to see the signs of the Helena limestones.

Figure 6.22 The surface of Venus looks like a flat volcanic plain where the Soviet Venera 14 spacecraft came to rest in 1982. The lander survived only 60 minutes in the harsh environment before succumbing to temperatures greater than the interior of an oven and pressures nearly 100 times the pressure at sea-level on Earth. The arm reaching out from the spacecraft was spring-loaded and designed to test the compressibility of the surface (a key in understanding what it is composed of). Unfortunately, the arm landed on the camera's discarded lens cap (New image processing by Ted Stryk).

shallow seas dry up, losing water to evaporation, and every day becomes hotter and more humid.

Water is a very potent greenhouse gas so the more humid the air, the warmer the surface and the more water evaporates. In this way a positive feedback[4] is introduced: higher temperatures produce more evaporation, putting more water into the air, which can hold more water vapor, causing even higher temperatures. A runaway greenhouse effect ensues and the oceans of the Earth evaporate. Under such a Sun, the Earth's carbonate rocks bake off their trapped carbon dioxide gas and the greenhouse effect increases until the combined effect of both gasses produces a planet as hot as Venus. Increased ultraviolet light from the Sun eventually splits the water vapor into hydrogen atoms that subsequently escape to space. The oxygen that remains chemically reacts with minerals on the surface. Once complete, the Earth's remaining atmosphere contains virtually nothing but carbon dioxide at levels nearly 200,000 times its current amount and our transformation into Venus is complete.

Is this what happened to Venus in our Solar System's distant past? Analysis of Venus' atmosphere by the European Space Agency's Venus Express spacecraft has found evidence that Venus once had just as much water as the Earth. Water contains two hydrogen atoms for every oxygen. A hydrogen atom is simply a single proton surrounded by a single electron. But about 1 in every 50,000 hydrogen nuclei also contains a neutron, producing an isotope of hydrogen called deuterium. Since deuterium is twice as heavy as hydrogen, when it bonds

[4] Here the term 'positive' means that the feedback mechanism feeds on itself. Hotter temperatures create conditions that create even hotter conditions and the temperature grows mercilessly without control. A 'negative' feedback is one in which the feedback mechanism halts the conditions that caused it, and thus becomes self-regulating.

with oxygen it forms a molecule called heavy water. When heavy water is returned to its component atoms by the ultraviolet energy of sunlight, less of the newly freed deuterium manages to escape away to space as compared to its lighter hydrogen cousin. In 2007 researchers using Venus Express detected 150 times more deuterium in Venus' atmosphere as we have in our own. The implication is that vast amounts of water (literally an ocean's worth of water) were once present in the Venusian atmosphere and that it has since been destroyed by the Sun's ultraviolet light.

The image that fires the imagination is that perhaps Venus, like Mars, was once habitable by our standards. If so, one wonders for how long? Was there time for any life to evolve before the climate changed irreversibly for the worse? Is it possible that for at least a little while there were three neighboring worlds in our Solar System on which water flowed down mountain valleys to empty into three different planetary seas?

Evidence for whether this was ever a reality would almost certainly rest in the chemistry and composition of the rocks found on Venus today. Just as the layers in the rock walls of Glacier National Park reveal floods, droughts, seashores, ocean bottoms, and glaciers, so too could Venus reveal its past through its surface. From radar waves bounced off Venus' surface it appears it is a landscape of hard dry rock with evidence of no more than a few meager centimeters of soil or dust to cover them. The rocks themselves are almost certainly volcanic. A survey of the planet made by radar from the orbiting Magellan spacecraft shows signs of volcanism of massive proportion. On the very largest scales are towering mountains like Beta Regio rising three miles (5 km) above the average planetary radius (the closest approximation we have to a 'sea level') where the upwelling interior has stretched and broken the crust into great radial rift valleys.

On much smaller scales are such delicate features as 'pancake' domes only a half mile high (0.8 km) where thick viscous lava oozed out of fissures in the ground and solidified in rounded mounds like thick batter on a hot frying pan.

Figure 6.23 In the 1990s, NASA's Magellan spacecraft in orbit around Venus sent radar waves down through the thick atmosphere to bounce off the hidden surface. Rough and rocky surfaces with sharp edges reflected more radar waves than smooth surfaces (a fact that is at the heart of stealth technology). Combined with elevation information from an onboard altimeter, planetary scientists constructed three-dimensional models of the Venusian surface (although the elevation information has been exaggerated). Here rough, radar-bright, lava flows extend hundreds of kilometers over a smooth, cracked plain surrounding the 5 mile (8 km) tall volcanic peak of Maat Mons (NASA/JPL).

In between these scales are circular coronae where magma hotspots like those under Yellowstone National Park welled up, stretching the surface, and then sank back down again leaving behind a complex series of concentric ridges and valleys like a collapsed cake fresh out of the oven.

But one of the most important signs of planet-wide volcanism is the scarcity of craters. In maps of the surface, there are nowhere near as many craters as we see on the Moon or Mars. Like the Earth, some process must weather and erode the Venusian surface. But unlike the Earth there are no oceans to cover crater floors, there are no rivers to wash them away, there are no tectonic plates to slide under others, raising mountains and wiping the slate clean.

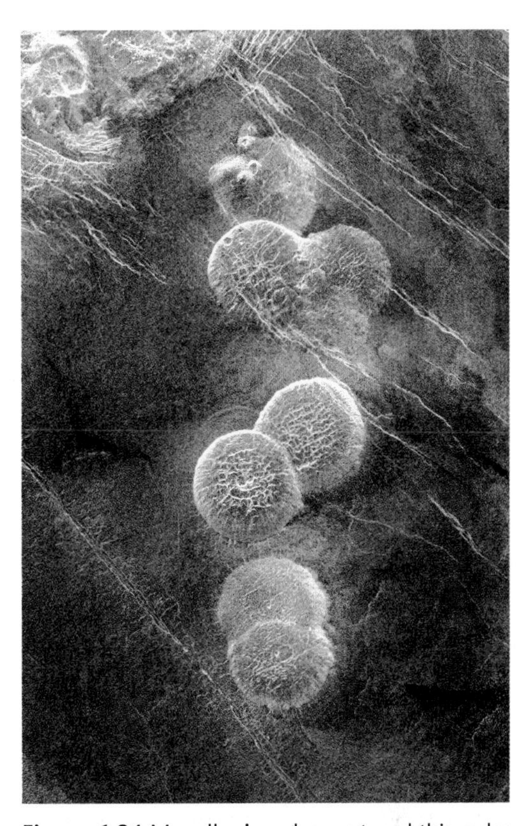

Figure 6.24 Magellan's radar captured this radar image showing where thick lava oozed up out of fissures in the ground to slowly spread outward across the surface before solidifying. Cracks formed as the surface cooled. Each ''pancake'' dome is about 16 miles (25 km) in diameter and rises 2,500 ft (750 m) above the dark, smooth plains (NASA/JPL).

There are some craters present, some of them quite large. But the fact that Venus does not look like the Moon means the planet was resurfaced somewhere around 600 million years ago to such an extent that what happened to the surface before is now nowhere to be seen. Walk across Venus today and what you would see (in the millisecond before you died) would look very much like the lava fields of Hawai'i Volcanoes National Park, minus the surrounding rain forest and tourists in flip-flops.

Also absent on Venus are any signs of tectonic plates like the Earth. There are no massive continents, or low flat ocean floors. If Venus is so similar to the Earth in size, mass, and composition, why does the planet's geography appear so different? One hypothesis goes back to the lack of water. Part of the lubrication that keeps the Earth's plates sliding along on top of one another is the water in our planet's mantle. Rock without water would form a crust too thick and strong, with too much friction preventing the system of sliding, folding, buckling and spreading that moves the plates across the Earth today. As Venus lost its water and its atmosphere superheated, so too did its interior as the internal heat that drives terrestrial plates no longer had an outlet. Such a buildup of heat, it is hypothesized, is what led to the periodic eruption and massive volcanic overturn of the entire Venusian crust. Venus is a planet that rips itself apart and turns itself inside out every half billion years or so, all as a result of the tragedy of its atmosphere.

The water on Venus is gone forever. Venus has gone down a road from which it can never recover. So too on Mars, the atmosphere that was stripped has led Mars on a path from which it cannot return on its own. Their climates have changed for good, yet we do not blame these changes on ancient Venusian industrialists or chastise Martians for their long-gone propensity for gas-guzzling sport utility vehicles. Their climates changed, but they did so on their own.

The Earth's climate has also changed on its own. Our planet's changes, though significant to us, have been relatively small compared with our planetary siblings. The most recent major change was the end of the last ice age about 10,000 years ago when the global average surface temperature was only a few degrees lower than it is today. Observations of deep-sea sediments washed there by cycles of glaciation and thawing on land show that during the last 800,000 years the length of time between each new ice age has averaged about 100,000 years. Over the last two million years glaciers have covered great swaths of North America 20 times. Every glacial feature you see within Glacier National Park today was caused by average global temperatures just a few degrees cooler than those today.[5]

Over the last *800 million* years, the Earth has undergone even larger temperature changes leading to more extreme ice ages on even longer timescales. During these times, it is thought, the Earth may have experienced great snowball periods where glaciers reached the equator and the seas froze to a depth of over half a mile (about a kilometer), followed by hothouse periods where the glaciers

[5] The primary cause for these past periods of glaciation may be long term fluctuations in the Earth's orbit, referred to as Milankovitch cycles. Over a period of 100,000 years the Earth's orbit gets gradually more and less elliptical changing how much energy the Earth receives at closest and farthest approaches to the Sun. Right now there is only a 3% difference between these distances, but in the distant past the difference was larger. Another change is the tilt of the Earth's axis. Today the Earth is tilted by 23.5 degrees with respect to the Sun. Evidence indicates that the Earth's tilt changes by up to 1.5 degrees in either direction over a period of 41,000 years. As the Earth's tilt becomes more extreme, so do the seasons. Hotter summers prevent ice building up in winter, shrinking glaciers and ice caps. A smaller tilt allows less light to reach the poles, cooling them and causing ice sheets to expand. The final change is the direction that the Earth's north pole points. Over a period of 26,000 years the direction our axis points in space draws out a great circle in the sky. This direction affects when the seasons occur. Right now northern hemisphere summer actually occurs when the Earth is farther from the Sun than it is when it is winter. In 10,000 years or so the position will be reversed and northern hemisphere seasons will become more extreme. Notice, however, that all of these changes due to gravitational tugs on the Earth by the Sun, Moon and other planets create changes over tens of thousands of years. None of these orbital effects produce rapid changes on the order of only hundreds of years.

melted completely and no ice-caps remained. But even during these enormous swings, still small on an astronomical scale, the Earth's climate has stayed stable to the point that temperatures were never so low that all our liquid water became locked up in ice, nor so high that it all evaporated into the atmosphere.

The remarkable thermostat that has kept us in this goldilocks climate, even while the Sun has increased in brightness by nearly 30% over the history of the planet, is due to the unique combination of gasses in our atmosphere, minerals on our surface, and abundant seas of liquid water.

While carbon dioxide is continuously removed from the atmosphere through the respiration of plant-life, it also dissolves in liquid water droplets, leaving the atmosphere as a mild form of carbonic acid rain. Where this rain falls to land, the weak acid slowly eats away at rocks and minerals in the landscape. The beautiful arches and hoodoos of Utah and the American southwest are a product of the global carbon cycle. Rivers and streams carry these eroded minerals down raging canyons to the seas where they settle to the bottom and react with carbon dioxide dissolved in sea water to produce calcium carbonate rocks: limestone.

The slow movement of continental plates sliding over the lower, denser, ocean-bottom plates forces the seafloor crust down into the Earth's hot mantle. Here, high temperatures melt the ocean crust, releasing the carbon dioxide into frothy, bubbling magma that rises and expands to explode violently at the

Figure 6.25 Eons of rain, snow, and ice have eroded a fairy-tale landscape in the colorful rock layers of Utah's Bryce Canyon National Park. A flashlight records the passage of time from a hiker descending into the canyon's maze (T. Nordgren).

surface from the summits of volcanoes. All around the Pacific Ocean is a 'Ring of Fire', volcanoes powered by the melting of the Pacific crust as it is slowly covered by the continents to west and east. With the eruptions of volcanoes like Mt. St. Helens in Washington State, the carbon dioxide is released back into the atmosphere and the global carbon cycle is complete.

It is this remarkable cycle that sets up a negative feedback mechanism that allows the Earth's temperature to remain so remarkably stable over millions of years. If, for example, temperatures fall too much, less moisture can stay suspended in the air and the atmosphere dries out. Rainfall decreases and with it less carbon dioxide is removed from the atmosphere. As volcanoes continue to erupt the concentration of carbon dioxide in the atmosphere begins to increase. With the increasing concentration of greenhouse gasses, the planet warms back up. Raise the temperature too much though, and the amount of moisture the atmosphere can hold increases too. With more water in the atmosphere there is now more rain. With increased rainfall, more carbon dioxide is removed from the air, reducing its concentration of greenhouse gasses. The temperature falls.

But the cycle is not instantaneous. It requires volcanoes and the motion of continental plates that move at a not quite glacial pace of one to two inches per year (about the same rate that your fingernails grow). Models of our planet and its atmosphere show that the timescales for stabilizing carbon dioxide levels through the negative feedback mechanism of the carbon cycle is on the order of a hundred thousand years. Thus ice ages naturally occur, followed over hundreds of thousands of years by warm periods.[6]

Given these natural changes, the power of volcanoes, the enormous size of the Earth, and the awe inspiring energy of the Sun, it really is hard to believe that there is anything human beings could do to significantly affect our atmosphere or the climate. But the power of life to alter a planet's climate is all around us. The very air we breathe, the oxygen that gives us life and colors the sky blue, is the waste product of life. Like the rocks on Mars, oxygen in the atmosphere quickly oxidizes iron turning the landscape red. In order to be a major component of our atmosphere it must be continuously replenished by life.

High in the Helena Formation limestone are one billion year old fossils revealed by the retreating Grinnell Glacier. The fossils look like concentrations of giant concentric circles; what they are are stromatolites and they record some of the earliest known forms of life on Earth. A stromatolite is a large cabbage-shaped knob of limestone created by the respiration of masses of aquatic cyanobacteria (commonly called blue-green algae). To see these stone remnants of ancient bacteria colonies is to look back nearly a quarter of the age of the Earth. During

[6] On Venus, there can be no carbon dioxide cycle or thermostat. Without rain or liquid oceans there is nothing to remove carbon dioxide from the atmosphere. With no plate tectonics, there is no means for recycling crust back into the interior. The only geologic process still at work on Venus is volcanism, which only serves to put more carbon dioxide into the atmosphere, forcing the greenhouse effect in one constant direction.

Figure 6.26 Bacteria colonies, forming structures called stromatolites, were one of the original forms of life on Earth over a billion years ago. Back then they were found in shallow seas all over the world and were the dominant form of life. Today, stromatolites are only found in a couple places on Earth; these are part of the Hamelin Pool Marine Nature Reserve in Western Australia. Notice the ripples on the seafloor and compare them to those recorded in the mudstones of the ancient seafloor in the mountains of Figure 6.7 (Warwick Hillier).

Figure 6.27 Grinnell Glacier huddles between the melt waters of the lake and the shadowed rock wall that marks the Continental Divide. In the foreground, the glacier's retreat uncovered the fossilized remains of billion year old stromatolites. These colonies of blue-green algae, like those in Western Australia, were shaped like giant cabbages. Repeated glaciation planed the rock flat revealing the concentric algal cross-sections. Plate tectonics have lifted the remains of these aquatic life forms until they now sit under a cloudless sky 6,500 feet (1,980 m) above sea level (T. Nordgren).

the time these simple organisms were alive in shallow seas they transformed the Earth's atmosphere from one composed largely of carbon dioxide (which the cyanobacteria breathed in from the seawater) to an atmosphere composed largely of oxygen (which the cyanobacteria breathed out). Their pollution changed the world and set the planetary stage for animal life as we know it

The evidence that we are doing what the stromatolites did is measured in the concentrations of atmospheric carbon dioxide. Air bubbles trapped in glacial ice tells us what the composition was of our atmosphere over the last several hundred thousand years. The deeper the ice, the farther back in time we can sniff the air of the planet that once was. Measurements show periods of atmospheric carbon dioxide fluctuations that vary between about 200 and 250 parts per million (ppm, meaning molecules of carbon dioxide per million molecules of air). These variations in concentration of about 25% occur over periods of 100,000 years. As predicted by the models of the carbon cycle, with each increase in atmospheric carbon dioxide came an increase in globally averaged temperature.

Beginning about 10,000 years ago, the Earth's atmospheric carbon dioxide levels once more began to increase, pulling the planet out of the last great ice age and ending the reign of the giant glaciers that carved the U-shaped valleys

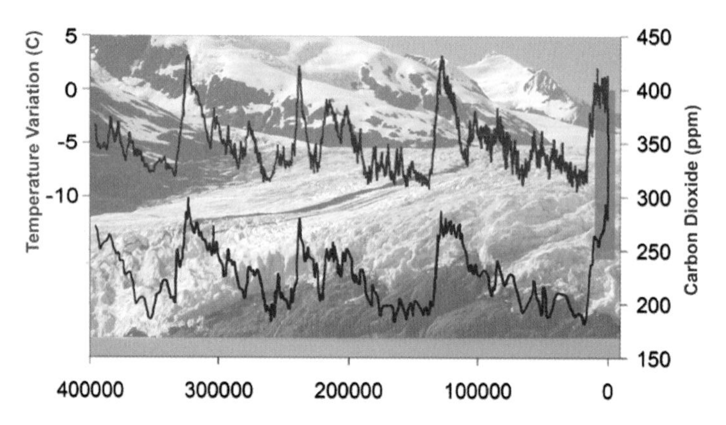

Figure 6.28 Carbon dioxide concentrations in parts per million (ppm) in the atmosphere are shown over the last 400,000 years. Ice cores in Antarctica reveal tiny bubbles of Earth's atmosphere, trapped when the snows that made the ice first fell. Temperature variations over this same time period are determined by changes in the ice itself. As basic chemistry and the carbon cycle predicts, when there's more carbon dioxide in the atmosphere, temperatures increase. Atmospheric carbon dioxide levels during the last 10,000 years (the grey box) are shown in greater detail in the top panel of Figure 6.29. The background image is the shrinking Portage Glacier south of Anchorage, Alaska ((carbon dioxide ice cores) Laboratoire de Glaciologie et de Geophysique de l'Environnement/Arctic and Antarctic Research Institute; (ice temperature) Laboratoire de Glaciologie et de Geophysique de l'Environnement/ Laboratoire des Sciences du Climat et de l'Environment /Arctic and Antarctic Research Institute; (glacier image) T. Nordgren).

around me here in Glacier National Park. Concentrations of carbon dioxide steadily increased, until about 1850 when, with the rise of industrialism in the northern hemisphere, the rate of increase itself increased. Atmospheric carbon dioxide levels have increased half again over the last 200 years, faster than at any time, and surpassing any levels seen, in the last 400,000 years. The U.S. Environmental Protection Agency predicts that if observed trends continue, atmospheric carbon dioxide levels measured in the 1990s will double by 2030. And with the carbon comes the heat.

I saw my first glacier when I moved to Alaska as a boy at the age of 10. Portage Glacier was my 'neighborhood' glacier just an hour's ride down the Seward Highway from Anchorage. I vividly remember standing with my brother on the edge of the glacial lake as the howling, frigid wind nearly blew us into the water. In the distance we could see Portage Glacier winding down the valley and into the water where giant icebergs gathered from the front of the floating ice sheet.

Twenty five years later I returned to find a modern visitor center built on the spot where my brother and I once stood. Panoramic windows looked out onto Portage Lake, except there was no longer any glacier to be seen. In less than my lifetime it had receded so far around the surrounding mountains that it was no longer visible from the visitor center that had been built for it. Today, a thirty-minute boat ride takes you around the bend to where the ice now stands, much

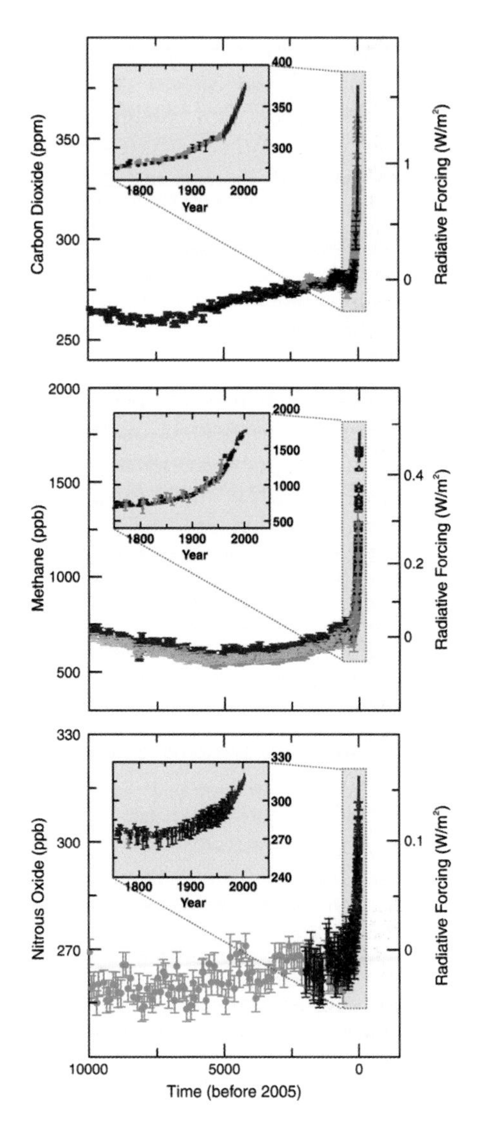

Figure 6.29 Changes in greenhouse gases from ice core and modern data. Atmospheric concentrations of carbon dioxide, methane and nitrous oxide over the last 10,000 years (large panels) and since 1750 (inset panels). Measurements are shown from ice cores (symbols with different colors for different studies) and atmospheric samples (red lines) (Intergovernmental Panel on Climate Change Fourth Assessment Report, Climate Change 2007, Synthesis Report (WGI Figure SPM.1)).

of its base now resting on rock. Before too many more years it will no longer even reach the lake that bears its name.

This is a story repeated all over the world and throughout Glacier National Park. Ask any long time visitor or park employee and they will tell you stories of what the glaciers looked like when they first started coming here. Out of 150 glaciers estimated to have existed in 1850, over two-thirds had disappeared by 1968. Within a decade those will be essentially gone as well.

Daniel Fagre is Research Ecologist and Director of the Climate Change in Mountain Ecosystems Project for the Northern Rocky Mountain Science Center of the U.S. Geological Survey at Glacier National Park. Since 1991 he's studied changes in the alpine ecosystems of the northern Rockies. At first, when it came to topics such as 'climate change' and 'global warming' Fagre was skeptical. "The idea that we could change the energy balance of the planet? It was like someone saying 'Hey, let's change the planet's orbit.'"

Today is a piercing blue day, like many in the mountains. It's cool in the shade, but surprisingly warm in the sunlight, and above it all the sky rises to heights of electric blue. Fagre has hiked with me up the long trail from the Many Glacier Hotel on the shores of Swiftcurrent Lake to where Grinnell Glacier sits in its cirque beneath the rough and ragged teeth of the Continental Divide. After hiking five miles (8 km) through a summer day even more beautiful than all the postcards promise, we pause high up on the inner rim of the ancient glacial gorge, and Fagre reaches out to the valley with a wave of his arm.

"George Bird Grinnell," he tells me,

Figure 6.30 This pair of photographs shows significant changes over just 56 years within Glacier National Park. Here an ice cave in Boulder Glacier was a popular location for pack trips as evidenced by the park visitors and their cowboy guide. In 1988, Park Ranger Jerry DeSanto re-photographed the same scene, dramatically illustrating the disappearance of the cave and the glacier itself ((1932) George Grant, courtesy of GNP archive; (1988) Jerry DeSanto, USGS).

"the first editor of what would become *Field and Stream* magazine was one of the first easterners to campaign for preservation of the region. In 1885 he was out here on vacation and tramped around the whole area. Exploring down in the valley below, he came across the aquamarine lakes that are so unique to glacial landscapes. Knowing enough about glacial geology he knew a glacier had to be nearby and so he set out to find it in 1887, a difficult prospect in those days, when the valley and mountain slopes were still covered in snow for much of the year." Fagre, pauses to chuckle, "Grinnell *discovered* the glacier by the means most folks used back then, he hired a local Native American guide who knew where the glacier was."

At that time there were no trails and so the two men and their horses just bushwacked straight up the valley floor. Eventually, they managed to climb, with all their equipment right up the series of waterfalls that runs down the steeply sloped valley's head. There, beneath the cirque where the glacier sat, they were right at a level with where we are now. "There Grinnell said he saw the edge of the glacier, an ice wall a thousand feet high (300 meters) and several miles wide. The entire cirque was one big glacier with just a tiny rim of rock where the waterfall is now."

Today it is mostly a lake. We can't even see the glacier from where we are now. Fagre turns back to me, "Only about 10% of the glacier remains today." Starting around 1920 and lasting a little over two decades, a series of droughts and hot summers caused glaciers all over the park to start retreating. Eventually they did so by as much as three hundred feet (100 meters) a year.

As of 1966 when aerial surveys were finally completed within the park, researchers found from the positions of moraines that there were 150 places where glaciers had definitely been as of the time of their greatest extent in 1850. As of 2008, there are only 26 still remaining in the park; 84% of the original ice cover is now gone. Fagre looks back up the valley at the glaciers that still cling to

Figure 6.31 Three photographs reveal the changes in Grinnell Glacier from near the head of Lake Josephine. The top photograph reveals the enormous extent of the glacier from soon after its "discovery" by George Bird Grinnell in 1887. Notice the great wall of ice that covers nearly the entire bowl at the head of the valley. The next two photographs from 33, and 120 years later are taken from a slightly different, but similar location, near the present day footbridge on the trail that leads up to the glacier. By 1920, significant shrinkage of the glacier had already begun, but the glacial ice masses within the cirque were still combined. By 2007, Grinnell Glacier is no longer visible from this vantage point. Only the now separate Salamander Glacier remains in the view ((1887) Lieutenant Beacon, GNP archives; (1920) T.J. Hileman, GNP archives; (2007) T. Nordgren).

shady spots in the rocks: tiny Gem Glacier that hangs beside a window in the thin arête at the continent's midline, and larger Salamander Glacier that was once part of Grinnell Glacier but has since broken off as the larger glacier receded back into the shade. "That's a reflection of climate change for the last century, and it's almost twice the global average," he said. "The upper elevations, the mountains, world wide, have all experienced this. The higher you go in elevation it still gets cold, it's just that it gets less cold than it used to."

Starting in 1992 Fagre began working with Myrna Hall at the State University of New York, on a project to model the future growth of glaciers within the park and what effect their change would have on the ecosystem that has grown to take advantage of their presence. They analyzed a hundred years of temperature and precipitation measurements from communities surrounding Glacier National Park. They measured the retreat of glaciers based on dating moraines and tree-rings in trees that grew as glaciers retreated. From these data, they found that they can explain 92% of the glacier ice melt by looking at only two things: average summer temperature and average winter precipitation. In addition they found that there was a 25 year lag in response to glaciers: cooling in the 1950s led to minor advances of several glaciers in the 1970s. However, now that once larger glaciers in the 1950s have become small cirque glaciers in the 1990s, the response time has shrunk to as little as a decade.

To predict what would happen in the future, they applied their models to the U.S. Environmental Protection Agency's projections for average summer temperatures through 2050. Based on current trends, these projections anticipate that the Earth's atmospheric carbon dioxide concentration will double by 2050 resulting in a global average temperature increase of 4.5° F (2.5° C). Because a warmer atmosphere can hold and release more moisture, one potential consequence of the EPA's projections is an increase in winter precipitation. If so, the increase in winter snowfall would soften the effects of higher summer temperatures and the glacial retreat might slow. What they discovered was just the opposite. In 2003 they published their results with the startling conclusion that by 2030 the glaciers would be history.

"We had some trepidation about the claim in our paper. There were all the jokes about what would you call Glacier National Park after all the glaciers were gone. And it just seemed a bit radical; glaciers have been here in these mountains for 7,000 years and possibly longer, and we're saying it's all going to go away in just a few decades? We felt like we were sticking our necks way out, and we were, but in the wrong direction."

Fagre, pauses and takes a moment to explain the sometimes glacial progress of the scientific process. "We began the model described in the 2003 paper using the data we had through 1990 and then projected the glacier status for each decade after that. In 2008 Myrna Hall looked at the 17 years of new data to compare them with the predictions we had made. It turns out when we look at the glacial parameters we have now – the size, extent, and rate of melting we see now – we are about 10 years ahead of when we thought we would be. It turns out we were too cautious. What this is telling us is change is happening much faster; it's much more pervasive and to a much greater extent. Today, cold years with lots of snow are merely what normal years used to be. We have now surpassed the hottest period for which there are any records. When you consider these glaciers have been here, in one form or another, for 7,000 years, and now, on the basis of our models, we are saying that the glaciers will be essentially gone by 2020, we are looking at a unique 7,000 year event occurring in the next 12 years."

We both look up at where Grinnell's glacier rests out of sight. By the end of the decade the glaciers will be gone. We are at a pivotal moment in geologic history. It is this generation that will witness the end of titanic forces that have shaped the landscape in one form or another for tens of thousands of years. But so what?

Beyond alpine aesthetics, why should we care? Back in his office Fagre tells me that we are all intimately tied to glaciers in a wealth of ways we don't often realize. Half of all water we use comes from mountains; so most of humanity is closely tied to what happens there. "The irony is that the mountains are changing faster than any other place except the Arctic and Antarctic, and we are tied to mountains for our water on a global basis. Here in the arid American west, 85% of the water we are dependent upon for growing our food, watering our lawns, filling our swimming pools, putting ice in our drinks, all comes from the mountains. And glaciers are a barometer for what's going on here. You can *see* them, where you can't see carbon atoms. They can be photographed and everyone can recognize that when it's hot, they melt. They are also politically very impartial; they aren't conservative or liberal, Republican or Democrat. They just sit there in their relation to the climate and most importantly they don't adapt. Plants and animals, as temperatures start to get warmer, they can try to change behavior to try and keep up. Maybe a plant blooms or an animal migrates a little earlier or a little later and for a time masks that there is a problem. But glaciers have nothing they can do; they're the perfect barometer for what is going on."

In Hall and Fagre's 2003 paper they wrote that because of the time it takes for glaciers to form, flow and melt, they do not respond easily to year-to-year variability, but rather provide an excellent measure of decadal climate trends. From the global retreat of glaciers we thus have direct evidence of real long-term climatic changes and not just temporary fluctuations of a warm summer or two. Glacier retreat in natural, protected ecosystems like Glacier National Park therefore serves as an early warning system, much like caged canaries in a coal mine, providing the first examples of climate and ecosystem changes on otherwise pristine environments.

We now know that by using temperature and precipitation proxies, such as tree-ring widths, borehole temperatures, layers of ice in glaciers, and growth patterns of coral, the mean global temperature is higher now than it has been at any time in the last 1,000 years with a temperature increase over the past two decades that is larger in both rate and magnitude than any seen during the last millennium. Based on study of growth and shrinkage patterns of the alpine glaciers within the park, other researchers estimate that the warming in alpine regions during the twentieth century may be as great as any century for the last 10,000 years.

While the Earth is self-regulating, and by itself has self-correcting feedback mechanisms, in each of humanity's actions over the last ten thousand years we are forcing changes that have self-*amplifying* feedbacks. The rise of our industry occurred at a time when much of the northern hemisphere was coming out of a period of relatively minor cooling called the 'Little Ice-Age.' Carbon dioxide levels and temperatures were increasing, the modern alpine glaciers of the world were beginning to retreat, and to this we added to the air the carbon waste of our industry. Coal, oil, and natural gas are all carbon compounds that like carbonate rocks are ways in which atmospheric carbon dioxide has been locked up in the Earth's crust. Like all green plants today, ancient plants breathed in carbon dioxide in order to live. They released the oxygen to our atmosphere and retained the carbon in their bodies. Over millions of years of burial, compression, heating and time, the carbon in the decayed plant material became the 'fossil' fuels we are currently digging up. As we release the carbon back into the air through our modern lives, we are adding back in only a few hundred years what the planet spent millions of years removing.

On top of this, our clear-cutting of lush, dense, forests for agriculture removes a source of natural carbon dioxide extraction from the atmosphere. In the Amazon rainforest, much of the clear-cutting for animal grazing is accomplished by fires. Soot, from forest-clearing and the burning of coal in factories[7] has slowly settled on ice fields in the arctic. The black of the soot absorbs the Sun's energy where previously white snowfields reflected it. Like black asphalt on a summer's day, the sooty surface heats up, increasing the melting of ice. As ice sheets melt, the dark rock, or darker ocean waters that are revealed absorb even more heat compared to the previous highly reflective ice, and the amount of heat absorbed by the Earth continues to increase. A positive feedback develops as the bright melting snows expose dark rock or ocean waters that as they absorb increasingly more solar energy, heat the planet and melt even more reflective ice.

Through these amplifying feedback processes, planetary scientists conclude that the forces we humans apply to our climate, when applied at critical times and in critical ways, really do have the ability to change our planet's climate for the worse. Our planet, just like Venus and Mars, goes through natural climate swings. We are lucky in that unlike either of those planets whose swings have taken them far from where they once began, our self-regulating carbon cycle keeps our swings in check. But like a parent pushing a child on a swing, push at the wrong time, and the angry child goes nowhere, push in the right way, at the right time, and the child goes higher than she ever could have on her own.

The power of science is in its ability to make predictions and to improve its models by the repeated testing of those predictions. A number of the climate scientists today working on modeling and predicting the effects of human-induced climate change developed the underlying physics of their models by studying the greatest test-case for greenhouse gasses: Venus.

Models predicting the effects of each of the many contributors to global warming (clearcuts, soot, decreased reflection from albedo, etc.) have subsequently been compared to the observations of what's actually happened. As in Fagre's model, the predicted rise in global average temperatures and subsequent effects on the ground have been confirmed – or found to be too conservative, rather than too extreme. Unless something changes, either in our planet or ourselves, the predictions for the coming century are seriously disquieting.

[7] Percival Lowell, the foremost observer of Mars at the dawn of the twentieth century, wrote that one of the reasons he fled the American Northeast in order to locate his telescope in the wilds of 1800s Arizona Territory was to escape the effects of great sooty clouds that hung over the industrialized cities of the East blotting out the light of the stars.

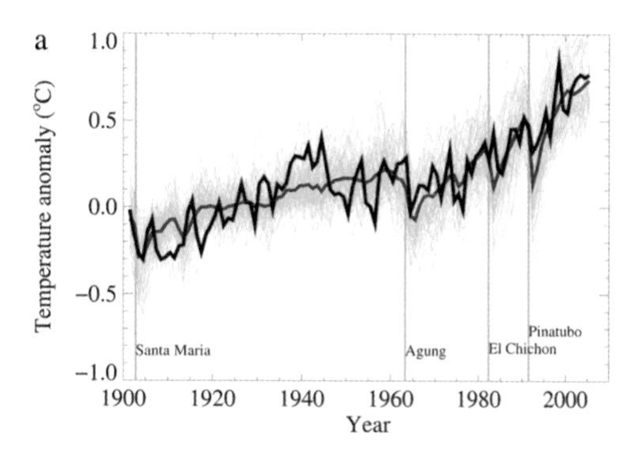

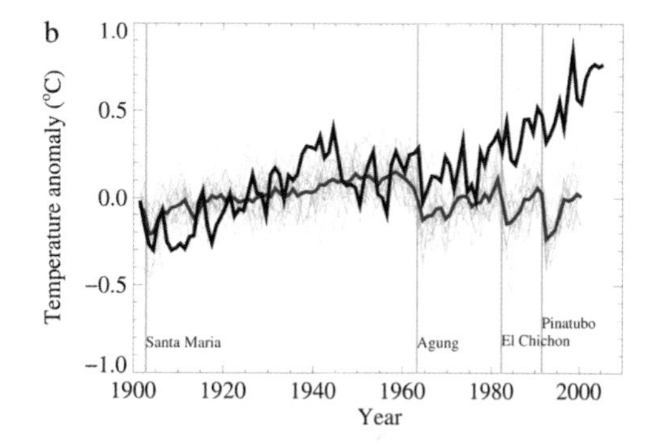

Figure 6.32 Comparison between global mean surface temperature anomalies (changes relative to the period 1901 to 1950) from observations (black) and different climate models. The models in (a) include both natural effects forcing climate change as well as man-made (anthropogenic) effects. The models in (b) only take into account natural forces. For models with anthropogenic and natural models: individual simulations are shown in yellow, while the mean is shown as a thick red curve. For only natural forcing models, the individual models are shown in light blue with the thick blue curve representing the mean. Vertical grey lines indicate the timing of major volcanic events. The key result is that models including only natural effects upon the world's climate do not accurately reflect the observed changes. Only those models that also include the influence we humans are having on the planet, do (Intergovernmental Panel on Climate Change Fourth Assessment Report, Working Group I Report, ''The Physical Science Basis,'' 2007).

As temperatures rise, growing seasons of plants and animals will be altered, affecting balances that have been in place for thousands of years. Many cool weather species will see their habitats shrink and vanish just as surely as if they'd been paved over with cities. Disease-carrying insects (like malaria-carrying mosquitos) will advance into regions that have never been exposed to them before. The elevation at which trees can grow will extend up and over many mountains in the Sierras and Rockies to such an extent that the alpine tundra and their meadows of wildflowers will be bulldozed under by an onrushing wave of foreign grasses and trees. No more fields of blue Explorer's Gentian; no more meadows of pink Stonecrop. And alpine swaths of pale Pearly Everlasting won't be.

With the melting of glaciers worldwide, including the great ice sheets covering Greenland and Antarctica, sea levels will rise as all that fresh water makes its way from mountaintop to ocean. Coastlines will be inundated, displacing billions of people. And for those living inland, say goodbye to the safety net that glacier meltwater offered when the last of the winter's snow melts in the height of summer. Rivers, streams and waterfalls that used to run year round, will now dry up in the warmer months. Everyone downstream, and that includes all of us, will feel this loss. West of the Mississippi River and Hudson's

Bay, everyone depends on the Rocky Mountains for water, and all of North America depends on the American West for food. In many ways, large and small, we are all fed by glaciers.

And while we may look back at the record of previous atmospheric changes with optimism that life has obviously survived, it is important to know that a billion years ago when carbon dioxide was far more prevalent in our atmosphere than it is today, stromatolites were the dominant form of life on Earth. Thanks to the oxygen atmosphere that their waste gasses produced, oxygen breathing organisms colonized the world, and stromatolites are now virtually extinct. A billion years from now I wonder who or what will be thanking us for what we have wrought during this century.

Over the last 50 years, the robotic exploration of the planets has shown there are no other hospitable planets in our Solar System. A hundred years ago, well educated science fiction writers could stir the imagination with tales of princesses on Venus and fighting men on Mars. As a result of our spacecraft we now know any princesses would be poisoned, crushed, and burned, and any fighting men would be asphyxiated, irradiated and frozen. And these are the most Earthlike planets we have yet found. While one way of looking at the history of NASA might be as one long crush of childhood hopes and dreams, of turning a living solar system into a dead one, the natural consequence of these cold hard facts is that now more than ever we understand the Earth is precious. Look around you at the trees, the waterfalls, the snow-capped peaks and arched over all is a vault of protective blue. The difference between life and lifelessness on this one single habitable world is our atmosphere, no thicker than the skin of an apple in comparison to its size. That's all there is between us and the vacuum, that single skin of blue between us and the all encompassing, infinite dark.

We now know the only way we will ever find another habitable planet will be if we make one ourselves. To transform such a world in our own image is a process called terraforming. Of all the planets we might terraform, the most likely is Mars. To do so, we would need to first increase the amount of greenhouse gasses in the atmosphere. We might seed its icy polar caps with some sort of dark material to make them less reflective and absorb more heat from the

Figure 6.33 An astronaut on board the International Space Station took this photograph of the Moon through the pale blue atmosphere that shrouds our planet (Image Science and Analysis Laboratory, NASA Johnson Space Center, "The Gateway to Astronaut Photography of Earth").

Figure 6.34 On April 6, 2008, the Japan Aerospace Exploration Agency (JAXA) spacecraft named Kaguya (Selene) captured the full Earth rising over the airless lunar limb (JAXA/NHK).

Sun. As they melt they will expose more of the dark red surface, further absorbing the Sun's warm glow and raising the planet's temperature. Evaporating water and carbon dioxide ice would once more raise the atmospheric pressure and increase the planet's greenhouse effect. Rock glaciers and subterranean ices could melt and liquid water could flow once more across the surface. Such a task, the engineering of a world, would be the most highly planned and precisely controlled scientific and engineering undertaking in all of human history.

And yet, ironically, this is exactly what we are currently doing to our own world, but here we are doing it with as little planning, monitoring, or control as possible. Every country on Earth with a coal-powered plant and anyone around the world who drives a car is a freelance engineer in the terraforming of our own planet. The problem of course, is that we are already terraformed. What we are doing to the planet and its natural systems – to the only Earth we have – needs a new name to reflect the accumulated effects of our actions. What we've already begun on this planet is most accurately described by the world we are attempting to become: we are Venusiforming the Earth. Though there is no plan, no organizing body, no group of brilliant engineers in Houston or Los Angeles, the experiment has already begun, and the initial data reports are in. The question is now: will we cancel this project, or wait to see whether all the predictions prove true?

See for yourself: greenhouse planets

Venus in evening and morning

Point a small telescope, or 20 × power binoculars, at Venus, and you will see its sunlit disk, as well as the fact (first seen by Galileo) that it goes through phases like the Moon. When Venus is about to pass behind the Sun, or has just come out from behind it, you will see its fully lit side and the disk will be small (Venus is nearly on the other side of the Sun as seen from Earth). When Venus has begun to pass between the Earth and Sun, or has just finished doing so, you will see its mostly unlit, night side, and Venus will appear as a large crescent (Venus is now closer than any other planet to the Earth). When Venus 'rounds the corner' between near and far sides of the Sun as seen from Earth, we see half a disk, as we see equal parts day and night on Venus. The following table lists the approximate dates that Venus is visible in evening and morning, as well as during what months it appears at its different phases.

The 'Position' column gives its position in the sky (east and west, and angle up from the horizon in degrees) 30 minutes after sunset for that month when Venus "rounds the corner" and is visible as a half disk. At this time, Venus is at its greatest apparent distance from the Sun. Positions are shown for the mid latitudes of the United States.

Table 6.1 Venus as an evening and morning star

Venus Visible in Evening			Venus Visible in Morning		
Full to Half	**Pos.**	**Half to Crescent**	**Crescent to Half**	**Pos.**	**Half to Full**
Feb–Aug 2010	WSW 20	Aug–Oct 2010	Oct 2010–Jan 2011	SE 30	Jan–July 2011
Oct 2011–Apr 2012	W 45	Apr–Jun 2012	Jun–Aug 2012	E 35	Aug 2012–Jan 2013
Apr–Nov 2013	SW 15	Nov 2013–Jan 2014	Jan–Mar 2014	SE 20	Mar–Sept 2014
Nov 2014–Jun 2015	W 35	Jun–Aug 2015	Aug–Oct 2015	SE 40	Oct–May 2016
July 2016–Jan 2017	SW 35	Jan–Mar 2017	Mar–Jun 2017	E 20	Jun–Dec 2017
Feb–Aug 2018	WSW 20	Aug–Oct 2018	Oct 2018–Jan 2019	SE 30	Jan–July 2019
Sept 2019–Mar 2020	W 45	Mar–Jun 2020	Jun–Aug 2020	E 35	Aug–Feb 2021
Apr–Oct 2021	SW 15	Oct 2021–Jan 2022	Jan–Mar 2022	SE 20	Mar–Sept 2022
Nov 2022–Jun 2023	W 30	Jun–Aug 2023	Aug–Oct 2023	ESE 40	Oct–May 2024
July 2024–Jan 2025	SW 35	Jan–Mar 2025	Mar–May 2025	E 20	May–Dec 2025

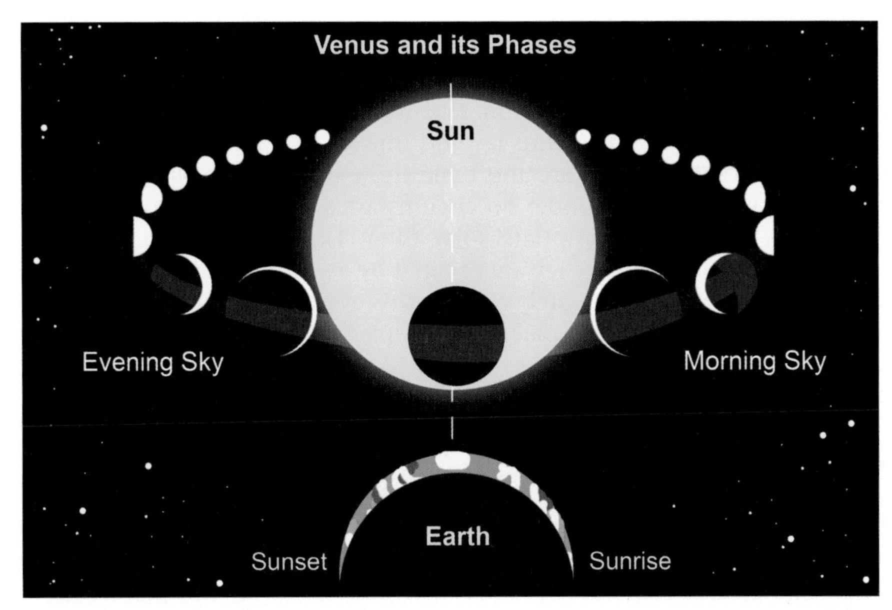

The sizes and phases of Venus as seen from Earth. While the apparent sizes of Venus, over the course of its orbit are correct, they are not to scale with the Earth, the Sun, or their distance from the Sun. When Venus is to the left of the Sun (as viewed from the northern hemisphere) Venus is the Evening Star. After passing between us and the Sun, Venus becomes the Morning Star (T. Nordgren).

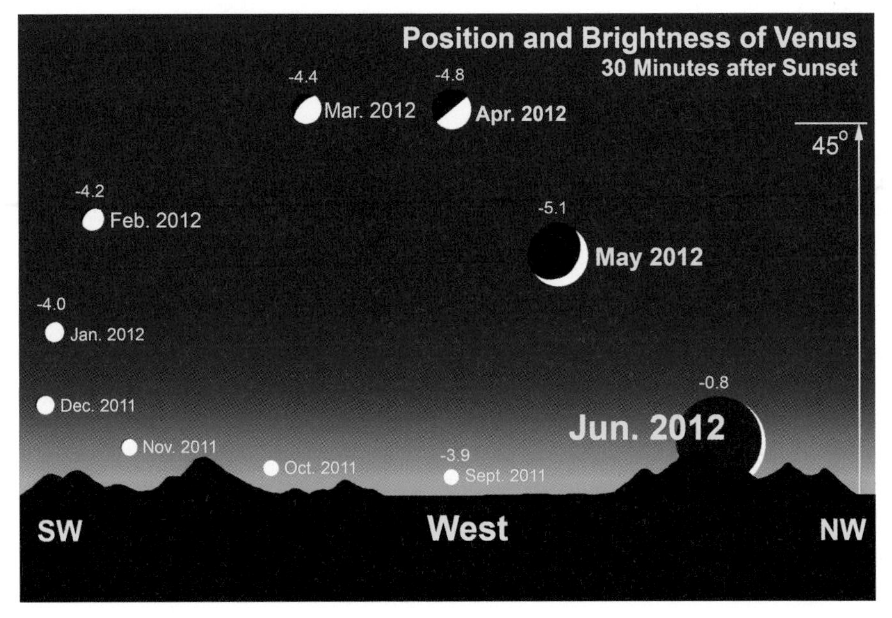

The position, phase, and magnitude (brightness) of Venus is shown at one month intervals. The relative sizes of the disk are correct, but not to scale with respect to the horizon or their position in the sky (T. Nordgren).

Transit of Venus

On June 6, 2012, as Venus passes from evening to morning star it crosses directly between the Earth and Sun. Venus last did this in 2004 but will not do so again until 2117. The following figure and table show the apparent motion of Venus across the disk of the Sun. Under no circumstances should you look at the Sun without the aid of an appropriate solar filter. Do NOT point a telescope or binoculars at the Sun unless they are covered by a solar filter designed especially for that purpose. One way to safely observe this event is to make a small pin-hole in one end of a shoe-box and place a white piece of paper on the inside of the opposite end. Point the pin-hole at the Sun and it will project an image of the Sun and the shadow of Venus, onto the screen at the other end.

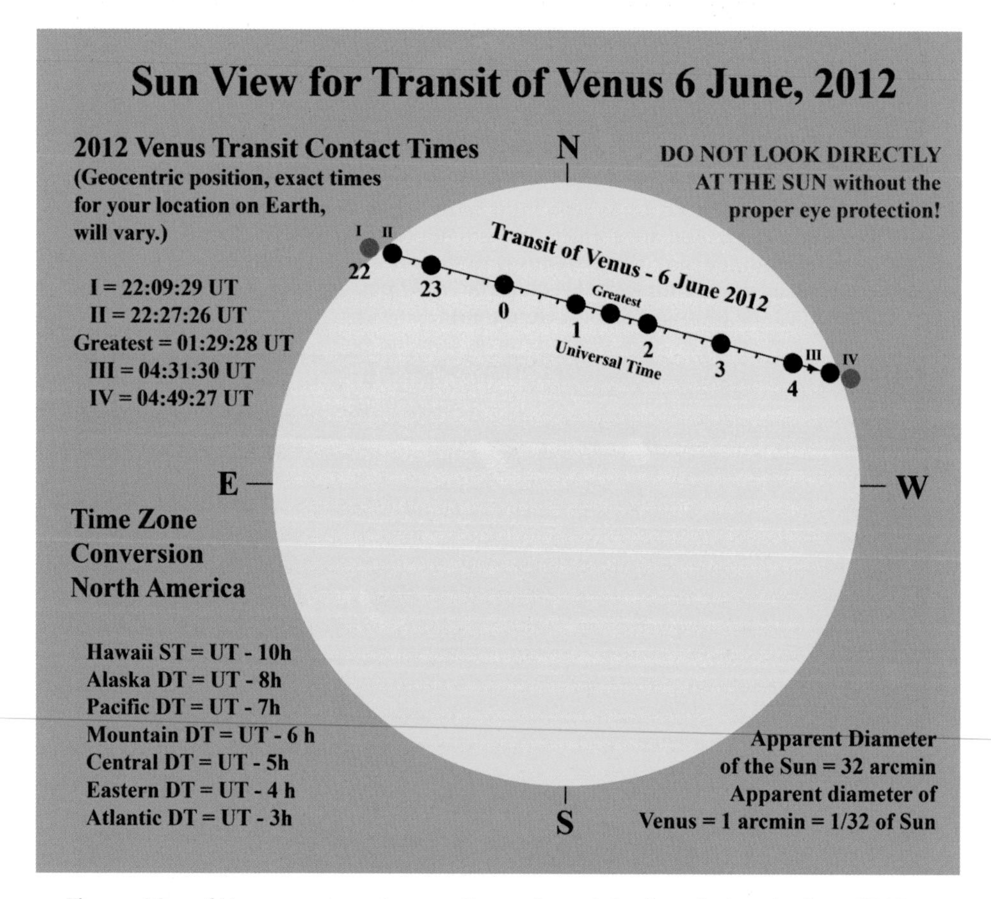

The position of Venus as viewed across the surface of the Sun during the June 2012 transit (Fred Espenak, NASA Goddard Space Flight Center).

Table 6.2 Visibility of the 2012 transit of Venus

Day	Location	Visible
6 Jun 2012, afternoon	Hawaii	Entire Transit Visible
6 Jun 2012, afternoon	Alaska and N. Canada	Entire Transit Visible
6 Jun 2012, afternoon	North America	Transit in Progress at Sunset
6 Jun 2012, afternoon	Central America, N.W. South America	Transit in Progress at Sunset
6 Jun 2012, night	S. and E. South America	No Transit Visible, Night
7 Jun 2012, night	Portugal, W. Africa	No Transit Visible, Night
7 Jun 2012, morning	Great Britain, Europe	Transit in Progress at Sunrise
7 Jun 2012, morning	Mid. East, E. Africa	Transit in Progress at Sunrise
7 Jun 2012, morning	Indian Ocean, Central Asia	Transit in Progress at Sunrise
7 Jun 2012, morning	Japan, Koreas, E. China	Entire Transit Visible
7 Jun 2012, morning	E. Australia	Entire Transit Visible
7 Jun 2012, midday	W. Pacific	Entire Transit Visible
6/7 Jun 2012, night	Antarctica	No Transit Visible, Night

Transit data courtesy of Fred Espenak, NASA Goddard Space Flight Center.

Grinnell Glacier repeat photography, 1900–2007

For the last several decades, the U.S. Geological Survey has documented the changing conditions in Glacier National Park by re-photographing the rapidly shrinking glaciers. The following two photographs show Grinnell Glacier, one of the most easily accessible glaciers, taken from the same position over 100 years apart. The photograph from 2007 clearly shows how the original glacier in 1900 has split in two, with Grinnell Glacier now invisible from this location, while the upper portion, called Salamander Glacier, is still found mostly in shade. Smaller Gem glacier is located high up on the rock wall to the left. Visitors to this same spot will be able to document the final disappearance of these remaining glaciers (expected to happen by 2020). The photographs were taken from the well maintained park trail leading from the Lake Josephine boat dock (in the Many Glacier section of the park) to the base of Grinnell Glacier itself (7.8 miles round trip, with 1600 ft elevation gain) at GPS coordinates of: 48° 46' 26".378 N., 113° 42' 11".269 W. *Note that the best photographs for comparison are taken in August. Photographs taken in June and July may still show recent snow in the area, obscuring the glacier's current extent.*

Two photographs, taken in 1900 and 2007, showing Grinnell Glacier, one of the most easily accessible glaciers ((1900) F. E. Matthes, courtesy USGS Archives, (2007) T. Nordgren).

Further reading

Mars – A Warmer, Wetter Planet by Jeffrey S. Kargel (2004)
Springer Praxis, ISBN 1852335688

*Venus Revealed: A New Look Below the Clouds of Our Mysterious Twin Plane*t by David Harry Grinspoon (1998)
Basic Books, ISBN 0201328399

Venus in Transit by Eli Maor (2004)
Princeton University Press, ISBN 0691115893

Geology Along Going-To-The-Sun Road by Omar B. Raup (1983)
Falcon Press Publishing, ISBN 0934318115

U.S. Geological Survey's Glacier National Park Repeat Photography Project
http://www.nrmsc.usgs.gov/repeatphoto/overview.htm

The NASA Goddard Institute for Space Studies, perhaps the leading U.S. research site for global climate change
http://www.giss.nasa.gov/

How to prepare to see the 2012 transit of Venus
http://www.transitofvenus.org/

7 Autumn Moon

The Man who has seen the rising moon break out of the clouds at midnight has been present like an archangel at the creation of light and the world.
Ralph Waldo Emerson

The cosmos isn't just for those who live under pristine skies in unspoiled corners of the mountain-west. Far more people live east of the Mississippi River than west of it and half the world's population now lives in cities and their suburbs. When I made plans to travel to the Great Smoky Mountains in eastern Tennessee for this book, everyone I told returned the same blank stare. It was as if they were trying to find a way of drawing my attention to the fact it was called the "Smokies" for a reason, as if I just hadn't quite noticed the name. When even the Park Ranger with whom I arranged this trip had the very same reaction, I knew I had to justify why this park was included in my plans. After all, if the skies aren't clear, and the lush green woods within the park aren't exactly a terrestrial analog of anything we're aware of on any other planet, then why on Earth am I even here? What does an astronomer see in a place like Great Smoky Mountains National Park?

I'm here in late October and the mountains are in full fall splendor, radiating waves of yellow and orange. It's one of those gloriously perfect fall afternoons when the park is at its busiest with folks having come from all over the southeast to see the Smokies' colors. Seasons are an astronomical phenomenon but one that most people misunderstand. When asked about the reason, the popular response is that summer and winter are due to the Earth's different distances from the Sun. If it's winter, we must be farther from the Sun than in summer. But if this were the case, then every location on Earth would experience the same season at the same time. But while the nights here in North America are just now beginning to get chilly and the lush greens of summer give way to the golds of

Figure 7.1 Clouds blow through Great Smoky Mountains National Park (T. Nordgren).

Figure 7.2 Fall colors burst out along Newfound Gap, the lowest route crossing the Smokies (T. Nordgren).

fall, my friends in Australia are just now seeing the first flowers of spring and looking fondly towards the lazy days of summer. While we in the north sing songs of white Christmas, those south of the equator are treated to Christmas in summer and swim-suits.[1]

The seasons are actually the result of the Earth's tilt on its axis. As our planet orbits the Sun, its axis of rotation is tilted from the vertical by an angle of 23.5 degrees. This angle keeps our north pole pointed towards the distant star Polaris, which we therefore call the North Star. During June and July the Sun and Polaris are off in the same direction and our north pole (and thus northern hemisphere) tilts towards the Sun. At these times the Sun rises in the northeast, passes high overhead when its light shines nearly straight down on us, and then sets to the northwest. It's summer.

Six months later, when the Earth is on the opposite side of the Sun, our north pole still points towards Polaris but now this means it tilts away from the Sun. The Sun rises towards the southeast, never gets very far above the southern horizon at noon, and sets to the southwest.

Now that the Sun no longer shines straight down on us and is instead low to the south, all its warming energy arrives at an angle. Every ray of light is spread out over a larger area than it was in June, and every spot in the northern hemisphere receives a little less energy than it did in summer. In addition, by rising in the southeast and setting in the southwest, the Sun spends less time in the sky - the days are shorter - and so both effects conspire to render the days cooler than in summer. Winter has come to the north.

Leafy trees sense the shorter days and, once nights are long enough, they begin to store up energy for the coming winter by blocking the circulation of

[1] Actually the Earth's orbit is almost perfectly circular, with a maximum change in distance of no more than 3%. To add insult to injury, the Earth is right now actually closer to the Sun in December than it is in June (though this slowly changes over thousands of years).

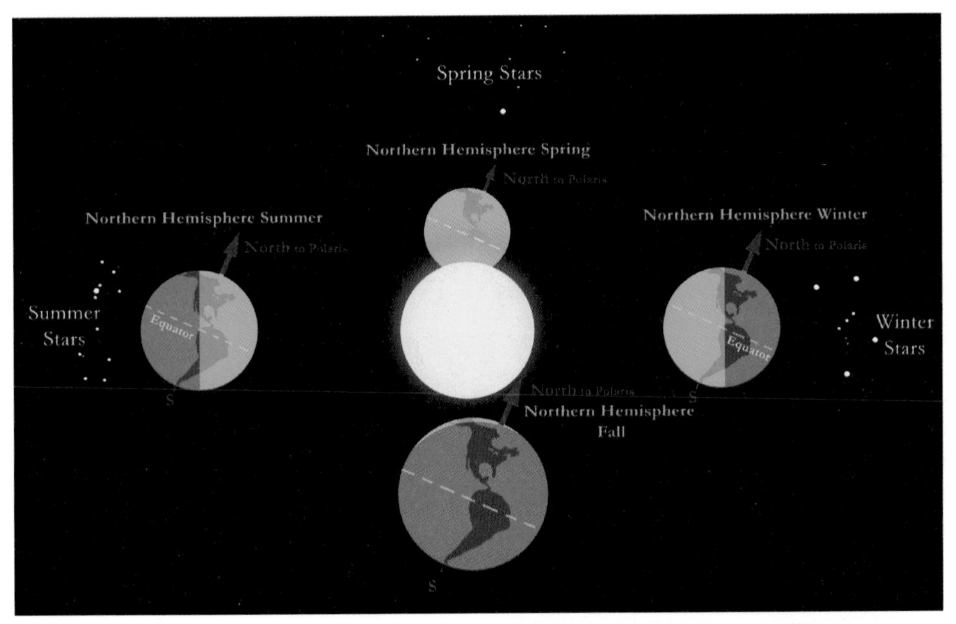

Figure 7.3 The Earth's tilt gives rise to the seasons as it orbits the Sun. Different hemispheres of the Earth receive more or less direct sunlight depending on whether they are currently tilted towards or away from the Sun. Also, depending upon which side of the Sun the Earth is on, different stars are visible at night (T. Nordgren).

Figure 7.4 The hardwood forests along the Roaring Fork Motor Nature Trail are ablaze with color as the longer nights of fall take over from the lazy days of summer (T. Nordgren).

nutrients to the leaves. Since the green chlorophyll that gives leaves their color must be continually replenished, this halt in the flow allows the leaf's natural pigments to emerge. Soon meadows, valleys and hills, from New England to the Smokies are alive in the fiery colors of a fall.

Because we are on the opposite side of the Sun in winter than we are in summer, different stars are visible at night at different times of year. The rise of old familiar constellations, not seen for nearly a year, is a way many cultures throughout history have kept track of the coming seasons. Follow the Appalachian Trail north through Great Smoky Mountains National Park and it leads you the entire length of the mountain

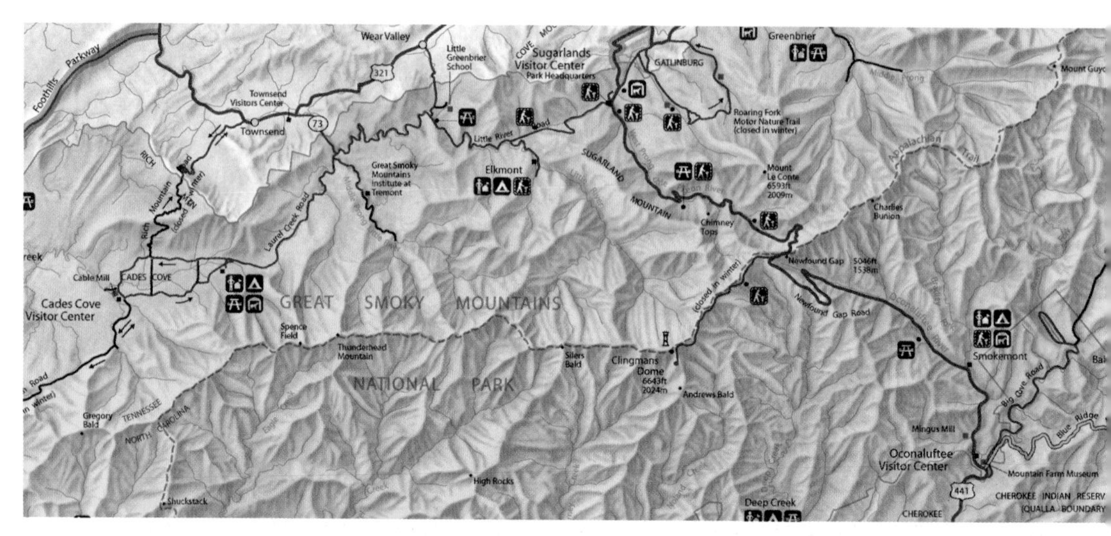

Figure 7.5 A close-up of the park map showing the Smoky Mountains and the places I visited within the park (NPS).

chain, all the way up to Maine. That far north, the Pole Star is higher overhead than it was in the south, and many more stars are visible all night long as they circle around Polaris. One such constellation of stars is Ursa Major, the Great Bear. In European tradition the bear includes the familiar Big Dipper with the handle for its tail (which has always baffled me given that bears have only extremely short tails).

For many American Indian tribes of North America, from the Micmac, Abenaki, and Iroquois of Maine and the American northeast back down the Appalachians to the Cherokee of North

Figure 7.6 Each night the stars in the north circle around the North Star, never setting beneath the horizon. Here in Maine, outside Acadia National Park, the North Star is always high in the sky (although hidden here behind the graceful stone carriage road bridge over Duck Brook) (T. Nordgren).

Figure 7.7 Maples turn red with the coming of fall (T. Nordgren).

Carolina and Tennessee, these same stars also form a bear. However, instead of a tail, the three stars that make up the handle of the dipper are three hunters in eternal pursuit. Audrey Salvatore, an amateur astronomer in Suffern, New York (along the northern portion of the Appalachian Mountains) has interviewed a number of Micmac tribal members who recounted for her their narrative of the three hunters.

> The first is called Robin, the marksman, the second is Chickadee, the cook, and the third who lags a little behind is Blue Jay. Look north in early spring, you will see the Bear climbing out of her den, the constellation of Corona Borealis. She was very hungry after her long winter nap, so she started to climb high in the sky in search of food. Chickadee spotted her but he was too small to hunt her alone so he called Robin to help. But first he had to make sure he had his cooking pot right next to him. Blue Jay lags behind gathering wood for the fire to cook the bear. All three chased the Bear all summer long as it got fatter and fatter in the sky. By the time Autumn came the Bear saw the Indians following her, she turned and reared up to fight the three. Taking careful aim, Robin shot an arrow and Mother Bear fell over on her back. By this time, Robin had waited long enough to eat some bear fat. In his eagerness, he jumped on the Bear and became covered with blood. Robin jumped up quickly and tried to shake the blood off. Although he shook most of the blood off there was one spot on his

chest that would not come off. Chickadee shouted, "You will have a red chest as long as your name is Robin." And so it is to this day. The blood that Robin scattered fell all over the maple trees and that is why the maple leaves turn red in the Autumn. Chickadee started to cook the Bear and as he stirred it, some of the yellow fat spilled over, and that is why some leaves turn yellow in the fall. Blue Jay was very tired so he stayed a little bit away, and was happy to get the scraps of the Bear. That is why you will see the Blue Jay today following the hunters and eating what they leave. All winter the Mother Bear's skeleton lies on her back while her spirit enters a sleeping bear in the cave, to emerge again in the Spring to start the hunt. And the story continues...

Micmac oral tradition, transcribed by Audrey Salvatore

Through star tales of this kind the behavior of the world around us is made sensible and we understand the purpose behind what we see. To those with sharp eyesight, the bright middle star in the Big Dipper's handle (Mizar) has a faintly visible companion (named Alcor), that together become Chickadee and his cooking pot. The constellation of Corona Borealis gives rise to new celestial bears each year, while the hunters' actions bring reason to the turning of the trees and the changing of the seasons.

Today we tell a different story for why the leaves and seasons change. One benefit of the scientific method is that it allows us to make predictions based on the consequences of our stories. The Earth is not alone in having an axial tilt and so it should not be alone in having seasons. In fact, we now know nearly every other planet in the Solar System goes through its seasons just as we do on Earth.

Mars' axis of rotation is currently tilted to the plane of its orbit by almost exactly the same amount as the Earth. In addition, its orbit is much more elliptical than the Earth's and so it really does get significantly closer and farther from the Sun. These effects combine to make its seasons more extreme than ours. Southern hemisphere summer comes when Mars is closest to the Sun and generates tremendous winds that flare into dustdevils and occasionally spawns globe-spanning sandstorms that blot out the sky. They give a whole new meaning to the hazy days of summer.

There is evidence that Mars' tilt periodically changes by tens of degrees over periods ranging from a few hundred thousand years to several million years in a complex cycle that produces dramatic changes in the planet's climate. When Mars' tilt is small, the poles get very little light while the Sun is always high above the planet's equator. The poles get so cold that water vapor and carbon dioxide (both greenhouse gasses) freeze out of the atmosphere and fall as snow. As they do, the atmospheric pressure drops so low that, with the drop in temperature, liquid water can no longer exist on the Martian surface. This is the Mars we see today. But let the planet's tilt increase and each pole gets more sunlight during summer and the greenhouse gasses re-enter the atmosphere, raising the temperature and atmospheric pressure enough that perhaps rain could fall and

The Song of the Stars
We are the stars which sing
we sing with our light
We are the birds of fire
we fly over the sky
Our light is a voice
We make a road for the spirits
for the spirits to pass over.
Among us are three hunters
who chase a bear
There never was a time
when they were not hunting.
We look down on the mountains
this is the song of the stars.
The Great Bear Ursa Major
"The Dipper"
Robin
Chickadee (Mizar)
Cooking Pot (Alcor)
Blue Jay

Figure 7.8 Watie Akins, an Abenaki Elder in Bangor, Maine, sang for me the song of the stars once told by his people. Here the Great Bear once again rises into the sky where the three hunters catch him turning the forests bright with color. A good test of eyesight is if you can spot Chickadee's faint cooking pot in the 'handle' of the Dipper (T. Nordgren).

seas form. Under these conditions life as we know it could flourish, but once the tilt decreases, the process reverses and Mars is once more locked in a deep freeze. Given such conditions, Mars is a difficult place for life to arise and survive.

Compared to Mars' wild swings, fluctuations in the Earth's axial tilt are tiny: no more than a degree or two over a 41,000 year cycle. While these small changes are one reason for long-term glacial periods, the change in our planet's tilt has never been so dramatic that all our oceans froze solid or all our seas boiled away. This climatic stability is one of the principal reasons our planet has remained habitable for life as we know it. So on a beautiful crisp autumn afternoon like today, I know I can look overhead and thank the Moon for the life I see around me.

The Earth is unique among the planets. We alone, have a moon so large that by comparison we are a double planet. Jupiter is over 300 times more massive than the Earth, yet its largest moon, Ganymede, is only twice as massive as our own. Our enormous Moon exerts daily forces upon the Earth that, over the long term, render our planet right for life.

The difference in the Moon's gravitational pull between near and far sides of the Earth creates the tidal water bulges that we experience twice a day as the Earth rotates under them. While anyone visiting the seashore has seen this type of tidal force in action, a much more subtle form of tidal force has stabilized the Earth's axial tilt. Even without the tidal bulges, the Earth itself is not a perfect sphere; like most people in middle age, the Earth bulges out a bit at the equator. The Moon therefore exerts a stronger gravitational pull on this wider mid-section than it does on the more distant parts of the planet's 'sphere.' Over time, any tendency the Earth's axis may have had to wobble and gyrate has been damped out and stabilized by the Moon's persistent tug. It's hard to go wild with a child forever tugging at your hip.

Figure 7.9 A first quarter Moon rises above the autumn woods on a sunny October day in the Smokies (T. Nordgren).

It might seem odd that the Moon could be responsible for such beneficence. Our language intimately associates the Moon with insanity and fear. How many of us have heard that the lunatics come out whenever there's a full Moon? Lunacy is from Luna, the Roman moon goddess and thus the Latin name for our moon. And while no close look at actual police records ever shows that lunatics prefer any one phase of the Moon over any other, our lore still tells of werewolves and vampires bringing terror by the light of a full Moon. Even so, several thousand years of poetry and literature attest to the beauty and power that the Moon has over us.

> Ignorance is the night of the mind, a night without moon or star.
>
> Confucius

> May you have warm words on a cold evening, a full moon on a dark night and smooth road all the way to your door.
>
> Irish blessing.

> Summer ends, and Autumn comes, and he who would have it otherwise would have high tide always and a full moon every night.
>
> Hal Borland (American author)

Perhaps it's no wonder then that full Moon hikes are some of the most popular evening programs the national parks provide. While we humans may still harbor an hereditary fear of the dark, when animals much larger and more vicious than we roamed the night where we slept; by the light of a gloriously full Moon we are able to hold our species' long memory at bay and take pleasure in the night, provided there's a knowledgeable Park Ranger for a guide.

The Moon is the one celestial sight upon which we probably spend the most time gazing: the Sun is so bright that most of us wisely spend very little time staring at its features, and the stars have become a thing of the past in the urban settings in which many of us live. Our Moon's close proximity makes it the one extraterrestrial body whose features we can see without a telescope and so after a lovely autumn evening regarding its strange markings, each of us can claim to be a planetary astronomer.

Figure 7.10 Full Moon hikes are a popular activity in many national parks (Ron Warner).

What do we see when we look at the Moon? Perhaps we notice its changing shape. Over a period of two weeks we see the Moon rise roughly an hour later each night and as it gets farther and farther away from the Sun in the sky we see more and more of its sunlit face. In the dark markings that the changing phases reveal, different people see different things. Some see a face: a Man in the Moon. Some see a rabbit, while for others it's the profile of a lady with a jewel at her throat, when the phase is just right.

Whatever form the markings take, they never change shape nor go away, and every month as the Moon's phase increases

Figure 7.11 The Clementine spacecraft captured this image of the full Moon. In a planetary psychology test, different apparitions are visible to different people. In the upper right is the Man in the Moon, while at lower left is the shadow of a rabbit. To the lower right is the profile of a dark-haired lady with the diamond pendant of Tycho's crater at her throat (Clementine Full Moon Mosaic (NRL)).

they're slowly revealed once more. The Cherokee, who live here in the Smokies, tell a tale of the Sun and Moon that makes sense of these sights.

> The Sun was a young woman and lived in the East, while her brother, the Moon, lived in the West. The girl had a lover who used to come every month in the dark of the moon to court her. He would come at night, and leave before daylight.
>
> Although she talked with him she could not see his face in the dark, and he would not tell her his name, until she was wondering all the time who it could be.
>
> At last she hit upon a plan to find out, so the next time he came, as they were sitting together in the dark of the âsi, she slyly dipped her hand into the cinders and ashes of the fireplace and rubbed it over his face, saying, "Your face is cold; you must have suffered from the wind," and pretending to be very sorry for him, but he did not know that she had ashes on her hand. After a while he left her and went away again.
>
> The next night when the Moon came up in the sky his face was covered with spots, and then his sister knew he was the one who had been coming to see her. He was so much ashamed to have her know it that he kept as far away as he could at the other end of the sky all the night. Ever since he tries to keep a long way behind the Sun, and when he does sometimes have to come near her in the west he makes himself as thin as a ribbon so that he can hardly be seen.
>
> James Mooney, *Myths of the Cherokee*, 1898

At the same time that the Cherokee told these stories, back when Europeans were first exploring the east coast of North America, the prevailing story among European intellectuals was that the Moon was a celestially pure and perfect sphere without blemish. The markings you and I see every night were merely a reflection of the impure corruptible Earth or strange differences in density of the smooth, featureless, ethereal lunar material. In either case, the heavens were separate from the Earth: They were perfect and we weren't. There was a reason they were called "The Heavens."

Figure 7.12 Thin waxing (i.e., getting bigger) lunar crescent is visible in the west after sunset. Waning crescents (getting thinner) are visible in the east before sunrise (T. Nordgren).

Then in November of 1609, Galileo pointed his telescope at a thin ribbon of Moon after sunset and sent our understanding of the Universe down a radically new path.

> From observations of these spots repeated many times I have been led to the opinion and conviction that the surface of the moon is not smooth, uniform, and precisely spherical as a great number of philosophers believe it (and the other heavenly bodies) to be, but is uneven, rough, and full of cavities and prominences, being not unlike the face of the earth, relieved by chains of mountains and deep valleys.
>
> Galileo Galilei, *Siderius Nuncius*, 1610

In 2009 we celebrated the International Year of Astronomy to commemorate the 400th anniversary of Galileo first pointing his telescope at the sky, but there is pretty good evidence that Galileo wasn't the first to do this, that in fact he wasn't even the first to look at the Moon. In July of that year, fully four months before Galileo, Thomas Harriot an English mathematician looked at the Moon through his own simple telescope. From his drawings we know that though he saw what Galileo would later see, there is no record that he drew any conclusions from what was visible in his eyepiece, that he understood their importance, or that he made any effort to tell anyone else.

Galileo did all of these and it was important that he did because this discovery marked the beginning of modern astronomy. For the first time, technology revealed something new to our senses that hadn't been known before; it expanded our concept of the Universe and required a great sifting between those previously held beliefs that could explain what was seen, and those that

Figure 7.13 Galileo Galilei's drawing of the third-quarter Moon published in his *Siderius Nuncius* (the *Starry Messenger*) in 1610.

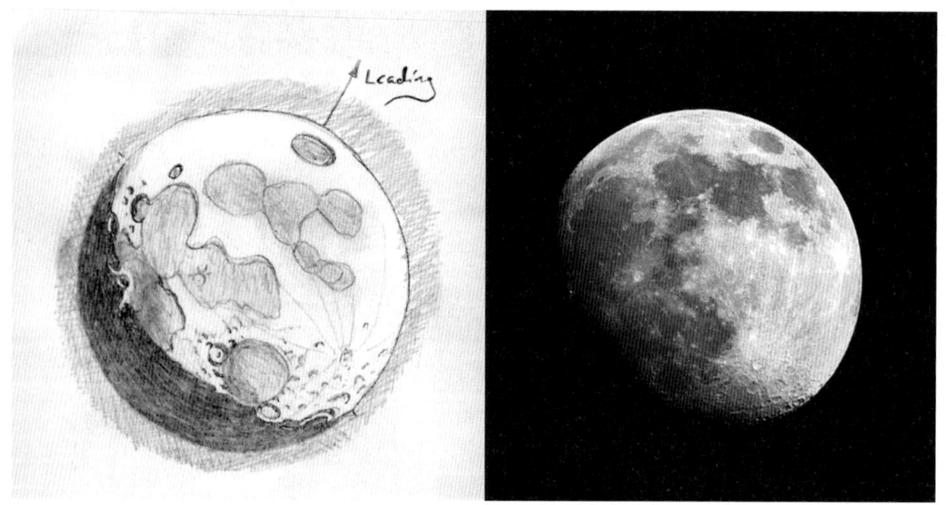

Figure 7.14 The photograph at right was made through a modern computer-controlled 10-inch (25 cm) diameter telescope, while the drawing at left was made through a simple replica of Galileo's telescope with a magnification of 20 ×. The two images were made at very nearly the exact same phase although three years apart. Notice the bright ray crater Tycho to the lower right in each picture (T. Nordgren).

couldn't. Just as important, anyone could see these new features for themselves; the physical Universe wasn't just revealed to a select few, it was there for anyone to see.

Go ahead, see what Galileo saw. A simple pair of 16 × magnification binoculars will show you everything he saw in just as much detail. The flat white lunar disk with dark grey markings becomes a spherical world with jagged mountains, deep shadowed valleys, smooth dark plains, and innumerable circular craters.

Through his eyepiece Galileo grasped that the Moon was a place, not a perfect

Figure 7.15 High lunar hills catch the light of a rising Sun along the quiet 'coastline' of the Sea of Serenity. Astronaut Ronald Evans captured this photo as he made his way alone around the Moon (R. Evans/NASA/LPI).

heavenly sphere, not an abode of angels, but a simple place like the Earth. And if the Moon could be a place like the Earth, then that implied the Earth shared something in common with the Moon. Rather than the Earth being a single, corruptible world, separate from the purity of the heavens above, the Earth now shared form and phenomena with the Universe of which we were an inseparable part. That single sight through a simple scope, now available for anyone to see, was the greatest world-altering discovery in the history of astronomy.

Today, through any small telescope or pair of binoculars, the familiar markings reveal themselves into two main types of terrain: smooth dark plains called maria (Latin for 'seas') and bright, mountainous highlands, ragged with craters in nearly infinite number and size.

While Galileo may have seen many lunar sights that reminded him of home, craters are pretty alien to the Earth. There are two principal processes that make giant circular holes in the ground. One comes from within; the other comes from without. Volcanoes leave circular craters where explosive eruptions blow open holes in the tops of mountains. Crater Lake in southern Oregon is a 6 mile (10 km) wide roughly circular mountain crater left behind by the eruption of Mt. Mazama, 7,700 years ago. Today clear blue water fills this caldera to depths of 1,943 feet (592 m), making it the deepest lake in North America and the seventh deepest in the world.

However, if you've ever seen a bullet-hole in a lonely, back country street sign, you've seen the other activity that leaves circular features behind: impacts.

Figure 7.16 Crater Lake is photographed in winter by astronauts onboard the International Space Station. Wizard Island is visible within the collapsed caldera lake (Image Science and Analysis Laboratory, NASA Johnson Space Center, 'The Gateway to Astronaut Photography of Earth,' http://earth.jsc.nasa.gov/).

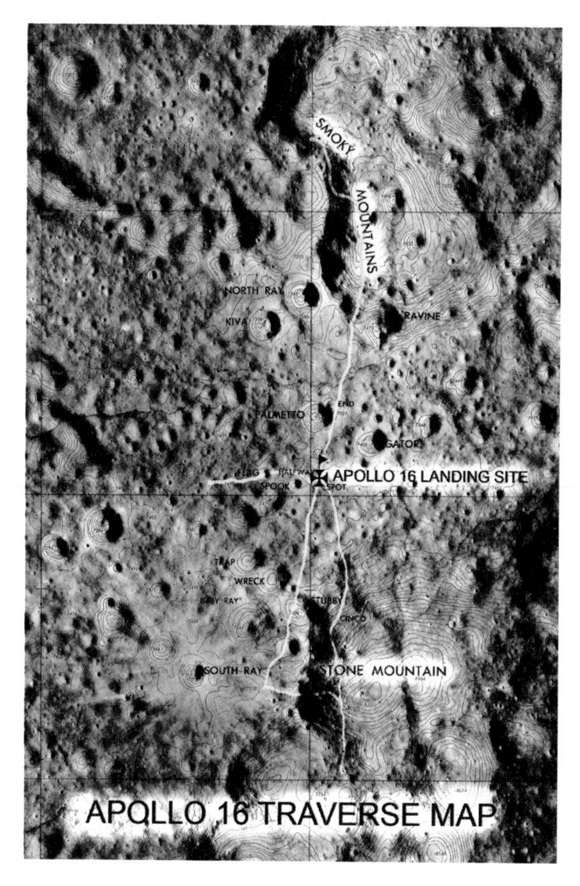

Figure 7.17 Topographic map showing the Apollo 16 landing site between the Smoky Mountains to the north, and Stone Mountain to the south. The route the astronauts drove in their lunar rover is drawn in yellow. The Smoky Mountains tower 500 feet (152 m) above the surrounding plain (elevations in red are in feet) (NASA/LPI).

Bombard the landscape with something heavy or explosive and bowl-shaped holes in the ground are the result. Many cities in Europe were littered with man-made impact craters at the end of World War II.

So which is it on the Moon, impacts or eruptions? The origin of lunar craters and maria was one of the principal scientific goals of sending men to the Moon in the 1960s. The first two Apollo missions to land on the Moon were designed to do so, on the sensibly flat smooth plains. The rocks they returned showed the dark maria are composed of volcanic basalt. They are enormous flooded basins where the lava long ago hardened. This type of rock is rare in the Smokies and much of the east coast, but travel to the western U.S. and big, blocky, black rocks found along many road-cuts will attest to the presence of ancient lava flows.

Starting with Apollo 14, the third expedition to land on the Moon, astronauts began to explore regions closer and closer to the lunar highlands looking for evidence of actual volcanoes. On the basis of telescopic observations from Earth it was thought that some of the features in the lunar mountains west of Mare Nectaris (Sea of Nectar) were volcanic in origin. On April 21, 1972, John Young and Charlie Duke in Apollo 16 became the only two human beings to ever explore the lunar highlands, doing so in a landing spot between two mountains named after prominent landmarks of the southern Appalachians: Stone Mountain in Georgia, and the Smoky Mountains where I now stand.[2]

[2] Along with Ken Mattingly, who stayed in lunar orbit while Duke and Young walked on the Moon, all three Apollo 16 astronauts had either lived or gone to school here in the southeast United States.

Figure 7.18 Panoramic image showing the Apollo 16 traverse to North Ray Crater, a 0.6 mile (1 km) diameter crater visible at left. Behind the crater at left of center is the summit of the Smoky Mountains. The lunar rover is visible to the right (NASA/LPI).

While Duke and Young had come looking for a volcanic origin, what they found was overwhelming evidence of eons of impacts. Everywhere they looked the rocks they brought back were nearly as old as the Solar System itself and in all that time the only change to affect them was the shock of repeated blasting by billions of years of planetary meteorite strikes. The flat-floored lunar valleys that were set amid the mountainous peaks (much like Cades Cove here in the Smokies) were composed entirely of rocks thrown there by unimaginably larger impacts elsewhere on the Moon. While the dark maria do tell a story of ancient volcanism, the results of Apollo 16's exploration of the Smoky Mountains reveal that the rest of the Moon is a long story with only a single theme: planetary bombardment.

Figure 7.19 Real Moon rock (in front of a fake Moon) on display at the Griffith Observatory in Los Angeles, California (T. Nordgren).

In large part, the history of our Solar System *is* a history of planetary violence. Our Solar System, the collection of one star, eight planets, and over 170 moons all began 5 billion years ago as a spinning disk of gas and dust (just like those we have begun to see around many other newly formed stars). Through gravity's persistent attraction, small things are forced together to become big things. And while many impacts between proto-planets probably resulted in their mutual destruction, in time, gravity will draw these pieces back together, building ever larger bodies in the end.

The planets we see today, their

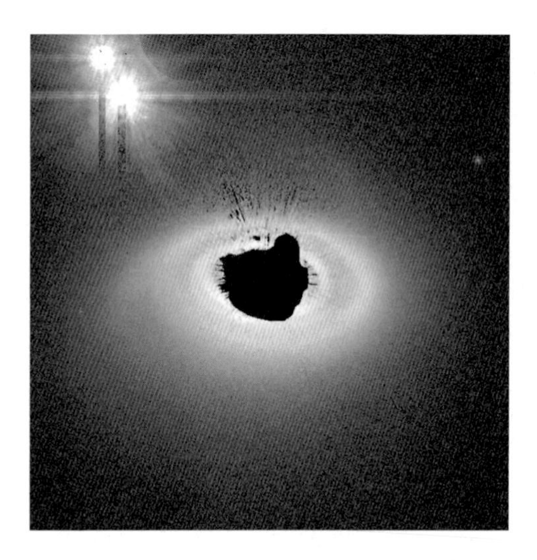

Figure 7.20 NASA's Hubble Space Telescope captured the faint reflection of sunlight off a disk of dust around the distant star HD141569A. New planets may be forming in the gaps this image reveals. The star is 320 light-years away in the constellation of Libra and has been blocked out in this image (since otherwise its glare would overpower the light of the disk, much like a spotlight overwhelms a firefly (NASA, M. Clampin (STScI), H. Ford (JHU), G. Illingworth (UCO/Lick), J. Krist (STScI), D. Ardila (JHU), D. Golimowski (JHU), the ACS Science Team and ESA).

compositions and characteristics, are dependent upon those early building blocks in the solar nebula. Close to the Sun, where its heat vaporizes water and other elements are gasses, the only building blocks available to form solid planets are rocks and metals. Not surprisingly, those are the principal components of the small solid bodies of the inner Solar System.

But those planets that formed farther from the Sun, where the temperatures are dramatically lower, formed out of bits and pieces that could also include ices. With more material with which to work these planets grew much larger and their gravity was able to retain much of the hydrogen and helium gas that made up the original solar nebula. They formed the large gaseous planets of the outer Solar System, and their moons are composed largely of ice and rock.

Many of those bits of leftover debris that crossed paths with the established planets eventually got swept up in the final stages of planet formation. But many other pieces on stable orbits still remain, most notably in a vast belt of rocky debris between Mars and Jupiter. We call these objects asteroids. Far from the Sun, comets are the left over icy material that orbit in the deep darkness beyond Neptune in what is now called the Kuiper Belt.[3]

For 4.5 billion years these asteroids and comets have continued their largely isolated existence too small for most geologic activity to take place, no volcanism, no tectonics, no wind or water to cause erosion.[4] They are thought

[3] The Oort Cloud is a second reservoir for comets located a thousand times farther out than the Kuiper Belt. It is a spherical cloud of ice balls that originally formed in nearer the Kuiper Belt but were flung out of the Solar System's disk by the gravity of Jupiter and the other large planets.

[4] In 2005 close-up images of the icy nucleus of comet Tempel 1, made by NASA's Deep Impact spacecraft, revealed layered deposits, apparent landslides and abundant craters. Geological forces had been at work on a comet which is hardly what anyone expected and proof that every time we explore some place new, our world opens up with new possibilities.

Figure 7.21 An artist's conception of the nearby Epsilon Eridanii planetary system where two planets orbit just outside each of two asteroid belts (much like Jupiter orbits just outside our asteroid belt). The inner belt is at very nearly the same location as the asteroid belt in our own solar system, while a reservoir of comets is thought to exist at about the same location as our system's Kuiper Belt. Epsilon Eridani is located only 10 light-years away and is easily visible low to the south in winter from southern locations (NASA/JPL-Caltech).

to be much as they were when the planets first started to form. This is why astronomers find these objects so fascinating; they alone are samples of what we are made of and from whence our planet came.

To understand why this is so rare, look around you at the mountains and trees, river and rocks; none of these show us the original building blocks of this planet. Take for instance one of the most visited spots in the park: Abrams Falls in Cades Cove. This afternoon I sit beside the pool into which the falls spill; I let my senses fill with the roar of the water and the rush of the wind through bright yellow leaves. There, eating my lunch I take stock of what I can see and what it can tell me about the early days of this planet.

The sandwich in my hand is only a day old. The bread of which it is made is no older than a week; the peanuts are no older than a year. The moss at my feet is perhaps a few decades old; the fallen tree upon which I rest may have first grown no longer than a century or two ago. The rocks are a bit older, but when compared to the age of the Solar System, not by much. Their rounded shape is the result of thousands of years of raging rivers that tumbled and polished boulders ripped out of deep mountain crevasses. The rock itself is a grey-colored sandstone made up of crystals and grains that weathered off of other rocks before being layered, compressed, solidified, and folded into its present upturned shape. While the bending and folding took place as two ancient continental plates collided a quarter of a billion years ago, the sand grains that first formed them are over a billion years old, but even that only reaches back a quarter of the way to the origin of the planet.

The Earth is old, even if what makes up the Earth isn't, so studying what we find here tells us only so much. Our photo-album for our planet is missing all the baby pictures. But walk along the surface of an asteroid or icy comet, and we finally get to see snapshots of those earliest days.

For better and worse, it isn't always necessary to go all the way out to the asteroid or Kuiper belts to examine these objects. So many left over chips from those early days still stream through the Solar System that the planets literally inhabit a celestial shooting gallery. Every month near-Earth asteroids cross our planet's path and every year an icy comet plunges in towards the Sun from its frigid outskirts, making quite a show as it quickly vaporizes from the heat.

Both NASA and JAXA (the Japan Aerospace Exploration Agency) have launched spacecraft to orbit near-Earth asteroids. What they've found is that asteroids can be complicated places with some asteroids a single dense object (like a solid rock) while others are the equivalent of a loose pile of rubble held together by its mutual gravitational attraction. While no spacecraft mission has yet returned a sample of an asteroid, we still have many samples from tiny bits and pieces that enter our atmosphere every year.

Here is perhaps a good time for some terminology: when a tiny chip off an asteroid is floating through space it is called a meteoroid. When a meteoroid enters our atmosphere and friction with the air causes a bright glow as it streaks across the sky, it's called a meteor (which folks commonly call a shooting star or falling star). Most meteors are no more than a grain of sand or small pebble in size. If a meteoroid is larger though, then the meteor might not burn up completely in the atmosphere and instead manage to strike the ground. Any charred black rock that survives the impact is called a meteorite.

From looking at the composition of meteorites we know that they and the asteroids from which they come are composed largely of rock and some metals like iron and nickel. Look up on

Figure 7.22 On Valentine's Day 2000, NASA's Near Earth Asteroid Rendezvous spacecraft went into orbit around an asteroid named Eros. At 21 miles (34 km) long and 8 miles (13 km) wide and tall, Eros is larger than most other asteroids that pass near the Earth, but in other respects is thought to be a pretty typical asteroid (NASA /JPL/JHUAPL).

Figure 7.23 A meteor streaks passed the full Moon and is reflected in a lake one misty November morning in Greenwood, South Carolina (Blake Suddeth).

Figure 7.24 This is perhaps the most famous depiction of any meteor shower. Made in 1889 for the Adventist book *Bible Readings for the Home Circle,* it portrays the spectacle of the 1833 Leonid meteor storm and was based upon the original account of Joseph Harvey Waggoner, a minister on his way from Florida to New Orleans (Adolf Vollmy/Library of Congress).

any clear night when you can see the stars and odds are you will see a meteor ending its existence in a blaze of glory, continuing our planet's formation by adding its mass to that of the Earth on which you stand.

About a dozen times a year your chances of seeing a shooting star increase dramatically. On these same few nights every year, meteor showers produce as many as a hundred or more streaks across the sky every hour. The darker the location, the more you will see. Some meteor showers, especially those before electric lights began to light up the sky, have gone down in history as meteor storms. The shower that happens every year on or around the night of November 17th is called the Leonids and on a cold clear November night in 1833 the Leonids lit up the sky in what became known as "The Night the Stars Fell." For those who were awoken in the pre-dawn hours by the sound of startled neighbors and the flash of light into darkened rooms, over thirty thousand meteors per hour blazed across the sky and it was thought that the world was ending and Judgment Day had come.

Samuel Rogers, a traveling preacher passing through Antioch, Virginia on the night of the storm recounted what he saw nearly 50 years later in his autobiography, *Toils and Struggles of the Olden Times*:

> I heard one of the children cry out, in a voice expressive of alarm: "Come to the door, father, the world is surely coming to an end."

> Another exclaimed: "See! The whole heavens are on fire! All the stars are falling!" These cries brought us all into the open yard, to gaze upon the grandest and most beautiful scene my eyes have ever beheld. It did appear as if every star had left its moorings, and was drifting rapidly in a westerly direction, leaving behind a track of light which remained visible for several seconds.
>
> Some of those wandering stars seemed as large as the full moon, or nearly so, and in some cases they appeared to dash at a rapid rate across the general course of the main body of meteors, leaving in their track a bluish light, which gathered into a thin cloud not unlike a puff of smoke from a tobacco-pipe. Some of the meteors were so bright that they were visible for some time after day had fairly dawned. Imagine large snowflakes drifting over your head, so near you that you can distinguish them, one from the other, and yet so thick in the air as to almost obscure the sky; then imagine each snowflake to be a meteor, leaving behind it a tail like a little comet; these meteors of all sizes, from that of a drop of water to that of a great star, having the size of the full moon in appearance: and you may then have some faint idea of this wonderful scene.
>
> Elder Samuel Rogers, 1880

Clear skies that night over much of the continent produced similar sights and fears that were repeated in cabin and teepee. Von Del Chamberlain, an astronomer specializing in Native American ethno-astronomy, writes that for American Indians out on the Great Plains there were:

> "Shooting stars like snowflakes in a storm." That was the phrase used by Oglala Teton Indians of the American Plains to name the year that other Plains Indians called, "Storm of Stars Winter," "Winter of the Falling Stars," and similar names. One record stated that the Kiowa ... were wakened by a sudden light. Running out from the tipis, they found the night as bright as day, with a myriad of meteors darting about in the sky. The parents aroused the children, saying, "Get up, get up, there is something awful going on." They had never before known such an occurrence, and regarded it as something ominous or dangerous, and sat watching it with dread and apprehension until daylight.
>
> Von Del Chamberlain, 1998

In 1898, Harriet Powers, a former slave from Athens, Georgia recorded the celestial event in brightly colored fabric in a quilt she made for the Reverend Charles Hall, chairman of the board of trustees of Atlanta University. Powers had never learned to read nor write so her quilts were her way of passing on the stories that were important to her life. Though time has caused the colors to fade, we can still see the power of her artistry and religious faith in the multiple panels that chronicle such Biblical stories as Adam and Eve, Jonah and the Whale, and the Crucifixion of Christ.

Figure 7.25 Born a slave in 1837, Harriet Powers depicted religious and astronomical themes in her quilts. In the central appliquéd square with grey background, orange meteors fall from the sky inducing terror and turmoil in the people below. The panel depicting the 'red light night' is at the bottom, second from the left. (Photograph © 2010 Museum of Fine Arts, Boston; Harriet Powers, American, 1837–1910 *Pictorial quilt* American (Athens, Georgia), 1895–98; cotton plain weave, pieced, appliqued, embroidered, and quilted, 68⅞ × 105 in. (175 × 266.7 cm), Museum of Fine Arts, Boston; Bequest of Maxim Karolik 64.619).

Right in the middle of these scriptural stories, in the very center of the quilt, is an appliquéd tableau of the Leonid meteor shower of 1833. Powers' own description of her scene, given upon delivery of the quilt reads, "The falling of the stars on Nov. 13, 1833. The people were frightened and thought that the end had come. God's hand staid the stars. The varmints rushed out of their beds." Its central location and the fact that Powers hadn't even been born the night the meteors fell, attests to the apocalyptic importance attached to the spectacle that was witnessed that night.[5] Today her story and her quilt hang in the Museum of Fine Arts, in Boston, Massachusetts.

[5] In the lower left portion of the quilt is a panel Powers describes as "The red light night of 1846. A man tolling the bell to notify the people of the wonder. Women, children and fowls frightened but God's merciful hand caused no harm to them." It is believed that this is a reference to a series of meteor showers Powers herself saw as a child on the nights of August 10 - 11, 1846. If so, then at least some of these shooting stars were Perseid meteors which still awe and delight observers to this day.

In a slightly less ominous vein, in 1934, almost exactly a hundred years after the 1833 shower, Frank Perkins and Mitchell Parish wrote the song 'Stars Fell on Alabama,' "We lived our little drama/ We kissed in a field of white/ And stars fell on Alabama/ Last night," which today, in honor of the song's title, has produced the only American license plate motto to memorialize an astronomical event.

Prior to the 1833 meteor shower it wasn't really known what meteors were. Some thought they were purely atmospheric phenomena, perhaps the result of flammable gasses and electrical discharges. Subsequent work by astronomers revealed that meteor showers are closely related to comets. Since each meteor is typically the fiery end to an interplanetary grain of sand, a meteor shower must be the result of a veritable river of sand through which our planet passes at the same point in its orbit every year. One of the principal discoveries of nineteenth century astronomy was that these celestial sandstorms are the calling cards of comets.

Comets are great mixes of rock and ice; "dirty snowballs" is the description astronomer Fred Whipple used in 1950. While these ice balls have their origin in orbits far from the Sun, every so often a random collision or gravitational influence of a passing planet sends one on a long plunge into the glowing heat of the inner solar system. When the frozen mass falls inward past the orbit of Mars, light and heat from the Sun begins to vaporize the comet's ices. Jets of dirty, dust-laden water vapor rocket millions of tons of these volatile compounds into a

Figure 7.26 Comet Hale Bopp above Cayuga Lake in upstate New York during the spring of 2007. The comet was so bright, stargazers could see it amid the glare of city lights, and it was visible for months as it passed around the Sun (T. Nordgren).

cloudy coma that envelops and hides the ever warming nucleus. Eventually, sunlight pushes the dust particles outward into a bright tail pointing away from the Sun. At its largest, a comet's tail can span the distance between the planets making them, for a brief time, the largest objects in the Solar System, even larger than the Sun itself.

Every comet produces two tails: dust reflecting the light of the Sun produces the large, bright, yellow-white tail that curves out away from the Sun while a second blue tail is composed of individual atoms stripped of their electrons and blown straight back from the Sun by the particles of the solar wind.

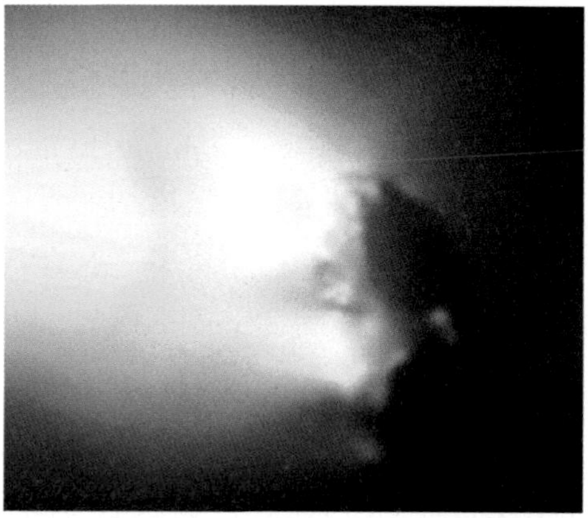

Figure 7.27 The icy nucleus of Halley's Comet is photographed by the European Space Agency's Giotto spacecraft as it flew through the comet's coma in 1986 (ESA, Courtesy of MPAe, Lindau).

As a comet passes around the Sun, its gradual evaporation leaves bits of itself behind that continue to orbit the Sun in a dusty stream. Where the ribbon of grit intersects the Earth's orbit, a meteor shower ensues whenever the Earth passes through it. This is why meteor showers occur at the same time each year. This is also why the Leonids are particularly spectacular every 33 years when its parent comet, Tempel-Tuttle, comes around and lays down a new swath of dust. The Leonids of 1966 were probably even more spectacular than the storm of 1833, while 33 years later, in 1999 and 2001, sky watchers were once more treated to a spectacular display. Seven years later in 2008 the Leonids were once more only an average occurrence.

Figure 7.28 Comet Hale Bopp from hazy skies in upstate New York. A comet's tails always point away from the Sun. The bright yellow tail is sunlight reflecting off dust ejected off the nucleus of the comet by the vaporizing ices. The blue tail is ionized gasses blown away from the Sun by particles of the solar wind (T. Nordgren).

Figure 7.29 The Spitzer Space Telescope captures this infrared image of the broken pieces of Comet 73P/Schwassman-Wachmann 3 as it sails along the trail of dust it left on previous orbits around the Sun. With each pass around the Sun, the comet left more of itself behind until, in 1995, it began to splinter apart. Each of its major pieces is a comet now, complete with a tail pointing away from the Sun. In 2022 the Earth is expected to pass close to this dusty trail resulting in a new meteor shower (NASA/JPL-Caltech).

The best time to see a shower is in the early morning hours after midnight and before dawn. Between these times you stand on the shadowed part of the Earth facing forward in our path around the Sun. Like riding in a car driving through a snowstorm, it's the passengers in the front seat that see the snow streaking toward them and perspective is the reason the snowflakes all seem to radiate outward from a point in front of the car. This point in a meteor shower is called the radiant, and the constellation in which it appears to be located is what gives the shower its name. After midnight, the constellation containing the radiant will almost always be rising in the east. To see as many shooting stars as possible, simply sit back and look up. Meteors will appear all over the sky, but every meteor that is part of the shower will always be shooting away from the constellation after which they are named.[5]

Every year we see the Perseids radiate from Perseus under late summer skies. Over two thousand years ago these meteors prompted the story of Zeus courting a young maiden by appearing as a golden shower. The fruit of their union was the hero Perseus who killed Medusa and whose constellation produces the most widely watched meteor shower each year on warm August nights. Most people think this shower is the best, as it is the only one the public typically hears about, but the most consistently spectacular shower is actually the Geminids of mid-December. The meteors of that shower are actually connected with an asteroid (3200 Phaethon), but it is probable that this is just a comet that long ago boiled away all its ice, leaving behind nothing but a rocky corpse.

Between the Perseids of summer and the Geminids of winter, are the

[6] You do not need a telescope to see meteors or a meteor shower. In fact, this is the worst thing you can do. Meteors will be all over the sky, while a telescope only allows you to look at one tiny part. The darker the sky, the more meteors you will see. The best nights are when there is no Moon. To tell if you are someplace dark enough to see a good shower, a simple rule of thumb is to look towards Polaris and see if you can see the stars that make up the handle of the Little Dipper. If you can, then you are in a fine location. If you can't even see Polaris, then you might want to find someplace darker.

Figure 7.30 Over a hundred meteors stream out of the constellation of Gemini in this composite image made over the course of three nights in December 2007. The Geminid meteor shower has become one of the more reliably brilliant showers in recent years, but because of cold temperatures very few people take the time to enjoy it. Notice that while meteors appear all over the sky in this fish-eye image, they all trace back to a single point, the radiant, in the constellation of Gemini (Berkó Erno).

Orionids of cool autumn nights. October's shower, that issues from Orion, is the result of the most famous comet of all, the one that Edmond Halley first realized passes by the Earth every 76 years. Halley's Comet won't return until 2061 but every October 20–22 you can see for yourself a part of its past. And while these are the meteors I will see tonight, my fondest memories are of the Leonids in November, because every year they occur on my birthday. So in a very real sense, I was born under a comet.[7]

Though we no longer view the Leonids, or any strong shower, as the end of days, it is important to understand the danger that comets and their inner Solar System cousins, the near-Earth asteroids pose. As of my writing this, there are

[7] While I am certainly not making any claims of comparison by noting this, Samuel Clemens (Mark Twain) was born the same year that Halley's Comet made an appearance in the sky. As he, himself said, "I came in with Halley's Comet in 1835. It is coming again next year, and I expect to go out with it The Almighty has said, no doubt: 'Now here are these two unaccountable freaks; they came in together, they must go out together." And so they did.

Table 7.1 Annual meteor showers

Meteor Shower Name	Date Range	Peak Date	Constellation Radiant	Peak Hourly Rate	Comet
Quadrantids	28 Dec–7 Jan	3–4 Jan	Quadrans Muralis(1)	10–60	?
Lyrids	16–25 April	22 April	Lyra	10	Thatcher
Eta Aquarids	21 April–12 May	5–6 May	Aquarius	10	Halley
Southern Delta Aquarids	14 July–18 Aug	28–29 July	Aquarius	15–20	?
Perseids	23 July–22 Aug	12–13 Aug	Perseus	50–80	Swift-Tuttle
Orionids	15–29 Oct	21 Oct	Orion	20	Halley
Leonids	13–20 Nov	17–18 Nov	Leo	10(2)	Tempel-Tuttle
Geminids	6–19 Dec	13–14 Dec	Gemini	50–80	3200 Phaethon

All data from Peter V. Bias' book *Meteors and Meteor Showers*. See his website at http://Meteorshowersonline.com.
Notes: (1) Obsolete constellation near the handle of the Big Dipper. (2) With Comet Tempel-Tuttle's return every 33 years this number can go up to several hundred an hour. The last outburst was in 1998-2002.

1,054 Potentially Hazardous Asteroids (PHAs) known. These are defined as those rocks larger than about a football field that come closer to the Earth than 20 times the distance to the Moon.

To get a sense of the danger, look back at the Moon. Consider that every crater is the record of an impact at some point in the last 4.5 billion years. The only impact crater you might be able to make out with your unaided eye is Tycho, a relatively fresh crater in the southern hemisphere with long bright rays that stream out in all directions and make the crater easily visible.[8] Fittingly then, it is named for the last great astronomer before the invention of the telescope.

While you can see Tycho's rays easily enough without a telescope when the Moon is full, the crater itself is best seen through binoculars or a telescope a few days before, when shadows still give a sense of the crater's full depth and dimension. It measures 53 miles (85 km) across, wider than Great Smoky Mountains National Park is long, and with a depth of nearly 3 miles (5 km),

[8] For those who see an upturned woman's profile in the Moon's dark markings, Tycho is the diamond jewel at her throat.

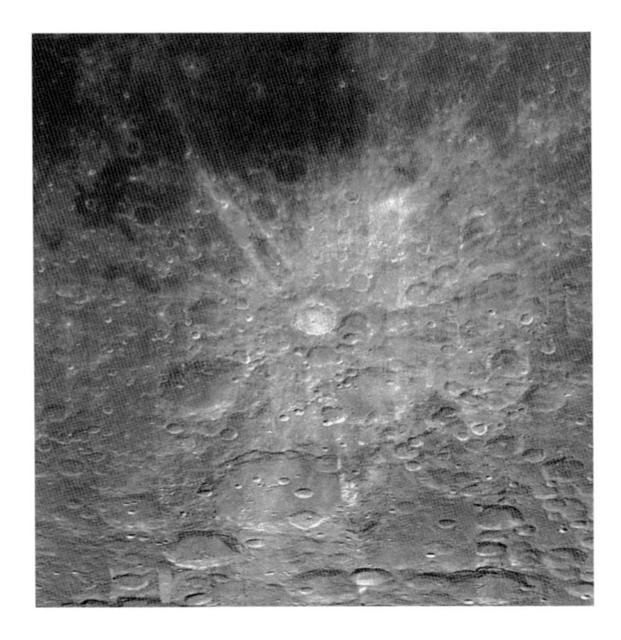

Figure 7.31 Because Tycho is one of the youngest impact sites we can see on the Moon, the rock and debris it ejected is still bright against the darker lunar surface. Tycho is thought to be less than 100 million years old based on samples of the rays brought back by Apollo 17 (NRL/Clementine).

Tycho is three times deeper than the Smokies are tall. For those of you on the west coast, this makes the crater three times deeper than the Grand Canyon and three times wider than the canyon at its widest. And yet a scar that would be a major tourist attraction on Earth is just one of umpteen-million craters, big and small, that cover the surface of the Moon.

Perhaps the most obvious sign of asteroid impacts are the large dark features that are the most easily seen from the Earth. Radioactive dating of Moon rocks brought back by the Apollo astronauts revealed that roughly 4 billion years ago the Moon underwent a period of extraordinarily heavy bombardment. During this time impacts on the Moon, and presumably the Earth, Mars, Venus and Mercury, were so intense that enormous impact basins were hollowed out of the crust. A billion years later, dark molten magma would well up out of cracks in the crust and pool in these low lying basins making it easier to see their boundaries. Three billion years later human beings would look up in wonder at the patterns they created and call them the maria, the Man in the Moon.

Ever since this time of heavy bombardment, debris has rained down on the Moon and planets at a low but more or less continuous rate producing those few craters we see on the lunar plains. And even though their numbers are small, some craters are quite big, such as the 62 mile (100 km) diameter crater Copernicus in the middle of Oceanus Procellarum, the Ocean of Storms. It's the scarcity of craters that confirm the maria are younger than the surrounding lunar surface. The number of craters you see on any particular body in the Solar System therefore serves as a clock revealing how old or young the surface may be.

The planets and moons themselves are all about the same age, 4.5 billion years

Figure 7.32 Ronald Evans, in Apollo 16, looks down on the flooded impact basin that is Mare Crisium. Notice how the dark volcanic rock that welled up and filled the huge circular bowl, overflowed parts of the surrounding rim, burying the older surface to the upper right. What few craters exist in Mare Crisium, only occurred after the basin was flooded (R. Evans/ NASA/LPI).

old, but because of the geological processes that have gone on there since, some have younger surfaces than others. It's as if the worlds in our Solar System were back-country roads. A mountain road built back in the 1930s when this park was created would be heavily potholed by now. But repave a segment every now and then and there will be parts where you can drive a car without damage. So while the entire road itself will be nearly 75 years old, different sections will appear younger by an amount revealed by the smoothness of the ride. On the Moon, the lunar potholes producing Mare Imbrium, Nectaris, and Serenitatis (the Seas of Showers, Nectar, and Serenity) were filled about 3 billion years ago, with the last of the remaining maria filled about a half billion years later. Since then the lunar surface has simply been left to bake under the Sun and accumulate its scars with the years.

What did the Moon do to deserve this bombardment while we on Earth were saved? In reality it did nothing; there is nothing special about the Moon, no propensity for impact, no magnetic, or gravitational potential that draws in the debris. For every asteroid that has ever scarred the Moon, the Earth has been hit by just as many.[9] The question is not why the Moon has so many craters, but why the Earth has so few.

[9] In fact, while every square mile of the Earth's surface has been hit by the same number of meteorites as every square mile of the Moon, because the Earth is so much larger than the Moon, we have actually been a celestial target far more often.

Figure 7.33 Craters are found on virtually every moon and planet with a solid surface. University of Redlands student, Samantha York, has crocheted and quilted a representation of the Valhalla impact basin on Jupiter's moon Callisto (Samantha York).

The scarcity of Earthly craters is naturally explained by the wind and the water, the mountains and the streams around me. The Earth is an active place where volcanoes erupt, continents collide, oceans are deep, and wind, rain and ice, erode. Of all the exotic places in the Solar System, the Earth has the youngest surface of them all because our planet is constantly recreating itself.[10]

Still, the Earth is not wholly without extraterrestrial mar. Interstate 40 skirts the northern boundary of the park as it cuts through the Appalachians on its way west across the country. Seventeen hundred miles down the smooth multi-lane hardtop, outside Winslow Arizona, I-40 paved over the much older byway of historic Route 66. There, set amongst the tacky charm of roadside hucksterism is the original out-of-this-world road-side attraction: Meteor Crater.

Meteor Crater, also known as Barringer Crater, is a pretty big hole in the ground by Earthly standards. It's 600 feet (183 m) deep, nearly a mile (1.2 km) wide, and has a rim that rises 150 feet (45 m) above the surrounding redrock plains of Northern Arizona. When Europeans first came across the crater in the

[10] OK, Jupiter's moon Io probably has an even younger surface than ours as its globe girdling volcanism constantly recovers its surface resulting in the one other body in the Solar System with no impact craters.

1800s its uniqueness was obvious, set as it was between the barren plains of the Navajo reservation and the ancient snow-capped volcano of the San Francisco Peaks. In 1891, G. K. Gilbert, the pre-eminent geologist of the day and head of the U.S. Geological Survey, was drawn to the Arizona Territory by reports of iron-nickel meteorites around the giant depression. Gilbert had already speculated on the possibility of extraterrestrial impacts, but because the mechanics of planetary collisions were not well understood, he hypothesized that a hole that big should have an equally giant iron meteorite buried beneath its surface. Today we know that at the interplanetary speeds at which asteroids and Earth collide, a very small body can make a very large hole and the explosive force of a large enough impact is enough to shatter the meteorite into a million small fragments. When Gilbert's magnets turned up no giant mass of iron he announced the only viable hypothesis was that the crater was the result of exploding volcanic steam; the abundance of nearby meteorites, was merely a coincidence.

Figure 7.34 Astronauts onboard the International Space Station captured this photo of Meteor Crater on the plains of northern Arizona. Canyon Diablo is the meandering canyon to the upper left (Image Science and Analysis Laboratory, NASA Johnson Space Center, 'The Gateway to Astronaut Photography of Earth,' http://earth.jsc.nasa.gov/).

A decade later, Daniel Moreau Barringer, a Philadelphia mining engineer, former lawyer, outdoor enthusiast, and hunting companion of Teddy Roosevelt would be convinced the crater really was the result of an impact. All around the crater the rock that had been ejected in its formation was laced throughout with bits of iron meteorite. If the meteorites had fallen first, they should be buried beneath the ejected rock, if they came afterwards, they should be sitting on top. Only if a single event was responsible for both, would they be found as they were. No matter what Gilbert's results might have been, Barringer was sure the enormous iron mass must be somewhere beneath the crater floor.

After securing the mineral rights to the land, Barringer created the Meteor Crater Exploration and Mining Company and spent the next 20 years looking for the iron meteorite he estimated weighed 10 million tons. For two decades he battled quicksand and skeptics. At first his battles were with those who believed, like Gilbert, that the crater owed more to volcanism than cosmic encounters. Eventually though, his fight shifted to those who accepted a meteoric origin, but

Figure 7.35 Visitors stand on an observation platform that hangs out into the bowl of Meteor Crater (also known as Barringer Crater). I took this photo standing high up on the rim, amid the rocks and boulders ejected there during the explosive impact. The lunar crater Tycho is almost a hundred times wider with bright ejecta that stretches for over a thousand miles across the Moon's surface (T. Nordgren).

based upon lunar observations, thought that the crater's size was due to the explosive power of the impact, and not simply the large size of the impactor.

If this latter point were true, which we now know it is, then the fortune Barringer had spent would have been for waste. When in 1929, yet another mine shaft failed to find any iron, and renewed calculations showed there should be little to no iron left from any impact, the board of directors pulled the plug. Two months later Barringer was dead of a heart attack.

In 1963 Gene Shoemaker, a geologist with the astrogeology branch of the U.S. Geological Survey (which he founded in nearby Flagstaff, Arizona), conclusively proved Barringer Crater was caused by a meteor impact by looking at the composition of its minerals and the structure and layering of its various features. The only naturally occurring explanation for the required high pressures and temperatures comes from beyond the Earth, and is the result of the explosion created when the Earth and an asteroid collide.

Today Meteor Crater is owned and operated by Barringer's descendents as a scientific and tourist destination.[11] The techniques Shoemaker used to identify

[11] While it is designated as a national natural landmark, it is not part of the national park system, so don't try to use a national parks pass to enter there. Even so, the privately run visitor center is well worth a stop on any road trip down Route 66.

Figure 7.36 On the rim of Grand Canyon, just off I-40 and only a few dozen miles from Meteor Crater, a piece of the iron meteorite that formed the crater sits on display in Verkamp's Curio Shop. Prior to 2008, the Verkamps had operated the shop for as long as the Barringers had operated at the crater (today the Verkamp shop is a park visitor center). A sign on the meteorite chunk says it was found in Canyon Diablo, weighs 535 pounds (243 kg), is 92% iron, 8% nickel, with trace amounts of gold, silver, platinum, and diamond (T. Nordgren).

Meteor Crater as an asteroid impact, have now positively identified 150 other impact craters around the world, and have been used to subsequently study craters on the Moon. The Apollo astronauts even trained in Meteor Crater in order to get experience with what they could be expected to see when they got to the Moon.

The impact that created Meteor Crater released as much energy as 2.5 megatons of TNT, about 150 times the force of the atomic bomb that destroyed Hiroshima.[12] While this would have obviously been a very bad day for anything grazing in the neighborhood we know from the Moon that the Earth must have been hit by far worse many times in its past. What must one of those have been like?

In my rambles through the Smoky Mountain woods I have come upon salamanders, frogs, lizards, wild turkeys, elk, and a bear. One animal I didn't see,

[12] During the Cold War it was a fear of some astronomers that if the Earth should be hit by a large enough meteor, even one smaller than what created Meteor Crater, the United States or Soviet Union would mistake the explosion for a surreptitious nuclear first strike by the other side. As a result, a very unlucky asteroid encounter could be the spark to ignite an accidental nuclear war.

Figure 7.37 This is what happens when Route 66 entrepreneurship and astronomy collide on dusty back roads (T. Nordgren).

however, was a dinosaur. The very lack of giant, ferocious man-eating dinosaurs (with the exception of the tourist trap attractions of Gatlinburg and Pigeon Forge just outside the park's boundary) is due to these planetary impacts. My very presence here, and the wildlife I hope to see or avoid, is thanks to astronomical collisions.

Sixty-five million years ago life on Earth looked very different than it does today. Reptilian-looking creatures, big and small (but mostly really big) had been the dominant animal throughout land, sea, and air for 180 million years. Think about that: dinosaurs ruled this planet for three times longer than we mammals have had it to ourselves. All over North America, dinosaur fossils are found in layers of rock, provided the rock is older than 65 million years. In North Carolina and western Tennessee fossil remains have been found of duck-billed dinosaurs thought to have lived about 80 million years ago. But in all rocks younger than 65 million years, any trace of these globe-spanning creatures is utterly absent. What's more, 65 million years ago, at the boundary that marks the end of the Cretaceous and start of the Tertiary periods (called the K-T boundary for short) as many as 75% of all the Earth's animal and plant species appear to have suddenly become extinct.

Dinosaurs, like astronomy, are something every kid loves at some point when they're growing up. By the time I was 12 I could name all the really cool dinosaurs, and to make matters even more awesome, whatever happened to them was a complete and utter mystery. How neat is it that in the years since then an asteroid impact has turned out to be the chief culprit in their extinction? My childhood interests have come full circle.

The evidence for dinosaurs' astronomical demise comes from many directions. In layers of rock around the world that span the date of their extinction, a layer of brown clay is found that is rich in the rare mineral iridium (at least compared to the soil above and below it). While iridium is not normally found on the Earth's surface, it is much more common in meteorites. The 65 million year old layer of clay also contains grains of shocked quartz (as if the quartz had been hit by a great force) and tiny tear-drop shaped glass rocks called tektites that are known to form when molten material solidifies as it falls through the air. In addition to these, quantities of soot are found in many sites around the globe as if great fires had burned the landscape during its formation. When these pieces of evidence are paired with a recently discovered crater nearly 120 miles (200 km) across partially buried off the coast of the Yucatan Peninsula near the town of Chicxulub (after which it's been named) the evidence is overwhelming that a massive impact took place 65 million years ago.

Figure 7.38 Cartoon by the author illustrating an alternative idea for how an asteroid might have wiped out the dinosaurs (T. Nordgren).

Imagine what it might have been like to be alive here in the American southeast on that fateful day the asteroid hit. For millions of years the 9 mile (15 km) wide mass of rock had been floating in space, orbiting the Sun on a path that brought it near the blue-green planet.[13] Many times it passed quite close to the Earth, or where the Earth would be just a few days or months later. But now the two trajectories cross at just the wrong time.

Today is a fine summer day and the sky is a drowsy, hazy blue. Over the horizon, coming in out of the south, the giant rock streaks through the thin atmosphere. The meteor's surface chars with the heat of atmospheric friction but inside it remains the same subarctic cold of the space in which it's been immersed for over ten thousand centuries. In less time than it takes to read this paragraph the asteroid slams into what will become the Gulf of Mexico over a thousand miles away from where I now stand.

In an instant, the energy of a hundred million hydrogen bombs is unleashed creating a fireball that erupts over the curvature of the Earth, shining three times brighter than the Sun. The skin on my face blisters and swells with second degree burns while many of the leaves around me burst into flames. All those poor plants and animals within 60 miles (100 km) of the impact are immediately incinerated by a fireball 400 times the intensity of the Sun.

The impact itself creates a crater nearly 120 miles (200 km) in diameter and

[13] Bill Bottke, a planetary scientist in Boulder Colorado, has found evidence that two large asteroids collided 100 million years ago in the asteroid belt. Based on the orbit of modern asteroids, the position of the potential collision was at just the right location in the belt that debris would have been sent spiraling into the inner Solar System. The asteroid that killed the dinosaurs 65 million years ago, as well as the asteroid that created Tycho Crater on the Moon, may have their origin in that collision. In fact, the near-Earth asteroids we see today may be the last remaining remnants of that long-ago event.

sends massive shockwaves through the Earth that reach me five minutes after impact with a rumble that shakes the ground where I stand. I stumble, but don't fall. The still smoldering trees around me shake wildly, sending countless winged creatures into the sky.

Ten minutes after the impact, thick molten rock from the impact rains down out of the sky over North America killing most life on the continent instantly. A little over an hour later, I'm no longer in any position to care about the 400 mile per hour (650 km/hr) air blast that races across my burnt and buried corpse, knocking flat the now raging inferno that is the forest where I once stood.

While I and all life on the continent is wiped out within the first few minutes, the rest of the planet follows suit pretty quickly. The hot, molten chunks of the Yucatan Peninsula that were blasted into space, rain back down all over the planet for the next hour or so. The sky overhead literally bakes the planet's surface as the re-entering rock heats up and radiates energy in every direction. For as long as material falls out of the sky the surface of the Earth is as hot as an oven and any creature that isn't able to bury itself deep in the ground, in dark holes and cool mud, or far down in the bottom of lakes or seas, dies in those first few hours after the impact.

For those that survive the initial hours, death comes slower. Sulfur compounds blasted into the atmosphere react with water vapor, producing concentrated acid rain that kills those aquatic species that survived in lakes and seas. Dust from the impact and smoke from the globe-spanning fire-storm blot out the Sun. Darkness and winter reign over the entire planet for months regardless of the local season. Green living plants and algae that depend on sunlight for food wither and die and nothing new grows for up to a year. Those animals on land and sea that depend on these plants for their food die soon thereafter. Carrion feeders have a field day at first, but soon they too starve.

Though probably 99% of all living organisms died as a result of the catastrophe, the one percent that survived included small furry animals that were able to burrow deep and live off food they had stored, not needing much to eat because of their small size. When the dust finally settled, they found themselves in a world without predators where nearly every ecological niche suddenly had room to go forth and prosper. 65 million years later they became us; we are only here because the dinosaurs are not.

Who knows what the world would look like today if orbital paths had been just a little different? There is preliminary evidence that many mass extinctions in our planet's past may be the result of unlucky collisions. In each instance, our world changed.

Look back at the Moon again. Even the Moon and the season's themselves may be the greatest sign of our planet's bombardment. The rocks the Apollo astronauts brought back show our sister world contains almost no metals like the Earth, or any other elements (like water) that are easily vaporized. But the rock that remains is very similar to Earth's mantle, our thick upper crust. This mix of similarities and differences with the Earth originally confounded ideas of where the Moon came from. For example, if the Moon formed in orbit around the

Earth, then it should have been built out of the same stuff we were and so have the same composition as the Earth.

Since the end of Apollo, planetary scientists have proposed an explanation that fits all the new data that had been brought back from the Moon. They theorize that 4.5 billion years ago, in the final days of planet formation, the nearly complete Earth was struck by another young planetoid roughly the size of Mars. By this stage in the Earth's formation, the Earth would have been fully differentiated as all the high density metals sank downward forming our core, leaving lighter density rock forming our planet's mantle. A glancing blow by another differentiated body would have sent a spray of the Earth's rocky mantle

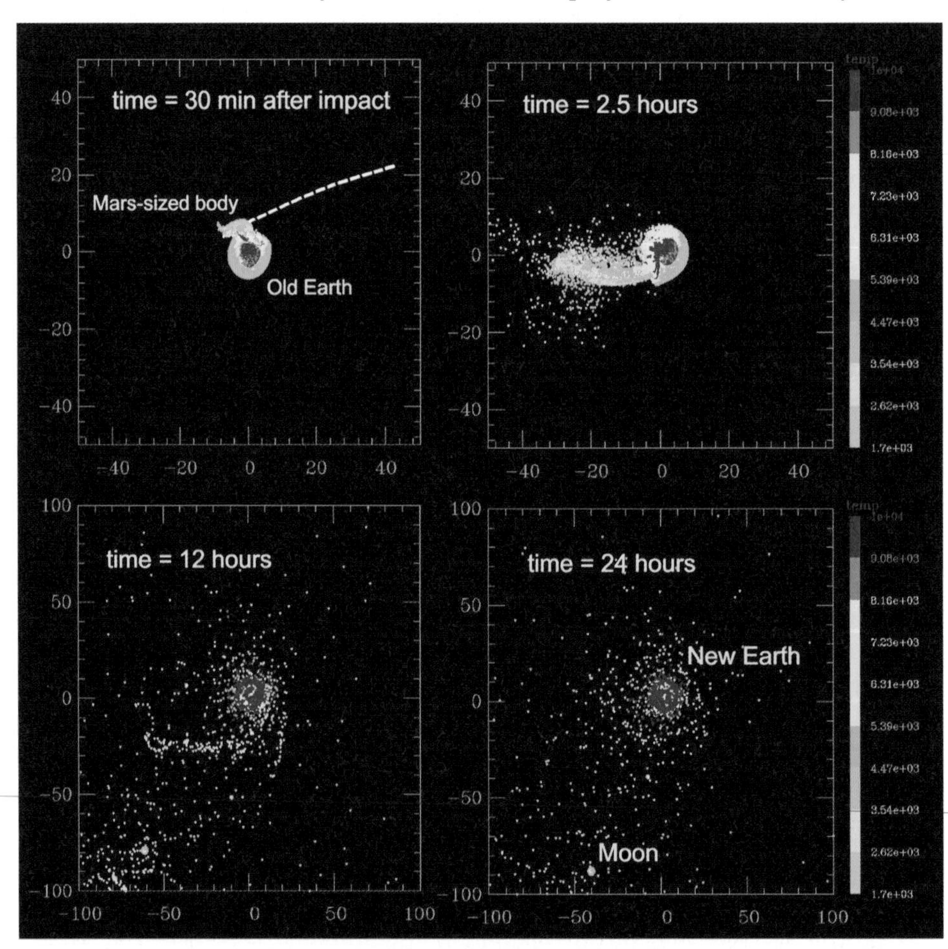

Figure 7.39 Robin Canup in Boulder Colorado successfully simulates the formation of the Moon as the result of a glancing collision between the early Earth and a Mars-sized planet. Colors show the temperature of the heated material. The original Earth begins with a mostly molten metallic core and a cooler rocky exterior. After only 24 hours the new Earth (a mix of both original planets) is largely complete with a new massive Moon (R. Canup/Southwest Research Institute).

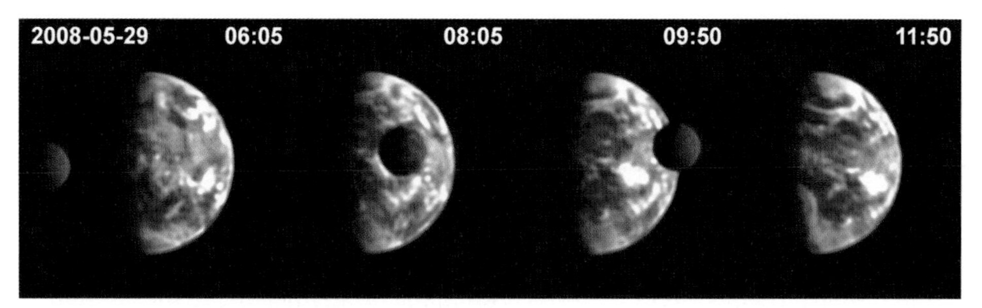

Figure 7.40 NASA's Deep Impact spacecraft that sent a probe hurtling into a comet nucleus, in order to see what it was made out of, turns back to look at the distant Earth and Moon from a distance of 31 million miles (50 million km). Africa is visible on the sunlit side of the Earth in the first frame. Five hours later South America and the east coast of America come into view with the sunrise (Donald J. Lindler, Sigma Space Corporation/GSFC; EPOCh/DIXI Science Teams).

Figure 7.41 The Solar System is still a dangerous place. In 1994 a comet that had once passed too close to Jupiter, and broken up into nearly two dozen pieces, re-crossed Jupiter's path, with each piece hitting the giant planet one after another. Every telescope in the Solar System, from backyard amateur telescopes to mountain-top observatories to the Hubble Space Telescope and cameras onboard the Galileo spacecraft on its way out to Jupiter all witnessed explosions erupting out of the Jovian clouds (top frame) that would have dwarfed the Earth with its aftermath (lower frames) (R. Evans, J. Trauger, H. Hammel and the HST Comet Science Team and NASA).

into orbit around the coalesced remains of the two previous worlds (the Earth we know today is the sum of these two once separate worlds).

In the aftermath of the collision, the newly-formed Earth now had an axial tilt (or at the very least, a new one) giving rise to seasons. Meanwhile, within a day of the collision, the ring of debris orbiting the new planet would have begun to contract to form our Moon. This moon would lack most metals and be made mostly of rock just as the Apollo astronauts found. From that day forward the Earth and its Moon formed the Solar System's sole double planet system.

Since then, the Earth's tidal forces have slowed the Moon's rotation so that it completes one turn on its axis for every trip around the Earth. In this way the Moon always keeps the same face pointed towards us. Nearly every moon in the Solar System is tidally locked like this to its planet. But tidal forces from the

Moon also work on the Earth. They raise the ocean tides we experience at the seashore while also slowing our rotation by 1.7 milliseconds per day per century. Eventually the Earth will turn once on its axis for every trip the Moon makes around us and the two planets will be locked together keeping their same faces forever facing one another. When that day comes, one whole hemisphere of the Earth will lose its view of the Moon while for the other hemisphere it will remain a constant fixture, motionless in the sky. It's impossible to say which hemisphere will win the Moon as our continents are constantly moving across the face of the world. Indeed, the distribution of the continents on Earth actually plays a role in the fate of the Moon.

The Earth turns once every 24 hours, faster than the time it takes the Moon to complete one trip around it. Friction between the waters of the tides and the Earth's seafloor does two things: 1) it slows the rotation of the Earth, and 2) it drags the tidal bulge forward with the Earth, ahead of the motion of the Moon. From the mass of this watery bulge our Moon feels a forward-pulling gravitational attraction. The extra gravitational energy raises the Moon into a higher orbit that increases its distance from the Earth below. The result is that as the Earth slows, the Moon spirals away.

The rate at which the Moon recedes therefore depends upon the ocean's friction with the Earth. Place a lot of bumps and barriers all around the Earth, let's call them continents, and the net frictional force is large: the rate increases at which the Earth slows and the Moon recedes.

Forty years ago Apollo astronauts placed reflective devices on the Moon. Powerful lasers fired from the Earth, reflect off these mirrors and the total travel time from the Earth to the Moon and back, tells us the precise distance to the Moon. These simple measurements reveal the Moon is moving away from us at a rate of 1.5 inches (3.8 cm) per year. At the Moon's current distance from the Earth, run the movie backwards, taking into account that the rate of recession increases the closer the two get, and our calculations require the Moon to come spiraling out of the Earth as little as 2 billion years ago. But we know the Earth and Moon are older than that.

The rate at which the Moon is currently receding is therefore much too fast if everything else we understand about the Moon, the Earth, the Solar System, and basic physics is correct. But if at some point in the past all the continents were bunched together in one single mass, then the total frictional force is less: the Earth fails to slow as quickly, and the Moon's outward spiral slackens. The only way the Moon can be where it is, receding as it is, and be older than the rocks that make up its mountains and maria, is if the Earth has periodically had supercontinents.

The evidence for periodic bunching of the continents is wonderfully all around me. The Great Smoky Mountains, and the Appalachians of which they are a part, formed 300 million years ago when the continents of North America and Africa collided to form the supercontinent of Pangaea. Like the Himalayas, which are a result of the Indian sub-continent plowing into Asia, the combined force of the continents' collisions drove the Earth's crust upwards building an

Figure 7.42 The center of the Milky Way rises above the southern extent of the Appalachian mountain range through Great Smoky Mountains National Park (T. Nordgren).

Appalachian range that probably towered higher than any other mountains on Earth today. Who knows how many Everests once stood where we stand now?

About 175 million years ago, Pangaea split and the now familiar continents began to go their separate way. The forces building the Appalachians ended and eons of erosion have been at work ever since wearing them down to dust. The Great Smoky Mountains we now see are nothing more than the heavily weathered bones of those once great mountains while the coastal plains that make up the eastern seaboard are their washed away remains. But look around now and in the gorgeous gently rolling, mist filled mountains is evidence of the Moon's evolution overhead. Even if we had no other evidence for Pangaea, the Moon I see over these mountains that are its legacy tells me it must have been so.

> Soon as the evening shades prevail,
> The moon takes up the wonderous tale,
> And nightly to the listening earth,
> Repeats the story of her birth
>
> Joseph Addison

This is what I see when I look around me at these Smoky Mountains. The roll of the hill, the light in the trees, and the color of the leaves; the wild turkeys in the grass, the farmer in the field, and the Cherokee people and their history; all of these are astronomical in origin or influence. All around me is astronomy.

And as the Sun finally sets, and the Park Ranger leads a procession of campers out into the meadows of Cades Cove where the local Smoky Mountain Astronomical Society has set up their telescopes, I am excited to see that the stars are beautiful overhead. I don't know what people were warning me about; this is a wonderful place for astronomy.

See for yourself

Moon phases

Watch the position of the Moon change from day to day, apparently getting larger as it gets farther away from the Sun in the sky. Each night right after sunset, stand in the same spot at the same time and notice the position of the Moon. Each day it will move farther east. This is the Moon's actual orbital motion around the Earth. It takes a little less than a full month (month comes from the word, moon) to go once around and come back to where it started. As the Moon gets farther away from where the Sun is setting, you will see more of its sunlit hemisphere and less of its night side. These are the Moon's phases. New Moon is when the Moon is nearly between us and the Sun and so we see only its dark, unlit side. A week later we see equal parts lit and dark (like a **D**) and this is called First Quarter. A week later we see the fully lit hemisphere and the Moon is full. One week after that and the Moon is back to being half lit and half dark (now like a **C**) and this is called third quarter. One week later and we are back to seeing the fully dark side and new Moon again. When the Moon's phases get fatter, it is waxing; when the phases get thinner, the Moon is waning.

Moon features through binoculars or 'scope

Look at the Moon through binoculars or a spotting 'scope. A magnification of as little as 16× will show nearly all the features Galileo saw. Notice the bright, rough and cratered highlands versus dark, smooth lowlands (maria). Look near first quarter, and along the terminator (the line dividing the lit side from the dark) you are seeing sunrise on the Moon. Shadows are long, and lots of detail is visible. Avoid full Moon when sunlight shines straight down on the Moon and all shadows and detail disappear.

You can see how rough the cratered highlands are compared to the smooth maria by looking at how the terminator cuts across the two regions. Galileo first noticed this in 1609. Notice how many of the lunar maria (most obviously Mare Crisium) are circular, and thus reveal their origin as giant volcanically flooded impact craters. The farside of the Moon, which we never see, has almost no maria at all. The highlands where Apollo 16 landed are best seen at First Quarter, seven days after new Moon.

Meteor showers

Meteors can be seen any night of the year from dark locations. There are about eight major meteor showers each year. While each peaks during one or two nights, meteors from these showers are often visible for several days before and after. During showers, observers in dark locations may see as many as a few hundred meteors an hour. Wherever you happen to be, showers are best viewed after midnight. Find a position that allows you to see the most amount of sky

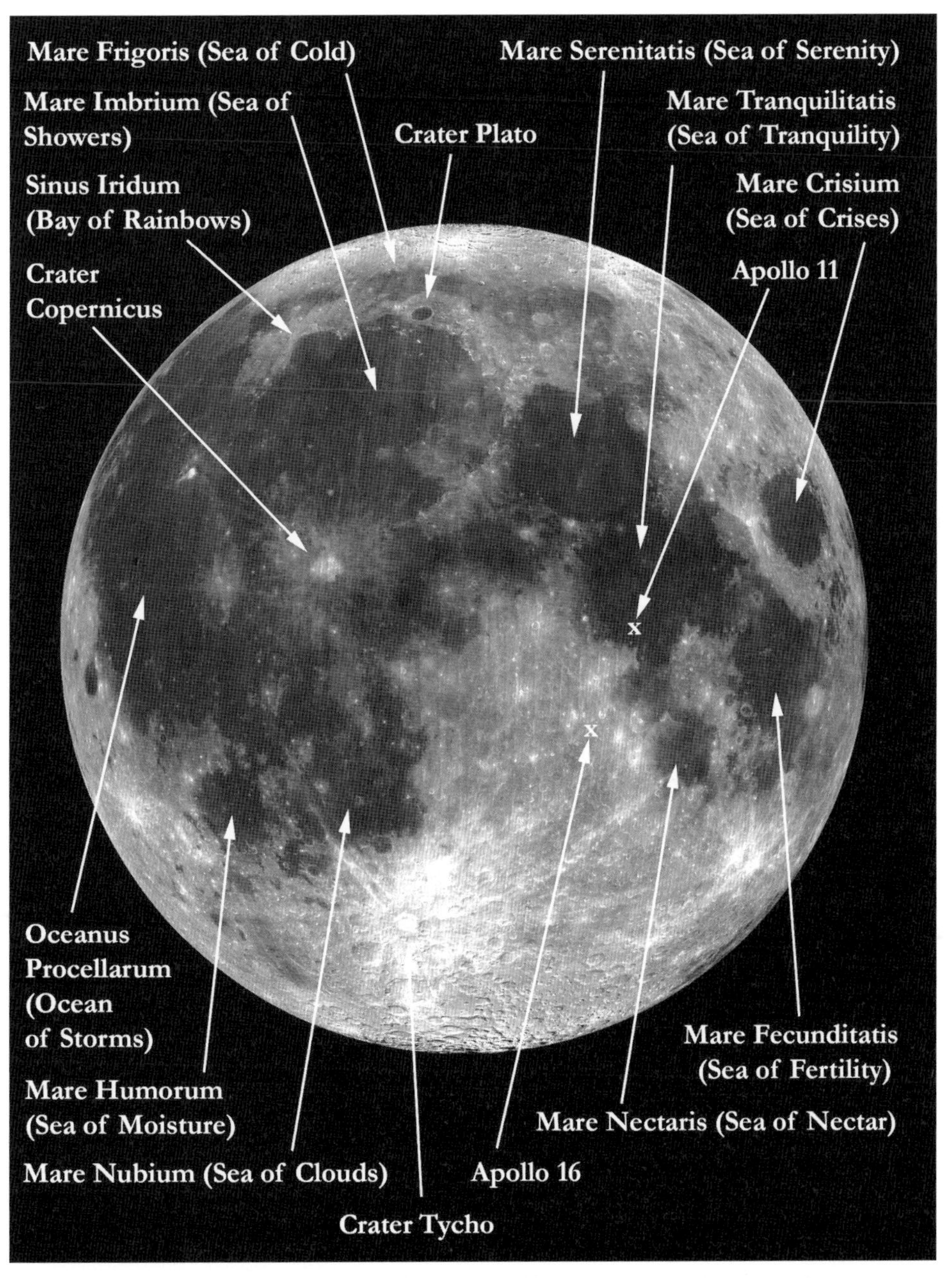

An annotated mosaic image of the full Moon showing the main features that may be discerned with the naked eye, binoculars or a small telescope (Clementine Full Moon Mosaic (NRL)).

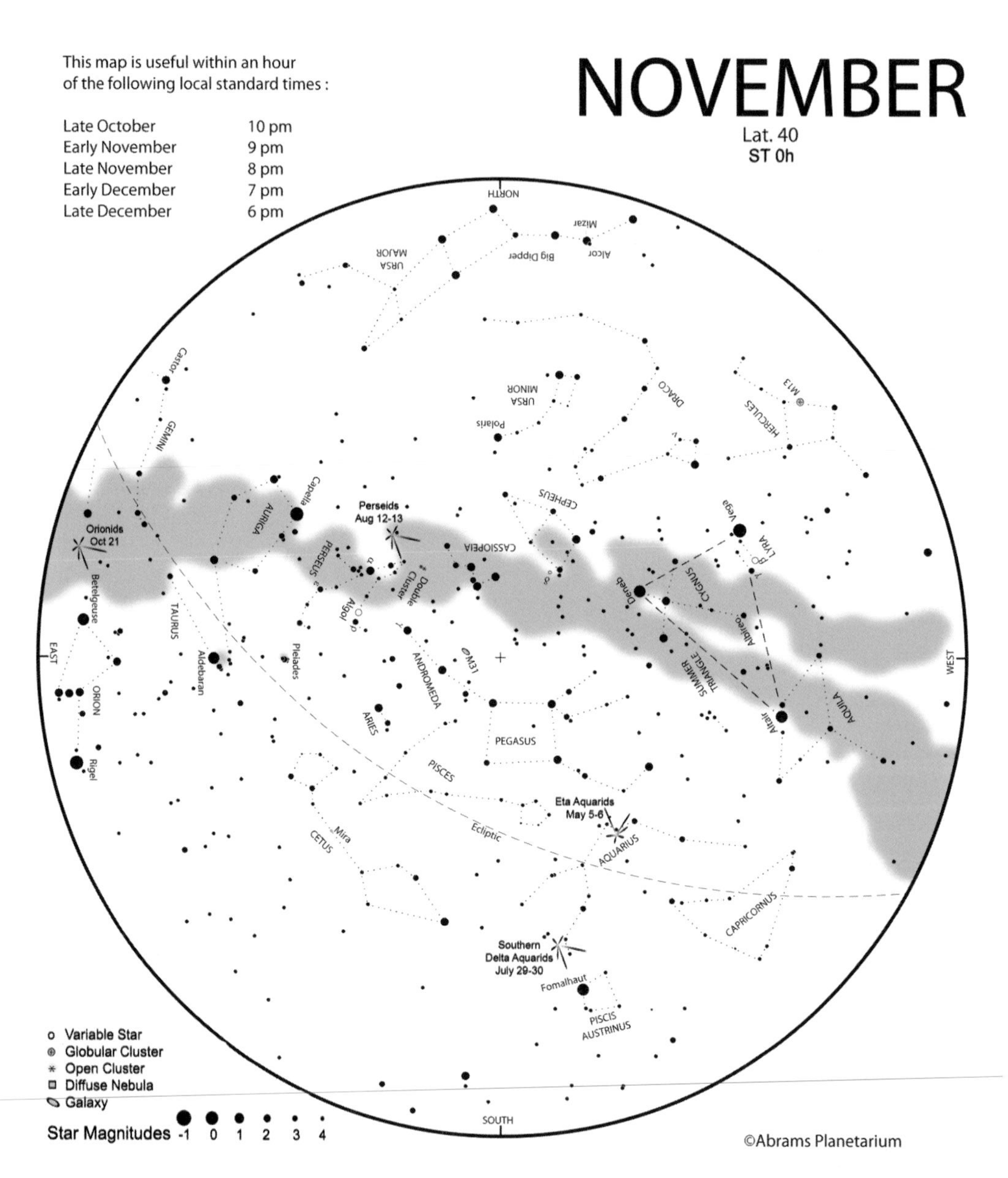

Hold the star map above your head with the top of the map pointed north. The center of the map is the sky straight overhead at the zenith. The meteor shower radiants from Table 7.1 are shown on the November and May star maps to give you an idea of their position relative to the constellations. Since meteor showers are best seen after midnight, many of these constellations will not be visible on the corresponding month's evening star map. To see what the sky looks like after midnight for a particular meteor shower, do the following: 1) Use the November and May star maps to find the radiant of a particular shower (for instance, the Perseids in the constellation of Perseus); 2) Note

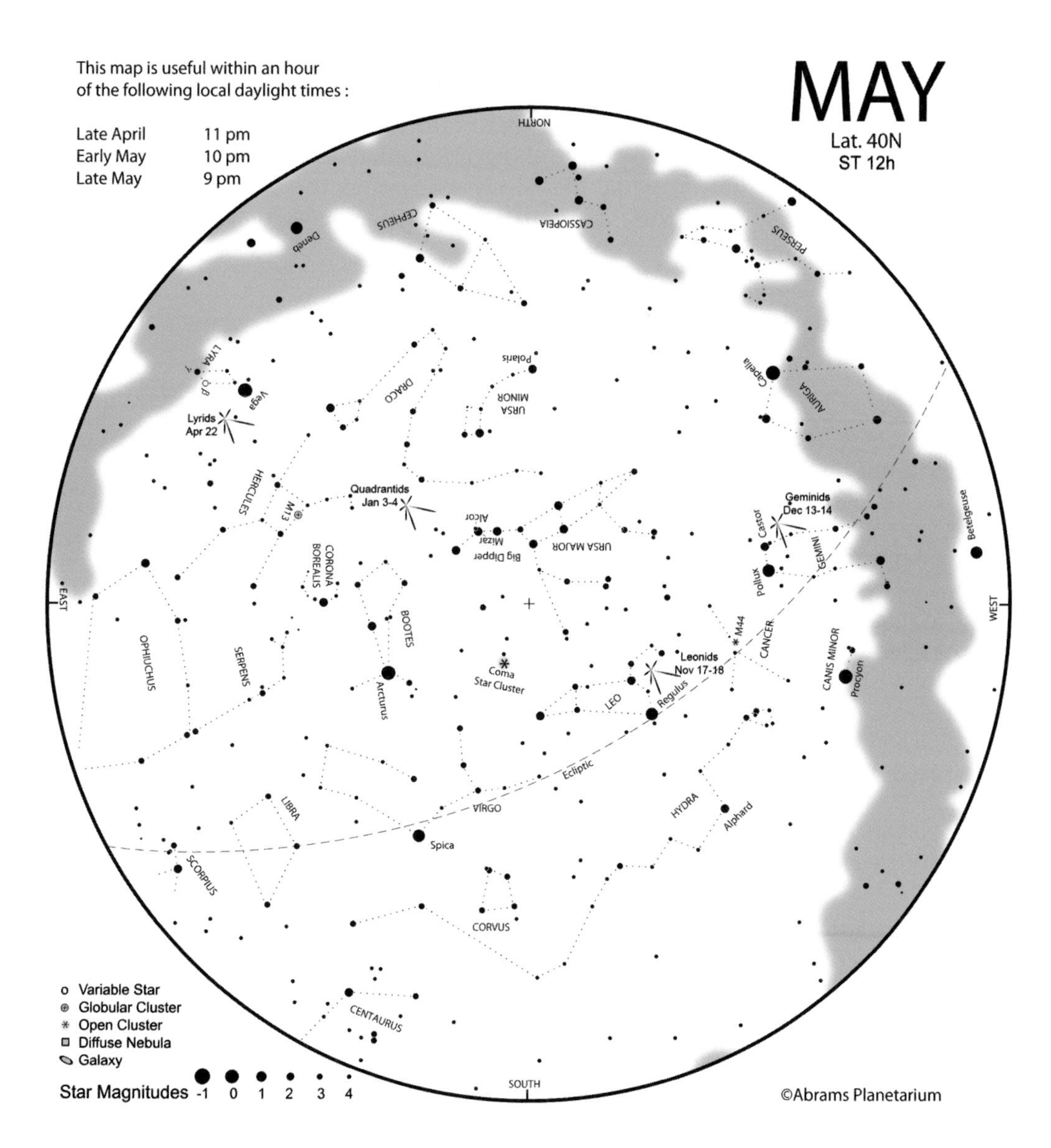

the date of the shower (in this case, August 12-13); 3) Consult the star map for August. Notice that in the upper left it says this map shows how the sky will look at 10:00pm during early August (when the shower is at its peak). 4) To see what the sky will look like at midnight you will need to find a star map that shows what the August sky will look like two hours later. Notice that in the upper right of the August map it states that this view is for a Sidereal Time (star time) of 18 hours, "ST 18h." 5) To find a view of the sky for two hours later, check the monthly star maps for one with a Sidereal Time of 20 hours, "ST 20h." This turns out to be the map for September. By consulting the September star map you will see the constellation of Perseus is just rising to the northeast; this is the view at midnight after which showers are usually at their best.

and then sit back and wait. After midnight the radiant will generally be towards the east so meteors that are part of the shower (as opposed to normal random meteors) will be predominantly shooting towards the west and will all trace back to the constellation after which the shower is named. The table in the chapter lists the different showers while the all-sky maps show the relative positions of radiants with respect to their constellation and include each shower's date of maximum.

Further reading

The Moon Watcher's Companion: Everything You Ever Wanted to Know About the Moon, and More by Donna Henes (2004)
Da Capo Press, ISBN 1569244669

Comet by Carl Sagan and Ann Druyan (1997)
Ballantine Books, ISBN 0345412222

Coon Mountain Controversies: Meteor Crater and the Development of Impact Theory by William Graves Hoyt (1987)
University of Arizona Press, ISBN 0816509689

Stitching Stars: The Story Quilts of Harriet Powers by Mary E. Lyons (1993)
Macmillan Publishing, ISBN 0684195763

Sky and Telescope's online Moon observing articles
http://www.skyandtelescope.com/observing/objects/moon/

Meteor Shower Online Information
http://meteorshowersonline.com/

NASA's Asteroid Watch Program: detecting and tracking impact hazards
http://www.jpl.nasa.gov/asteroidwatch/

University of Arizona, Earth impact effects simulator
http://www.lpl.arizona.edu/impacteffects/

8 Our cosmic connection

But the stars throng out in their glory,
And they sing of the God in man;
They sing of the Mighty Master,
Of the loom his fingers span,
Where a star or a soul is a part of the whole,
And weft in the wonderous plan.

Robert W. Service 'The Three Voices'

It's five in the morning the week of the winter solstice and the temperature is lower than the hour. Three of us are making a trek that will eventually take us up the far western wall of the canyon, and all of it under the light of a thousand shining stars. The trail we follow winds through scrub brush and over frozen washes, past a pictograph of a thousand-year-old stellar spectacle and up onto a lonely plateau where we'll find half-buried stone walls and ceremonial sites that have lain there largely undisturbed for eleven hundred years. We make this trip in the cold and the dark to be there at sunrise on this week of the shortest day of the year. What we hope to see at that moment is perhaps what brought the first people here so long ago, and what, as is evidenced by our very presence, still brings them today: a connection to the cosmos around us.

In the far northwestern corner of New Mexico, set quietly amid a country of dry washes and maze-like badlands, is a low wide canyon, the heart of a national park to which no paved road leads. The 13-mile (21-kilometer) dirt road from Nageezi is an alternately dusty, muddy, wash-boarded, or rutted white-knuckle adventure that never fails to transport me back in time with every jolt. At road's end is the isolation of Chaco Culture National Historical Park. Those few who find their way here are here by no accident; something about the park draws them, as it has drawn me regularly over the last decade. In winter, the temperatures plummet and what few visitors the park sees drop to almost zero. So, while today Chaco Canyon is nearly empty of people and feels like the far end of nowhere, a thousand years ago it was the center of the universe in the desert southwest. And it's a universe that we can still faintly see today.

Very little about the canyon is exactly how it was when the Chacoans, the ancestors of many of today's southwestern puebloan people, were here. Today the canyon is a parched ravine, carpeted in greasewood and saltbrush. Over a thousand years ago when the first stone construction began, the climate in the southwest was slightly less arid and stands of pinyon and juniper dotted the landscape. The immense ruins that cover the canyon floor today, and are the

Figure 8.1 Snow covers old stones in Chaco Culture National Historical Park. Fajada Butte stands silently at the entrance to the canyon beyond the tumbled walls of a 'Great House' (T. Nordgren).

chief draw for the adventurers who make it this far, are no more than the skeletal remains of 'Great Houses,' the purpose of which is still a subject of debate.

Even the geography of the valley has changed due to the continuous work of wind and water on the canyon and its contents. A thousand years of dry blowing sand has partially buried the gigantic great houses, while a thousand spring thaws have cracked sandstone blocks from off cliff walls to tumble down on the ruins below.

With all that has changed though, the one thing I can see today that is virtually identical to what the Chacoans saw are the patterns of the sky. The subtle shifting patterns of Sun, Moon, stars and planets are all but indistinguishable from what the builders of the great houses saw, and through the sky we are offered one of the few windows into a world of what used to be called the Chaco Anasazi, the ancestral Puebloan people of Chaco. For within this canyon are innumerable walls, doors, windows, petroglyphs and pictographs, all of which silently yet insistently say that here there were astronomers.

Figure 8.2 Aerial view of Pueblo Bonito taken by Anne Morrow Lindbergh in 1929. She snapped the photo as her husband Charles Lindbergh flew low over the canyon for a survey of ancient ruins in the American Southwest. In 1941 a massive section of the canyon wall (visible in the center of the photo) collapsed, destroying or damaging 65 rooms, including a section of the highest wall visible here (Anne Morrow Lindbergh, Courtesy Museum of New Mexico, Negative #130198).

The eleven great houses within Chaco Canyon were once the center of the Chacoan world, which at its height in about 1100 A.D. extended over 60,000 square miles (160,000 square kilometers) of the Colorado Plateau in today's states of New Mexico, Arizona, Utah and Colorado. The silent ruins preserved in such widely separated national parks and monuments as Mesa Verde, Aztec Ruins, Hovenweep, and Canyon de Chelly were all built over time by the culture that had its ceremonial center in this canyon. Today their descendents, the Pueblo Indians of Hopi, Acoma, Zuni, and others, still regard the valley and its lonely buildings as sacred.

The canyon itself is not remarkable by the standards of the southwest: low, flat-topped walls are less than a mile wide and extend for only a dozen miles or so. It's only when you reach the center of the canyon's extent, where the silty grey Chaco and Fajada washes mingle, that the valley's most impressive geologic feature is found. Here Fajada Butte stands alone, a single block of sandstone cut off from the distant Chacra Mesa. It's the first sentinel seen by any visitor to the canyon and beyond its flanks are the great houses with names that still call out to me of far off exotic mysteries: Hungo Pavi, Chetro Ketl, Pueblo Bonito, Kin Kletso, and Casa Rinconada.

Pueblo Bonito (Spanish for Pretty Village) is the largest of these great houses and contains over 600 rooms, many stacked three and four stories high in an area over three acres (1.2 hectares) in extent. Altogether it is as large as the Roman

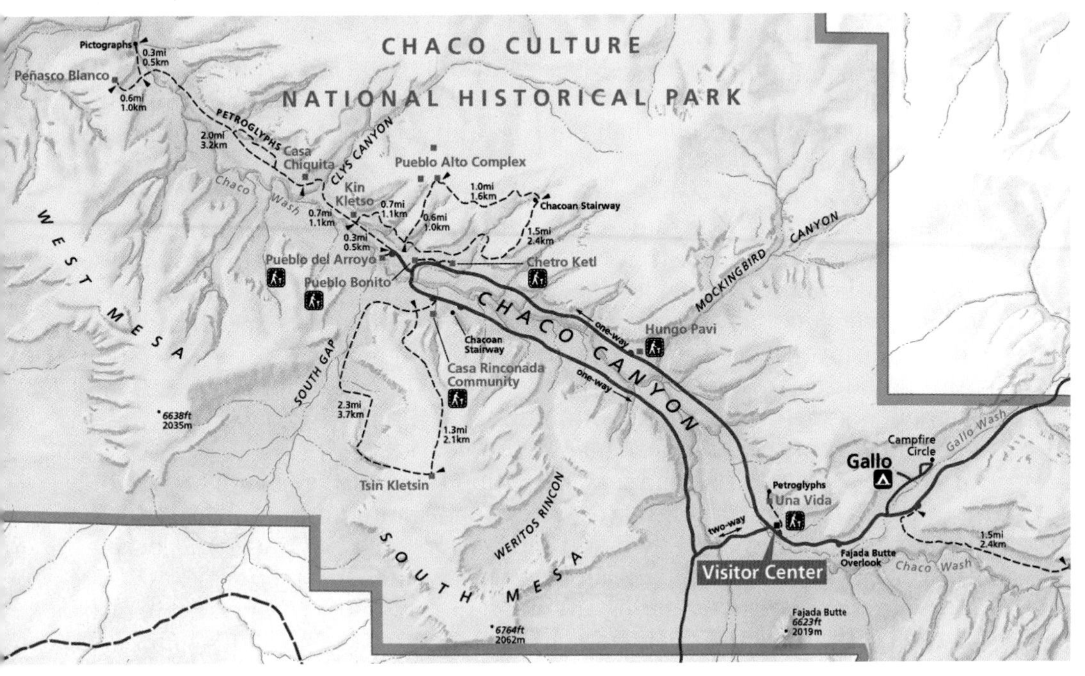

Figure 8.3 Map of the central portion of Chaco Culture National Historical Park showing the positions of the great houses in relation to the rest of the canyon (National Park Service).

Figure 8.4 A pre-1941 photo of a Park Ranger standing on the overlook above Pueblo Bonito. Today much more of the sprawling great house has been excavated so that the curving back wall now stands three to four stories above the surrounding ground. The stable off to the right was once the living quarters for archaeologists working in the park and acted as a kitchen and trading post (where many precious artifacts undoubtedly left the canyon). These structures have been removed and perhaps only foundations remain (to be unearthed by future archeologists) (George A. Grant, Courtesy National Park Service).

Coliseum, and until the mid 1800s was still the largest building in the continental United States. But as impressive as Pueblo Bonito and the other houses are today in their eerie, half-buried beauty, at their height they must have been awe inspiring.

Walk around these enormous buildings and one is struck by the power of intention in this place. The canyon was never truly inviting by our standards, yet people chose to come here. The thickness of the lower walls in the great houses shows that the builders intended them to reach their ultimate heights of several stories. Each building appears meticulously planned for its specific location within the canyon, even though actual construction took place over generations – three hundred years, in the case of Pueblo Bonito. It is from the precision of their buildings, and other works within this valley, that we know the sky was an important part of the Chacoan's plans.

That the Chacoans would pay attention to the sky is not a startling hypothesis. On a practical level, the rising and setting of the Sun defined the length of the day and the time that could be worked. In addition, the sky is the source of the heat that bakes the canyon walls in summer and brings violent, yet life sustaining rain showers during afternoon storms. In winter the sky brings the wind and snow that chill the bones and freeze the ground as solid as granite. As an agrarian society in a climate that was always harsh, the Chacoans needed to know the timing of the seasons, to anticipate the changing conditions and make the best use of the short growing seasons.

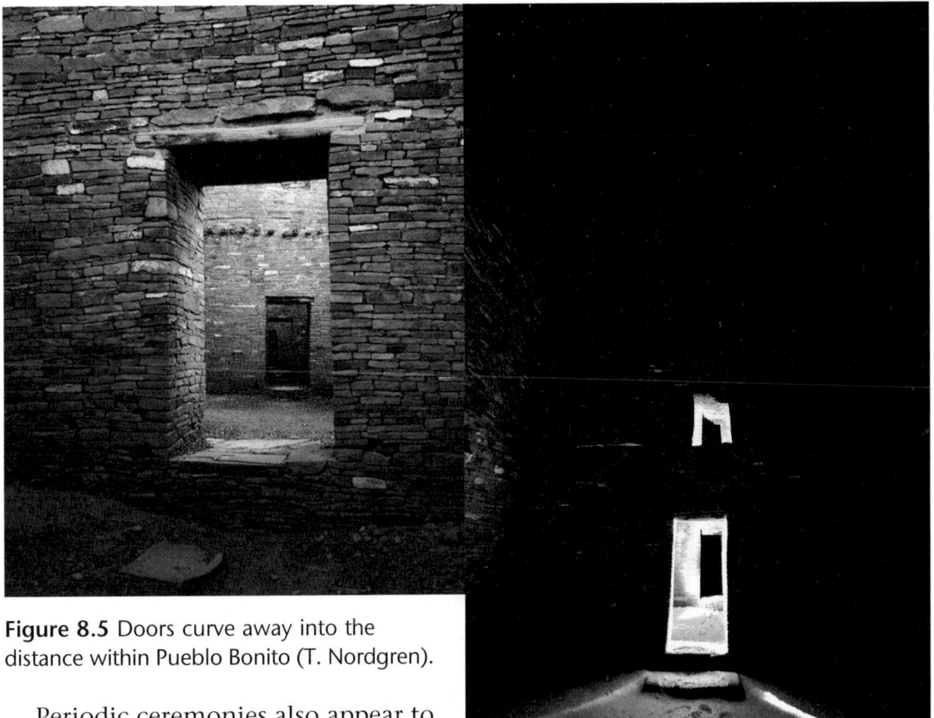

Figure 8.5 Doors curve away into the distance within Pueblo Bonito (T. Nordgren).

Figure 8.6 Stars shine above empty rooms in a Chaco great house (T. Nordgren).

Periodic ceremonies also appear to have been an important essence of Chacoan life. During these special times of elaborate ritual it was vital that people be brought in from large distances to help build the great houses that are such an important part of the canyon. Such ceremonies needed an accurate astronomical calendar to assemble the masses as they probably were held in the depths of winter on the shortest day of the year, or at the closest full Moon, during the period of no agricultural activity.

Today we don't have to notice the sky very much. Inside our houses, thermostats keep the temperature constant and the rains off our heads. While the Sun's motion still defines the length of the day, indoor and outdoor lighting make the night as bright as the day and divorce us from the need to finish our work by the time the Sun sets. Food is available from the grocery store all year long and if you want to leave town for a holiday all you need do is consult the calendar on the wall.

A month ago, when I first arrived here, I walked alone through Pueblo Bonito the morning after a silent snowfall. My path took me through a semicircular maze of square windowless rooms and past deep circular kivas undisturbed by any footprints. The kivas are believed to be ceremonial gathering places, similar to those the Hopi and other pueblos use today. The square rooms, on the other hand, may be apartments, storage rooms, or simply support structures for the larger building (the intent of their architecture is a mystery). Set amidst all of these rooms is a large central plaza bisected by a single wall that runs perfectly north and south. Beside it, aligned in exactly the same way, is a great kiva larger than any other in the house.

There are literally an infinite number of ways these walls and kivas could be oriented, if the orientation were random. Yet these structures at the heart of this great house are perfectly aligned along the Earth's true north and south axis, and they are not the only ones. How does one do this by design?

Figure 8.7 The central North Wall within Pueblo Bonito (T. Nordgren).

I know their alignment because I used a compass, but the compass only tells me the alignment to Earth's magnetic north pole, not the true pole around which the Earth turns. The north magnetic pole slowly drifts around thanks to the Earth's molten iron core (today the magnetic pole is located in Greenland). Knowing where it currently is provides the offset, the magnetic declination (also

Figure 8.8 The great kiva at the center of Pueblo Bonito is aligned towards the true north axis of our planet. Today, the North Star, Polaris lies almost directly over the pole and so in this two-hour-long photo is the one star that does not appear to move as the Earth turns. For observers in the northern hemisphere, it will always point us north (T. Nordgren).

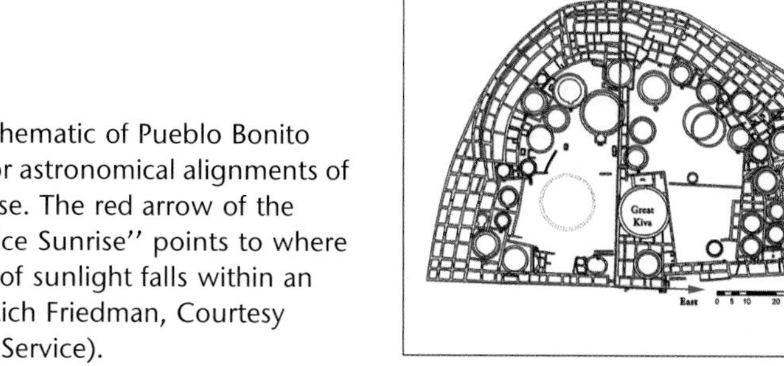

Figure 8.9 Schematic of Pueblo Bonito showing major astronomical alignments of the great house. The red arrow of the ''Winter Solstice Sunrise'' points to where the rectangle of sunlight falls within an inner room (Rich Friedman, Courtesy National Park Service).

known as the magnetic variation) found on most topographic maps. The magnetic declination (or variation) is the horizontal angular difference between True North and Magnetic North and it allows a hiker or an engineer to compensate for what she sees on her compass. Currently, the difference between true north and magnetic north is 10.5 degrees along the horizon as seen from Chaco. This is almost equal to the width of your fist and thumb held out horizontally at arm's length. But, the Chacoans had neither compasses, nor topographic maps.

Today an astute sky watcher knows that the North Star, Polaris, happens to be located off in space nearly directly above the Earth's north pole. Look to the north and it alone appears to the eye to stand still as the Earth's rotation carries the rest of the celestial sphere around it. But the direction the Earth points in space slowly precesses; it traces out an enormous circle on the sky, completing one cycle every 26,000 years. A thousand years ago, Polaris was five degrees away from where it is now, about the length of my thumb held at arm's length. This is far enough away so that it too traced out a noticeable circle in the sky; any structure aligned to it would have no guarantee of pointing exactly north.

No, the only way to locate true north is to be a Sun watcher. Halfway through each day the Sun must pass from the eastern sky where it rises to the western sky where it sets. In the northern hemisphere, north of 23 degrees north latitude (a line called the Tropic of Cancer) this passage always takes place south of the zenith, the point straight overhead. Plant a stick in the ground perfectly straight up and down and when the Sun is at the highest point of its trip across the sky, its shadow will be at its shortest and will always point north.

Watch the Sun rise or set only once a week during the spring or fall, and another pattern of the Sun's motion emerges. During spring, the Sun rises each morning progressively farther north along the horizon compared to where it rose the day before. Around the time of the spring equinox, when the Sun rises nearly due East, the change can be as much as the Sun's full diameter each day. After only 30 days, the Sun will rise almost 15 degrees farther north than where it did the month before (this is the width of your outstretched hand).

Eventually this northward motion slows until on June 21st, the day of the summer solstice, it stops. Solstice literally means "Sun stands still." On this day the Sun rises and sets as far north as it will ever get. Here the Sun "stands still" as it appears to rise from nearly the same spot for four or five days in a row until finally the motion reverses, and each day the Sun rises a little farther south. Six months later, on the morning of the winter solstice, it rises as far south as it will ever get and its motion halts before changing direction once more.

This pattern of motion along the horizon holds equally true for sunset, and you don't need to be in Chaco or the American southwest to see this. If you have a far horizon with many stationary features, be they mesas and buttes, or barns, apartment buildings and trees, you can tell the date by noting where the Sun rises and sets.

Why does the Sun do this? Imagine the Earth's equator projected out into space tracing a line across the sky like that shown on the sky maps in this chapter (called the celestial equator). If you stood at the equator, you would see this line rise from the horizon due east, pass through the zenith and set due west. Any star, planet, Sun or Moon that happens to lie along the projection of the Earth's equator will be carried along that imaginary line as the Earth turns on its axis. If an astronomical object happens to be located in the sky north of the celestial equator (the distance north is known to astronomers by the archaic-sounding term *declination*) then as the world turns, it will rise north of east and set north of west. The same goes true for any object to the south of the celestial equator; the farther south the declination of the Sun, for instance, the farther south it rises.

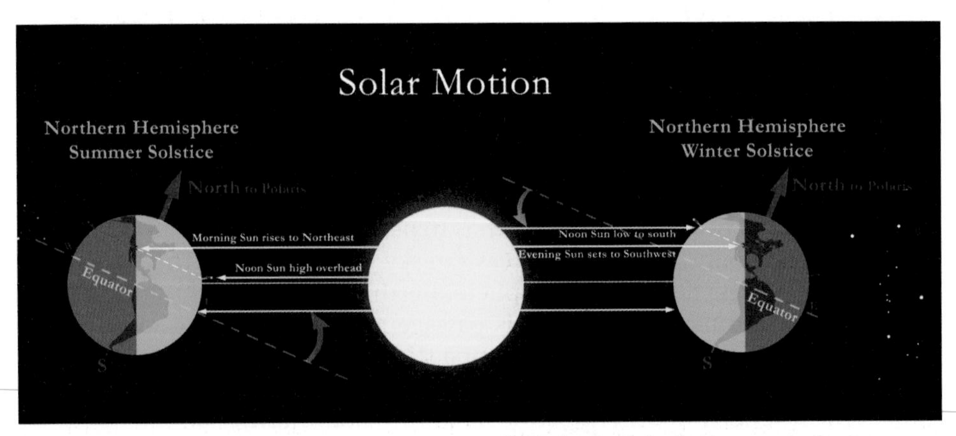

Figure 8.10 In northern hemisphere summer (left) the Earth is tilted towards the Sun as it points towards Polaris. The Sun is therefore at a higher north declination compared with the celestial equator (the dashed-line projection of the Earth's equator). Around the summer solstice, the Sun thus rises north of east, passes high over the tiny figure of a person standing at the latitude of Chaco Canyon at noon, and sets to the northwest. Six months later (right), the northern hemisphere points away from the Sun which has a large southern declination compared with the celestial equator. Around the winter solstice the Sun appears to rise in the southeast, never passes very high above the southern horizon (viewed by our tiny figure at the latitude of Chaco) and sets towards the southwest (T. Nordgren).

Because the Earth tilts on its axis relative to the Sun, sometimes the Earth's northern hemisphere points more towards the Sun (northern hemisphere's summer) and the Sun is farther north in the sky than the celestial equator. At these times the Sun has a northern declination and this is why in summer the Sun rises to the northeast, passes high overhead, and then passes below the horizon to the northwest. Six months later the position is reversed; the northern hemisphere points away from the Sun and so the Sun has a large southern declination. The Sun therefore rises in the southeast, sets in the southwest, and never attains an altitude very high above the southern horizon. Since the Sun's declination changes as the Earth's tilted axis and equator move around it, so too does the Sun's rise and set positions change, north and south, over the course of the year.

Back in Pueblo Bonito I continued my silent and lone exploration by stepping out of the plaza and walking eastward along the southern wall. Standing at the southeast corner there is evidence of a sunwatching shrine where each day the Sun rises behind a different part of a series of downward steps formed by the overlapping cliff-tops of the canyon wall. Observers standing here at dawn see the Sun rise against these various features in so regular a pattern that they can easily use this as a calendar to announce the approach and arrival of ceremonial days. In fact, the Hopi do this in their communities today.

After October 29, however, the Sun rises southward over a section of featureless plateau. Perhaps not coincidentally, on the very morning the Sun rises beyond this flat expanse, a sliver of sunlight first makes its way through a curious diagonal door set high up on the corner between two walls behind me. Through this door the Sun casts a glowing rectangle of light into a small room. Each morning in December as the Sun moves farther south, the golden rectangle moves farther north until on the days immediately around the winter solstice the first light of dawn shines perfectly into the room's far corner.

We know from a few protected chambers in other great houses that plaster used to cover the masonry walls of these ruins. Rich decoration still adorns one such room in nearby Chetro Ketl. If

Figure 8.11 Sunrise from the solar observing station at Pueblo Bonito's southeast corner. Here, three weeks after the summer solstice, the Sun rises high on the mesa rim. As summer turns to fall and winter, the Sun rises progressively farther south (to the right). Steps and notches in the mesa wall provide a reference calendar until the Sun reaches the featureless expanse at the extreme right around October 29 each year (T. Nordgren).

Figure 8.12 Sunrise the morning of the winter solstice shines the first light of dawn through an odd, diagonal door, perfectly illuminating the far corner of a room within Pueblo Bonito (T. Nordgren).

Figure 8.13 The great kiva of Casa Riconada sits surrounded by winter snow on its solitary hill (T. Nordgren).

any plaster once covered these walls it would have been easy for the local solar observer to paint a calendar, saying for one such mark: "When the rising Sun shines on this spot the time for the winter solstice ceremony is exactly 10 days away." Sadly, any painted plaster which once covered the stonework in this tiny room is long gone, and so we are left to simply speculate about what might have been.

It is at the very least intriguing that there should exist a window to help mark time starting exactly on the date that the horizon fails in its ability to do so. And while there are other corner doors in Pueblo Bonito that show no such alignment today, the alignment through this door certainly works and it is not unreasonable that Chacoans would have noticed and made use of it even if it wasn't the original intention.

Still, one way to tell if this was the corner door's intended purpose is to ask the following question: If the Chacoans built a winter solstice observing site, then perhaps somewhere else they built a similar site to observe the summer solstice?

Across the wash from Pueblo Bonito, sitting atop a quiet hill where there is a clear view to the summer solstice sunrise, there are the remains of the great kiva Casa Rinconada. Unlike the other kivas in the valley that are all components of larger great houses, Casa Rinconada sits alone. Sixty feet (18 meters) in width and 12 to 16 feet (4 to 5 meters) deep, it is the largest of all the kivas in the valley and is awe inspiring in its geometrical perfection. Perfectly round to within four inches (10 cm), the stairways within its two massive T-shaped doors at either end are aligned exactly north and south.

Today the kiva is open to the air, but a thousand years ago it was enclosed with a log roof supported by four massive posts. The post-holes are still visible amidst the grass that now grows here and they describe a square whose sides are aligned with north, south, east and west.[1] Astronomer Ray Williamson, a pioneer in the field of archeoastronomy and one of the first to see astronomical significance in Casa Rinconada, recounts a story of the 'first kiva' told by the people of Acoma Pueblo in his book, *Living the Sky: The Cosmos of the American Indian*:

> When they built the kiva, they first put up beams of four different trees. These were the trees that were planted in the underworld for the people to climb up on.... The walls represent the sky, the beams of the roof (made of wood of the first four trees) represent the Milky Way. The sky looks like a circle, hence the round shape of the kiva.

As the Acoma are one of the puebloan peoples that trace their ancestory back to the Chacoans, creation stories like this one support the cosmological significance of this greatest of kivas in the heart of Chaco culture.

Within Casa Rinconada's circular structure is another astronomical alignment that occurs every year

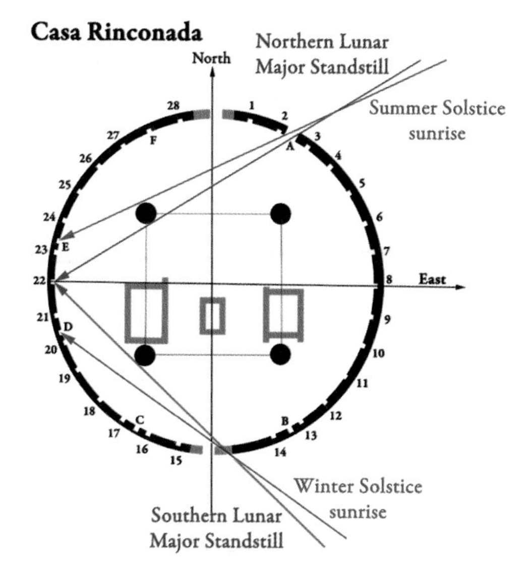

Figure 8.14 Schematic diagram of Casa Rinconada showing the major astronomical alignments. Each upper niche is numbered, while the larger, lower niches are lettered A – F. The four (now missing) posts are shown in their proper positions with lines joining them aligned with the cardinal directions (Modified from R. Williams, *Living the Sky*).

[1] Visitors to Aztec Ruins National Monument, Chaco's sister park, can walk through a fully reconstructed great kiva, believed to be very similar to what Casa Rinconada would have looked like. Aztec Ruins NM is located outside Aztec, New Mexico, almost exactly due north of the great houses of Chaco Canyon.

Figure 8.15 A two-hour exposure through the southern doorway of Casa Rinconada shows the kiva's alignment to the Earth's true north pole. The window through which the summer solstice and northern major lunar standstill light shines is visible to the right of the northern T-shaped doorway (T. Nordgren).

on the summer solstice and today draws visitors from all over the world. On this morning, the rising Sun shines in through an opening in the high eastern wall and casts a square beam of light across the space of the ceremonial room and into a large niche at the bottom of the opposite wall.

The circular kiva is ringed with 28 niches, small square openings, located above a stone bench that runs the circumference of the enclosure. Beneath these regularly spaced niches are six larger niches, four on the west side, and two on the east. It is into one of the lower western openings that the rising Sun shines on the days immediately around the summer solstice.

Was this alignment intended as a marker for the summer solstice? It's hard to say. Much of the kiva we see today was actually reconstructed during excavation in the early 1930s. All but one side and one corner of the high eastern opening were built anew from the stones found tumbled from the ruined walls. There is evidence that there used to be a room outside the small window through which the Sun's morning light now passes. Did this block the sunlight or was there another window for the Sun? No one knows. In addition, there is evidence that the niches themselves used to be covered and so perhaps no light would have fallen inside them. Is this important to their purpose? No one can say.

The Sun is not the only astronomical object that casts light and shadow and moves across the horizon and sky with clockwork precision. While the tilt of the

Figure 8.16 The ruins of Casa Rinconada soon after excavation began in the 1930s (George A. Grant, Courtesy National Park Service).

Earth carries the Sun north and south along the horizon with the seasons, the Moon rises in a similar pattern that is repeated each month.

When the Moon is full, it is on the opposite side of the Earth from the Sun so that we see the entirety of its sunlit side. If it is winter in the northern hemisphere, the Sun has a southern declination and sets in the southwest. At the moment it does so the full Moon must be opposite the Sun and therefore rising in the northeast. This means the Moon must have a northern declination.

Half a month later, when the Moon has moved half way around the Earth, the new Moon occurs when the Moon lies more or less between the Earth and Sun. The Moon is invisible at this time as we are looking entirely at its shadowed side and it rises and sets in unison with the Sun. However, a couple days after new Moon, we see a hair's sliver of the lit portion and are treated to a beautiful thin crescent setting just after sunset. At this moment, the Moon is still nearly in between the Earth and Sun and so from our perspective it appears very close to the Sun in the sky. Since it is winter and the Sun has a southern declination, so too must the Moon.

Over the course of a single month, the position of the rising and setting Moon therefore travels first south then north against the horizon. The very path the Sun takes along the horizon in one year, the Moon repeats every month. On top of this monthly cycle, which you can pick out for yourself over just a few nights, there is a second more intriguing cycle through which the Moon passes more slowly. The full range north and south over which the Moon rises gradually expands and contracts over the course of 18.6 years.

Because of this slow cycle, the Moon doesn't go perfectly between the Sun and Earth each time it is new Moon; if it did, we would see a total solar eclipse each month at the moment of alignment. The reason this doesn't happen is that the Moon's orbit is tilted by 5.1 degrees relative to the orbit of the Earth around the Sun. Nearly every month during new Moon, the Moon passes uneventfully above or beneath the Sun as viewed by Earthly observers. So while the Moon's declination changes north and south just like the Sun's, it changes by a slightly different amount.

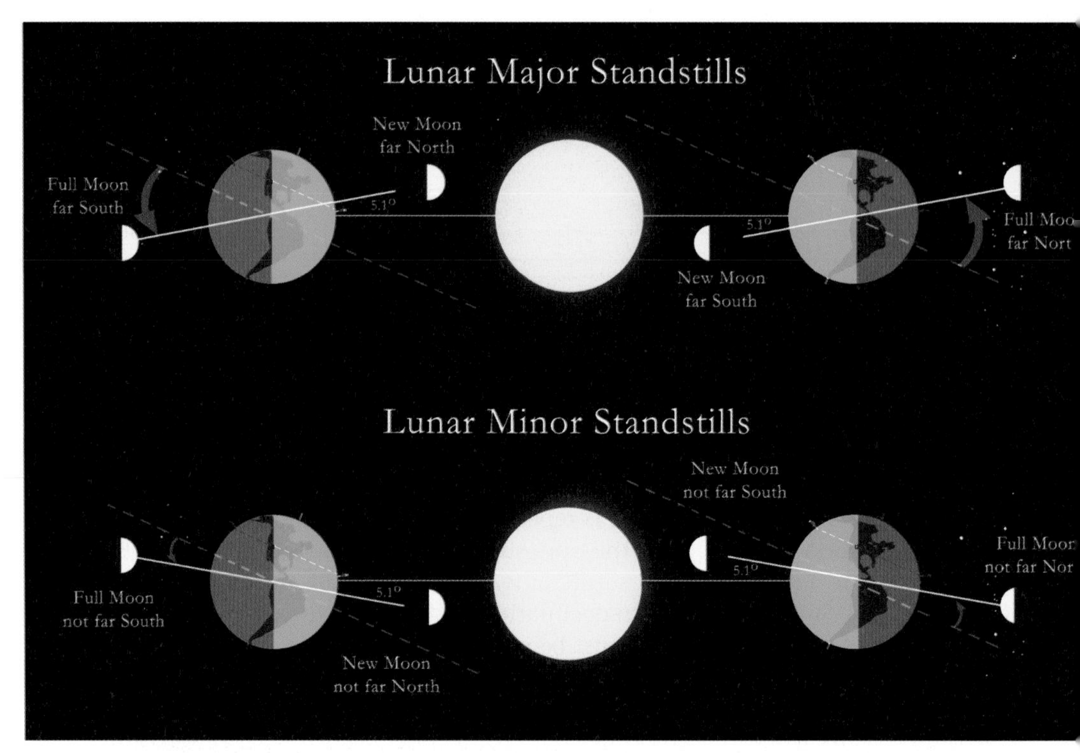

Figure 8.17 The Moon's orbit is tilted at a constant 5.1 degrees with respect to the ecliptic (the solid yellow line showing the plane of the Earth's orbit around the Sun). However, the direction the Moon's orbit tilts slowly changes over an 18.6-year period. When the Moon's orbit is tilted opposite the Earth's tilt (top) the Moon rises and sets at its maximum extent north and south along the horizon. The farthest north and south the Moon rises is called the major standstills and occur when the Moon is at its greatest north and south declination compared with the line of the Earth's equator (dashed white line).
9.3 years after major standstill, the Moon's orbit is now tilted the same direction as the Earth, and the north and south declinations of the Moon are much less. These are the minor lunar standstills during which the Moon's motion along the horizon is now rather small (T. Nordgren).

Sometimes, the Moon's orbit is tilted so that it adds to that by which the Earth is tilted and the full Moon has a maximum declination north or south greater than that ever reached by the Sun. In summer, when the Sun rises at its northernmost point on the horizon, the full Moon rises farther south than the Sun ever reaches on its annual path (the Moon is said to be at its southern major standstill). Six months later, when in the depths of winter the Sun rises at its southernmost point, the full Moon once again rises farther north than the Sun ever reached in summer (and is now said to be at its northern major standstill).

But the direction the Moon's orbit points gradually changes over an 18.6-year period (just as the Earth's pole traces out a 26,000-year circle). As a result, 9.3 years after the rising Moon sweeps out its largest path along the horizon, the direction that the Moon's orbit tilts is the same as the Earth, and the Moon's declination swings through its smallest range. During these times, the full Moon rises nowhere near as far north or south as the Sun and its monthly motion is at a minimum (these are now called northern and southern minor standstills).

I'll be honest; with all our modern-day distractions, I never noticed this cycle when I was growing up. In the entire time I have been alive I have lived only long enough to experience two full lunar cycles. Frankly, as a child I was paying attention to other things, and eighteen years is a long time to notice the position of the rising and setting Moon change from year to year. This is especially true for someone in a culture that hardly notices the Moon even goes through phases.

Today it takes an unusual person to spend the time outdoors needed to really understand what the Moon is doing in the sky. Ron Sutcliffe, an engineer and adjunct professor with a passion for naked-eye lunar observing is just such a person. Sutcliffe normally works up in southern Colorado at Chimney Rock where a Chacoan era great house sits precariously on a narrow mountain ridge to the west of two massive rock towers. From the kiva at the center of the ruins the full Moon at its northern major standstill is just visible rising in the narrow gap between the towers every 18.6 years. While some putative ancient astronomical alignments may seem to border on coincidence, the placement of the large ceremonial kiva on the lonely ridge-top in just the right position to see this rare lunar apparition seems unquestionably planned.

Figure 8.18 This photo shows the view from the central kiva at Chimney Rock, Colorado. Superimposed is a diagram showing the farthest north the full Moon ever rises compared with the Sun on the summer solstice. Over 18.6 years, the northernmost full Moon-rise varies so that only during the three years around the major lunar standstill does the light of the full Moon shine between the two stone pillars. The winter of 2007 marked the last year this occurred, and won't be seen again until 2024 (Ron Sutcliffe from his book *Moon Tracks: Lunar Horizon Patterns*).

Figure 8.19 The full moon rises through the towers of Chimney Rock during northern major lunar standstill in 2006 (Ron Sutcliffe).

During the month leading up to the winter solstice of 2007, the last year the Moon would appear at nearly the same position as its northern major standstill, Sutcliffe and I were both working at Chaco Culture NHP, and we spent a number of evenings, as the park staff often do after hours, sitting around and talking about our interests and passions. It was here I learned why Sutcliffe was down in Chaco. When the full Moon is at its northern major standstill, its first light at dusk shines in through the same window in Casa Rinconada that plays such an interesting role with the Sun on the summer solstice. Now, however, the square of moonlight shines on one of the 28 smaller, higher, niches that ring the great kiva.[2] The niche it illuminates is the niche that is set exactly due west of the kiva's center and thus faces directly east. For all other lunar rises when the sky is dark the Moon is too far south and the moonbeam falls to the north of this niche. Six months later, however, when the Moon is at the southernmost major standstill, it shines in through the great T-shaped south door and illuminates the exact same niche. For all other Moon rises it is too far north, and its moonbeam falls to the south of this eastern most niche. Only on this one year of lunar major standstill that happens every 18.6 years, will the two moonbeams illuminate the same niche six months apart.

Figure 8.20 During the winter solstice, the rising Sun shines through the southern door of Casa Rinconada illuminating one of the lower niches within the kiva (and mimics what happens with the rising Moon during the southern major standstill). The reason there are multiple squares of light is because of the shadow from a recently installed fence blocking the southern doorway (T. Nordgren).

[2] The fact that there are 28 niches is interesting in relation to the Moon. It takes the Moon 27.3 days to move fully around the Earth. An observer can see this by watching the Moon's position relative to the stars and seeing how long the Moon takes to come back to the same position with relation to them. This is called the Moon's sidereal period. Because the Earth is in constant motion around the Sun, the Moon has to travel a little farther than one full orbit to come back to the same position relative the Sun as seen by an Earthly observer. The time between two successive full Moons (or two successive new Moons) is therefore a little longer: 29.5 days. This is where we get the concept of a *month*. Is it significant that these two periods are almost equal to the number of niches? As always, it is intriguing.

If Sutcliffe's hypotheses are confirmed, consider the care such an architectural feat requires. Being aware of the solstices and equinoxes is common sense if you are an agrarian people living in a harsh and unforgiving environment. Careful attention to the rising of the Sun for a year or two is sufficient to pick out the trends. But consider the motivation and care required to pick out a nearly 20-year period of the Moon? The Moon brings no rain. Lunar standstills signal no time for planting. And to pick out a cycle of nearly two decades with no written language, when surviving to see no more than two or three full cycles is the norm is a stunning achievement.[3] It is therefore highly impressive – or highly improbable – for such a culture to have designed a circular structure where window, door, niches, and radius all conspire for annual alignments of the Sun and 18.6-year alignments of the Moon. Which it is, impressive or improbable, is still open to question.

Intriguingly, there are other ancient markers that may record the exact same celestial phenomena. Fajada Butte sits just a few miles east of Casa Rinconada. In the summer of 1977 Anna Sofaer was an artist participating in a field school to document the rock art of Chaco. She and Jay Croty, a rock art expert, had climbed to the summit of Fajada Butte to investigate a petroglyph site consisting of an elaborate spiral chipped into the cliff face beneath three massive vertical stone slabs. Arriving shortly before noon, they found a beam of sunlight slipping between two slabs to cast a thin dagger of light that perfectly pierced the spiral's heart. Having recently studied astronomical markers created by the Maya and other peoples of the Americas, Sofaer knew she had stumbled upon something special: a celestial marker to celebrate the Sun's apex in its annual travel around

Figure 8.21 High atop Fajada Butte, three slabs of sandstone rest against a rock wall. At noon of the summer solstice a thin dagger of sunlight shines between two of the slabs to perfectly pierce a large spiral petroglyph on the rock wall (D. Ford, Courtesy National Park Service).

3 It is not unreasonable to suggest that petroglyphs and pictographs form a written language that could be used to keep track of long-duration events like those noted here. Archeoastronomer Kim Malville has documented a series of pictographs near a tower window at Mesa Verde National Park that appear to illustrate both the monthly position of the Moon and (as if in the form of a running tally) the 18.6 year period of major and minor lunar standstill.

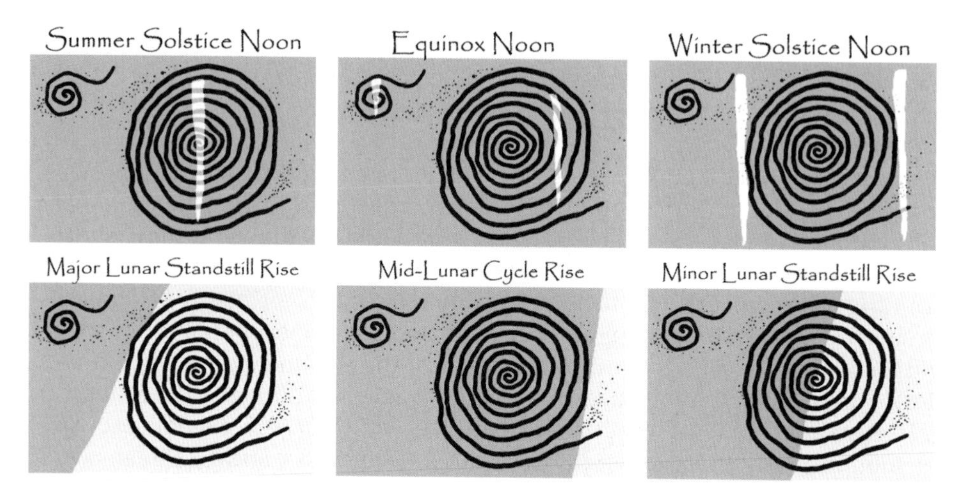

Figure 8.22 Simplified diagram of the play of light and shadow against the "sun dagger" petroglyph. At noon on the solstices and equinoxes, the Sun casts one or more daggers of light across the two spirals. When the full Moon rises at major and minor standstills it casts a shadow that either pierces the center of the large spiral or brackets it on the left. In between major and minor standstill it casts a shadow at full moon rise that brackets the spiral on the right (T. Nordgren, modified from The Solstice Project).

the sky. Sofaer has spent the better part of the last 30 years studying that ceremonial Sun station high in the New Mexico sky. To date, the 'Sun Dagger' (as it has come to be known) has been shown to mark noon around the time of both solstices and the equinoxes. In addition, a shadow from the rising Sun perfectly bisects the spiral around the 15th of May which may have been a ceremonial date for spring planting. There is also conjecture that the full Moon's shadow plays across the spiral at both major and minor standstill and that the nine and a half turns of the spiral may signify the approximately nine years between the two lunar events.

The complexity and beauty of the spiral and slabs high atop the lonely mountain is unquestionably moving. There is a tingle to the senses when we cannot help but recognize that this is a place of intense ceremonial importance. How I wish I could stand there a thousand years ago when the canyon was alive with activity and see what took place as the solstice approached. On each of those dates when special Sun Watchers mounted the ramps and stairs that led to this isolated site, the precise movement of Sun and shadow across carefully etched stone, in a familiar pattern that was annually greeted with great expectation, must have been a tremendously reassuring sign that all was well in the Chacoan Universe.[4]

[4] Sadly, you can no longer visit the summit of Fajada Butte. There is evidence that in the mid 1980s one of the slabs slipped due to erosion from all the visitors who came to see the site. The park service has had to close off all further access to avoid any more unintentional damage.

And these are just a fraction of all the hypothesized astronomical alignments, symbols, and observation stations within the canyon. For every one that has been identified there may be dozens, if not hundreds, that will never be found or recognized. But the question arises: are all these celestial alignments real? Sure, we can show the alignment works, but did anyone purposefully design them to be this way? Were they intended or just a coincidence that was never even recognized by the canyon's original inhabitants? On which side of the thousand-year gap between us and the Chacoans does the astronomical ingenuity lie? We can only speculate.

The evidence is tantalizing and certainly sends a shiver down the spine. But with literally thousands of walls, doors, windows, kivas, petroglyphs and pictographs it is surprisingly easy to have alignments in fact that were in no way intended in reality. With enough patience I know I can find an alignment of some building, wall or petroglyph that marks the sunrise of my birthday, made a thousand years before I was born to see it. To proclaim, as some do, that the Chacoans had an unimaginable astronomical insight beyond all modern scientific comprehension, is ultimately as disrespectful of them as claiming they were idiots and could only have built these wonders with the help of aliens. Ultimately, if we are to learn something meaningful about ourselves and our ancestors, it is important to learn what they really knew, what they really observed, and what they really held sacred.

To determine the reality of these alignments we should do more than just pay attention to a single set of silent stones, we should look for supporting evidence that these astronomical events were actually of importance to the people who made them. If it was, then it stands to reason that if they constructed multiple similar sites then all should mark the same or similar phenomena. In the words of the archeoastronomer and physicist Bradley E. Schaefer "A word of ethnography is worth a thousand alignments." Sadly, there are no written records left to us from the Chacoans. When they left the canyon for the last time they carefully packed up their belongings, walled up their houses, and set fire to their kivas.

Figure 8.23 Summer monsoon thunderstorms lend an aura of power and mystery to the canyon (Jeff Swartz).

But we can consult the children of the Chacoans as some of them still practice their ancestor's traditions. We do know they must have watched the sky since the Hopi and other pueblos still use the interplay of the Sun, sky, and horizon to mark the passage of the seasons and the

Figure 8.24 The Sun sets behind Fajada Butte shortly before the winter solstice. As viewed from this spot, the Sun sets in the notch at the summit where the three slabs of the Sun Dagger site are found. The setting Sun is photographed at 3 minute, followed by 2 minute, then 1 minute intervals right as it disappears. There is nothing special about the spot where I took this photo, just a lonely section of trail where my calculations told me I should stand to see this sight (T. Nordgren).

timing of celebrations.[5] In addition, since Sofaer first found the Sun Dagger, several more sites (with excitingly similar, though less intricate, design) have been found throughout the southwest. Taken together, these confirm that whatever else may take place within Chaco Canyon, atop Fajada Butte, the silent play of nearly a thousand summer solstice sunbeams and ten thousand full Moon shadows across a beautiful lonely spiral has certainly been no accident.

Today many visitors flock to the canyon eager to feel some connection to the cosmic energy they believe was once harnessed here.[6] The possibility of ancient astronomical knowledge fills us with awe and wonder. But why do we feel this way? According to archeologist, W. James Judge:

> Consider for a moment the fact that Native Americans lived very close to nature in this area for at least 12,000 years before the Caucasians arrived. Now, you can't live that close to a phenomenon for that long without observing something about it.... My hunch is that what we are witnessing here is a "rediscovery" of the sun's path by modern Anglo populations.... This rediscovery is a function solely of our having been very effectively insulated from our environment for a long time.... Thus, in my view, the amazement we display over prehistoric solstice markers is largely a function of our own recently acquired ignorance about the environment.
>
> *Astronomy and Ceremony in the Prehistoric Southwest*
> edited by Carlson and Judge, 1983.

[5] What exactly they, and other pueblos, practice and what beliefs have made their way down through a thousand years of oral tradition is not clear; history has made them disinclined to share the knowledge they hold sacred.

[6] During one recent total lunar eclipse my students and I saw the campground filled to overflowing as visitors flocked in over the rough, dusty road. It was an absolute madhouse. I shudder to think what would have ensued had the road been paved like local business leaders want. While paving the road to Chaco may bring visitors by the thousands, it would be at the cost of destroying what is most special about the park.

In addition I wonder if it isn't also a reflection of how people have always felt a need to believe we have some sort of connection to the heavens, to feel that the world isn't random and that everything happens for some larger purpose. The heavens change with a clockwork regularity in counterpoint to the messiness of our everyday lives. There is comfort in the orderly change of the Sun, Moon, planets, and stars and therefore comfort to be drawn from the possibility that their movements bring order to our lives? Many of those who feel this way regularly look for signs that the stars affect us and we affect them, and I understand why they do so.

Throughout all of human history the approach of winter solstice has been a stressful time. For some cultures, elaborate ritual and ceremony took place to appease the Sun in hopes it would halt its southward journey. If the ceremony, or the people, were found to be wanting, the days might continue to shorten and the Sun might just continue onward over the southern horizon to disappear forever and leave the world in coldness and dark. Only through our action could the Sun be reversed and the world brought out of winter. While most of us might claim that our cultures no longer believe this to be true, consider that one of the most sacred of Christian holidays takes place within just a few days of the winter solstice and features a star appearing in the East. Later in spring, the high holy day of Passover (usually followed closely by Easter) begins on the first full Moon after the Spring Equinox.

Even in secular circles, children still learn to wish upon a star, while millions of adults regularly read their horoscopes published in nearly every major newspaper. A complaint against modern science is that it seeks to divorce us from our cherished connection to the heavens, and at first glance this is certainly true. I am very clear with my students that there has never been any scientific evidence that astrology works in any way, shape, or form. Yet far from the stars being a simple instrument for telling us who we should love and what days are auspicious for business ventures, modern astrophysics reveals a much more powerful connection we share with the stars, a connection that has been going on for over five billion years and is responsible in nearly every way for making us who we are today.

It is with these thoughts that I am once more brought back to our early morning hike through the canyon. We are making our way out to Peñasco Blanco in the frigid predawn hours to test a winter solstice alignment marker between the far western pueblo and a small peak along the eastern horizon. If we are correct, then on the winter solstice the first rays of the rising Sun should just touch this distant point as viewed from a ruined kiva at the heart of the great house. If this alignment is true, it might explain why this great house is located where it is. Testing this hypothesis is at the heart of what it means to be a scientist.

Only a few hundred yards beyond a frozen wash (that by day will be several feet of slippery, grey mud) the trail reaches the low western wall of the canyon. At its base we cross under a red painted pictograph located high under a yellow sandstone overhang. Overhead a star, a crescent, and a hand print are revealed in

Figure 8.25 Possible pictograph of a supernova as it appeared next to the crescent Moon in the summer of 1054. The artist has placed his or her handprint here in a sign that is interpreted as marking its ceremonial importance. Located on the underside of an overhang they have been protected from nearly a thousand years of damaging sunlight like that which has faded the red "flames" extending from the carved circles on the rock wall below. In 1066 A.D. Halley's Comet made an appearance in the sky that was seen by people all over the world. Is the flaming circle a representation of that astronomical apparition? No one knows (T. Nordgren).

the light of our head lamps under a still dark sky. On this spot the view of the eastern sky is completely uninterrupted. Anyone standing here a thousand years ago, at the height of the Chacoan culture, would have seen a new star rise in the east next to a thin crescent Moon on the morning of July 5, 1054. Chinese astronomers recorded that on that day a strange new star appeared that was four times brighter than Venus and for 23 days it was so bright that it was visible in broad daylight. For nearly two years this strange new star joined the common constellations in the sky before finally fading away to invisibility.

It is this event, an event so rare that no human being in the last four hundred years has seen its equal, that some believe is depicted here under this quiet overhang. What ancient Chinese and Chacoan astronomers saw is the explosive destruction of a star. It's what we call a supernova. Like so many of the other astronomical connections in Chaco, what this pictograph panel actually represents is still the subject of vigorous debate. There is an historical account of a site among the Zuni that also features a pictograph panel of a star, moon, and hand, but it is said to mark a Sun-watching station (while other contemporaneous descriptions fail to mention any sign of hand, or alternately, a crescent). And though there are five or six other rock art locations around the

Figure 8.26 Every element in nature is just a matter of different combinations of protons, neutrons and electrons. In this periodic table of elements at the Griffith Observatory in Los Angeles, each element is on display with its atomic number (T. Nordgren).

southwest that feature a star and crescent, only Chaco's adds the feature of a hand, a symbol believed to mark a sacred ceremonial site.[7]

Whatever the reason this pictograph was painted, what is certain is that in this canyon, on this spot, a strange and startling sight was witnessed a thousand years ago that intimately links us to the stars above. Whether or not anyone chose to record it here, we are the offspring of the phenomenon they saw; we are the children of supernovae.

[7] As a further example of the magnitude of the debate over the meaning of this pictograph, between the three experts who reviewed this manuscript I received three separate arguments for and against various interpretations of this site, including Sun-watching station, the planet Venus, and two separate supernovae that would have been visible during a single Chacoan's life.

Take for instance, the red in the paint of the pictograph. It's due to atoms of iron and oxygen that we call rust. Both of these atoms are found within me, in the red blood cells pounding in the pulse I feel in my veins after the long cold hike to this spot. At the turn of the last century scientists found that every element we see in the world, the wood of the tree, the sandstone of the rocks, the air we breathe is all made of atoms and those atoms are made from only three simple building blocks: protons, neutrons, and electrons. The cottonwood tree in the wash is primarily carbon, a collection of six positively charged protons, glued to six neutrally charged neutrons forming a tight bundle called a nucleus. Around this nucleus is a dizzying quantum cloud of six negatively charged electrons held in place by the force of their equal but opposite charge to the nuclear protons. The sand in the sandstone is made primarily of silicon with an atomic number 14, meaning 14 protons, 14 neutrons and 14 electrons. Similarly, the oxygen in the air I'm breathing is no more than an atom with an atomic number of eight. Eight negative charges held to a nucleus of eight positive charges is all it takes to be called oxygen.

The simplest element of all is hydrogen, one proton and one electron. Hydrogen, carbon, oxygen and nitrogen (the most common element in our atmosphere) combine in a myriad ways to make up the molecules required for all organic life on Earth. It is strange beyond almost all words that the only difference between the yellow sand whose roughness gently scratches my outstretched finger tips and the cold crisp, life-giving air I breathe into my lungs is nothing more than an extra six protons, neutrons and electrons, and that that is nothing more than an atom of carbon.

The power and beauty of the atomic structure of all matter is that we are all related at an atomic and sub-atomic level. If you have the power to combine one oxygen atom and one carbon atom you literally make an atom of silicon. The dream of the alchemists to turn lead into gold is a physical reality, provided you are willing to pay for the energy to do so.

And that is the difficulty. To make silicon you must combine the positively charged nuclei of two atoms. As with love, opposites attract in the sub-atomic world, and like charges repel. As an analogy, attempt to force two magnets together in the wrong way and one is met with futility and strain. But reverse one magnet and the two fly together. Forcing the two positively charged nuclei together requires enormous energy and the larger the nuclei (the greater the number of protons) the more energy required. There is one place in nature where the combination of heat and pressure is just enough to force the simplest of all elements together: the hearts of stars.

Look around you. The component atoms of everything you see, including yourself, began life in the center of a star. The astronomer Carl Sagan, said it simply and best, "We are star stuff." The story of how hydrogen, a mundane proton with a single electron, could eventually be transmuted into you and me and the world around us is at heart a simple story of gravity and time.

If I were an atom of hydrogen I would be simply one amongst a nearly infinite number of absolutely identical companions slowly drifting in great clouds

Figure 8.27 A dark cloud of dust hangs in front of glowing red clouds of hydrogen gas filling the space between the stars. Hydrogen gas emits red light when excited by ultraviolet light from stars, much like neon gas in beer signs glows reddish-orange when excited by electricity (NASA, ESA, The Hubble Heritage Team, (STScI/AURA) and P. McCullough (STScI)).

throughout the galaxy. Alone, I weigh next to nothing, yet with my fellow hydrogen atoms our cloud makes up nearly 2,000 times the mass of a single sun. Astronomers know this because of the light we give off.

You see this almost every night along a busy city street. Run an electric current through a tube of gas with an atomic number of 10 and you produce the wonders of late night bars, roadside attractions, and Route 66: the neon sign. Neon has an atomic number of 10. Every element gives off a unique spectrum of light when it is exposed to energy (be it in the form of electricity from the wall or ultraviolet light from hot young stars). Because of this, astronomers can look out into the Universe and by analyzing the light they receive from distant stars, nebulae, galaxies and quasars they measure the composition of the Universe.

Not surprisingly, ninety percent of everything out there is the simplest thing there is, hydrogen gas. Telescopes around the world attuned to the specific wavelength of light given off by hydrogen atoms reveal that the 'empty' space between the stars is awash with this gas. While we hydrogen atoms may be as light as it is possible for a single element to be, we still possess some mass and so I am attracted to every other atom out here through the power of gravity. Though the forces are small, the Universe has time to wait. Eventually gravity finds an eddy or knot in the galactic clouds where just a few more solar masses of us are found than elsewhere and the relentless pull of gravity builds like an avalanche to draw more gas into the growing mass. Slowly I pick up speed as I fall into the developing maelstrom.

With growing speed my fellows and I increase in temperature. Temperature, at its most basic, is nothing more than the average speed of an object's atoms. The colder something is, the more lethargic the atoms. Absolute zero in the Kelvin temperature scale of physicists (–459° F or -273° C) is the temperature at which all atoms stop. Heat anything, and you cause its atoms to speed up.

As air in a balloon heats up, the excited atoms crash into the balloon's fabric

Figure 8.28 There is nothing empty about space. Here are identical views of the same constellation. On the left is Orion seen in visible light. Glowing red gas is visible in places from clouds of hydrogen. On the right, the exact same view is seen in the infrared light of glowing gas and dust. The brightest spot on the right corresponds to the Orion Nebula, the bright red "star" within the sword hanging from Orion's Belt on the left. Betelgeuse, the bright orange star in the upper left of the visible image, is no more than a faint white spot in the upper left of the infrared image ((visible) T. Nordgren; (infrared) IPAC, JPL-Caltech, NASA).

pushing it outwards. Hot gasses produce an outward pressure. As the growing cloud of which I am a part collapses and warms, we begin to see what will eventually become a relentless tug of war between two forces: gravity pulling us in, and thermal pressure pushing us out. But gravity never gives up, and we are still far too cold to put up much of an outward fight. Mass builds upon mass, and the cloud continues to shrink.

Over time, denser pockets form within my cloud and it fragments into what will become an entire cluster of hundreds of individual stars. The collapse of my individual clump takes nearly 10 million years to produce a spherical ball so dense and hot that we hydrogen nuclei are finally forced together hard enough to fuse and form the second simplest of all elements: helium.

It takes four protons coming together in an elaborate chain of collisions to produce one helium nucleus.[8] Add up the mass of the helium nucleus, however, and you find that it is slightly smaller than the mass of the four hydrogen nuclei

[8] During this chain of reactions two of the protons are converted into neutrons producing helium with an atomic number of 2.

Figure 8.29 Hubble Space Telescope mosaic of the Orion Nebula. Newly formed stars in the very brightest, inner portion of the nebula light up the entire cloud. The Orion Nebula (also known as M42) is only 1,500 light-years away, making it the nearest star-forming region to the Earth. This mosaic is made up of 520 individual Hubble images taken in 5 different colors to make a single picture that covers an area on the sky the same size as the full Moon. This nebula is an amazing sight through even a small amateur telescope and looks strikingly similar (though fainter and without any color) (NASA, ESA, M. Robberto (STScI/ESA) and the Hubble Space Telescope Orion Treasury Team).

that made it. This missing mass is converted into energy by the amount of Einstein's $E=mc^2$ (where m is the missing mass, c is the speed of light, and E is the energy released). Though the missing mass is very small, the speed of light is very large, and thus the energy released is enormous. Only with the onset of nuclear fusion is the inward push of gravity balanced by the outward pressure of heat. At that moment of Einsteinian ignition a new star turns on.

The light from these new stars lights up the hydrogen clouds out of which

Figure 8.30 The Pleiades cluster of new stars is passing through wisps of gas and dust that reflect the hot young stars' blue light. On even a clear dark night, most of us are only able to see the brightest stars visible here (NASA, ESA, AURA/Caltech).

they and their siblings are born. The Orion Nebula which you can faintly see with your unaided eye as a fuzzy 'star' in the winter sky is just such a cloud.[9] Through even a small telescope it is revealed as one of the most beautiful objects in the night sky. Out of a single cloud like the Orion Nebula, a thousand stars may emerge. Every star in the sky, including the Sun, began life with its siblings in this way. The seven sisters of the Pleiades are just such a cluster of young stars in the winter sky.

In the heart of a star where hydrogen fusion is taking place, the delicate balance of gravity and heat results in a feedback loop that keeps the star happy for millions if not billions of years. Here in the core the combined weight of 10^{57} fellow hydrogen atoms above me (that's a 1 followed by 57 zeros) fuses us together to produce the energy that heats the interior and keeps the crush of that mass at bay. If the temperature in the core should ever drop, the weight above will push in more forcefully and the rate at which we fuse goes up, thus increasing the energy produced, and once more raising the temperature. If the temperature rises too much, the thermal pressure pushing out overcomes the gravitational pressure pushing in, and the rate of fusion decreases, thus dropping the temperature back down. For as long as there is hydrogen in the core of the star, the delicate balance of gravity and fusion holds. Since hydrogen is the most common element in the Universe, hydrogen is the most common element in stars, and the quantity of hydrogen fuel for fusion's energy is vast.

The nuclear energy that bubbles upward in a star, heats the surface, producing life giving warmth for any family of planets. If you are a civilization on one such planet you therefore have plenty of time to watch me rise and set and follow my patterns. I'll give you light in summer, should your planet be tilted, and in winter you can offer me elaborate ritual incentive to bring back my life-giving warmth.

If I am part of a small star like the Sun, we hydrogen nuclei will fuse happily for nearly ten billion years. The difference in time between civilizations on such

[9] An analogous object you can see in the summer sky is the Lagoon Nebula in the constellation of Sagittarius. Chapter 1 shows a picture of what you can see through any small telescope.

a planet, where one may build stone observatories of light and shadow to watch my motion through the sky, while the later builds observatories of steel and glass to plumb the depths of my core, is no more than an hour in my star's life. As of today, we are only half way through our comfortable life and residents of any such planet can sleep comfortably knowing there will be many more days to come.

For stars much more massive than the Sun – and theoretically they can be as much as 150 times greater – the inward crush of gravity is so strong that the nuclear fuel is fused at an accelerated rate. Stars no bigger than a dozen times larger than the Sun exhaust even their enormous stores of hydrogen fuel in as little as 10 million years but they shine with the light of 10 thousand suns while they do so. Compare this with stars less massive than the Sun where the weak force of gravity produces only an anemic rate of fusion. Dimly shining out there in the darkness are small, cool stars that have been steadily glowing since the formation of the Universe almost 14 billion years ago.

Look up at the night sky. On a clear night you can see at least some stars from all but the heart of the brightest cities. The first thing that you notice is that they are not all the same brightness. In the starry darkness of a national park the range in brightness is astounding. Astronomers since the time of Hipparchus in 140 B.C. have classified this range in brightness by saying the very brightest stars are of the first magnitude. Those about two and a half times fainter (at about the limit of the human eye to pick out a difference in brightness) are stars of the second magnitude. Two and a half times dimmer still are third magnitude stars and so on until we come to the seventh magnitude limit of what the typical human can see under the very darkest of skies. The result is a scale that has confounded astronomy students for centuries: the dimmer the star, the greater the magnitude.

Although the magnitude scale has come down to us largely unchanged, modern astronomy has identified stars with zero magnitude (Vega) and very bright objects with negative values of magnitude (Venus at its brightest can be -4, while the Full Moon is -13, and the Sun at noon is –26).[10] So when we look at the night sky and see the brightest stars of zero and first magnitude, can we conclude that they are the massive stars while the dim ones are all small and anemic? Not quite.

The distance to a star plays an enormous factor in how bright it looks: distant stars are dim. However, space is vast, and there are a lot more stars very far away from us than there are stars close by. Pick any bright star in the sky at random and odds are it isn't very close. To look as bright as these stars do means they're

[10] Every difference of five magnitudes means something is a hundred times brighter or dimmer. Since the noonday Sun is about 30 magnitudes brighter than the faintest star you can see with the naked eye, this means there is a range of 10^{12}, a trillion times in brightness ($100 \times 100 \times 100 \times 100 \times 100 \times 100$) between the faintest thing and brightest thing we see in the sky.

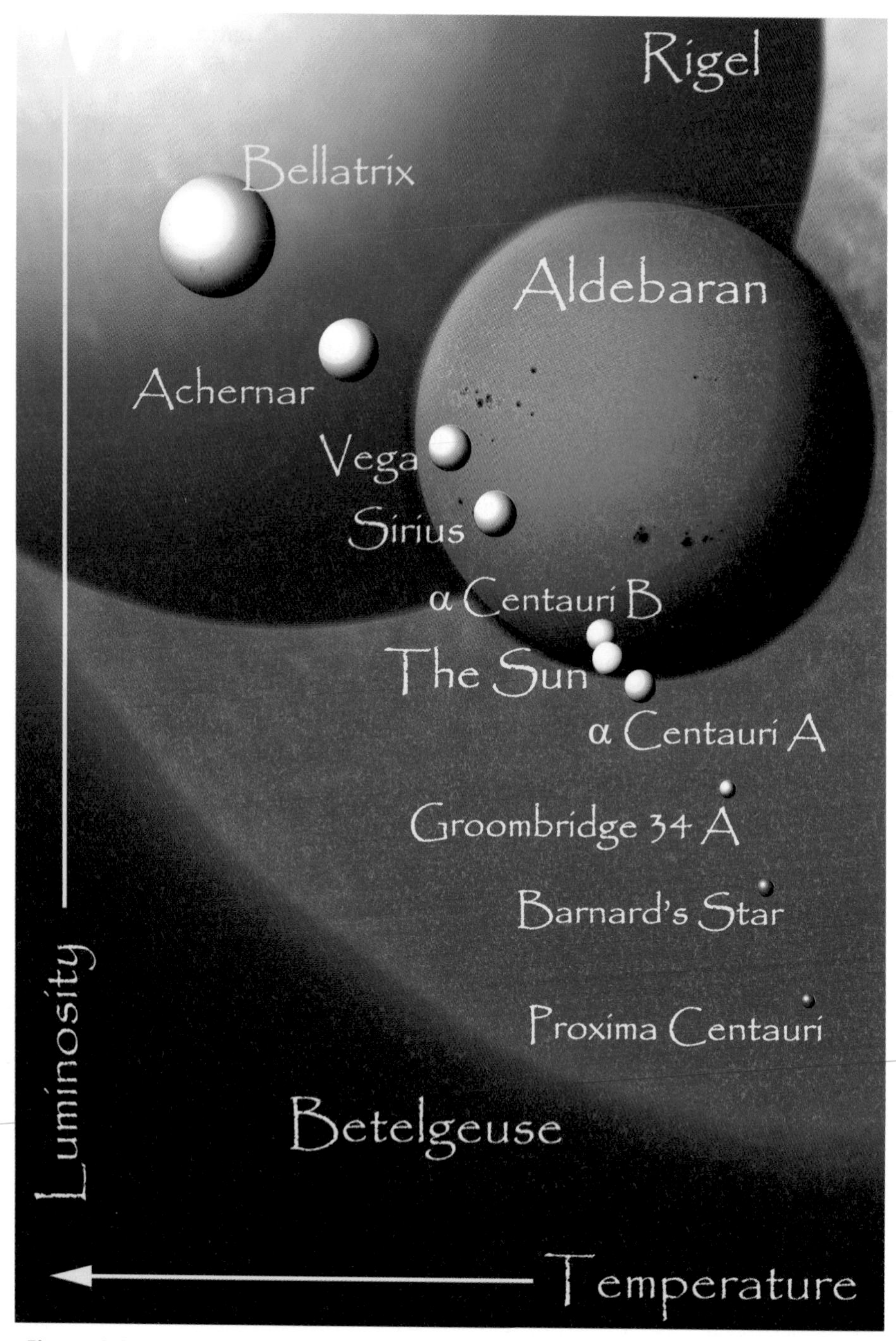

Figure 8.31 A collection of bright and nearby stars is shown with approximately true relative sizes. Hotter stars are shown to the left with more luminous stars (those that are truly brighter, taking into account their actual distance from us) at top (T. Nordgren).

typically among the galaxy's most luminous. So, on average, the brightest stars that make up the constellations you normally see, especially if you view the sky from your home in the city, do in fact tend to be the most massive stars burning quickly through their hydrogen fuel.

But look closely at the stars and another quality soon comes into play. Stars are not all white. There are red stars, orange stars, yellow stars, white stars and blue stars. The colors tell us their surface temperature. Place a metal bar in a fire and at first it doesn't glow at all. In actuality it is glowing, but in the infrared where your eyes can't see it. Let the metal get hot though, and soon it glows a dull red. As the temperature increases the bar gets brighter and the color changes from orange, to yellow, and finally if a fire is hot enough, bright white then blue. The bluest stars are therefore the hottest stars, and what this reveals about the energy escaping through the stellar surface gives you clues to their hidden interior. The colors are a window to the hearts of stars.

In summer, the heart of Scorpius is the cool red star Antares (the Rival of Mars, because of its ruddy color). In winter, the shoulder of Orion is the great orange star Betelgeuse with almost half the surface temperature of the warm, yellow Sun. Compare this with Rigel the hot blue star diagonally across Orion's belt, whose

Figure 8.32 The stars and constellations of evening in summer rise above Casa Rinconada. Labeled are stars, clusters and nebulae showing nearly every stage of stellar evolution from birth (Lagoon Nebula) to old age (red supergiant Antares). See the 'See for yourself' section at chapter's end (T. Nordgren).

Figure 8.33 The stars and constellations of evening in winter rise above Fajada Butte. Labeled are stars, clusters and nebulae showing every stage of stellar evolution from birth (Orion Nebula) to old age (red supergiant Betelgeuse), and death (Crab Nebula). See the 'See for yourself' section at chapter's end (T. Nordgren).

surface is nearly twice as hot as the Sun, and the color of stars quickly becomes apparent.

If all stars were at the same stage in their lives, contentedly fusing hydrogen into helium, then you could simply look at the colors of the stars in the sky and tell which stars were massive and which were stars like the Sun or smaller. While some may forever debate whether nature or nurture is the ultimate guide of what a person may become, for stars the matter is settled. Everything a star will do and become is set at the moment of its birth by the mass that it contains.

As stars burn through their nuclear hydrogen fuel I am converted from hydrogen into helium. Were I on Earth, I would weigh less than the surrounding atoms of oxygen and so I would float up and away. In a star, the condition is reversed and I now weigh more than the surrounding hydrogen and so I sink slowly into the heart of the star where I and my friends over a few billion years gradually choke the nuclear furnaces with atomic ash. For a star like the Sun, this process will take another five billion years.

When that time comes, the nuclear fires are momentarily banked and gravity seizes its opportunity to once again push inward while the opposing heat of fusion is absent. As we fall we give up heat, just as we did in the original collapsing cloud that made the star grow hot in the first place. The liberated heat ignites hydrogen fusion in a shell around the core. Since this extra nuclear energy is high up in the depths of the star, its thermal pressure causes the surface of the star to swell, but does nothing to counter the crush of gravity compressing its center.

In the core, the ever increasing pressure and heat one day reaches a point where even we heavy helium atoms fuse. When three helium nuclei come together they produce a nucleus of carbon where, just as before, the mass of the carbon nucleus is slightly less than the mass of the helium that made it. The missing mass is converted into the energy that heats the star and halts the compression of gravity.

On the outside, the surface bloats and cools as it swells outward away from the additional interior heat (much like a camper seated too far from a roaring bonfire who feels the surrounding cold creep between her shoulder blades). The Sun, like all other stars in their time, will go through this process. As its surface swells and cools it loses its sunny yellow color and grows a corpulent red. Its expanding bulk means scorched death for the planets of the inner Solar System. First Mercury, then Venus is engulfed by the bloated Sun. The star that gave Earth life for so long may very well swallow it too, though the planet's surface had long since become a charred and broken cinder. On that day there will be no more sunrises and sunsets, no more seasons or solstices. Everything humanity is or ever was will perish unless it has found somewhere else to go. You have less than five billion years to figure out how to do so. The Sun is now a red giant. Aldebaran, the orange eye of Taurus the bull that Orion pursues is just such a giant in the winter sky.

Eventually the nuclear core chokes with this new heavy carbon ash and the heat from fusion fails. Gravity wins again and once more presses the carbon atoms irresistibly inward. What happens next depends on the star's mass. If it is small like the Sun, then the mass of the star is too low for the constricting pressure to force any new atomic nuclei together and no new fusion processes occur. As the stellar heart falls inward a point is reached at which quantum physics says that the electrons in the core cannot be pushed any closer together and a new form of pressure, an electron degeneracy pressure, halts any further contraction of the stellar core.

The additional heat that's released gently blows the thin outer layers of the star off into interstellar space. Ultraviolet light from the exposed hot core, now a beast called a white dwarf, illuminates this expanding gas like those roadside neon signs. These planetary nebulae (so named because to early astronomers with poor telescopes they resembled the small disks of planets) are some of the most beautiful objects in all the Universe. A gorgeous tombstone, the Ring Nebula near Vega is a warning in the summer sky of what fate holds in store for the Sun someday.

While white dwarfs and planetary nebulae are the fate of stars like the Sun, the fate of the most massive stars is of even greater importance for what it reveals about the past. For these stars which, like Betelgeuse, may be 20 times more massive than the Sun, the inward crush of gravity is strong enough to force the remaining helium to fuse with the carbon ash; the result is atoms of oxygen. Over the next million years this process will be repeated many times as after depleting each stage of nuclear fuel, the core collapses until it can fuse the remains of the previous reaction. Stars like these are red supergiants like Betelgeuse in Orion.

While helium fused to become carbon for maybe a few hundred thousand years, carbon fuses with helium to become oxygen for maybe only a few hundred

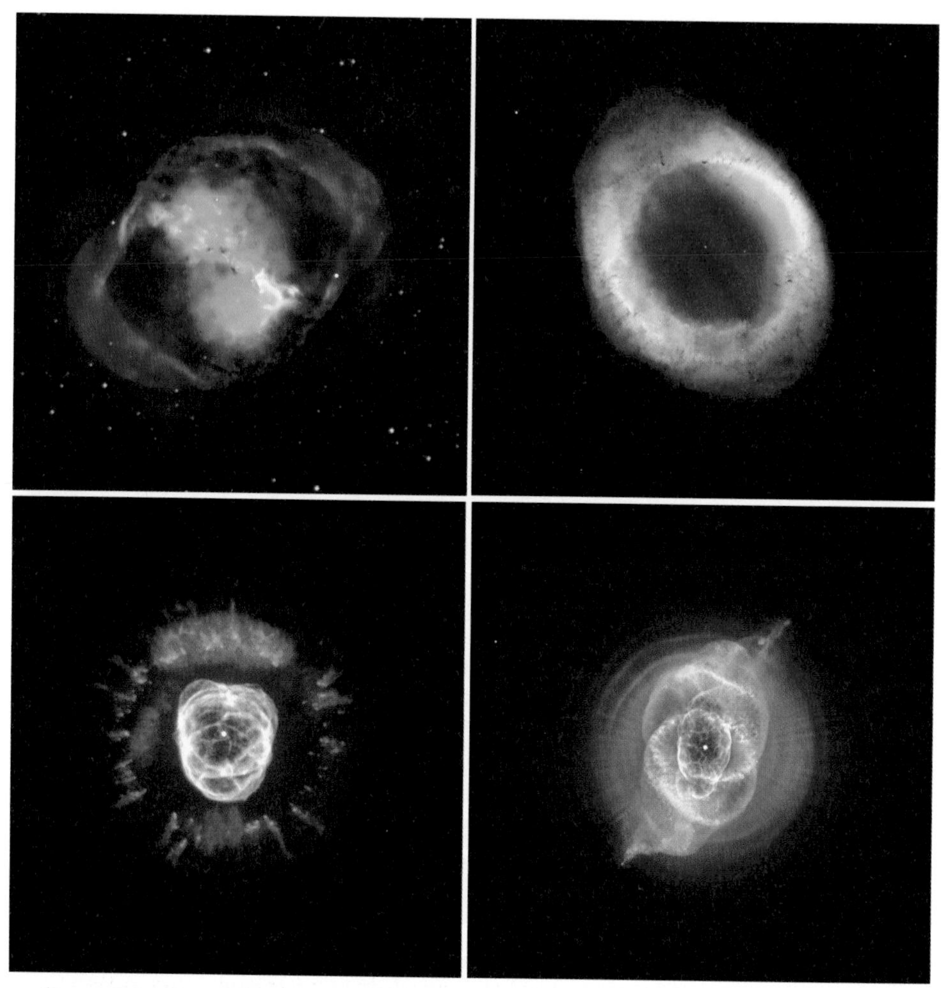

Figure 8.34 Four different planetary nebulae are shown. Each is the death of a star like our Sun where the outer atmosphere is gently blown off revealing the hot, exposed core of the star (now called a white dwarf). Different elements in the expanding clouds light up with different colors from the ultraviolet light of the white dwarf found at the center of each. Clockwise from upper left is the Dumbbell Nebula, Ring Nebula, Cat's Eye Nebula, and Eskimo Nebula ((Dumbbell) Chase Ellis/U.S. Naval Observatory, (Ring) Hubble Heritage Team (AURA/STScI/NASA), (Cat's Eye) NASA, ESA, HEIC and Hubble Heritage Team, (Eskimo) Andrew Fruchter (STScI) et al., WFPC2, HST, NASA).

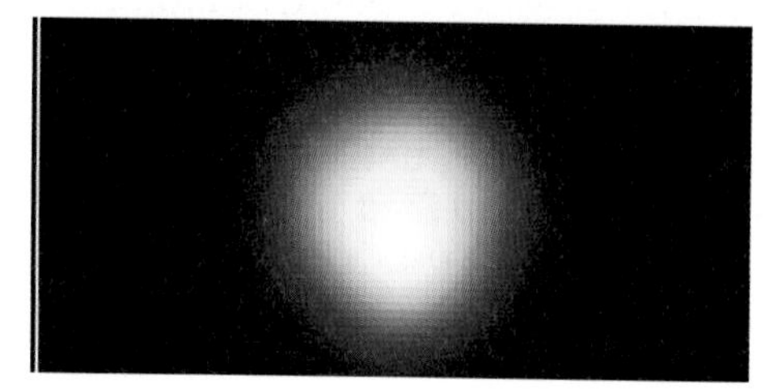

Figure 8.35 This is the first and only Hubble Space Telescope image of the surface of another star. Betelgeuse is expected to go supernova any time within the next hundred thousand years or so (Andrea Dupree (Harvard-Smithsonian CfA), Ronald Gilliland (STScI), NASA and ESA).

years. Each new reaction burns for less time than the one before as the star runs through ever newer and larger atomic reactions. For less than a year, oxygens fuse together to produce silicon (and more helium) while silicon atoms spend the last day of the star's 10-million-year life fusing together to produce iron.

In every one of these reactions, the resulting element always weighs less than the ones that made it. In each case the missing mass produces the energy that heats the interior and produces the outward pushing 'radiation pressure' that balances gravity and prevents the star's collapse. In that final day, however, the last day of the star's life, the iron that is made is so stable that any attempt to fuse it to produce something else actually takes energy away from the star. To fuse iron would be to cool a star, not to heat it. The star has reached a dead-end from which no escape through fusion is possible.

After millions of years in which I began as the lightest stable element in the Universe, I find myself in the star's final day as one of its most massive. On that day gravity crushes we iron atoms together so tightly that, just as with a white dwarf, electron degeneracy pressure briefly holds the mass of the star at bay. But with every new nuclear reaction that takes place above, more iron rains down on us below until, at the final moment of the stars life, our combined weight is too much for even the degenerate pressures of quantum physics.

In an instant, negatively charged electrons are forced inward into their iron nuclei where they fuse with positively charged protons to produce neutrons with no charge at all. Atoms are mostly empty space; they're a cloud of electrons with a microscopic nucleus 10,000 times smaller, hidden at its center. At the moment the electrons disappear, all this empty space is revealed, and the bottom drops out. Now that I'm a neutron I can be packed a billion times smaller than the former electrons would allow and suddenly we're on an elevator a billion floors up whose cables have been cut. Gravity runs rampant until we finally reach bottom with such force and energy that two things happens. Some neutrons 'stick,' forming a tight ball of neutrons, a single atomic nucleus, two and half times the mass of the Sun in a perfect sphere no larger than a small city. The quantum property of neutron degeneracy pressure (neutrons packed as tightly as physics allows) is now all that holds against gravity.

The other event is that the heat and energy of that final collapse sends bits and pieces of the stellar interior crashing outward through the star at just the moment the upper floors realize that the bottom's gone. I, and the other subatomic nuclear debris, ride the wave of one of the most energetic explosions the Universe ever sees and are accelerated to near the speed of light (and yet gamma-rays traveling *at* the speed of light, pass us as if we were standing still). Astronomers on Earth call us cosmic rays and we flood the galaxy from nearly 15 billion years of cosmic evolution.

As I rush outward through the star's insides, I see a nightmarish cascade of collisions and nuclear reactions that could never have taken place within the heart of a stable star. Iron nuclei fuse with passing oxygen forming selenium with an atomic number of 34. Two iron atoms fuse to make tellurium with an atomic number of 52, and somewhere a chain reaction of iron and other new elements combine to produce a nucleus with an atomic number of 79: gold. Look at a periodic table and every element after iron is rare because these are the elements that are only ever formed in the final cataclysmic instant of a massive star's death: a supernova.

But you don't need a telescope or a chemistry set to examine the final seconds of a star. Is there any iron in your belt buckle? Do you wear any gold on your fingers or hanging from your ears? Do your glasses have titanium in their frames? If so, then you wear about you a sample of a supernova.

For a brief moment this massive star's death lights the galaxy with the combined radiance of a hundred billion stars. The expanding wave of its stellar remains seeds the interstellar clouds with an alphabet soup of actium, barium, and carbon, xenon, yttrium and zirconium. The pressure of its passage gives these gasses that very first kick of instability that allows gravity a toe-hold causing the collapse that forms new stars. And when new stars and planets do form, they will now contain within them the silicon, oxygen, carbon, iron, and gold that are the elemental legacy of all the stars which have ever come before.

My life in a star will then begin again....

In the constellation of Taurus sits the Crab Nebula. It is the remains of the supernova perhaps recorded on the overhang I see above me (now that my imaginative wanderings have come full circle). Astronomers have measured the speed with which the stellar debris is expanding and at its current size the mathematics confirms that it must have exploded at about just the right time to be what ancient astronomers saw here almost a thousand years ago. You can still see the Crab Nebula from Chaco using the park's own modern observatory.

During the summer, rangers conduct evening astronomy programs while twice a year Chaco is home to star-parties put on by local astronomy clubs. And though the days can be warm, volunteers use specially equipped telescopes to show visitors the Sun that has always meant so much to this canyon. Thankfully, it appears that there will always be astronomers here.

Figure 8.36 Hubble Space Telescope image of the Crab Nebula (also called M1). This is the exploded remains of the star that was seen to go supernova in 1054 A.D. Within its twisted gas filaments are all the elements it ever created during its long life. These elements, which include among others, carbon, oxygen, and nitrogen (the organic elements need for life as we know it) will enrich the interstellar gasses out of which new stars and planets will someday form (NASA, ESA, J. Hester and A. Loll (Arizona State University)).

Figure 8.37 Thanks to generous private donations of money, time, and the building itself, Chaco Culture NHP is currently the only national park with its own modern observatory. Here stars circle over it, just as they circle over the kivas elsewhere within the park. Visitors are invited to take part in evening astronomy programs most clear nights using the park's telescope (National Park Service).

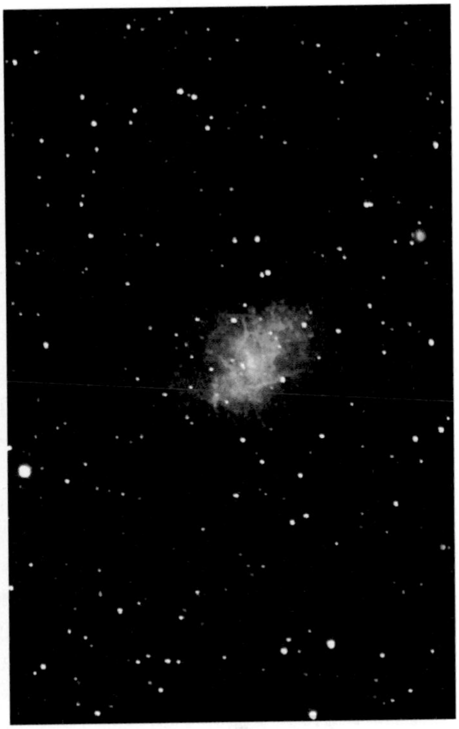

Figure 8.38 Here the remnant of the 1054 A.D. supernova is seen in an image from Chaco's observatory, made nearly a thousand years after it was first seen in the sky in the very same canyon. This image and the observatory that made it are thanks to the work and donation of John Sefick (National Park Service).

As I continue up the trail past the pictograph site, I take a moment to eat a handful of trail mix. The raisins it contains are high in iron which comes from the soil that is a component of the Earth from the days when it formed from an interstellar cloud. The raisins I eat are therefore the final stage from starlight to me. But the influence of supernovae in our lives doesn't end there. The cosmic rays produced in those explosions rain down on the Earth from all directions. Every once in a while one such proton traveling at nearly the speed of light hits a molecular bond in my DNA and changes it. If this particular strand of DNA is in a gamete used in reproduction, then this mutated gene will get passed on to the next generation.

In a world where far more living things, say coyotes, are born than can survive on the available food supply, any advantage spells the difference between a coyote that manages to pass on its genetic code, and one that doesn't. Nine hundred and ninety-nine times out of a thousand, a mutation is gibberish or harmful, and the coyote won't pass it on. But on that rare occasion when it isn't gibberish, where the change actually produces something beneficial that works slightly better than what was there before, the bearer stands a slightly better chance of surviving to mate, have puppies, and pass on its genes. The mutation is passed on to future generations. This is natural selection through random genetic mutation.

We humans have been trying this for only a few thousand years in a process called artificial selection and it's given us the difference between coyotes and cocker spaniels. But let nature do this for five billion years and it takes the building blocks of stellar evolution and gives us every living thing we see in the world today.

A half hour before dawn, as the stars begin to fade, we reach the ruins of Peñasco Blanco and my friends leave me to set up my camera for the coming sunrise. While I document any sunrise alignment visible from here, they continue on to a panel of beautiful rock art along a nearby mesa to check for any alignments visible from there. Alone against the crumbling wall of an ancient kiva, I look up to see Venus still shining in the growing twilight where waves of purple and pink begin to wash across the sky. The stone faces of half-buried walls are all turned to the east, and again I wonder about how many times someone may have stood where I am standing now on just such a day in order to see what I am about to see.

In fact, the very presence of people here on this continent may itself be attributed to supernovae. Physicists in Germany have identified a layer of rare

Figure 8.39 Pre-dawn light reflects off the walls of Peñasco Blanco (T. Nordgren).

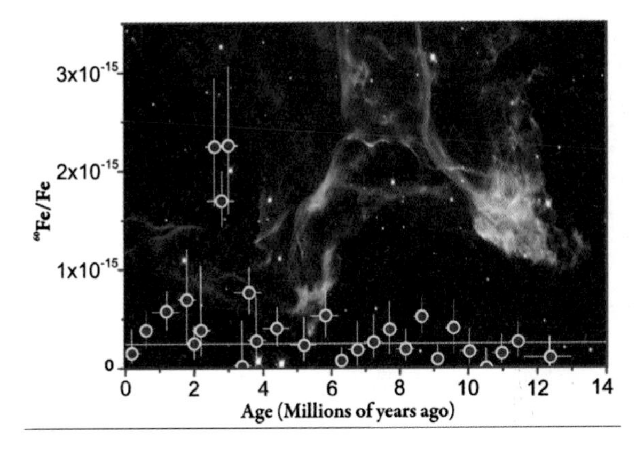

Figure 8.40 The abundance of an unstable iron isotope (^{60}Fe) measured relative to typical iron found on Earth in sea-floor layers of different ages. The average ratio of Iron-60 with time is shown by the thin white line. Three million years ago the abundance of Iron-60 jumped as a wave of new supernova elements is thought to have swept by the planet. The background image shows the light of elements created in another supernova captured by Hubble ((data) K. Knie and others, (background) NASA/STScI).

iron isotope in the sea floor laid down 2.8 million years ago (different isotopes of iron have the same number of protons, but different numbers of neutrons). Because this isotope decays into other elements on a timescale of millions of years, its presence in great abundance means it was not formed with the other elements that formed the Earth over five billion years ago. Rather, they suggest this iron must have been produced in a much more recent supernova, and then rained down upon the Earth to form a fine layer on the seafloor as the shockwave swept passed the planet.[11]

Since supernovae produce cosmic rays in addition to other elements, one result of a nearby supernova is an intense mass of charged particles flooding the Earth's atmosphere as the wave sweeps past. In addition to decimating our protective ozone layer, it's thought that increased levels of cosmic rays would act as seed points for the formation of clouds. For the hundred thousand years that the supernova's cosmic rays bombarded the planet, weather and climate patterns would change, altering, perhaps significantly, those plants and animals living at the time.

Imagine a supernova like that witnessed in 1054 A.D. but ten thousand times brighter – almost a hundred times brighter than the full Moon. As iron deposited on interstellar dust slowly wafts in from space, a tsunami of cosmic rays sweeps through the atmosphere, changing weather patterns, and the skies above Africa begin to change. Evidence confirms that 2.8 million years ago, Africa became a little more arid and the tree cover a little more sparse.

Our ancestors who still spent much of their time in and around trees notice the good days are a little less frequent; the forests a little less lush. The animals

[11] On average there should be one supernova every hundred years somewhere in our Galaxy. So over the five-billion-year history of our planet this process should have happened a number of times. As a child I recall a nearby supernova being one possible explanation for the extinction of the dinosaurs.

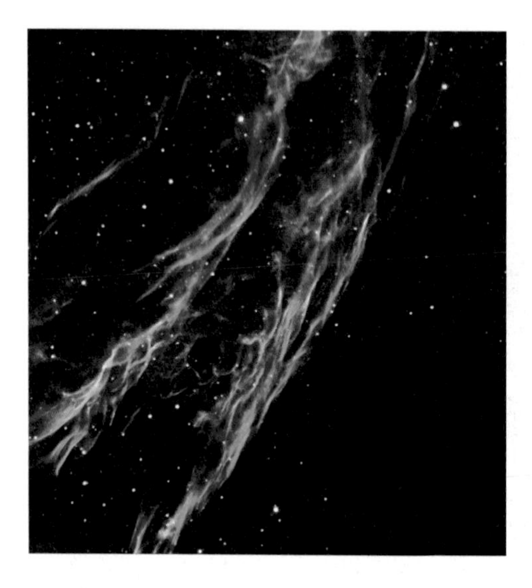

Figure 8.41 A portion of the Veil Nebula, a supernova remnant in the summer constellation of Cygnus, sweeps through space. Red light is from hydrogen atoms, while green light is produced by oxygen. Any interstellar clouds, stars or planets in its expanding path will have new elements (like Iron-60) settle upon them (Lua Gregory/U.S. Naval Observatory).

that live off the plants that live off the Sun become scarce in the old, familiar places under the trees. Those early hominids with a genetic make-up more suited to walking upright and capable of exploring the opening savannah would have had a much easier time passing on their genetic code to future generations.

Anthropologists today believe that 2.8 million years ago, the change in Africa's climate, whatever the cause, forced our ancestors down from the trees to finally walk erect. Our ancestors took the first steps that would eventually lead them out of Africa and around the world. We are all the descendents of those people; the Chacoans are descended from those who walked east, while I am descended from those who walked west.

By the light of stars, now lost to the coming light of day, I acknowledge their power and presence to make me who I am today. In every sense of the word I am who I am because of them. On a personal level they fill me with wonder and awe. On a physical level their life and death gave rise to my body, stimulated my evolution, and altered the environment that drove my ancestors.

It is at this moment, in the week of the winter solstice, that the final star upon which I owe so much finally appears. My genetic cousins the coyotes yip and howl around me as I stand here at Peñasco Blanco and see the first rays of the rising winter Sun stream down the long valley. Before me I see the sharp silhouettes of mountains and walls casting their long shadows, every one pointing at me. Today, on this spot that I have reached through the ruins of ancient astronomers, the sunrise is for me alone and I feel the power of my position on this canyon rim. From me to the Sun, and to the stars that made me, I am connected to everyone on this planet and we are connected through this Earth to ten million millennia of stellar evolution. The Navajo who live here now have a greeting they use when they meet one another. They say, "Yá'áát'ééh," or "It is good." But a Navajo I once knew told me that on another level it can also

Figure 8.42 Winter solstice sunrise bathes the canyon and our faces amongst the ruins of Peñasco Blanco. The intervals at which we captured the sunrise began at 2 minutes, then increased to 3 minutes (E. Dodd, G.B. Cornucopia, T. Nordgren).

mean "You are part of the Universe." Standing here now, a child of the stars, I know they speak the truth.

See for Yourself: The Stars

Personal Chaco solar calendar

On any of the solstices (June or December 21) or equinoxes (March or September 21) find an easily remembered outdoor location near your home where at the instant the Sun sets, it does so behind a prominent feature on the horizon (tree, wall, building, water tower, or mountain). Find a way to mark your spot. Return each week to the exact same place and see where the Sun sets. Take a picture of your western horizon (when the Sun isn't within the field of view) and make a mark on the image at weekly intervals for a year showing the sunset position; you now have a solar calendar that will allow you to know the date and predict the solstices.

Colors and temperatures of stars

If you look closely you can identify the different colors of stars noting which look reddish, bluish, or yellow/white. Colors allow us to take the star's surface

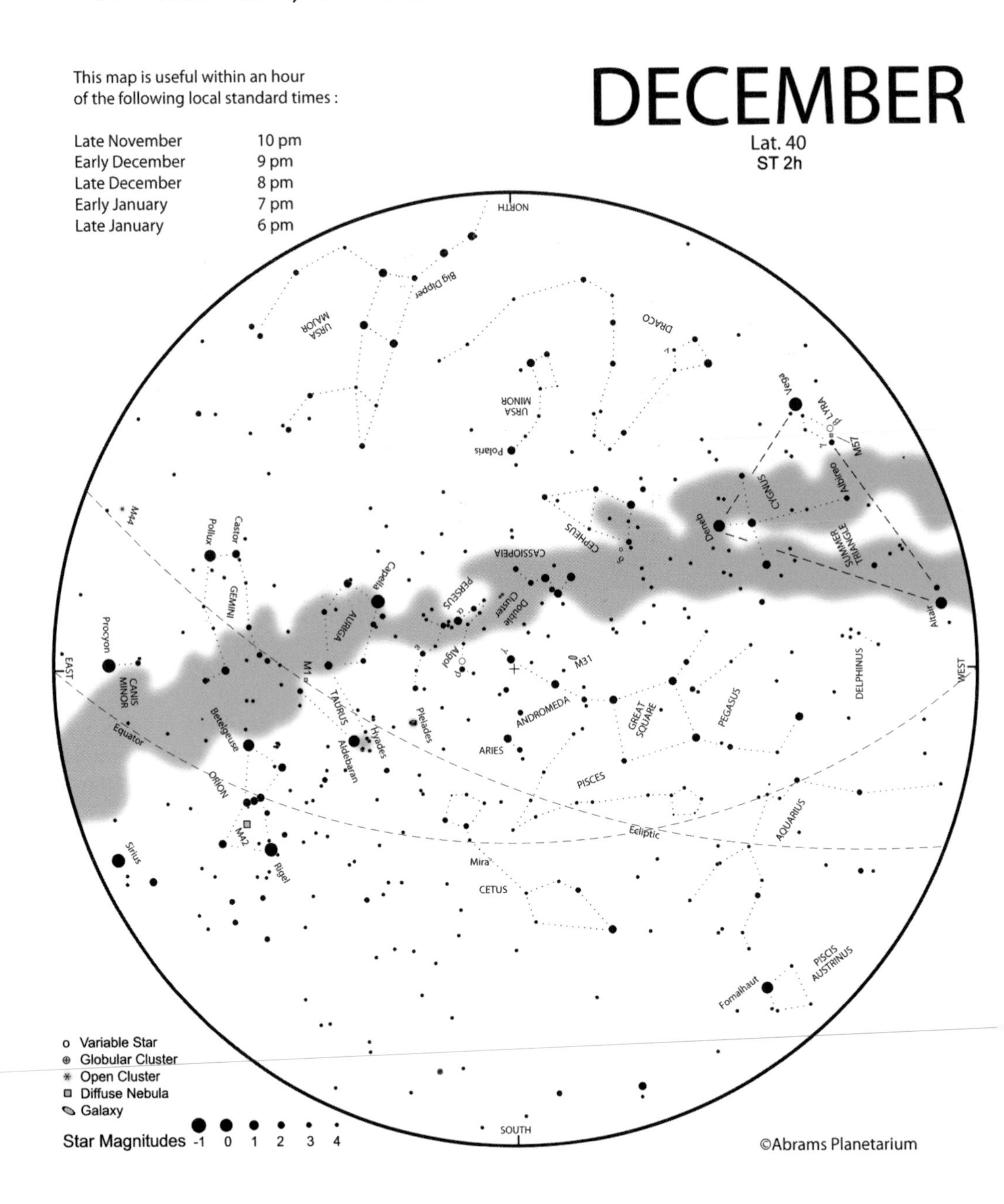

Hold the star map above your head with the top of the map pointed north. The center of the map is the sky straight overhead at the zenith.

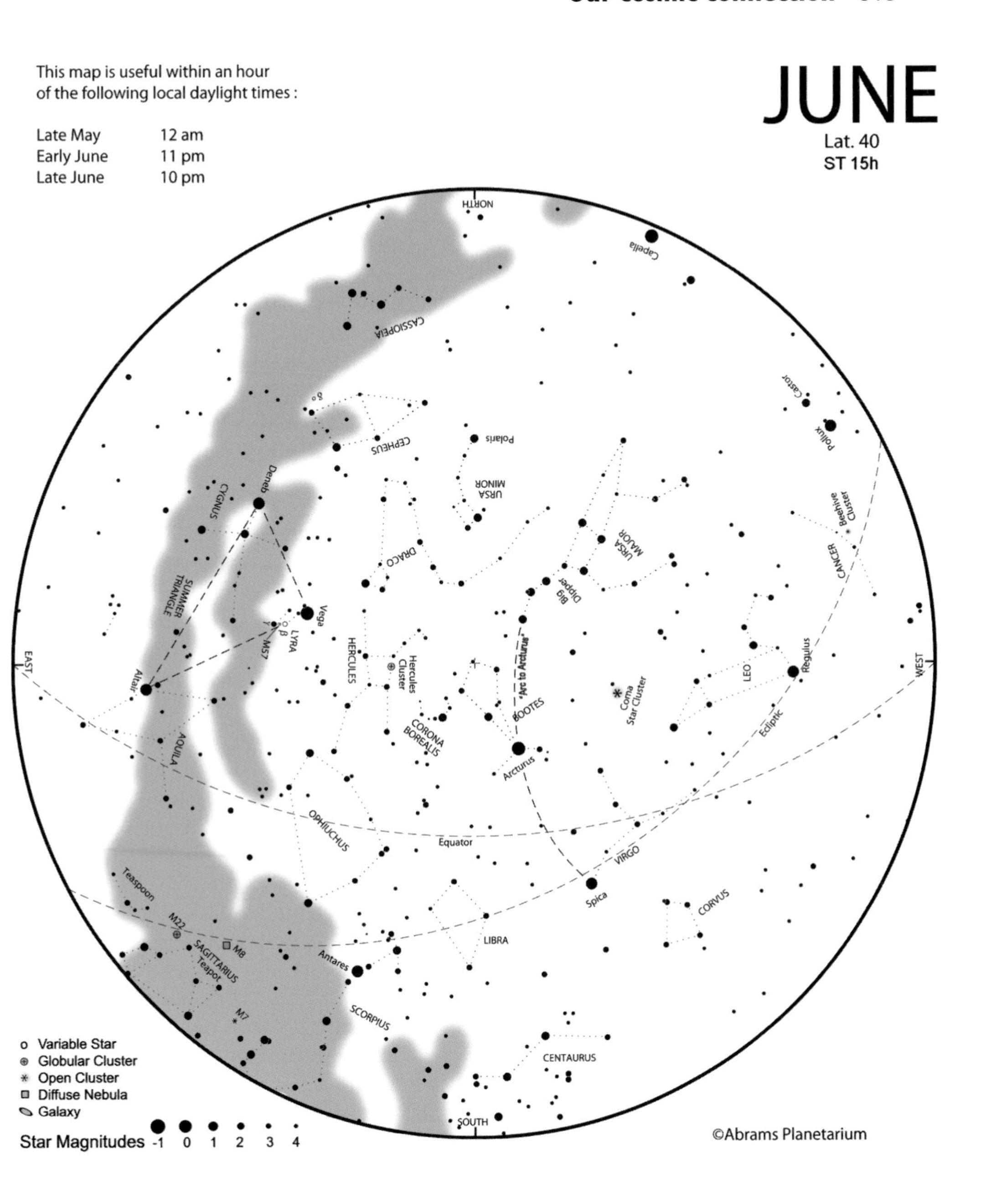
This map is useful within an hour
of the following local daylight times :
Late May 12 am
Early June 11 pm
Late June 10 pm
JUNE
Lat. 40
ST 15h
NORTH
EAST
WEST
SOUTH
Capella
CASSIOPEIA
CEPHEUS
Polaris
URSA MINOR
Castor
Pollux
Deneb
CYGNUS
DRACO
URSA MAJOR
Big Dipper
Beehive Cluster
CANCER
SUMMER TRIANGLE
Vega
LYRA
M57
HERCULES
Hercules Cluster
"Arc to Arcturus"
LEO
Regulus
Altair
BOOTES
Coma Star Cluster
CORONA BOREALIS
Arcturus
Ecliptic
AQUILA
OPHIUCHUS
Equator
VIRGO
Teaspoon
Spica
CORVUS
M22
M8
SAGITTARIUS
Teapot
LIBRA
Antares
SCORPIUS
M7
CENTAURUS
o Variable Star
Globular Cluster
Open Cluster
Diffuse Nebula
Galaxy
Star Magnitudes -1 0 1 2 3 4
©Abrams Planetarium

temperature (the cosmic equivalent of resting our hand against a forehead to see if we have a fever). On average, red stars tend to be about half the surface temperature of our Sun, while blue stars tend to have surfaces three to five times hotter than the Sun.

In winter skies, red stars are Aldebaran in Taurus and Betelgeuse in Orion. Compare Betelgeuse with the blue star Rigel at the opposite corner of Orion. Compare bluish Castor (one of the heads of Gemini the twins) with its yellowish sibling Pollux in order to see subtle differences in color.

In summer, red Antares in Scorpius looks like the red planet Mars after which it is named. Arcturus in Boötes ("arc" to it from the handle of the Big Dipper) is also redder than any of the stars that make up the Big Dipper which look noticeably blue/white in comparison. If you should have a small telescope, point it at Albireo, the southernmost star in Cygnus, the Northern Cross. What looks like one star to our unaided eye reveals itself to be two stars: one turquoise blue, the other bright yellow.

Stellar evolution in a single night (see Figures 8.32 and 8.33)

Winter (late fall to early spring): Find the constellation of Orion by looking for his belt of three nearly equally bright and equally spaced stars. Above Orion (if it is rising) or to its west is the constellation of Taurus the Bull which Orion hunts. Farther overhead, or to the west, is the small cluster of the Pleiades. In these three objects you can see every stage of stellar evolution for massive stars.

Hanging down from Orion's Belt is a line of three faint stars. The middle star looks slightly fuzzy compared to the other nearby stars. This fuzzy 'star' is the Orion Nebula (also called Messier 42, or simply M42), a cloud of hydrogen gas lit up by new stars being born. This is a stunning sight through a telescope or good binoculars.

The Pleiades are a small tight cluster of newborn stars. Called the 'Seven Sisters' six or seven stars are visible to most eyes. Some observers under dark skies can see many more. In reality there are hundreds of stars there; we see only the brightest.

Follow the line of Orion's Belt down and to the east and you come to the brightest star in the sky: Sirius in the constellation of Canis Major (Orion's large dog). Sirius is a star in the middle of its life fusing hydrogen into helium. It is the brightest star in the sky because it is the closest star to us that we can see with the naked eye in the Northern Hemisphere (only 8.6 light-years away). The light you see tonight took 8.6 years to get to you. If Sirius hasn't risen yet, find Bellatrix, the upper right (northwest) corner of Orion. Bellatrix is also a typical hydrogen-fusing star in the middle of its life. Both of these stars are hotter, brighter, and more massive than our Sun.

The V-shaped head of Taurus the Bull is a loose group of stars called the Hyades. In the same direction, but closer to us in space, is the red giant Aldebaran that forms one of Taurus' eyes. As a red giant star, Aldebaran has begun to die; it has run out of hydrogen fuel in its core and is now fusing helium to make the energy to support itself.

Betelgeuse, the bright orange star in the upper left (northeast) corner of Orion is a high-mass red supergiant star. It is in the very last stages of its life, fusing elements heavier than hydrogen and helium to support its weight. Betelgeuse is expected to go supernova any time within the next hundred thousand years or so. Perhaps it will tomorrow. When it does, it will be the brightest supernova in human history, perhaps as bright as the full Moon, and much brighter than the supernova of 1054.

To see the location of the 1054 supernova, go back to the V-shaped head of Taurus and follow both arms of the V upward until you come to two stars (one at the end of each arm) that form the horns of Taurus. Between these two stars, but closer to the one on the east, is the spot in space where Chinese and Chacoan astronomers saw the supernova of 1054. In its spot today is the Crab Nebula (also called M1), faintly visible through a medium sized telescope.

Summer (early summer to mid-fall): The Lagoon Nebula (M8) is a star-forming region above the north spout of the Sagittarius Teapot (see Chapter 1's 'See for yourself' section). Here, as with the Orion Nebula, newborn stars light up the cloud so bright it looks like a fuzzy 'star' to the unaided eye. Good binoculars or a small telescope show the cloud of gas (including a dark dust lane) and new-forming stars. To the immediate right (west) of Sagittarius, half way between the spout and the stinger at the end of Scorpius' tail, is a much brighter cloud that's visible to the naked-eye. This cluster of new stars, called M7, is only 220 million years old (twice as old as the Pleiades).

For a typical star like our Sun in the middle of its hydrogen fusing life, look to Vega or Altair, both are part of the famous Summer Triangle. They are two of the brighter stars. Each is slightly more massive than our Sun.

Find the Big Dipper to the north then following along the handle, 'arc' to the bright orange star Arcturus. This is a dying and bloated giant star fusing helium into carbon, very similar to Aldebaran in winter.

After stars like our Sun pass through the red giant phase of their life they form a planetary nebula where the outer part of their atmosphere expands out into space. The most famous of these objects is the Ring Nebula in Lyra. It is located between beta and gamma Lyra (see the finder-photo in Chapter 3). It is not visible with the naked-eye, but appears as a bright 'O' in a medium sized telescope.

On the southern horizon, is the red supergiant Antares. Antares is the heart of Scorpius and is the brightest (and reddest) star in the southern half of the sky. Antares is 15 times more massive than the Sun, 700 times larger, and is near the end of its life.

The Veil Nebula in Cygnus is a supernova remnant of a massive star like the Crab Nebula (though 10 to 20 times older). It's not visible to the naked eye, and hard to see through a moderate sized telescope unless in a dark location. Its brightest part is off the western arm of the Northern Cross and is one of my favorite things to see in a telescope.

Further reading

A Guide to Prehistoric Astronomy in the Southwest by J. McKim Malville (2008)
Johnson Books, ISBN 1555664148

Living the Sky: The Cosmos of the American Indian by Ray A. Williamson (1987)
University of Oklahoma Press, ISBN 0806120347

Origins: Fourteen Billion Years of Cosmic Evolution by Neil DeGrasse Tyson and Donald Goldsmith (2005)
W. W. Norton, ISBN 0393327582

Extreme Stars by James B. Kaler (2001)
Cambridge University Press, ISBN 0521402620

Richard Wetherill – Anasazi: Pioneer Explorer of Southwestern Ruins by Frank McNitt (1974)
University of New Mexico Press, ISBN 0826303293

Ron Sutcliffe's lunar cycle observer's guide (and links to his book *Moon Tracks: Lunar Horizon Patterns*)
http://www.moonspiral.com

Anna Sofaer's website: The Solstice Project
http:// www.solsticeproject.org/

9 Worlds without number

Colorado Rocky Mountain High,
I've seen it raining fire in the sky,
Shadow from the starlight is softer than a lullaby,
Rocky Mountain High,
Colorado.

John Denver

Since 1995 we have discovered over three hundred planets around other stars. Think about that; we now know of forty times as many planets outside our Solar System as inside it and the number increases with each passing year. What will we find if we ever go out there? Will there be new, exotic forms of life awaiting our scientific expeditions? Will there be other civilizations with whom we can communicate?

Figure 9.1 Artist's conception of newly discovered alien planets orbiting the nearby star Gliese 876 (NASA and G. Bacon (STScI)).

Say we *could* send explorers to these new worlds. If suddenly our concept of the habitable Universe expanded overnight as we looked out on this new frontier of unimaginable space, what would the reaction be? Imagine we elected a new President who was the foremost expert on these alien worlds. He'd be the head of America's leading scientific association and he'd conceive the expedition from within his own home. He'd personally choose the explorers who would make the journey, determine the skills that would be necessary, and enlist his friends (the best scientific minds in the country) to oversee their training. He'd be the one to send them off into the vast unknown with orders that they report back to him exclusively with all that they discover about the potential for new life, resources, and territory for us to expand and conquer.

Science fiction? No, history. Thomas Jefferson was elected the third President of the United States in 1800 at a time when a gentleman could be both a politician and scientist, and still make significant contributions to both. Not only had Jefferson authored the Declaration of Independence and been the nation's first Secretary of State under George Washington, he was also the foremost expert on the geography of the American West and was president of the American Philosophical Society, the leading scientific organization in the country.

Jefferson's passion for science was serious. He considered Sir Isaac Newton, the discoverer of gravity, one of the greatest men the world has produced, and might well have become a scientist himself. "Before I entered on the business of this world I was much attached to astronomy & had laid a sufficient foundation at the College [of William and Mary] to have pursued it with satisfaction & advantage," Jefferson told a friend late in his life.[1] In

Figure 9.2 In this drawing of Thomas Jefferson, the third President points to the Declaration of Independence while to his left are scientific instruments including a celestial sphere (Library of Congress).

[1] Donald Jackson's *Thomas Jefferson and the Rocky Mountains, Exploring the West from Monticello.*

Figure 9.3 Portraits of Meriwether Lewis (left) and William Clark (right), co-Captains of the Corps of Discovery. Reproduction of watercolor paintings by C.W. Peale in Independence Hall, Philadelphia, PA. Copyrighted 1903 (Library of Congress).

1962, President John F. Kennedy joked at a White House dinner honoring 49 Nobel Prize laureates, "I think this is the most extraordinary collection of talent and of human knowledge that has ever been gathered together at the White House - with the possible exception of when Thomas Jefferson dined alone."

Jefferson's scientific knowledge of geography, geology and the American West was perfectly placed to take advantage of his times. Three years after being elected President, Jefferson arranged the purchase of the Louisiana Territory from Napoleon Bonaparte for $15 million. In addition to the geo-political advantage of precluding French expansion into North America and opening up new territory for the young expansionist country, the Louisiana Purchase offered unparalleled scientific opportunity to reconnoiter terra incognita.

For the exploration of these new lands, the President chose his personal secretary, Meriwether Lewis, to head the Corps of Discovery. As its name implied, its primary goal would be scientific exploration and the scope of what Jefferson intended for Captain Lewis and his co-Captain William Clark, is best understood from Jefferson's personal orders:

> The object of your mission is to explore the Missouri River, and such principal streams of it, as, by its course and communication with the waters of the Pacific Ocean, whether the Columbia, Oregon, Colorado, or any other river, may offer the most direct and practible[sic] water-communication across the continent, for the purposes of commerce....
>
> Beginning at the mouth of the Missouri, you will take observations of latitude and longitude, at all remarkable points on the river... Your observations are to be taken with great pains and accuracy; to be entered distinctly and intelligibly for others as well as yourself; to comprehend all the elements necessary, with the aid of the usual tables, to fix the latitude and longitude of the places at which they were taken....
>
> You will ... endeavor to make yourself acquainted, as far as a diligent pursuit of your journey shall admit, with the names of the [Indian] nations and their numbers; ... Their language, traditions,

> monuments; Their ordinary occupations in agriculture, fishing, hunting, war, arts, and the implements for these; Their food, clothing, and domestic accommodations; Moral and physical circumstances which distinguish them from the tribes we know;... And, considering the interest which every nation has in extending and strengthening the authority of reason and justice among the people around them, it will be useful to acquire what knowledge you can of the state of morality, religion, and information among them...
>
> Other objects worthy of notice will be;
>
> The soil and face of the country, its growth and vegetable productions, especially those not of the United States; The animals of the country generally, and especially those not known in the United States; The remains and accounts of any which may be deemed rare or extinct; The mineral productions of every kind, but more particularly metals, lime-stone, pit-coal, and saltpeter; salines and mineral waters, noting the temperature of the last, and such circumstances as may indicate their character; Volcanic appearances; [and] Climate, as characterized by the thermometer, by the proportion of rainy, cloudy, and clear days; by lightning, hail, snow, ice; by the access and recess of frost; by the winds prevailing at different seasons; the dates at which particular plants put forth, or lose their flower or leaf; times of appearance of particular birds, reptiles or insects....
>
> Given under my hand at the city of Washington, this twentieth day of June, 1803.
>
> Thomas Jefferson, President of the United States of America

The Corps of Discovery set in motion a tradition of Federal patronage for American scientific exploration that would continue through the later Hayden and Powell expeditions to Yellowstone and the Grand Canyon, and ultimately to John F. Kennedy's pledge to send a man to the Moon and NASA's robotic missions surveying the Solar System and studying the stars.

That exploration is now entering a new chapter with NASA's 2009 launch of the Kepler spacecraft. It's designed to find new Earth-like planets around nearby stars and survey the summer sky for evidence of solar systems other than our own. The prime contractor that built this planet-hunting spacecraft is Ball Aerospace located in Boulder, Colorado just a short drive from the entrance to Rocky Mountain National Park. Today I'm taking a break from a NASA workshop in Boulder to go hiking in the park along the Continental Divide and to get a small sense of what Lewis and Clark must have experienced on their expedition through the Mountain West.

It's hard to imagine how little the scientific establishment of Jefferson's day knew about this region just two hundred years ago. They knew next to nothing of the geography, geology, botany, or the people already living here – and nothing at all about their languages or cultures. Read between the lines of

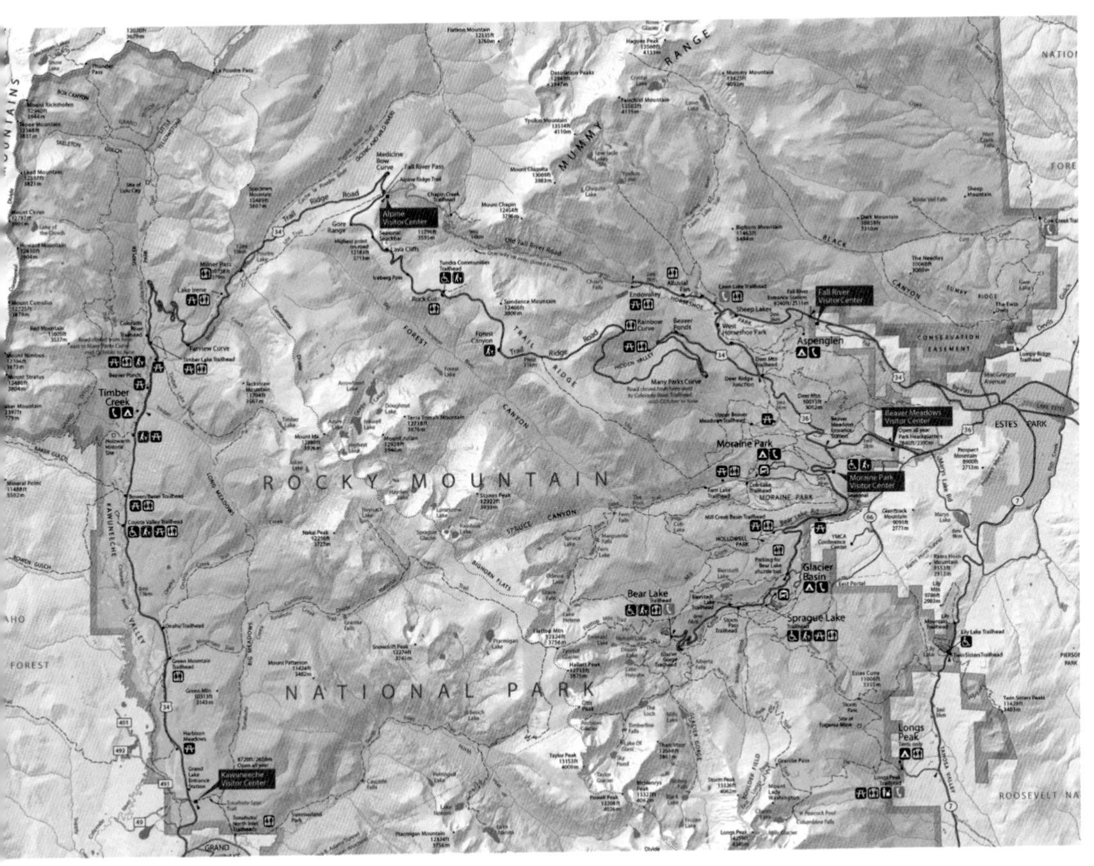

Figure 9.4 Detail of National Park Service map of Rocky Mountain National Park showing the Continental Divide outside Estes Park, Colorado (NPS).

Jefferson's letter and you see a territory in which the President of the United States could still hope to find herds of woolly mammoths, mountains of salt, and erupting volcanoes. The majority of this continent, from the Mississippi River to the Oregon Coast and all that it contained was a complete and utter mystery to Europeans.

One persistent question was the possibility of a Northwest Passage. Going as far back as Columbus, Europeans dreamt of finding a westward route to the markets of China and Japan. With the discovery of a brand new continent in the way, this dream turned into the search for an all-water route from the Atlantic to Pacific Oceans through the middle of the continent. "The belief that the Missouri [River] could be used to reach the Pacific dominated the geographical lore of the Northwest," writes John Logan Allen, a Lewis and Clark expert and chair of the geography department at the University of Wyoming. "It was the central ingredient of the geographical knowledge which, when interpreted in the light of early nineteenth century American thought, was to form the basis of

Thomas Jefferson's image of the Northwest and the chief objective of the Lewis and Clark Expedition.[2]"

For the majority of the last two thousand years, science was akin to philosophy. For the ancient Greeks, most notably Aristotle, it was enough that an idea was logical for it to be considered true. It made sense that two balls of differing weights should fall to Earth at different rates because the heavier ball had more of the stuff that wanted to return to its place of origin than the lighter ball. Never mind that it was possible to show that this is exactly not what happens, the idea was logical. Aristotle's writings were eventually used as support for Christian philosophy, so that the truth of a scientific idea depended not only on how well it agreed with Aristotle, but by extension, Scripture.

As an example of this way of thinking, consider that prior to the 1500s geography and theology could be intertwined disciplines. "Irregular landforms in general, and mountains in particular, were considered not part of the original creation but something that came later, a consequence of the Flood, put here by God as a reminder of His extreme displeasure at our evil ways," writes Gary Ferguson in *The Great Divide, The Rocky Mountains in the American Mind*. By 1609, however, when Galileo saw mountains hidden away on the Moon, ideas of their presence as a symbol of our sins eventually gave way to European Enlightenment ideals of reason and natural explanations (which led to very different ideas about the distribution of the world's landforms).

With increased recognition of the importance of empirical evidence, the hallmark of Enlightenment science, discoveries about the physical Universe could attain a level of acceptance independent of whether or not the ideas seemed intuitive, were argued with particular flair, or agreed with Scripture. The scientific method that arose out of this tradition can be briefly stated as follows: Observe the world, form a hypothesis to explain what you see, test the hypothesis, revise the hypothesis as the need arises, and repeat. Those hypotheses that survive repeated testing are given the exalted name of scientific theory. In the sixteenth century Tycho Brahe was a master of celestial observations and Johannes Kepler was a master mathematician at turning observations into testable hypotheses. He would be so successful that four hundred years later NASA's planet-hunting spacecraft would be named in his honor.

But what about the early stages of the scientific process? Where do the initial ideas come from? The early stages of the scientific method are some of the most chaotic and exciting. You observe. Look around you. See what the world has to offer and then hazard a guess. The better your initial observations, the better chance your initial hypothesis has of surviving the subsequent tests to live another day. The fewer the observations, or the more your 'observations' are colored by pre-conceived ideas without basis in fact, and the better your chance of Nature surprising you.

[2] From Allen's *Passage Through the Garden, Lewis and Clark and the Image of the American Northwest*.

American geography in the 1700s was a matter of scientific observation. European explorers had been mapping the east coast of North America for over two centuries and based on these observations they developed theories as to what the rest of the continent should look like. As early as the 1720s, writes Allen:

> ... British promotional literature on North America had introduced the concept of symmetrical geography. It was known that rivers which flowed westward into the Mississippi had their sources in mountains that were close to the Atlantic. It was further known that those same westward-flowing streams had interlocking drainage systems with the rivers that flowed east to the Atlantic. The same set of geographical conditions, according to the tenet of symmetrical geography, should apply to the western parts of the continent.

If symmetric geography was true then the mountains in the west should be like those in the east. For the Appalachian Mountains that run the length of the American east coast, anyone traveling by boat has a simple one day portage up and over a single ridge of roughly 3,000 ft-high (1,000 m) mountains before descending by river down the other side. The same would surely be true for anyone attempting to travel up the Missouri, cross the Rockies, and sail down the Columbia to the Pacific Ocean. Allen concludes:

> It is obvious that the true height of the Rockies, either in terms of their vertical rise or of their base height above sea level, was not even remotely understood. Confusion also ruled the American conceptualization of the location of the American ranges. Most sources agreed that the mountains of the interior were similar in structure to the Blue Ridge of Virginia, a single ridge or series of parallel ridges transected by rivers; the true nature of the Rockies as a broad and massive alpine region was not even dreamed of.

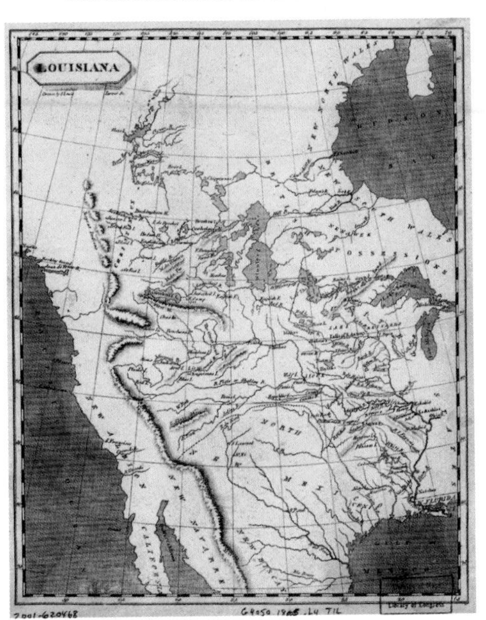

Figure 9.5 Map of the Louisiana Purchase and western North America by Samuel Lewis in Aaron Arrowsmith's and Lewis' *New and Elegant General Atlas* published in 1804. This map, using the best available data at the time Lewis and Clark set off, represents the prevailing ideas of western geography. In particular notice how the Rocky Mountains are represented as a single range of mountains, with sizeable gaps at the headwaters of the Missouri, nearly within sight of the Pacific Ocean (Library of Congress).

As an example of the meager evidence on which these theories rested, the British naval explorer George Vancouver, upon exploring the mouth of the Columbia River along the Oregon coast, saw in the distance a range of low mountains punctuated with the occasional snow-capped peak. What today we know as the Cascade Range with its isolated volcanic peaks, Vancouver assumed was the Rockies. Any explorers portaging across from the headwaters of the Missouri would therefore find a simple short trip downstream to the Pacific.

In reality the Cascades are simply one of the westernmost ridges of a vast mountain region. From the western edge of the Cascades to the eastern edge of the Rockies is 800 miles (1,300 km) of alpine landscape sitting at an average elevation of nearly 6000 ft (1,800 m) above sea-level. Far from being only 3000 feet above the surrounding landscape, the Rockies in Colorado and Montana regularly reach heights of 14,000 ft (4,300 m), with many passes at heights of 10,000 – 12,000 ft (3,000 – 3,700 m), four to six thousand feet (twelve to eighteen hundred meters) higher than the plains at their eastern feet.

From very few observations an entire theory of the west took shape that in every respect was utterly wrong. And so with visions of symmetric continents, and single-day portages over low mountain chains, Lewis and Clark and the Corps of Discovery set out to survey the interior of America. To do this the Captains became astronomers. According to Arlen J. Large, Lewis and Clark expert, member of the *Wall Street Journal* Washington Bureau, and amateur astronomer, "Lewis and Clark could not ignore their sky. It was a sky that most Americans today could not imagine, unpolluted by factories and automobiles, far from glowing cities, black and transparent to starlight from the glittering constellations. More important, the captains *needed* the sky to tell them where they were, and to show the world where they had been after they got back."

Figure 9.6 Even in June snow still covers the high elevations within Rocky Mountain National Park along the Alpine Ridge Trail (T. Nordgren).

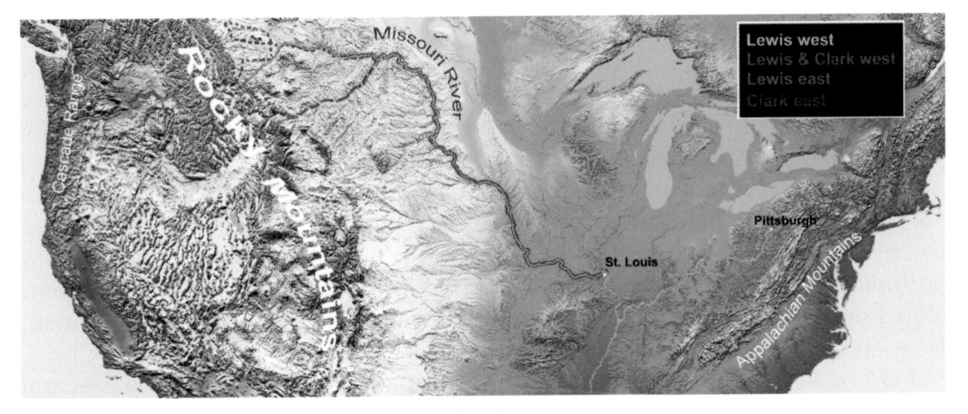

Figure 9.7 Topographic map of North America created by the Shuttle Radar Topography Mission on board the Space Shuttle Endeavour in 2000. Colors denote elevations ranging from green at sea-level to beige around 5000 ft (1524 meters) with grey and white even higher. The paths of Lewis and Clark both west and on their return trip (where both explored different routes through the Rockies) are shown. Compare this map of the country to that predicted in the map of 1804 and notice the differences between the western and eastern halves of the country (NASA/JPL).

Figure 9.8 Stars appear to revolve around the stationary North Star, Polaris, as the Earth spins in this two-hour exposure from Rainbow Curve along the Trail Ridge Road through Rocky Mountain National Park. The farther north you are, the higher Polaris is above the northern horizon (T. Nordgren).

Jefferson had Lewis trained in celestial navigation by the astronomer Andrew Ellicott in Philadelphia, the preeminent surveyor in the early 1800s. He taught Lewis that to find one's position on the Earth and survey the continent, one needs to first look to the stars. You can try this for yourself; go outside tonight and find the North Star: Polaris. This star is special in that just by chance it happens to lie in the sky almost directly over the north pole of the Earth. As the Earth spins, and everything else in the Universe rises in the East and sets in the West, they all trace out enormous circles across the sky with this star at their center.

Polaris' distance above the horizon tells you how far north of the equator you are. Imagine you were camping at the North Pole, ninety degrees north of the equator. The North Star would be directly above you. The angle between this star and the horizon would be almost exactly 90 degrees (Polaris is only half of a degree away from being perfectly above the North Pole). Move to the equator, and the North Star will lie almost exactly upon the northern horizon thus forming an angle of zero degrees with it. Turn around and hike northward half the distance to the pole, forty-five degrees north of the equator, and the North Star will be halfway up the northern sky again at 45 degrees. A trend develops. No matter where you are in the northern hemisphere, measure the angle from the northern horizon up to Polaris and it is the same as your angle up from the equator. This is your latitude. Sailors have been using this method for centuries to find their position North and South, including a variation that uses the Sun's altitude at noon; Lewis and Clark did the same.

Finding their latitude was easy; finding their longitude – the position East and West – was trickier. Because the Earth turns from west to east, the westerly position of a star in the sky is not just a matter of position on the Earth, it is also a matter of what time it is. Imagine I know it is noon on the east coast in New York (perhaps I have just gotten off a plane from there and have forgotten to change the time on my watch). At the instant my watch says noon, I look up and see that where I am the Sun is half way between the eastern horizon and the zenith overhead. From high overhead (where the Sun should be at noon) to where the Sun is now, is an angle of 45 degrees, one eighth of a full circle. I am therefore, one eighth of a full circle around the Earth away from New York. I must be on the west coast (measure my latitude as before and together I find I am in Los Angeles). Figuring out longitude in this way requires that I know the time in some reference location (e.g., my watch which I had not changed since leaving New York). But in the 1800s, clocks were not accurate or sturdy enough to survive rough sea voyages or long overland treks. What was needed was another way to tell the time back home, and for this astronomy was crucial.

For years astronomers had studied the orbit of the Moon in agonizing detail for just this purpose. To this day they still publish dense tables of the Moon's position, the dates of its eclipses and the positions of nearby stars as seen in the sky. All of these tables provide the times of these occurrences for someone in Greenwich, England. With this knowledge, the sky becomes the clock and Greenwich the reference. Observe an eclipse or the distance between the Moon and a star at a particular local time, and the difference between local time and the time calculated in the astronomical almanacs tells you how far around the globe you are.

This is precisely what Lewis and Clark did in their trek across the continent. "At night," writes Large in *Lewis and Clark: Part Time Astronomers*, "the captains would choose a bright star as a fixed point against which to measure the Moon's easterly motion in its orbit around the Earth. They were well enough acquainted with the sky to identify in their journals such first-magnitude beacons as Antares, Altair, Regulus, Spica, Pollux, Aldebaran and Fomalhaut as their target stars, depending on the season of the year. (Clark extended his bent for inventive spelling into the heavens, rendering the latter two as 'Alberian' and 'Fulenhalt.')"

When on the night of January 14, 1805 a rare total lunar eclipse swept across the American plains, Lewis was prepared to find their position that way as well, and wrote in his journal:

> Observed an Eclips [sic] of the Moon. I had no other glass to assist me in this observation but a small refracting telescope belonging to my sextant, which however was of considerable service, as it enabled me to define the edge of the moon's immage [sic] with much more precision than I could have done with the natural eye.

Determining all of these angles and positions took painstaking care. Comparing them with tables where complex calculations were required to interpolate the published positions and times to the exact time and date where you were making your measurements (while also standing on the shore of a raging river in the height of the spring flood season) was staggeringly difficult. While Lewis and Clark made their stellar observations with the intent of calculating their longitudes upon their return, in special cases like that of the lunar eclipse, Lewis attempted the laborious calculation there in the field. "Lewis' result, just over 99 degrees west of Greenwich," wrote Large, "was wrong by about a hundred miles. But such was the imprecision of the method that he was lucky to be so close." For the Corps of Discovery, the first GPS unit used to find their way West were three large editions of the British Astronomical Almanac (one for each year of the expedition).

Lewis and Clark finally saw the Rockies for themselves in central Montana near Helena, where the Missouri River enters into a chasm they named the Gates of the Mountains. Today, most travelers encounter the Rockies further south in Colorado, where Rocky Mountain National Park is the sixth most visited national park with 2.7 million visitors annually. Climb to the Continental Divide here or in Montana and the view is spectacular in its rugged remoteness. To east and west, row upon row of snow-capped mountains disappear over the horizon. What must a view like this have done to men who had set out thinking that they could simply carry their boats and

Figure 9.9 The eclipsed Moon is visible against the background stars. From timing the moment eclipses begin, or noting the position of stars relative to the Moon, a person can calculate his or her longitude using astronomical tables (T. Nordgren).

Figure 9.10 Cars drive across the Rocky Mountains in hours over distances that once took days. The Trail Ridge Road passes through Rock Cut turnout just below the Alpine Visitor Center at an elevation of almost 11,796 ft (3,595 m) above sea-level (T. Nordgren).

supplies from one river to the next in the course of a day or two?[3] The actual portage would take very much longer.

South of Rocky Mountain National Park the Rockies can be crossed between breakfast and dinner by means of Interstate 70. Within the park, the much more scenic Trail Ridge Road that leads up and over the Continental Divide takes a bit longer. It's the highest paved route through the Rocky Mountains, and tops out at the Alpine Visitor Center at an elevation of 11,796 feet (3,595 meters). Standing on the trail above the visitor center, the wind at this altitude is fierce. My hands sting where bare skin comes in contact with the cold air. With each labored breath at this altitude, my head fills with the smell of the air's crisp freshness yet burns my lungs with its icy purity. Every sense seems over saturated as above me an electric blue sky is as blinding bright as last winters' snow still resting on hillsides.

For the Corps of Discovery the hazards of hypothesizing too much from too little initial information resulted in the captains finding themselves ill prepared and too late in the fall season for what they had expected to find in these mountains. So it was that here in the Rocky Mountains, heavy with snow with the onset of winter in 1805, those theories of symmetric geography and dreams of a Northwest Passage finally died.

Modern visitors to these peaks are still surprised by the mountains when their

[3] One winter I crossed paths with the Corps' route through Montana and Idaho while I was in the middle of a snowstorm. In near whiteout conditions my truck passed in and out of the clouds as the road climbed up and down over row after row of mountains. As I descended down the back side of each ridge I kept hoping that I'd reached the last and each time I was disappointed to see yet another mountain before me. If this is how I felt in a nice warm truck going 25 miles per hour (40 km/hr), I can only imagine what Lewis and Clark must have thought making this trek on foot looking for the river that would take them to the Pacific.

previous experiences prove too limited for the conditions they find when they get here. At this elevation the atmospheric pressure is only two thirds that at sea-level. Hikers and day-trippers who have driven here in only a few hours from elevations seven to eight thousand feet (2,300 meters) lower, often experience shortness of breath, light headedness, fatigue and even chest pains. It is not uncommon for people to think they are having heart attacks here, and helicopter evacuations are not uncommon.

The mountains therefore still offer danger and difficulty, but only two hundred years later we've filled in the map with the most stunning alpine vistas a mountain-lover could hope to see. On a clear blue day – perhaps it is only the lack of oxygen – but I feel I can reach up and touch the sky. Later, after day gives way to dusk and the Sun goes down, I and other visitors to the park are once more faced with what we are no longer accustomed to seeing: stars much the same as they must have looked before electricity followed in the footsteps of westward expansion. During summer months the local Estes Valley Astronomical Society sets up telescopes in the Beaver Meadows part of the park on permanent piers the park has set up for anyone to use. At these summer star parties, park rangers, retired professional astronomers, and avid amateurs all show visiting eastern city dwellers the dark starry sky that was once a nightly occurrence for every generation of human being who lived out here until the end of the nineteenth century. Only a hundred years later the Universe above us, the last remaining frontier that still beckons those who look for the unknown and unexplored, is rapidly fading from view behind the glow of our own urban atmosphere. The people who live and work in these mountains are now the protectors of a quickly vanishing resource: our only window into the cosmos beyond us.

Two hundred years after Lewis and Clark, our robotic Pioneers, Mariners, and Voyagers have made the first reconnaissance of the planets in our Solar System. Eventually we will go beyond the Moon ourselves, but the question remains,

Figure 9.11 Park visitors gather on Friday and Saturday nights during summer months for ranger-led evening stargazing in Rocky Mountain National Park (T. Nordgren).

Figure 9.12 Telescope piers are provided at Upper Beaver Meadows Trailhead for visitors with telescopes. Stars are not the only sights visitors can expect to see (T. Nordgren).

Figure 9.13 The light of our galaxy reflects in the water of Bear Lake within Rocky Mountain National Park (T. Nordgren).

what then? Like easterners did with the American west at the dawn of the nineteenth century, we look beyond our Solar System at the thousands of stars we see every night in the sky and we wonder what is out there. Are there other planets and other solar systems like our own orbiting other stars? Are there other life-forms out there with whom we can communicate? What does our own experience in our own Solar System reveal about what we will find out there, and like those ideas of symmetric geography, what errors will our limited knowledge of this one planet and single solar system introduce into what we discover?

To truly take Lewis and Clark to the stars and answer the question of whether or not there is other life in the Universe, we first need to fill in those blank places on the celestial maps. Are there other solar systems out there; are we alone? Up until only a few years ago, our ideas of how solar systems could be arranged were based on hypotheses designed to produce the one example we knew: ours. Computer models showed that new stars should form at the centers of clouds of gas and dust that over time quickly settle into disks spinning around the young sun. Gravity pulls tiny dust grains into ever larger masses that through repeated collisions accumulate into solid objects. These growing bits of debris are called planetesimals and over time, the persistent action of gravity brings them together to form the planets we know today.

In close to a star where it's warm, planetesimals are made out of rock and metal, the things that are solid under high temperatures. Out far from a star where the temperature rapidly drops, molecules of water, carbon dioxide, ammonia, nitrogen and other 'gasses,' turn to ice and add to the list of solids out of which planetesimals can form. With more building blocks to play with, the outer solar system forms the seeds of big planets first. These icy masses are big enough that their gravity attracts and holds the light hydrogen and helium gasses swirling in the proto-planetary disk. The end result is that out there, planets can get very big, very fast.

In close to the star, the rocky masses don't have nearly as much to work with and their paltry gravitational pull fails to keep the lighter gasses from streaming away. To this day, let go of a helium balloon on Earth and it will float off into the sky. When the first nuclear fires within a star switch on, the solar wind that blows outward from its surface clears away the remaining gasses in the disk. As the last remaining planetesimals coalesce, the only remaining planets are those few that are on stable, mostly circular orbits.[4]

Alan Boss is a theoretical astrophysicist at the Carnegie Institution in Washington D.C. working on the question of where our Sun and planets came from. This picture of how solar systems form is largely based on computer models he developed over the last couple decades. In his book, *Looking for Earths, the Race to Find New Solar Systems*, Boss says that the upshot of these calculations is, "that the innermost giant planet should occur at about five times the Earth-

4 While also leaving a few well-defined belts of left-over debris, such as the asteroid and Kuiper belts.

Figure 9.14 Artist's conception of planets and asteroids in orbit around a nearby star. The ring of dusty debris is all that remains of the disk out of which the alien solar system formed (NASA/JPL-Caltech/T. Pyle/SSC).

Sun distance from its star. This distance is precisely that of Jupiter's orbit. Only terrestrial [small rocky, Earth-like] planets should be found inside this distance, as indeed is the case for our Solar System.'' From these results, he continued, "the calculations strongly supported the idea that we understood the basics of how the Solar System formed.'' But how can we know if this hypothesis is correct?

Finding planets around other stars is enormously difficult. Go out on a clear night and look up. The starlight you see is generated by nuclear fusion deep within the star heating up the stellar surface to thousands and even tens of thousands of degrees. Planets by comparison are cold. Even though the brightest 'stars' in our sky are often planets like Venus (40 times brighter than the brightest stars like Vega and Betelgeuse) they are bright only because they are close to us. Planets emit no light of their own that we can see with our eyes. The light we do see is reflected sunlight so actually seeing a planet around a distant star is as difficult as picking out a shiny golf-ball in the glare of a spotlight, a spotlight pointed right at us.

Astronomers have therefore long sought other ways to find these planets. If the planet itself is too small, too dark, or too close to the star to be seen, then we could instead turn our sights on the star itself. One of the white-lies that we astronomers tell students is that planets orbit stars. They don't; planets and stars orbit around their common center of mass. Put a large man and a small child on a playground teeter-totter (a beam balanced on a pivot) and the only way to balance the two so that one doesn't crash to the ground and send the other flying into orbit is to place the pivot very near (but not exactly under) the larger person. This pivot point is the center of mass between the two. For the Sun and Earth, the center of mass is so close to the Sun that it is actually inside it, but it is not at the Sun's center. As a result, our Sun does not sit still as the Earth goes around it, but

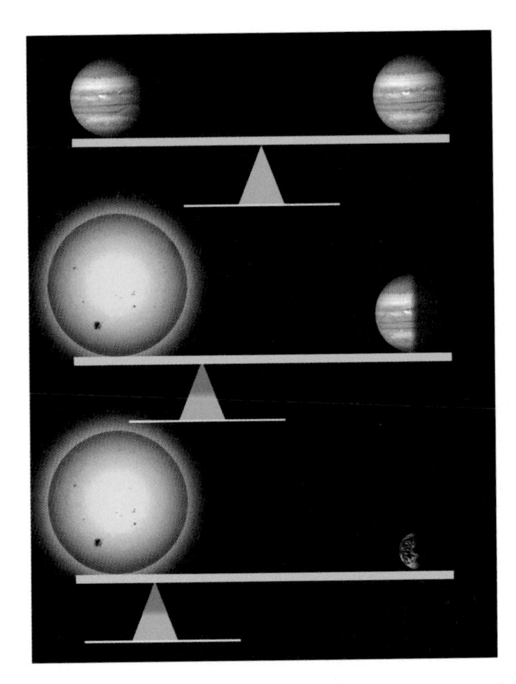

Figure 9.15 Three views of where the center of mass (the balance point) is found: (top) between two equally massed objects, (middle) between a star and a large gaseous planet, and (bottom) between a star and small rocky planet like the Earth. Sizes are not to scale (T. Nordgren).

rather the Sun travels around a small circle in space (while the Earth travels around in a big one) always keeping the center of mass between the two. The result is the Sun wobbles.

The larger and farther away the planet, the farther the center of mass is from the center of the Sun and the larger the wobble. Jupiter, the most massive planet in our Solar System is 300 times larger than the Earth and, just like our models 'predict,' sits out at a distance of 5 astronomical units.[5] Jupiter therefore causes the greatest wobble on our star and its effect is the one any aliens living on a planet around a nearby star would stand the best chance of discovering.

There are two ways to see a wobble. The first is to actually see the star move around on the sky. The precise measurement of a star's position on the sky is called astrometry and for the last couple thousands of years it's the most common type of astronomy that we've done. It's what Lewis and Clark were doing two hundred years ago as they crossed the country.[6] Unfortunately, the size of the wobble even Jupiter induces is microscopic when viewed from the nearest star like the Sun (alpha Centauri) and it only gets smaller and harder to see the farther away you are. In order to precisely measure really small changes in position you need to have a really, really big telescope, and look through really, really steady air (and ideally none at all). Giant space-based telescopes, many times larger than Hubble, are expensive and don't yet exist.

The only other way to find planets from the wobble they cause is to look for the change in the star's velocity. If a star is moving around in a circle, then, for all but a star whose orbit is face on to our observer, the star will sometimes move

5 An astronomical unit (AU) is the average separation between the Earth and Sun.

6 So that you don't have to reproduce the calculations Lewis and Clark did when you go on a road trip today, modern astrometrists measure the positions of stars so that Global Positioning Satellites can determine where they are in space, thus helping you know where you are relative to them.

towards us, followed soon thereafter by it moving away. Around and around, becomes towards and away.

If you've ever stood beside a road and heard the sound of the cars change pitch as they pass you by then you've experienced the physics that allows us to calculate the car's (or star's) velocity: the greater the change in pitch, the greater the velocity.

This effect (called the Doppler effect) occurs because sound (and light) are waves that travel at a constant speed (the speed of sound for one, the speed of light for the other). As a speeding car approaches us, passes, and then moves away its sound changes from high pitch to low pitch (this is the same as a change from short wavelengths to long wavelengths). The light given off by a passing star does the same thing, but where we experience the wavelength of sound as the pitch or tone, we experience the wavelength of light as its color. Short wavelengths are blue, long wavelengths are red.

When a star travels towards us it emits starlight, but this starlight is forced to travel at the speed of light and no faster. No matter how fast the star is moving towards us the light coming off the star is forced by physics to keep the same speed. Light waves coming off the star, therefore, get bunched up as they travel towards us. Shorter distances between waves means a shorter wavelength, meaning the light changes color: it gets bluer. We call this a Doppler shift and that the star's light is blueshifted.

Half an orbit later, as the star is no longer moving towards us but is now moving away, its light waves appear stretched out. Each succeeding crest in the light's wave comes later as the star speeds rapidly away. The waves are now longer and the color reddens. This is a redshift.

In both cases, the faster the velocity along our line of sight (called the radial velocity), the greater the Doppler shift and the greater the change in color.

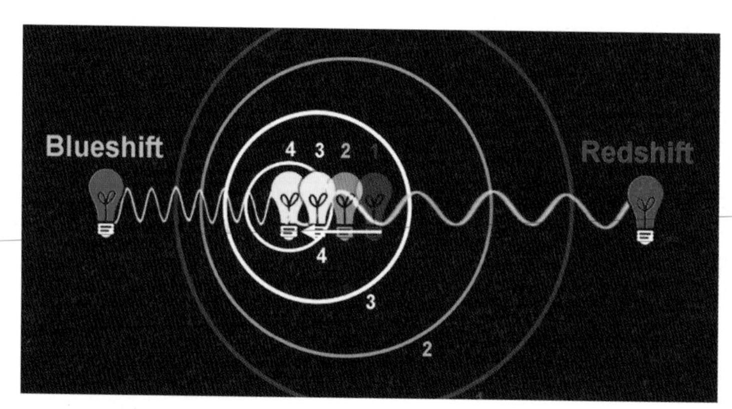

Figure 9.16 As a yellow light moves to the left, the waves it emits in front get bunched up because they can move no faster than the speed of light. To an observer in front, the yellow bulb's light is Doppler shifted to shorter wavelengths: it looks blue. An observer behind the bulb sees a similar shift to longer wavelengths: it looks red. The faster the speed, the greater the shift in wavelength will be (T. Nordgren).

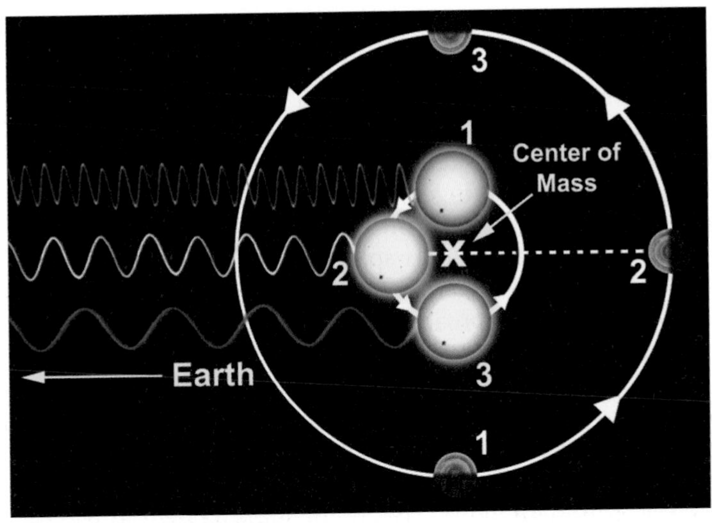

Figure 9.17 As a star and planet orbit around their combined center of mass, the bright star traces a small circle. The motion of the star around this circle requires it to move towards and away from observers on the Earth, Doppler-shifting its light. When star and planet are at time (1) the star is moving towards the Earth and its light is blueshifted. At time (2), the star and planet are moving perpendicular to the Earth, and there is no Doppler shift. When star and planet are at time (3) the star is moving away from the Earth, and its light is redshifted (T. Nordgren).

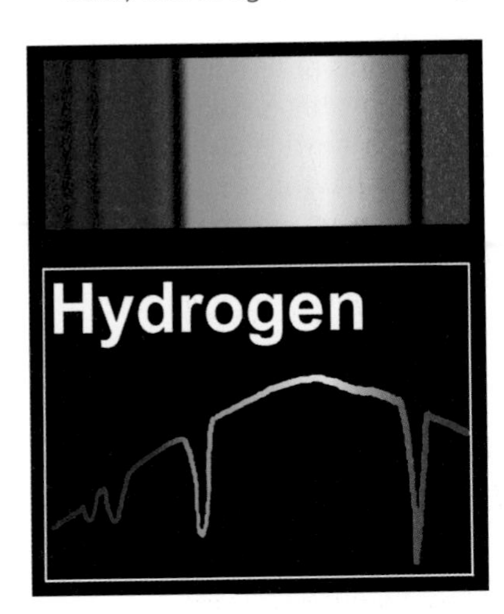

Figure 9.18 Diagram of a stellar spectrum. The upper figure shows the rainbow of light produced by passing a star's light through a prism. Dark absorption lines are observed. The bottom diagram shows the intensity of light as a function of wavelength (color) for the same spectrum. The dark absorption lines are deep troughs. By identifying the pattern of lines, we see the strong presence of hydrogen gas in the spectrum (T. Nordgren).

Watch a star wobble in a circle and for all but a face-on orbit (where the star no longer moves towards or away from us), the star's light will get bluer, then redder, then back again.

To precisely measure a change in wavelength you need to have a reference, something that tells you what the wavelength should be if the star was perfectly at rest. Pass sunlight through a prism and you spread out its white light into a spectrum that shows all the colors of the rainbow. Use a high precision prism, and buried within the rainbow are innumerable dark gaps (called absorption lines) where cool gasses in the outer atmosphere of the star have absorbed certain wavelengths of the light that streams by from out of the solar depths.

Every element in nature absorbs a unique pattern of wavelengths giving each element a special spectral fingerprint. Using these known absorption lines astronomers have identified elements like hydrogen and oxygen in stars clear across the Universe. When a star's spectrum is analyzed for known elemental fingerprints, if we see the fingerprint of hydrogen (for instance) shifted to the blue, we know the star is approaching us and we can precisely measure exactly how fast based on the size of the shift.

In reality this change is minuscule and not at all detectable by our eyes, but using this technique with sensitive detectors, we can now measure a star's velocity as small as three feet per second (1 m/s, the equivalent of a stately walk). This level of precision is achieved by passing the star's light through a chamber of iodine vapor at the telescope. The iodine absorbs light at a large number of wavelengths producing a large number of closely spaced parallel dark absorption lines, rather like a ruler. Against these reference lines (that are known to be at rest with respect to the observatory and Earth) the Doppler shifted lines in the spectrum of the star can be measured and compared.[7]

Jupiter causes our Sun's velocity to change by as much as 40 ft/s (12 m/s), so if we had any Centauri neighbors they could easily detect Jupiter's presence using technology no more advanced that what we have today.

From the time it takes our Sun to go through one period of red- and blueshift, as well as the maximum velocity it achieves (and thus how big a circle the planet can sweep out in the same period) any potential Centauri astronomers can calculate just how big Jupiter must be without ever having seen it. This is possible because four hundred years ago Johannes Kepler found that all planets follow certain specific orbital laws. One of these is that the orbital period, the orbital radius, and the total mass of the bodies doing the orbiting, all obey the relation:

$$P^2 = a^3/M_{total}$$

If you know the period (P) and the orbital radius (a) then provided you have some way of knowing how big the star is, you learn how massive any planet must be.[8] Wonderfully, this technique doesn't depend on how far away the star is from us. As long as the star is bright enough for us to capture its light and measure the wavelength change, we can find planets around stars as far away as we like.

[7] Prior to developing the iodine cell, astronomers used chambers of deadly hydrogen-fluoride gas for the reference source. Radial velocity searches like these were a dangerous business.

[8] The absorption lines in the star's spectrum also tell us the type of star it is and one thing the last hundred years has shown is that stars with similar spectra have similar masses. In this way we know the mass of the star when looking for the masses of potential planets.

Starting in 1981 the first team of astronomers with an instrument that could detect radial velocities as small as 50 ft/s (15 m/s) began observing nearby stars similar to our Sun for the tell-tale velocity signature of planets only a little bigger than Jupiter. Over the next decade as technology and radial velocity measurements improved, they would be joined by other groups. By the summer of 1995 they had found nothing.

That June at the annual meeting of the American Astronomical Society in Pittsburgh, Pennsylvania (the city from which Meriwether Lewis first set forth to pick up the other men for their trip cross-country) the planet hunters were glum. It was the first professional meeting I ever attended and I remember the mood well. After 14 years of searching, the original planet-hunting team had found nothing and were calling it quits. On top of that, another team led by Geoff Marcy in California had written to Boss in 1992 describing how they expected to find signs of Jupiter-mass planets at Jupiter distances within two or three years. As of that meeting three years later, they hadn't. No one had. [9]

Where were all the Jupiters and why weren't we finding any? One unpleasant conclusion was that planets are rare; that we are alone. Then again, perhaps all the results were saying was that Jupiters are rare and around each of these stars there might still be a family of perfectly lovely Earth-mass planets where life could evolve. But the potential lack of Jupiters has a profound effect upon the likelihood of this possibility.

Standing on the Continental Divide, I look north and south along the spine of the continent. Our planet may be called Earth, but it should have been called Water. Even in July snow still blankets some hillsides. Summer sun reflects off melting droplets watering high mountain tundra. What water doesn't nourish these high alpine meadows, gathers in streams and lakes that flow to the sea where they cover three-quarters of the planet's surface. The rocks of our crust, including the seafloor, are awash in water that provides the lubrication and low melting point that allows our continental plates to move and slide over one another. The very mountain range I am standing on is the result of collisions between the Earth's tectonic plates; without the water at my feet we wouldn't have the Rockies.

The same theories of planet formation that predict our planet should be made of metal and rock predict that water should not have been a major component of an early hot, dry Earth.

One explanation for this contradiction is that in the early days of the Solar

[9] Well, not quite. In 1992, radio astronomer Alex Wolszczan found evidence for three Earth-sized planets in orbit around a pulsar, a rapidly spinning neutron star that is the dead remnant of a once massive star. The planets around PSR 1257+12 were found using a variant of the Doppler method. Since they almost certainly formed in the aftermath of the supernova, one implication was that "the process of rocky planet formation is very robust," explained one planet searcher. As a result, he therefore expected "to find Earth size planets everywhere." But as of 1995, no one had found anything else.

System, our planet was pummeled by left over icy debris from the outer Solar System. Massive comet impacts literally rained water down on our planet that mixed with what water was already here to create a planet possible for life. But thanks to Jupiter, the bombardment didn't last. Computer models show that Jupiter's massive gravitational presence eventually swept up the remaining comets and flung them far out of the Solar System much like an angry child tossing toys. This is a process that continues still as at least two comets have been swallowed by Jupiter in the last 15 years (the most recent impact occurring in the summer of 2009).

Because of Jupiter, the Earth experiences major extinction-level cometary impacts only once every 100 million years or so, long enough for life, including intelligent life, to get a toe-hold between encounters. While the last major impact may have said so-long to the dinosaurs, 65 million years ago, it opened the door for mammals and eventually us.

Remove Jupiter from the Solar System and models say that comet impacts should unleash global devastation every hundred thousand years: too short for any intelligences like us to get started. In a solar system without a Jupiter, any planets with life may have produced nothing more than their own version of hardy cockroaches.

Those then were the choices in the summer of 1995: either no other solar systems, or no other solar systems with protective Jupiters. Either way we might very well be alone in this Galaxy. Either way, the repercussions would be profound. Five months later, everything changed.

In the November issue of the prestigious journal *Nature*, Swiss astronomers Michel Mayor and Didier Queloz announced the discovery of a planet at least half as massive as Jupiter orbiting a star in the constellation Pegasus.[10] The star, 51 Pegasi (51 Peg for short) is a normal star in every way like our Sun and its spectra showed the unmistakable sign of a small unseen planet tugging it around in a perfectly repeating circular orbit.

The only problem was that our theory of planet formation said a planet like Jupiter should orbit 5 AU from its star, taking a decade or more to complete a single trip around. The planet around 51 Peg takes only 4 days, requiring it to orbit at a distance of only 0.05 AU. This is a hundred times closer than physics says it should and seven times closer than the planet Mercury in our own system.

[10] The Doppler method has one major disadvantage: it produces only a lower estimate of how massive the planet might be. With the Doppler method we detect only the radial velocity of the star, the velocity along the line of sight to us. A star with a very small radial velocity might be circled by a very small planet. But, if the orbit of the planet and star is oriented so it is almost face-on to us, then very little of its orbital motion will be along our line of sight. The real orbital velocity could therefore be quite large and therefore due to a very big planet. Fortunately, this orientation is not all that likely to happen by random chance. On average, the mass of the planet will be closer than not to the lower possible limit.

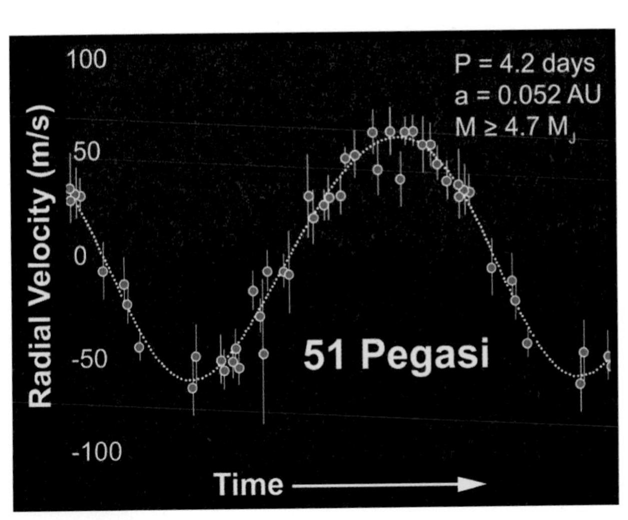

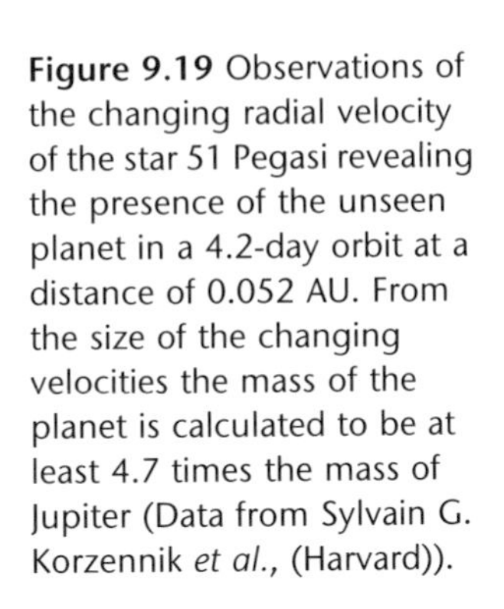

Figure 9.19 Observations of the changing radial velocity of the star 51 Pegasi revealing the presence of the unseen planet in a 4.2-day orbit at a distance of 0.052 AU. From the size of the changing velocities the mass of the planet is calculated to be at least 4.7 times the mass of Jupiter (Data from Sylvain G. Korzennik *et al.*, (Harvard)).

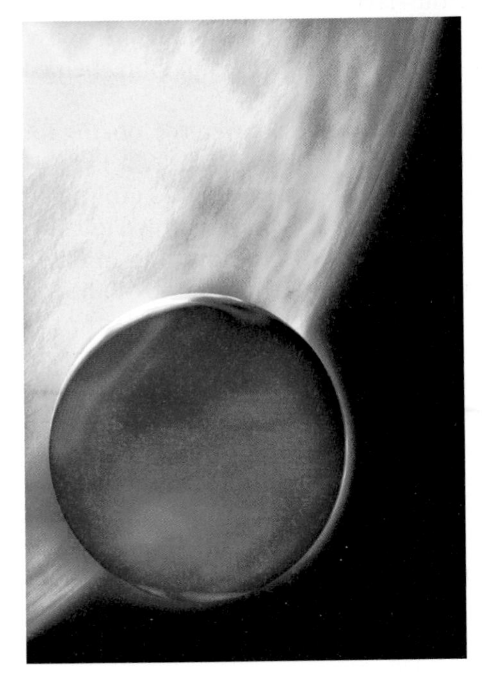

Figure 9.20 Artist's concept of a gaseous planet like that around 51 Pegasi (NASA, ESA, and G. Bacon (STScI)).

Where did that come from? Bill Cochran is an astronomer at the University of Texas at Austin. He and his colleague Artie Hatzes have been using the 107-inch (2.7-meter) diameter telescope at the McDonald Observatory in west Texas to look for planets since 1987. I first met Cochran at a conference on extrasolar planet searches in 1998, only a little while after the discovery of that first planet. Cochran told me later, "Prior to 51 Peg, we expected all planetary systems around main-sequence stars to look roughly like our own. We had one example of a planetary system and, guess what, all of our models succeeded in producing that one example. The discovery of 51 Peg was a real shock. There was any number of papers that tried to explain it away, (for example, as stellar pulsations, or star spots). It really took a major reset of our attitudes to realize that once again Nature was far cleverer than we had been."

None of the other planetary searches, some of which had been going on for over a decade, had found anything like this. Mayor and Queloz, had been searching for only a little over a year. As it happened, the reason they were first was a result of what astronomers expected and therefore how they planned their searches. Initially, there wasn't a lot of hurry to a project of this kind. After all, the theoretical astrophysicists said that Jupiter-sized planets, the ones most likely to be detected, could only form far from their stars on orbits that would take a decade to show any sign in the gradual change from blueshift, to redshift and back again.

"The expectations we had for what solar systems looked like governed how the surveys were designed." Cochran explained, "If you were looking for Jupiter with expected fluctuations of 12 m/s with a period of 10 years, and you had low errors on your observations (3 m/s) you didn't think to observe any one star very often."

In Northern California, Geoff Marcy and Paul Butler had been using a 10-foot (3-meter) telescope at the Lick Observatory to search for planets since 1987. For them, the task of calibrating their data, of converting the star's spectrum into very precise velocities, was a laborious process. If, according to Boss, you "were looking for Jupiter clones and expected to find companions with periods of a decade or so, then with that sort of expectation, in fact, you might as well not even bother to analyze your data until you have accumulated several years' worth. My *Science* paper had only helped to fuel this attitude.... The Swiss' absence of such preconceptions about the orbital periods of companions helped enable their discovery of 51 Pegasi b."[11]

Mayor and Queloz were ultimately successful because they weren't looking for planets at all. They were searching for objects called brown dwarfs, a kind of intermediate stage between stars and planets where a star has failed to grow large enough for nuclear fusion to ignite in its core. In the case where a brown dwarf is in orbit around a normal star there is no theoretical limit to how close the two might be and thus how short their orbital period. Expecting the possibility of changes on a nightly basis, Mayor had developed a spectrometer similar to the one that other planet-search groups were using but where the calibration was much simpler and radial velocities could be determined immediately.

After the Swiss team freed other groups from the constraints of what "wasn't possible," Marcy and Butler quickly confirmed the presence of 51 Peg's planet and within months had discovered two more so-called 'hot Jupiters' hiding away in the backlog of data on their computers. The planet around 70 Virginis was nearly seven times more massive than Jupiter with an orbital period of only 117 days, while the planet around 47 Ursae Majoris was two and a half times larger than Jupiter at a distance of 2 AU from its star. According to Boss, with the benefit of hindsight, had they not been biased towards finding solar systems like our own, "it was clear that Marcy and Butler easily could have beaten Mayor and Queloz to the punch," and been the first explorers to discover planets around a star like our own.

[11] The accepted convention in naming planets around other stars is that if you should find a previously unknown companion to a star like 51 Pegasi, the companion is named 51 Pegasi b, while the star itself is assumed to be 51 Pegasi A. For subsequent companions you work your way down the alphabet.

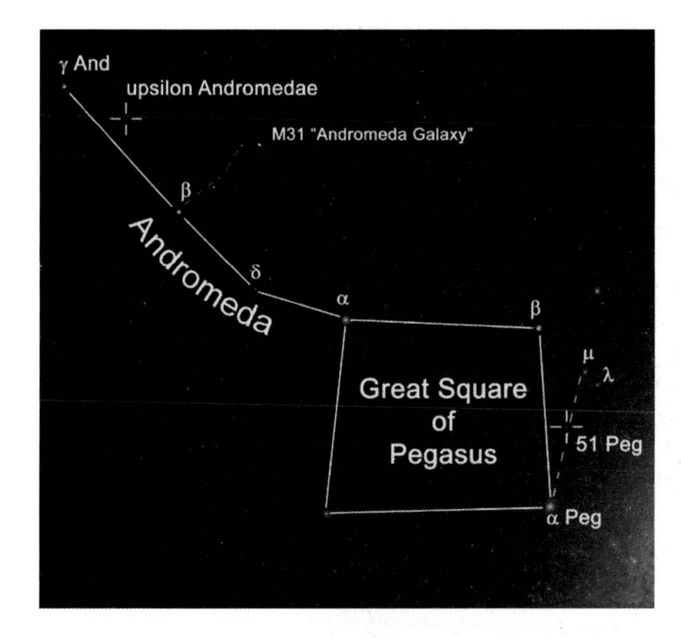

Figure 9.21 A number of stars with known extrasolar planets are visible at night to the naked eye. Two are shown here, 51 Pegasi (the first planet found) and Upsilon Andromedae (the first solar system), in the constellations of Pegasus and Andromeda (see the 'See for yourself' section for details). North is up (T. Nordgren).

But not everyone was willing to immediately say the conventional theory was wrong. 51 Peg and the other two planets might very well have been special cases, or maybe not even planets at all. "At first, it was just an oddball, a weirdo, an exception to the rule, a cosmic whim that was fun, an exotic circus animal that could astonish or even frighten you," Mayor would write about his discovery. "But the number of such exceptions multiplied: Tau Bootis b, 55 Cancri b, Upsilon Andromedae b." Over the next couple of years, Cochran and others would find so many examples of hot Jupiters that eventually it was obvious something groundbreaking had been discovered. "Theorists have no other choice than to go back to their drawing boards," Mayor would claim with complete justification.

The theory that eventually emerged is that solar systems can be much more dynamic places than previously thought. If a new star takes too long to clear out the gas and debris disk in which the fast growing Jupiter-like planets orbit, then gravitational interactions with the disk can cause the planet to slow down. As it does, the star's gravity pulls it spiraling inward towards the star. Any terrestrial planets on Earth-like orbits would face a perilous future as the massive planet goes rumbling by: potentially swallowing it whole, sending it crashing into the star, or ejecting it out of the solar system altogether. In the end, the rampaging planet stops its inward plunge only when it is close enough to the central star that the dust disk disappears or tidal forces from the star counter the slowing effects of the disk.[12]

[12] Recent models of planetary dynamics indicate that in our own Solar System, Uranus and Neptune once traded orbits. In the process they scattered numerous comets inward towards the Sun causing the period known as Late Heavy Bombardment that pummeled the Earth and Moon four billion years ago.

Figure 9.22 Artist's conception of a large gaseous planet orbiting within the disk of debris out of which its planetary system formed. Gravitational drag with the dust disk causes the massive planet to slowly spiral in towards its star resulting in the class of 'hot Jupiter' planets (NASA/JPL-Caltech/R. Hurt (SSC-Caltech)).

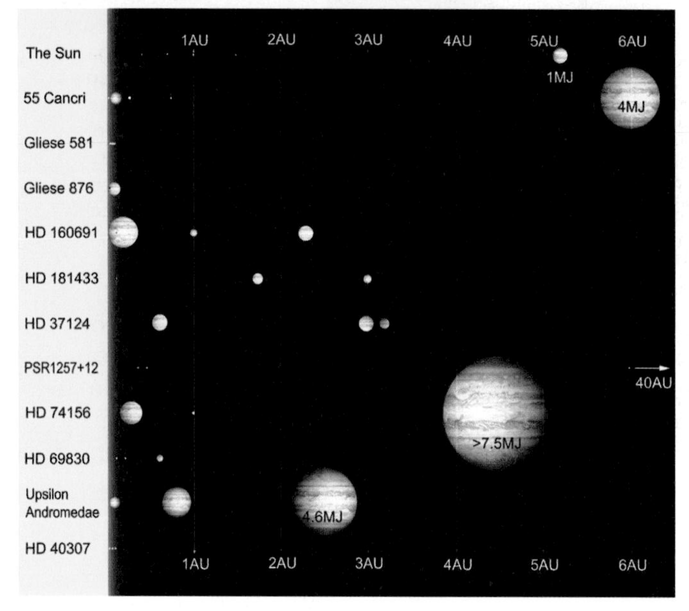

Figure 9.23 Diagram of known extrasolar planetary systems with at least three planets as of 2009. Our own Solar System is shown at top for comparison. The masses of the planets (in units of Jupiter mass, or MJ) are shown by the relative sizes in the diagram. The Earth orbits at a distance of 1 AU (T. Nordgren).

A whole new spectrum of possible planetary arrangements thus emerged in the late 1990s, showing that we may be just one among many possible configurations in the family of planetary systems. We now know of 360 stars in the sky that have at least one planet. And where we knew of only one solar system (our own) prior to 1995, today we know of 45 others.[13]

Here are a few of my favorites that also have the advantage of orbiting stars I can see with my own eyes:

Rho Cor Bor is 55 light-years away in the constellation Coronae Borealis (Northern Crown). The planet is one and a half times as massive as Jupiter and is

[13] The very rate at which new planets are being discovered requires that any tally specify not just a year, but a month (and on some occasions even a day). As of February 2010 the tally stands at 429 different planets.

Figure 9.24 Finder photograph for the extrasolar planet around the star Rho Coronae Borealis (ρ CrB) as seen above the Continental Divide in Montana. North is to the right (T. Nordgren).

in a nearly circular orbit only a quarter the distance between the Earth and Sun. Astronomers believe they know the true inclination of the orbital plane of the planet (and hence its true orbital velocity and mass) from observing a dust disk still in place around the star.

Pollux, also called beta Geminorum, is the second brightest star in Gemini and 34 light-years away. It's the slightly yellowish star compared to the bluish Castor that is the head of Pollux's twin. Pollux b is at least twice as massive as Jupiter and orbits out at about 1.6 AU, just outside of where Mars is in our own solar system. Pollux is a first magnitude giant star in our winter and spring skies so even urban observers can look up and see a star home to another potential solar system.

Epsilon Eridani is the closest star to us with a planet. Only 10 light-years away and visible in winter skies, epsilon Eridani has long been a target for planet hunters. A number of times over the years, astronomers had thought they'd found one, only to have it shown to be a tantalizing but ultimately spurious signal in noisy data. In 2000, Hatzes and Cochran found Epsilon Eridani b which from subsequent Hubble Space Telescope observations of the actual astrometric wobble of the star, reveals the planet is 1.5 times the mass of Jupiter. Epsilon Eridani b is in an elliptical 6.8-year orbit that takes it as close as 1 AU from its star, and as far out as five times farther away.

55 Cancri is a yellow sun-like star 44 light-years from Earth. Around it orbit at least five planets in the largest extrasolar system we currently know. The first, and closest is 55 Cancri b with an orbital period of just 15 days; it was one of the first planets found in the slew of discoveries that followed 51 Peg. The largest planet in the system is at least four times larger than Jupiter out at a distance of 6 AU (exactly where a Jupiter-like planet should be). The smallest planet yet detected there is 55 Cancri e which is at least 18 times the mass of the Earth but skims so close to its sun at a distance of 0.04 AU that it completes one orbit every 3 days.

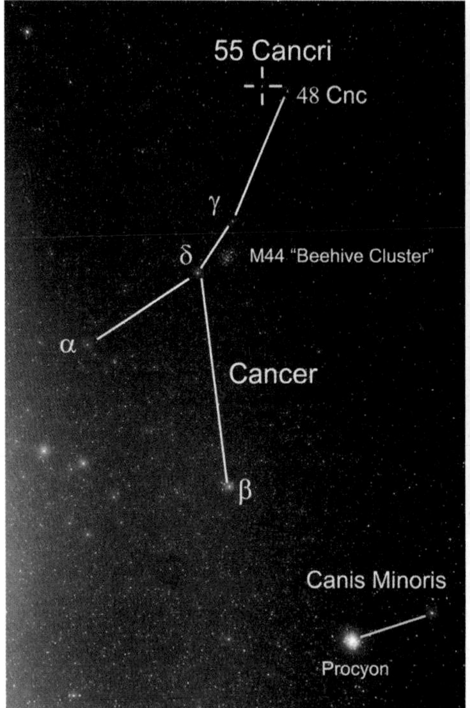

Figure 9.25 The star 55 Cancri hosts at least five known planets and is visible in this finder photograph for the constellation Cancer, the Crab. North is up (T. Nordgren).

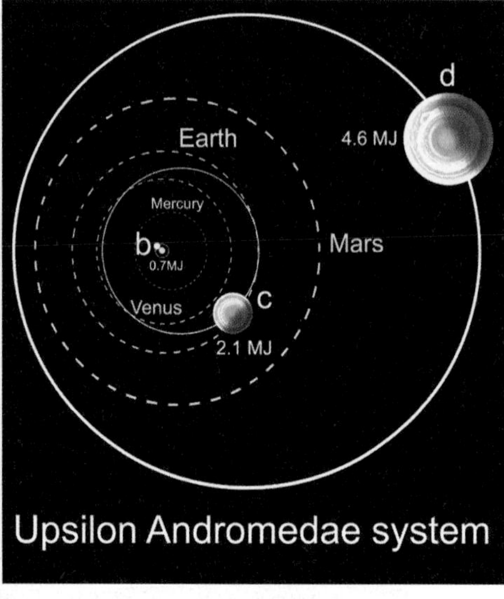

Figure 9.26 Diagram of the Upsilon Andromedae planetary system showing the orbits of the three known planets against the orbits of the four inner planets within our own Solar System (T. Nordgren).

Upsilon Andromedae is a star a little hotter than our Sun at a distance of 44 light-years with three known planets. Each of its planets is massive, yet all three orbit within 5 astronomical units of the star. The outer two planets are on highly elliptical orbits, something not originally expected to be possible in a stable solar system. In fact, Ups Andromedae c with a mass at least twice that of Jupiter's is on such an elliptical orbit that during its 241-day period it passes closer to its Sun than Venus, then loops farther out than the orbit of the Earth. Theoreticians now think that early in its past the gravitational interaction between the three massive planets probably sent them careening around their star to wind up on the eccentric orbits they now inhabit.

Fomalhaut, one of the brightest stars in fall's southern sky (and one of the stars William Clark had such difficulty spelling in his journal), is a young hot blue star twice as massive as our own Sun. Only 25 light-years away, Fomalhaut has long been known to have a warm dusty ring around it. In 2004 astronomers used the Hubble Space Telescope to capture an image of this dusty debris. Embedded within the inner edge of the dust ring their image revealed a bright spot that two years later had moved farther around the star. This is the first confirmed instance of an actual planet photographed in orbit around another star. It's a large planet to be sure, at least three times the mass of Jupiter and

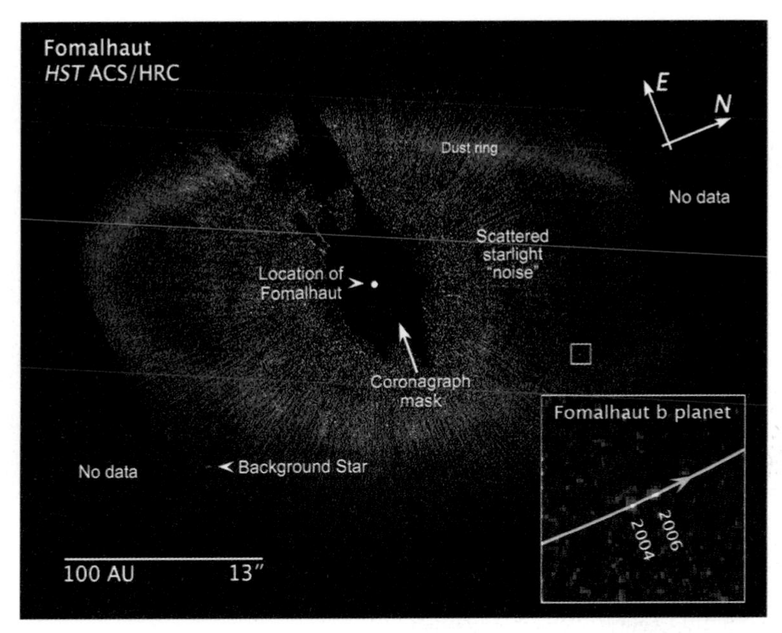

Figure 9.27 Hubble Space telescope image of the dust ring around the star Fomalhaut. The otherwise blinding glare of the star is blocked out revealing the faint structure of the disk as well as the light of a planet discovered in 2004 and observed a second time two years later (inset) (NASA, ESA, P. Kalas, J. Graham, E. Chiang, E. Kite (Univ. California, Berkeley), M. Clampin (NASA/Goddard), M. Fitzgerald (Lawrence Livermore NL), K. Stapelfeldt, J. Krist (NASA/JPL)).

orbits over twice as far away from Fomalhaut as Pluto does around our own star. With an orbital period a little under 900 years, Fomalhaut b has experienced no more than a single change of seasons in the entire 400 hundred years since anyone on Earth first pointed a telescope at the sky.

These worlds are just the tip of the iceberg. As new technology allows the planet hunters to precisely measure ever smaller changes in wavelength, ever smaller planets will continue to be found. Currently, the smallest known planet around another star is only five times larger than the Earth orbiting a small, cool red dwarf star.[14] Eventually the precision will be so good that they'll be found around normal stars like our own.

But consider that all these planets (with the exception of the recent image of Fomalhaut b) were found from no more than a little wiggle in the spectral line of a star. No planets were ever actually seen. In 1999, astronomers found the final proof that these objects were actually real planets. As with the lunar eclipse that Lewis observed along the Missouri River in the village of the Mandan, every once

[14] Because the host star, Gliese 581 is only a third the mass of our own Sun, planets in orbit around it have a much larger effect on its wobble around its center of mass.

in a while the orbital paths of suns, stars, and planets will align. When the Earth passes between the Moon and Sun, lunar eclipses occur; when the Moon passes between the Earth and Sun a solar eclipse takes place and the amount of sunlight reaching the Earth momentarily drops.

Eventually, after enough planets have been found around enough stars, probability says that one of them should periodically pass between its star and us, producing a stellar eclipse where the star's light momentarily dims by a miniscule fraction. These are called transits and in 1999, the first one was discovered just as the radial velocities said the planet should pass in front of the stellar disk.

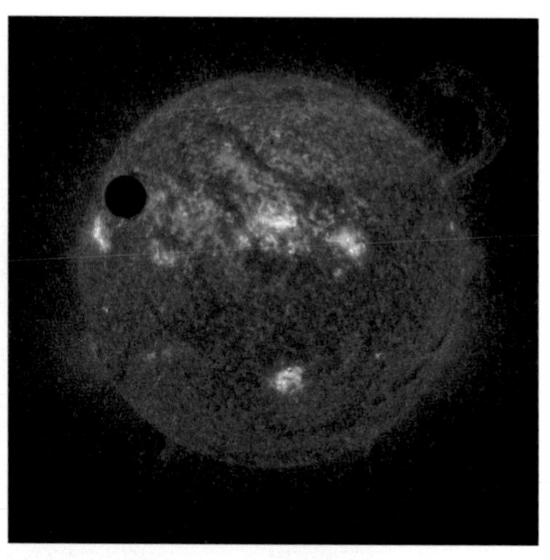

Figure 9.28 Artist's conception where a planet transits in front of its star blocking out a tiny fraction of the star's light ((solar image) NASA, ESA).

From the size of the drop in starlight we could tell how big the planet was. From the mass and size of the planet we could determine its density (unsurprisingly it was gas).[1] Half an orbit later the planet passed behind its star as viewed from Earth. If you record the spectral fingerprint of the star when the planet is eclipsed by the star, then record another when the planet comes out from behind, the difference between the two spectra reveals the elemental signature of the planet (even though at no point is its light ever distinguished from the star's). Gasses that have so far been found in their atmospheres include water vapor, sodium, carbon dioxide, and methane. We are sniffing the air of worlds around other stars.

In 2009 NASA launched the next phase in planet exploration. The Kepler spacecraft is a robotic mission to repeatedly observe the brightness of 170,000 stars in the direction of the Summer Triangle. It will spend three and a half years (and hopefully more) looking at this region of the sky for the telltale signs of tiny eclipses by transiting planets with a precision so great that we should be able to find planets as small as Earth, in Earth-like orbits around stars like our Sun.

The primary mission will last three years, so only planets in the equivalent of our inner Solar System will be detected over multiple passes in front of its star. Once they've been found however, then Cochran, a co-investigator on the mission team, will use the radial velocity method to look for any larger planets in

[15] A benefit of transiting extrasolar planets is that the orientation of the orbit is revealed and the mass of any planet is no longer just a lower limit, but actually the true, unambiguous mass.

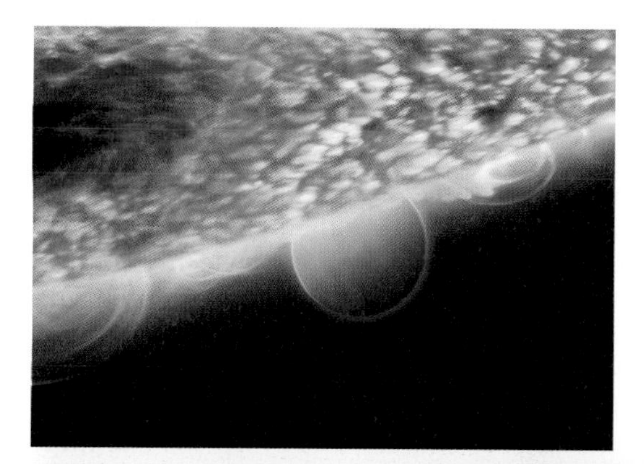

Figure 9.29 After transiting in front of its star, the planet is subsequently eclipsed by it as shown in this artist's conception (NASA, ESA, and G. Bacon (STScI)).

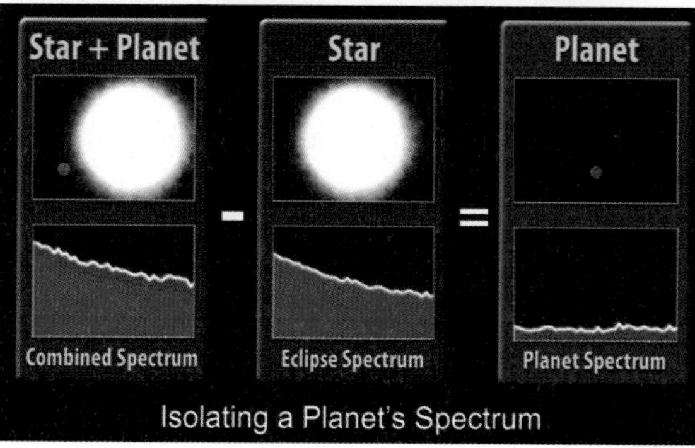

Figure 9.30 Diagram showing how astronomers acquire the spectrum of a planet around another star that experiences eclipses. In the left panel the combined light of planet and star is observed (bottom panel shows the contribution to the spectrum of star and planet). Astronomers then wait for the planet to be eclipsed by the star. The resulting spectrum shows only the light of the star. This spectrum (at middle bottom) is subtracted from the combined spectrum (bottom left). The result is just a spectrum of the planet (bottom right) (NASA/JPL-Caltech).

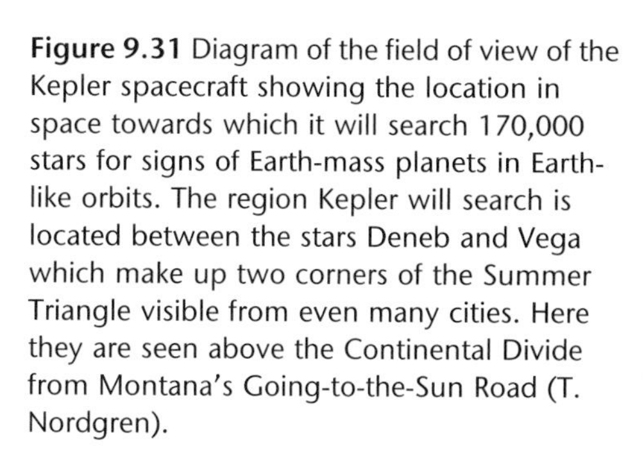

Figure 9.31 Diagram of the field of view of the Kepler spacecraft showing the location in space towards which it will search 170,000 stars for signs of Earth-mass planets in Earth-like orbits. The region Kepler will search is located between the stars Deneb and Vega which make up two corners of the Summer Triangle visible from even many cities. Here they are seen above the Continental Divide from Montana's Going-to-the-Sun Road (T. Nordgren).

more distant orbits. Soon we will have completed the first surveys of entire solar systems and finally know just how common or rare our collection of eight planets is. As unbelievably exciting as this prospect is, one of the most common questions I or any astronomer is asked is why it matters? When there is so much that needs fixing down here, why spend our time looking out there?

While there are many historical examples of world-changing discoveries coming from the most unlikely of places (where would modern electronics be if in the 1700s inquisitive people like Benjamin Franklin hadn't experimented with lightning and static electricity) there is a lesson we can learn from the national park where I now stand.

As American and European settlers spread west across the continent in the 1800s they did so to find land, freedom, new resources, and prosperity (all those things people still look for today). Yet, in the midst of that expansion, the federal government declared there were some places we would not develop, farm, mine, or 'civilize,' but leave as best we found for future generations. These are the national parks. As costly and counterproductive to the immediate needs and the wants of the people of the time as they were, can anyone argue that the peace, beauty, and reflection they have brought to the human spirit did not make them money well spent?

This is what I see in every picture from the Hubble Space Telescope or distant orbiter flying past the rings of Saturn. And just like geology and geography was only part of Lewis and Clark's charge, the ultimate astronomical question isn't the quantity of other Earths but the possibility of life out there looking back at us. Are we alone or are we part of a greater family of living, thinking, beings in the Universe? Imagine what we could have learned from those who were living on this continent before us if we had been truly intent on communication and not colonization, and that is only a microscopic fraction of what we could learn from a truly alien intelligence. The laws of physics make the colonization of other planetary systems no more than a distant dream, so for perhaps the first time in our history, we will be forced to communicate first with anyone we discover.

Figure 9.32 The Planetary Society is dedicated to inspiring the public with the adventure and mystery of space exploration. Through their projects and publications, the Society plays a leading role in the quest to engage the public and fuel support for exploring other worlds. Their very logo draws on humanity's long tradition of exploration and discovery (The Planetary Society).

But how likely is it that there *is* anyone to talk to out there? Frank Drake, an astronomer at Cornell University in the 1960s and '70s came up with a formula for at least framing this seemingly impossible question. His idea was to break down a complicated question with no clear answer into a series of smaller questions, each with at least some possibility of being addressed. What he and his colleagues came up with has come to be known as the Drake Equation, and it looks like this:

$$N = R^* \times f_p \times n_e \times f_l \times f_i \times f_c \times L$$

where N is the number of alien civilizations out there in our Galaxy with which we could possibly communicate. Each term on the right is some number or fraction that has a bearing on how big or small N could possibly be. For instance:

R^* is the rate at which stars are born in the Galaxy. From the size and age of the Milky Way, we can estimate this at about 40 stars per year.

f_p is the fraction of those stars with planets. Our searches for extrasolar planets shows this may be a very large fraction, perhaps almost 1. But let's be cautious: $f_p = \frac{1}{2}$.

n_e is the number of Earth-like worlds per planetary system. In our own Solar System we know perhaps four or five places where liquid water and the requirements for life could be: Earth, Mars, under the ice sheet of Europa, and possibly on Saturn's moons Titan and Enceladus. Let's remain conservative and say $n_e = 3$.

f_l is the fraction of those planets or moons in the solar system where life actually develops. The last decade of human exploration of Mars has all been aimed at identifying if life ever arose there. Evidence indicates that life on Earth evolved pretty quickly once the planet formed. While life may be inevitable, let's remain pessimists: $f_l = \frac{1}{100}$.

f_i is the fraction of those places with life where it actually evolves to intelligence. Obviously with each term in this progression we are less and less sure of what values to use and the answer we derive is simply a guess, but at the very least it will be an educated guess. We're intelligent, but we also know dinosaurs lived on this planet for 185 million years and never built asteroid deflection capabilities, so let's say f_i is one out of two.

f_c is the fraction of intelligent civilizations that develop the technology to communicate across interstellar distances. It does us no good if an agrarian race of truly enlightened beings evolves who eschew all form of technology. They may be admirable people from whom we could learn a lot, but there is no way we will ever learn it if we can't communicate with them across interstellar space. Using our one data point which is the modern Earth, we have three possibly intelligent species (us, apes, and dolphins) but only one of which is signaling our presence to extraterrestrials via televisions. Let's say $f_c = \frac{1}{3}$.

Multiply all these terms together and you have:

$$N = 40 \times \frac{1}{2} \times 3 \times 1/100 \times \frac{1}{2} \times \frac{1}{3} \times L$$
$$N = \frac{1}{10} \text{ technological civilizations per year } (\times L)$$

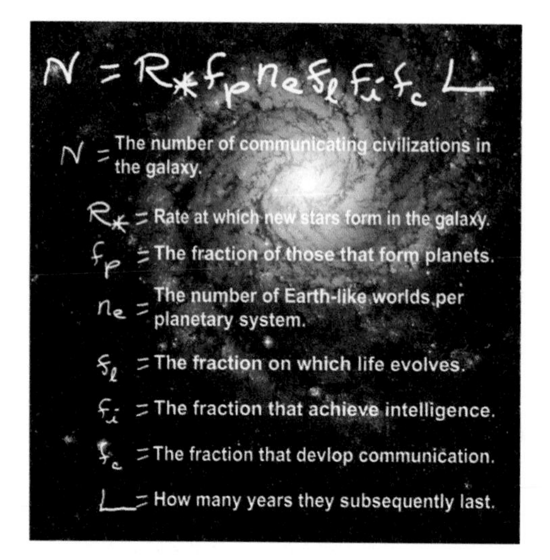

N = R* fp ne fl fi fc L
N = The number of communicating civilizations in the galaxy.
R* = Rate at which new stars form in the galaxy.
fp = The fraction of those that form planets.
ne = The number of Earth-like worlds per planetary system.
fl = The fraction on which life evolves.
fi = The fraction that achieve intelligence.
fc = The fraction that devlop communication.
L = How many years they subsequently last.

Figure 9.33 The Drake Equation for calculating the number of civilizations in our galaxy with whom we might be able to communicate right now ((background) Hubble Heritage Team).

This means that every 10 years one intelligent communicating civilization pops up somewhere in the Milky Way. But the number that is out there at any one time depends on how long they stick around. That's L, the lifetime of a technological civilization. Right now for us, it is about 100 years, the length of time we've had radio technology to signal our presence to the stars through TV reruns. If during the height of the cold war we'd annihilated ourselves in a nuclear holocaust, or if we ruin our planet during the next century due to ecological carelessness, then L might be no more than 100 years.

Let's say that's typical; that as soon as technological civilizations sprout they snuff themselves back out. Then N = 10. As amazing as this would be, unfortunately for us, the Galaxy is a huge place. At 100,000 light-years across and about 1000 light-years thick, there are about 10 trillion cubic light-years of space in our Galaxy. If all 10 inhabited star systems are spread randomly throughout the Galaxy then that implies that each civilization is alone within its own bubble of a trillion cubic light-years and the nearest civilization is about 10,000 light-years away. If all civilizations are like us, then no one ever lives long enough to hear from their neighbors.

But imagine we don't destroy our planet or ourselves. Say we overcome the tendencies that decimated indigenous populations, melted glaciers, and harnessed the power of the Sun to make hydrogen bombs. If civilizations can do that and L = 1 million years, then there are one hundred thousand civilizations out there in the Galaxy today waiting to hear from us and in the lifetime of our species there would be time enough for interstellar conversations.

In the end, even if we find we are alone in the Galaxy, with no one to talk to or learn from we will still have found those worlds that we will someday need to travel to. While our physics provides no practical way to do that yet, our astrophysics is quite clear that when our Sun begins to die in a little less than 5 billion years, we won't be able to stay here. If we should be so lucky that L is a very large number, then all those planets we discover out there will be the last hope for the survival of our species and everything it will have ever accomplished down here.

Currently none of these planets are Earth-like worlds with a solid surface orbiting in an Earth-like orbit at just the right distance for liquid water to be present for life as we know it. But Jupiter-like planets such as Upsilon

Figure 9.34 An artist's conception of Upsilon Andromedae c rising above the distant mountains as an alien expedition explores an unknown continent on their Earth-like moon (T. Nordgren).

Andromedae c do orbit within their star's habitable zone and the one thing the last 50 years of robotic space exploration has shown is that Nature surprises. Several moons of the great gaseous planets in our Solar System almost certainly have liquid water hiding under frozen protective crusts. Move Jupiter and its system of moons to the orbit of the Earth and provided there's enough volcanic outgassing for a heat-trapping greenhouse atmosphere, then Europa becomes a balmy south-sea world.

Imagine just such a moon around Upsilon Andromedae c where oceans and continents create a world like ours in miniature. Like all moons in our Solar System, tidal forces slowly synchronize its rotation so that one side stays forever pointing at its planet. On such a world, the planet never rises or sets, but stays forever hanging in the sky, visible forever by one hemisphere, forever invisible to the other. If there should be intelligent life there, any expedition of alien beings, born and raised on the anti-planet hemisphere setting off to explore the far reaches of their world will therefore come to discover more than their initial hypotheses predicted. In following some unknown river across a brand new continent, forever heading for the distant horizon, a day will come when a new planet rises, one they've never seen before. And on that day they will discover a new world, the first of many if they ever turn their exploratory view to the sky as we have just taken the first steps in doing ourselves.

If so, perhaps someday our two species will meet and we will learn to tell one another what we have discovered in the course of our travels.

See for yourself: other stars with planets

'See' an extrasolar planet

As of the summer of 2009 there are 42 stars visible to the naked eye around which we know other planets orbit. Many of these stars are visible only from dark locations on nights where the Milky Way is easily seen. Some, however, are visible even from the hearts of cities (in none of these cases are the planets themselves bright enough to be seen, even through a telescope). The following are six of the easiest stars to find that have extrasolar planets or planetary systems. The outlined boxes on the July and January star maps correspond to those photographs within the chapter that show finder-maps for hard to find stars. They are listed in order of increasing difficulty to identify.

Pollux or **Beta Geminorum (**β Gem): Magnitude 1.15, visible in evening from December to June (best in March). Gemini is best found by looking for the two stars Castor and Pollux that form the head of Gemini the Twins. They are often visible even from cities and are found just to the east of the prominent constellation of Orion. Pollux is east and slightly yellow relative to its close, blue twin, Castor. The planet Pollux b is at least 2.3 times the mass of Jupiter and at an average distance of 1.6 AU from its star (equivalent to the orbit of Mars) takes 590 days to complete one orbit.

Fomalhaut: Magnitude 1.16, visible in evening from September to December (best in October/November). During the fall, Fomalhaut is the brightest star in the southern sky. In the constellation of Pisces Austrinus (the Southern Fish), Fomalhaut is the only star that stands out and is often the only star visible from light polluted skies. If it's fall, and there is a bright star along the southern horizon, it's probably Fomalhaut. The planet Fomalhaut b is less than 3 times the mass of Jupiter and at an average distance of 115 AU orbits more than twice as far away from its star as Pluto does in our Solar System. One year for Fomalhaut b lasts 876 terrestrial years. Of all the stars listed here with planets, only Fomalhaut is not visible on the July or January star maps included with this chapter (see instead the star maps in Chapters 3, 4, 7, or 8).

Gamma Cephei (γ Cep): Magnitude 3.2, visible all year (best November). The constellation of Cepheus is found so close to Polaris that as the Earth turns, it never sets below the horizon as seen from the United States (except for Hawaii). Only during spring is Cepheus at its lowest. The constellation looks like a simple house with a square and a triangle on top for a roof. The peak of the roof is the star γ Cep and it is located almost 15 degrees away from Polaris, halfway between the North Star and the W of Cassiopeia. At arm's length, your hand with thumb and pinky-finger extended spans 15 degrees from tip to tip. If Cassiopeia isn't up and you can't find γ Cep from the star map, place your thumb on Polaris and rotate your hand around the North Star. γ Cep will be the only star of near comparable brightness that your pinky-finger crosses. Gamma Cephei b is a planet 1.76 times bigger than Jupiter at a distance of 2 AU from its cool orange-red star. The unseen gas planet takes 900 days to complete a single orbit.

Epsilon Eridani (ε Eri): Magnitude 3.7, visible in evening November to March (best January). The constellation of Eridanus represents a river that meanders through the winter sky to the southwest of Orion. The best way to find the star epsilon Eridani is to start at the bright blue star Rigel in Orion (the southwest corner of the constellation) and "star-hop" from one star to the next along the river as shown in the January star map in this chapter. ε Eri is only 10 light-years away, making it the closest known star with a planet. The planet is 1.5 times more massive than Jupiter in an elliptical 6.8-year orbit that takes it between 1 and 5 AU from its star (the difference between the Earth and Jupiter in our own Solar System).

Upsilon Andromedae (υ And): Magnitude 4.1, visible in evening August to March (best December). The December sky map and Figure 9.21 show how to find υ And along the string of stars that extends eastward from the Great Square of Pegasus. It is usually best to begin at α And, the northeastern corner of the Great Square in Figure 9.21, and star-hop eastward to υ And. Upsilon Andromeda has three known planets in orbit. They range from at least 0.71 to 4.61 Jupiter masses and have orbital periods ranging from 4 days to 3.5 years.

Rho Coronae Borealis (ρ Cor Bor or ρ CrB): Magnitude 5.4, visible in evening April to October (best July). The large C-shaped constellation of Corona Borealis is found in summer along the line joining the bright orange star Arcturus and the bright first-magnitude star, Vega. As Figure 9.24 shows, extend the two arms at the open end of the C and they come to a point at the dim star ρ CrB. The planet that orbits this star is a little larger than Jupiter and orbits in a nearly circular orbit only a quarter the size of the Earth's, completing one circuit of its star every 40 days.

51 Pegasi (51 Peg): Magnitude 5.5, visible in evening August to January (best November). Find the Great Square of Pegasus in the fall and early winter sky. Figure 9.21 shows that just west of the two stars that make up the short, western side of the Square are two close stars (μ and λ Peg). Almost directly in-between μ Peg and α Peg (which is the southwest corner of the Square) is the very dim star 51 Peg. This is the very first star like our Sun found with an extrasolar planet and it is visible only from dark locations with no Moon, and on nights when the Milky Way is also easily visible. See it and know that a planet at least half as massive as Jupiter makes one complete orbit every four days.

55 Cancri (55 Cnc): Magnitude 5.95, visible in evening January to June (best March). This is the hardest star in this list to find as it is just at the limit of what you can see from a dark location and it is in a constellation with no bright stars

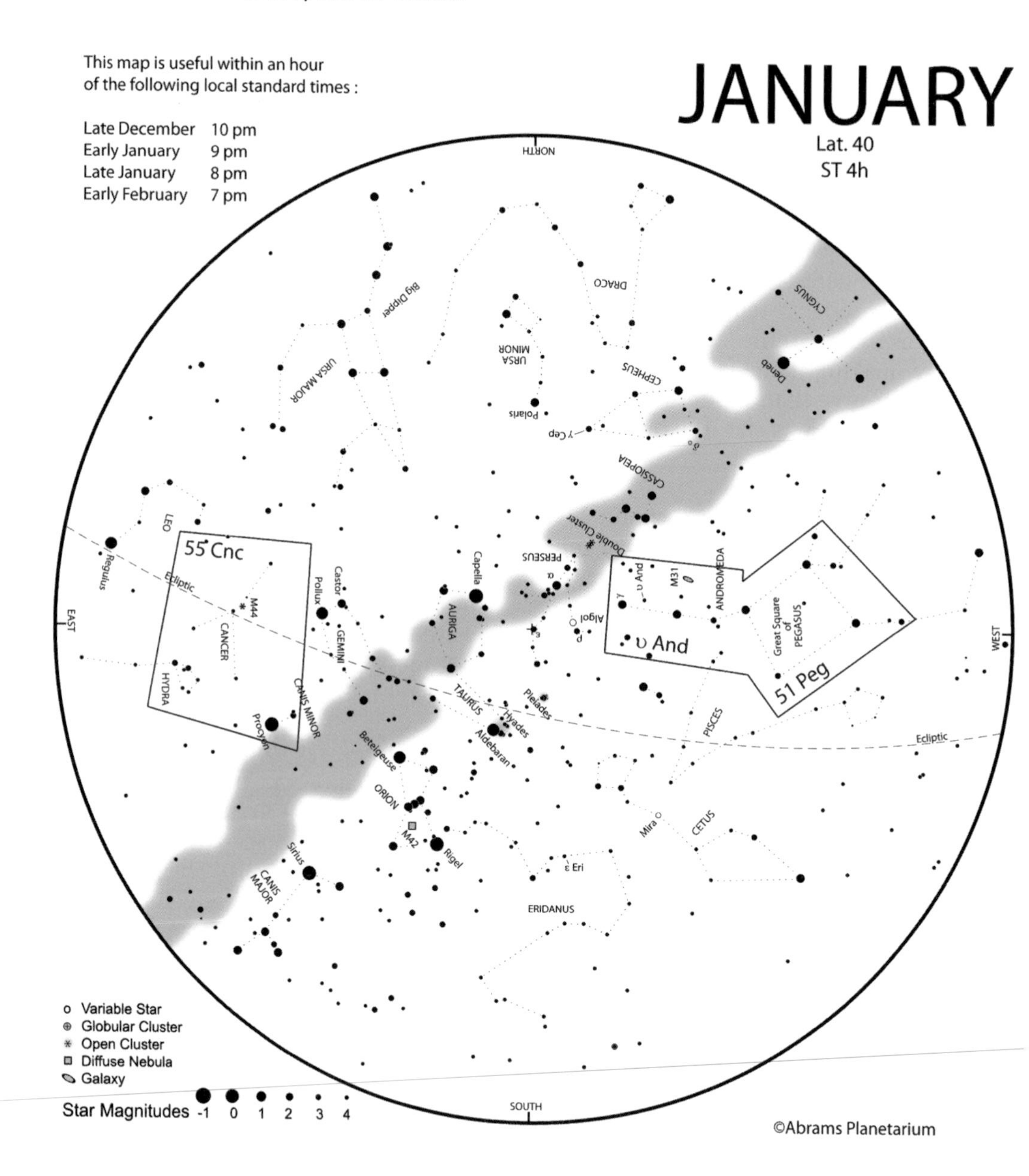

Hold the star map above your head with the top of the map pointed north. The center of the map is the sky straight overhead at the zenith. Boxes are shown outlining selected photographs for finding specific extrasolar planets.

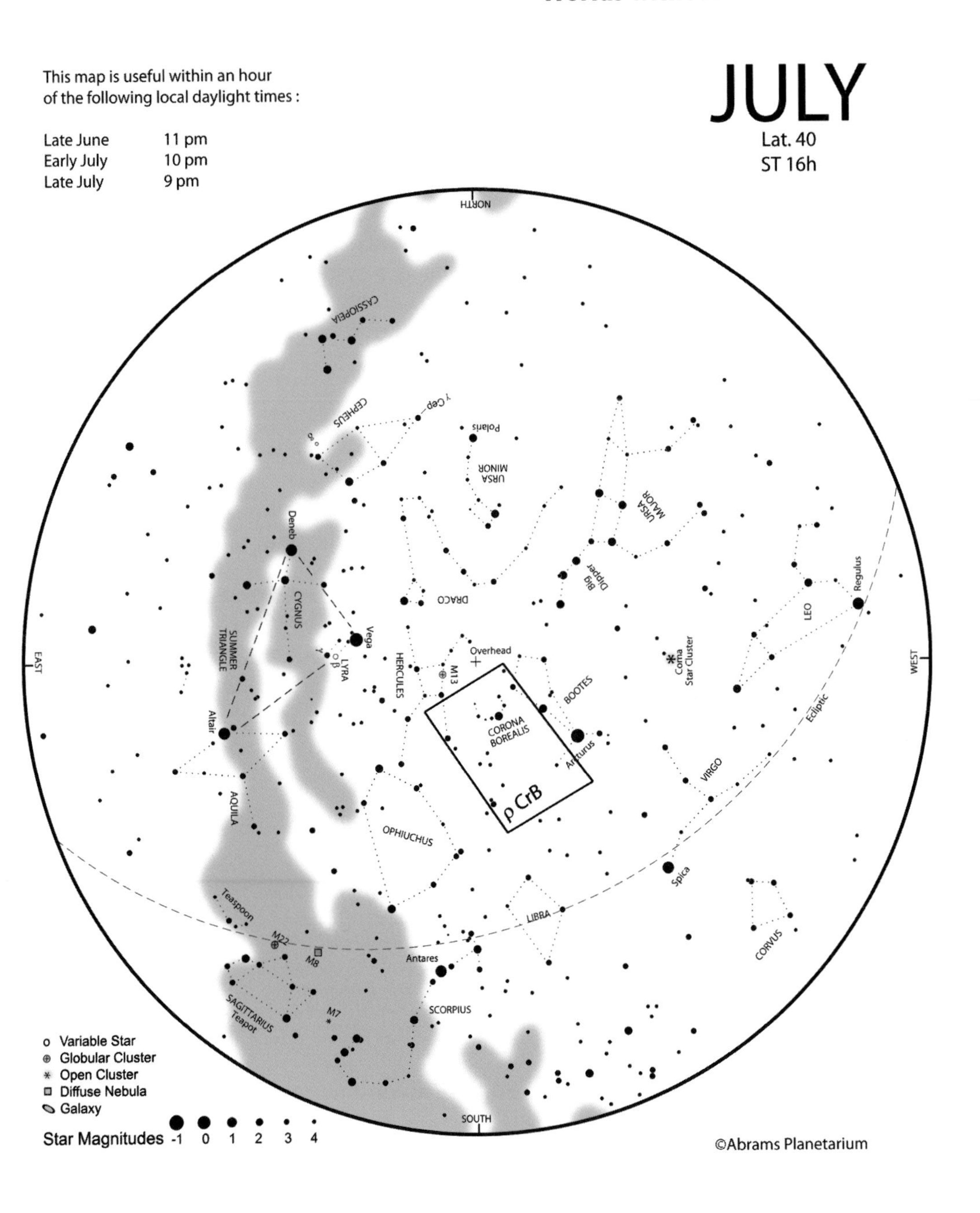
This map is useful within an hour
of the following local daylight times :
Late June 11 pm
Early July 10 pm
Late July 9 pm
JULY
Lat. 40
ST 16h
NORTH
EAST
WEST
SOUTH
CASSIOPEIA
CEPHEUS
γ Cep
δ
Polaris
URSA MINOR
URSA MAJOR
Big Dipper
DRACO
Deneb
CYGNUS
SUMMER TRIANGLE
Vega
LYRA
γ
β
HERCULES
M13
Overhead
BOOTES
CORONA BOREALIS
Arcturus
ρ CrB
Coma Star Cluster
LEO
Regulus
Ecliptic
VIRGO
Spica
Altair
AQUILA
OPHIUCHUS
LIBRA
CORVUS
Teaspoon
M22
M8
Antares
SCORPIUS
M7
SAGITTARIUS Teapot
o Variable Star
⊕ Globular Cluster
⁎ Open Cluster
▫ Diffuse Nebula
Galaxy
Star Magnitudes -1 0 1 2 3 4
©Abrams Planetarium

and therefore very difficult to spot. Half way between the constellation of Gemini and Leo there is a small fuzzy cloud that looks like a blurry star. This is the small cluster of stars called M44, the Beehive (see Figure 9.25). It is the heart of the Y-shaped constellation of Cancer the Crab. M44 sits between δ and γ Cnc. North of γ Cnc is the nearly equally bright star 48 Cnc. Just to the east of 48 Cnc is 55 Cnc and it will be just barely within your ability to see it from dark locations with no Moon. If you can also see the faint winter Milky Way then it is a good chance you will be able to see 55 Cnc. Around this star is a system of at least five planets ranging in mass from at least 18 times the mass of the Earth to four times the mass of Jupiter. It is the largest planetary system we know of beyond our own.

Further reading

Planet Quest by Ken Croswell (1999)
Oxford University Press, ISBN 0192880837

Looking for Earths: The Race to Find New Solar Systems by Alan Boss (1998)
Wiley, ISBN 0471184217

New Worlds in the Cosmos: The Discovery of Exoplanets by Michel Mayor and Pierre-Yves Frei (2003)
Cambridge University Press, ISBN 0521812070

Undaunted Courage: Meriwether Lewis, Thomas Jefferson, and the Opening of the American West by Stephen E. Ambrose (1997)
Simon & Schuster, ISBN 0684826976

The Great Divide: The Rocky Mountains in the American Mind by Gary Ferguson (2004)
W.W. Norton & Company, ISBN 0393050726

Thomas Jefferson and the Rocky Mountains: Exploring the West from Monticello by Donald Jackson (1993)
University of Oklahoma Press, ISBN 0806125047

Passage through the Garden: Lewis and Clark and the Image of the American Northwest by John Logan Allen (1975)
University of Illinois Press, ISBN 0252003977

Journals of the Lewis and Clark Expedition, an on-line searchable database
http://lewisandclarkjournals.unl.edu/

The Planet Quest Atlas of Extrasolar Planets
http://planetquest.jpl.nasa.gov/atlas/atlas_index.cfm

NASA's Kepler Mission Homepage
http://kepler.nasa.gov/

The Planetary Society
http://www.planetary.org

10 Far away and long ago: the Universe before you

The sky begins at your feet.
Park Ranger G.B. Cornucopia

"In the beginning there was only Tokpella, Endless Space. Nothing stirred because there were no winds, no shadows fell because there was no light, and all was still. Only Tawa, the Sun Spirit, existed, along with some lesser gods. Tawa contemplated on the Universe of space without objects or life, and he regretted that it was so barren. He gathered the elements of Endless Space and put some of his own substance into them, and in this way he created the First World. There were no people then, merely insect-like creatures who lived in a dark cave deep in the earth. For a long while Tawa watched them. He was deeply disappointed. He thought, 'What I created is imperfect. These creatures do not understand the meaning of life.'" [1]

In time Tawa created the Second World so that he might perfect all things that have life in them. When Spider Grandmother led the creatures to this new world they were changed into hairy forms that resembled dogs and coyotes and bears. But because they still had no knowledge of things they eventually became embittered and fought amongst themselves and Tawa was still not pleased.

Figure 10.1 Hopi pot showing a traditional migration pattern (T. Nordgren).

Tawa next created the Third World. When Spider Woman led the creatures there, their bodies changed into people and she said "Tawa has given you this place so that you may live in harmony and forget all evil. Do not injure one another. Remember that Tawa created you out of Endless Space, and try to understand the meaning of things."

1 From Harold Courlander's *Fourth World of the Hopis.*

Figure 10.2 The Grand Canyon stretches into the distances as seen from the Rim Trail just east of the historic El Tovar Hotel in Grand Canyon National Park. Snow has fallen well into the canyon from a recent winter snowstorm (T. Nordgren).

But even there the people fought and Powakas, or sorcerers, brought evil and conflict. Tawa was once more displeased and so told Spider Grandmother to tell the people that all those of good heart should go away from this place. The people left their homes and climbed a great bamboo tree that led up through a hole in the sky through which they entered the Upper World. There, in *this* world, they became all the different tribes of people we know today.

Among the many Hopi clans whose origin tale this is, the details have long since diverged in the telling and retelling over the centuries. Most believe the location of the opening through which they emerged into this world, the sipapuni, is long forgotten. But a number of clans from Third Mesa believe the sipapuni can still be found, protected beneath a protective cap of water near the place where the Colorado and Little Colorado Rivers meet in the heart of the Grand Canyon.

Grand Canyon is a fitting monument to origins and legends. Its enormity dwarfs words and renders simple size measures meaningless. And while western religious tradition ascribes divine beauty to the heavens above, the canyon's awesome spectacle renders it reasonable that the Hopi world of creation should be located within the earth below. Even some orthodox Protestant traditions find justification for their own creation story within the canyon. For them the canyon's expanse is evidence enough for Noah's flood and the Earth's creation only six thousand years ago.[2]

For myself, looking down from one of the many overlooks along the South Rim, I understand the need to find simple explanations for what I am seeing. It's truly difficult to comprehend how something as big and beautiful as this could be here. So while some may only be able to come to grips with the canyon by believing that it was created in a single cataclysmic event, while others may take the opposite tack and conclude that it is as eternal as the Earth, the middle road between these extremes is the scientific story that the canyon was created by the scouring action of simple water over long periods of time. And while I am a scientist, I recognize that even with all the evidence in the world supporting this origin, this is probably the least intuitively obvious and emotionally satisfying explanation of all. Our mind recoils at the lengths of time required for what we see here.

Figure 10.3 Light from late afternoon sun reflects off the Colorado River as it continues to cut its way through the depths of the Grand Canyon (T. Nordgren).

That's the problem with 'Deep Time;' it's a span of time so far beyond our ordinary experience or historical reference that, like a bout of vertigo on the canyon's rim, our mental faculties whirl and spin without means of support. But to understand the geology of the canyon absolutely requires we come to terms with Deep Time. Standing here on the rim, I cannot help but see in the canyon's depths a glorious example of the observations, hypotheses, models, and theories that when coupled with Deep Time (and in the case of astronomy, Deep Distances) lead to the scientific origin of not just the canyon, but the entire Universe beyond.

To begin to think about Deep Time, consider a fly whose entire lifespan is no more than a single day. To a fly, you and I are unchanging features of the

[2] No matter what the physical evidence for a 4.5 billion year old Earth, from fields as diverse as geology, chemistry, biology, physics, or astronomy.

landscape. While we may move around in a vain attempt to swat them, consider how little our physical form changes over the life of a fly. We don't grow taller, get thinner, or change shape. Our hair doesn't grow longer or naturally change color. We look virtually the same from the moment we wake up to the moment we go to bed, during which time the tiny fly has grown, lived, and died.

Relative to a fly, you and I live for 25,000 lifetimes. To put this into perspective, 25,000 lifetimes for people is a span of time 1.7 million years long. During that time, the Earth has most recently undergone 20 periods of glaciation, mammoths and saber-tooth cats flourished and became extinct, and the Colorado River began the final phase of carving that created the image of Grand Canyon we see today. This is Deep Time. Now imagine trying to describe to the fly this still relatively recent geological time span. 1.7 million years for a fly, is the equivalent of 43 billion years for us: longer than the canyon, the Earth, our Galaxy, or even the Universe have been around.

Figure 10.4 Two of John Wesley Powell's boats photographed during his second expedition down the Colorado River and through the Grand Canyon (July 30, 1872) (Grand Canyon National Park Museum Collection, Image No. 14642B).

When we stand on the rim of the Grand Canyon we are confronted with a sight that is superficially unchanged over the lifespan of humans. The canyon looks the same to me now as it did when my parents brought me here as a kid. Their view, in turn, was identical to the one all previous generations of visitors had whether they arrived here by plane, car, train, horse, or foot. Stand on the rim and you see what Teddy Roosevelt saw when he declared Grand Canyon a National Monument in 1908, when John Wesley Powell first explored its depths by boat in 1869, or when that distant ancestor of today's Grand Canyon tribes saw this canyon for the very first time maybe 10,000 years ago.

But the canyon *has* changed over the last 10,000, 100, or even 10 years, and by just enough to explain its origin over a timespan a thousand to a million times longer.[3] To get a

[3] Every year rockfalls within the canyon require work crews to repair trails, while in the process the canyon widens and deepens just a little more.

better feeling for this, let's examine two of the unimaginably slow processes that make the canyon that we see today. The first is deposition: the rock that we see in the canyon's walls had to come from somewhere. With the exception of the dark volcanic rock in the Vishnu schist and Zoroaster Granite at the very bottom of the canyon (whose hard steep sides make the narrow inner gorge) the bright colored layers of the canyon are the result of sediments: the slow settling of sand, silt, mud, and minerals.

Layering like this happens constantly, yet so slowly we never notice. Live for a while in the desert southwest and you get used to the wind that blows sand in through windows and cracks in the doors. For everyone who has ever had to sweep off the porch and dust off the furniture, you've battled the forces that created the rocks through which this canyon cut.

Let's do a simple thought experiment. Stop sweeping and let the sand and dust grow about your house. After only a year I can easily imagine a layer of sandy sediments a single grain thick: say 1/32nd of an inch thick (about 0.75 mm), and an identical layer is deposited for every year you fail to tidy things up. In one hundred years that's dirt three inches deep (7.5 cm) and after a thousand, the floor is buried nearly three feet down (0.75 m). The layers I see revealed in the canyon's walls are only 5,000 ft thick (1,525 m), easily due (by our calculations) to the gentle drift of sand and sediment after only two million years.[4]

But the canyon isn't made of loosely stacked sand; the weight of those layers presses and compacts the sediments beneath it. The internal heat and pressures on all that buried sand cement and solidify it from sand grains into sandstones. If, in our simple thought experiment, we were to imagine that each layer of loose deposits is eventually compressed to even a tenth its original thickness, then the two million years we calculated turns into twenty million years. The geological record indicates the horizontal layers we see within the canyon were laid down by sediments, off and on, over a period of 200 million years, easily accomplished by our simple assumption of windblown sand. But the deposition is only half the story. Without the erosion that cut away the canyon, all those multi-hued sandstones, limestones, and shales would be forever hidden beneath our feet. Besides, it's what missing that impresses, just as much as what's there.

Consider then a second thought experiment: how much time is needed to remove the canyon's missing mass? Grand Canyon is roughly 10 miles (16 km) wide, 277 miles (443 km) long and about 1 mile (1.6 km) deep. That's 2,770 cubic miles (11,340 cubic km) of dirt that the river has removed. If a single grain of sand is no more than 1/32nd of an inch (0.75 mm) on a side, then to dig our

[4] In addition, our simple model of sedimentary layering also reveals that the oldest material is always at the bottom. Unless some other processes come along to fold, warp, tilt, or bend them, the stack of layers serves as a history text of what's come before and for how long. For five years now, NASA's Mars rover Opportunity has been reading similar stacks of sedimentary rocks revealing the geologic history of a planet other than our own.

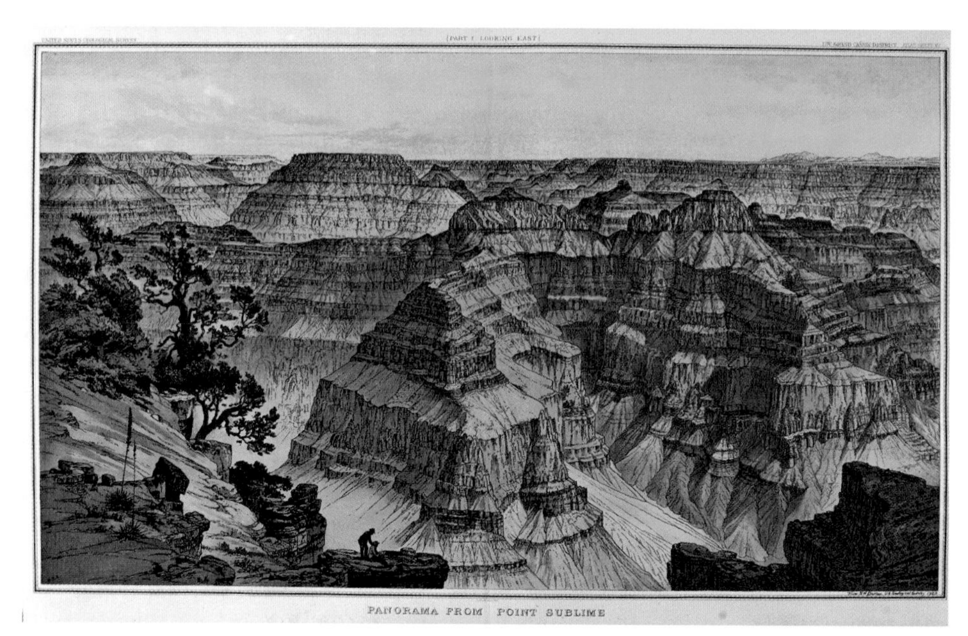

Figure 10.5 View of the Grand Canyon as shown from Point Sublime on the North Rim. This engraving by William Holmes is from the 1882 *Atlas of the Tertiary History of the Grand Canyon by Capt. Clarence E. Dutton USA for the Department of the Interior and the Geological Survey.* As a scientific illustration, Holmes' intent was to represent the geological layering and formations visible within the canyon (Grand Canyon National Park Museum Collection, Image No. 13572).

Figure 10.6 Time passes above the Colorado River within Grand Canyon National Park. As the Earth turns, winter stars rise in the east as the muddy, chocolate-brown waters of the Colorado continue their slow erosion of the canyon. In the distance the South Kaibab Trail Bridge spans turbulent waters rendered deceptively placid in this two-hour-long exposure (T. Nordgren).

canyon we need to remove 10^{22} grains of sand. Let's say each fluid ounce of water (0.03 liters) passing through the canyon were to remove just a single grain of sand (if you've ever seen how chocolaty brown the Colorado River is, you'll know this is probably an underestimate). Before the Glen Canyon Dam's construction upstream created Lake Powell, the Colorado flowed through the canyon at a rate of roughly 250,000 gallons (1 million liters) a second. At 128 fluid ounces per gallon, that's 30 million grains of sand removed each second, or a little over 10^{15} grains per year. At that rate it would take the Colorado 10 million years to carve the canyon we see today.

The geologic record shows the majority of the canyon was carved over the last six million years as the river cut through the Colorado Plateau that was lifted up in the mighty tectonic forces that also created the Rocky Mountains. So to within a factor of two, our simple explanation holds.

Sure, these thought experiments are crude; they grossly simplify the canyon's creation just as surely as if I were to tell the story of a man's life by simply saying he was born, lived, and died. The rock layers aren't all made of windblown sand, and forces of erosion are at work other than the single river. But these thought experiments serve the purpose of what scientists call an order of magnitude calculation. The amount of time our calculations require has the right order of magnitude (science-speak for the right number of zeros). In this case, they require lengths of time measured in millions of years. If our quick calculation had shown that the necessary depth of dirt laid down or dug out could not be done in a few million years, but instead required 10 billion years (or would happen in only 1000), we'd know something was wrong with our approximation. We'd know we had failed to take something important into consideration.

But our reasonable assumptions of one grain per year, and one grain per ounce, turned out to make sense: they are just enough to make the canyon we see today provided we are prepared to let the years add up. By and large, science requires Deep Time to explain everything we see around us, whether it is canyons, mountains, biological diversity, DNA, the chemical elements, the stars in the sky, or the origin of time itself. Wrapping our heads around this much time is hard.

It's a cold crisp winter morning today. A new year has just begun, and the first rays of a new day are just beginning to light the canyon's golden walls. I take in this view from the ice-laden summit of the South Kaibab Trail that I will follow all the way down to Phantom Ranch at the canyon's bottom along the Colorado River. Tonight I'll be giving an astronomy talk for the hikers camping at the ranch, and while the clear blue sky will be a joy to see from deep in the canyon, up here on the rim it produces temperatures well below freezing.

Still, as I take my first crunchy steps on the snowy 6.3-mile (10-km) hike to the bottom, I take off my gloves and reach out to the rock wall beside me. Every gentle scratch and scuff across my finger tips puts me in contact with the distant past. After a mile and half (2.4 km) I have descended a thousand feet (300 m) into the Earth and am astounded to think that the layer my finger traces follows the curve of a sand dune baking under the Sun of a long-vanished day. Here where

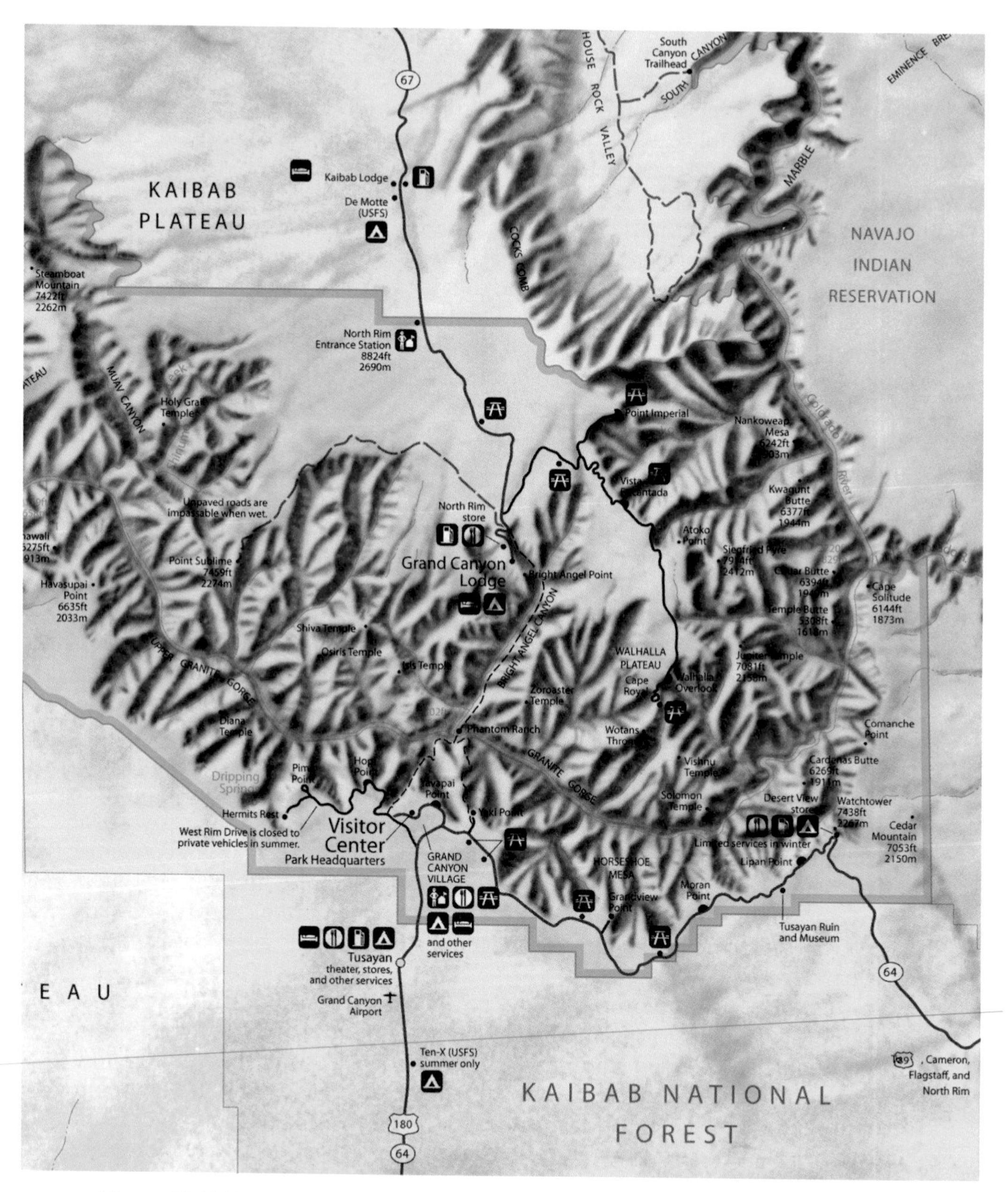

Figure 10.7 Close-up of the National Park Service map showing the south and north rims of Grand Canyon National Park. The South Kaibab Trail is the black dashed-line connecting Yaki Point along the South Rim to Phantom Ranch just north of the Colorado River at the bottom of the canyon. The Little Colorado River is the northeastern boundary of the park and flows westward into the Colorado at the intersection of the park and Navajo Reservation. Today the Hopi Reservation is located entirely within the much larger Navajo lands (NPS).

my fingers pass, a lonely lizard crawled up what seemed a momentary sand dune only to have it immortalized in cross-bedded swoops along the broad bright band of the Coconino Sandstone. As I descend through switch-backs along this prominent yellow cliff-face, the dunes that once covered the vanished mass of the canyon carry me ever farther into the past. At the average rate at which these sedimentary layers were deposited, each footstep I take forward along my path steps me 10,000 years backward in time.[5]

Figure 10.8 View from along the South Kaibab Trail deep within the canyon (T. Nordgren).

Time is something we astronomers have to come to grips with as well as geologists. Multiple lines of evidence point to a Universe that is 13.7 billion years old. At the rate of 10,000 years per footstep, I'd have to hike a trail 260 miles (416 km) long to walk back to the beginning of the Universe. And the time that our species has been here on Earth would take up no more than the first forty feet (12 meters) of the journey. With only a single step I'd leave all recorded human history behind.

In some ways geologists are fortunate compared with astronomers; geologists can hold their subject in their hand and actually touch the past like I touch the dark Redwall limestone that formed here 340 million years ago. For those astronomers who study the Universe beyond our Solar System we are forced to study everything by no more than the light we receive. Although we have little hope of ever reaching out and touching the things we wish to understand, we do have one advantage other sciences don't. We have a time machine.

Everything we see, we see because of light. It can be emitted directly, like light from the Sun or the stars in the night sky; or reflected, like light striking the sheer canyon walls or the snow on the rim. It takes time for light to reach our eyes, so we are forever seeing 'old' light, whether it has been traveling for milliseconds or millennia. The Universe we see around us shows how things looked when light first left its source. Light allows us to see back in time: the farther away we look

[5] From the South Rim to the Tip-Off point at the top of the inner gorge, the South Kaibab Trail covers 4.6 miles (7.4 km). Above this point, 250 million years of sedimentary deposits have led to the multi-colored layers we see. While each layer was put down over slightly different lengths of time, my passage through the entire section is done at an average rate of roughly 50 million years per mile (30 million years per kilometer), or 10,000 years per footstep.

Figure 10.9 A few of the layers that make the canyon grand. Included are the approximate ages of the layers (T. Nordgren).

the farther back in time we see. Unimaginable distance leads to unimaginable time: astronomical size itself is the window through which astronomers look to the earliest days of the Universe. And as with the Grand Canyon around me, even small steps inevitably take us to amazing destinations.

Phantom Ranch is a collection of cabins huddled within a narrow side canyon of the inner gorge along Bright Angel Creek. In 1922, Mary Elizabeth Jane Colter designed the buildings in a style that would become famous as National Park Service "Rustic." The rounded stones, polished by millions of years of water rushing through the canyon, coupled with rough-hewn logs brought down by mules, makes the ranch itself seem a natural part of the surrounding geologic wonder.

Dinner at the Ranch is a sociable time as everyone gathers around long plank tables and shares in the food like one big family related by the blood pumping in our veins from the long exertion needed to get here. I meet people at dinner that night from all over the world: a couple college kids on break, a family from Europe touring America, an octogenarian taking the time to do all those things she couldn't when she was younger. Imagine my surprise then when the fellow across from me recognizes my last name and tells me he was once my brother's college physics professor before he retired.

Figure 10.10 Cabins of Mary Colter's Phantom Ranch (NPS photo by Michael Quinn).

After sunset we all gather at the amphitheater set amidst cottonwood trees near where the creek and the Colorado River meet. Because of its size and distance from most large cities, Grand Canyon National Park has wonderfully dark skies. This is a fact not lost on park visitors that flock to evening talks like mine with such enthusiasm that the park has dramatically increased its ranger astronomy programs on the South Rim.[6] Tonight is a clear crisp night, perfect for stargazing, and although the rock walls of the gorge hem in our view of the sky, what we see will be enough. Though we've all walked many miles to be here, tonight we're going to explore even larger distances: we're going to take a tour deep into the Universe, starting from where we are right now in this little clearing beside the river.

Because of the enormous distances involved, ours will be a Power of 10 tour where every stop is 10 times farther away than the one before it. The "Power of 10" is simply the number of zeros after the number one, so one hundred miles, kilometers, or light-years is just ten to the power of two, or $10^2 = 100$). While we will start out small on scales with which we are all familiar from our everyday lives, our pace will rapidly lengthen as we eventually span the distances between planets, stars, and galaxies.

For this Power of 10 tour, our only concern is to know the distance of each step rounded off to the nearest whole power of 10 (the nearest order of magnitude). Fractions are inconsequential. All that is important to know is if something is 10 light-years away, or ten thousand. Given that our Galaxy is 100,000 light-years across, the difference between 1 light-year and 2.3 light-years will be utterly inconsequential.[7]

Look at your feet. The Earth is the nearest astronomical body to you. The average adult is roughly 65 inches tall (5 feet, 5 inches) so when you see the Earth you are looking over a distance of 65 inches. Unfortunately, the foot and inch aren't very good units for a Power of 10 tour since there are 12, not 10, inches in a foot, and I can never remember how many feet in a mile. But 65 inches is also about one and a half meters. If all one is interested in is a general order of magnitude, then we can round off our height to the nearest power of ten which is simply one meter ($10^0 = 1$). So on average, when you and I look down at our feet, we are seeing a distance of about 1 meter. The nice thing about this fact is that for those of us in the U.S. who were not raised on the metric system, one meter is simply person-sized. One meter is *you*. In the following trip through the

6 For those who like to explore at their own pace, Grand Canyon has an audio cell-phone tour that visitors can call to learn more about the canyon at various scenic spots along the South Rim. One of those stops in the tour specifically guides visitors to an exploration of the canyon's gloriously dark-starry night sky. You can call right now (there's even a phone at Phantom Ranch): After dialing 1-928-225-2907 press 4.

7 For those who want to know the most precise distances to the objects described in this Power of 10 tour, consult the 'See for yourself' section at the end of the chapter (which also includes instructions on how to find them).

Figure 10.11 Moonlight illuminates Grand Canyon in this view towards the North Rim from Yavapai Point. Because of the wonderfully dark skies at Grand Canyon, astronomy has become a major activity within the park with evening ranger programs, a seasonal night-sky stop on the park's cell-phone audio tour (call 1-928-225-2907 press 4), and an annual star party hosted by local astronomy clubs (T. Nordgren).

Universe everything will therefore be scaled not to just some arbitrary unit of feet, miles, furlongs, or kilometers, but rather it will be scaled to *you*.

We will start in this amphitheater at the bottom of the Grand Canyon.[8] The Earth, the third planet from the Sun and the planet we have done the most to explore, is one meter away, one unit of you.

From where I stand, the next power of ten (i.e., $10^1 = 10$, and so ten times farther away) is the boundary of the audience in this amphitheater. If, however, you are standing on the canyon rim as you read this, a distance of ten meters (or ten yous) away, encompasses the people around you: the crowd at an overlook after a tour bus has stopped.

For those of us at the canyon bottom, ten times farther than that (100 meters away) and we see across the Colorado River. For those of you at the top looking down, 10^2 yous only gets you to the bottom of the Kaibab formation, the very top layer of the canyon.

Look ten times farther away, approximately 10^3 meters distant, and those of us at top and bottom of the canyon finally see one another. The canyon itself is only about 2,000 people deep.

[8] Those not at the Grand Canyon can join up with this walk after the fifth step.

At 10^4 meters, we span the width of the canyon itself. To the nearest power of ten, about ten thousand yous laid end to end is all it would take to bridge the canyon, not a million, not a billion, just ten thousand: a typical state university student body could do it as a prank.

So what is ten times farther than that? At one hundred thousand yous (10^5m = 100 km) we come to the internationally defined edge of space. This is the height at which meteors enter the atmosphere and end their life in a fiery streak. The great irony of space, a place to which virtually none of us will ever go, is that if one could drive there it would only take an hour; you wouldn't even have to stop for gas. Virtually everything that makes life possible on this planet is contained within that distance.

By the time we turn our gaze ten times farther away, at one million meters (10^3 km), we have passed through the realm of the orbiting spacecraft. Look up on a dark night and watch for an hour and you will almost certainly be rewarded with a small faint spot of light moving slowly against the background stars. Use a website such as Heavens-Above.com to see when the International Space Station or Hubble Space Telescope will fly by overhead.

Figure 10.12 The Moon rises above the inner canyon near Phantom Ranch (T. Nordgren).

Ten million yous away (10^4 km distant) brings us to the beginning of geosynchronous orbit where communications satellites take 24 hours to orbit once around the Earth. Since this is exactly the same rate at which our planet turns, these satellites stay in very nearly the same spot over the Earth. All people with a satellite TV need do is point their dish at one spot in the sky, and be comfortable in the knowledge that orbital dynamics will keep the Food Network flowing to their homes. While it is very difficult to see these satellites with the naked eye, we see their effects daily.

At one hundred million yous (10^5 km) we turn our sights to the Moon. Looking at it rise over the rim of the canyon you are seeing the abode of human footprints a hundred million times farther away than your feet. This is the greatest distance from which any human being has been able to look back at us here on Earth, but we are still only eight powers of ten out into the Universe.

Figure 10.13 Park Rangers give evening sky talks most nights of the week, all year long. Even in the depths of winter a full Moon hike along the canyon's South Rim draws upwards of 50 visitors (T. Nordgren).

Figure 10.14 Mars as photographed by the Viking Orbiter in the late 1970s. Visible extending across the face of the hemisphere is Valles Marineris, a canyon network that, if on Earth, would extend from Los Angeles, California, to New York City. The Grand Canyon would be lost within a minor tributary of this grandest canyon in the Solar System (USGS/NASA).

Ten times farther than the Moon, at one billion (i.e., one thousand million) yous, we see nothing.

Ten times farther away, at ten billion meters, we see the end of an imaginary tower formed by standing every human being on Earth on the shoulders of every other. And still, there would be nothing there for the one on top to touch. The astronomer Carl Sagan described this mostly empty cosmos by writing that the only typical spot is "the vast, cold, universal vacuum, the everlasting night of intergalactic space, a place so strange and desolate that, by comparison, planets and stars and galaxies seem achingly rare and lovely." The Earth and Moon are awash in this nothingness.

At 10^{11} = 100 billion yous, we finally reach the Sun and other small rocky planets orbiting close in to its warm glow. Mars is overhead tonight, eleven orders of magnitude farther away than my feet. At only 11 orders of magnitude, the distance in meters is quickly becoming enormously large and astronomers begin to use a different unit of measure called the Astronomical Unit. One AU is the average distance between the Earth and Sun and so is about 1.5×10^{11} meters.

One way we know the distances to the planets is that we have bounced radio waves (radar) off their surfaces. We know that radio waves (a form of light) travel at the speed of light so measuring the time it takes to make the round trip tells us exactly how far away the planets are. And though Mars may be 100 billion times farther away than my feet, the light travel time between Mars and my eyes is only about 10 minutes. So while we are already seeing backward in time, it's not by what anyone would consider an exciting amount. The greatest effect of this light

Figure 10.15 Saturn, the ringed planet, as photographed by the Hubble Space Telescope in Earth orbit (NASA and The Hubble Heritage Team (STScI/AURA)).

travel time is to render impossible the real-time control of rovers on Mars by drivers on Earth.

Ten times farther away at 10^{12} meters (10 AU) our view carries us past the giant planets. Jupiter, Saturn, Uranus and Neptune are only 1 trillion times farther away from our eyes than our own planet is. Light from the Sun out there is 100 times fainter than the sunlight that falls all around us on a sunny day and temperatures plunge accordingly. From here on out, worlds with solid surfaces are made predominantly of ice. The Universe is a cold, dark arctic wilderness with only the occasional campfire of starlight where even a few planets are able to huddle for warmth. If Jupiter or Saturn is visible tonight, either would be the most distant thing you can see in our Solar System with the unaided eye.

On our way out to 100 AU (10^{13} m) our gaze pass completely through the Kuiper Belt, a region of icy debris from the Solar System's formation. Here is Pluto, once designated as the furthest planet. Composed largely of ice, with an orbit that carried it across Neptune's every so often, it was considered an oddball, unlike any other planet we knew of. Then, with the discovery of more (but smaller) icy objects, it became the largest known Kuiper Belt Object until a larger one was finally found in 2005.[9] Pluto is pretty boring to look at through a telescope, but it is the most distant object in our Solar System that you can see with a backyard telescope and a dark backyard.

Between 100 and one thousand AU from the Sun (10^{14} m) we enter a shadowy part of the Solar System. Everything we've examined so far (save the Sun) we've seen only by its reflected sunlight. We're now 1,000 times farther from the Sun than the Earth and out here the Sun's light is a million times fainter. As of 2009 only a handful of objects have ever been seen that orbit out at these distances. Sedna is one such object. Like Pluto, it is an ice world, but only 1,000 miles (1,600 km) across. Every 12,500 years its highly elliptical orbit brings it to within 76 AU of the Sun and then swings it out through this dark expanse to a distance of nearly 1000 AU.

Astronomers only found Sedna because it is currently at its closest approach where the sunlight is brightest. How many other worlds are there waiting to be

[9] It's been named Eris, after the Greek goddess of discord and strife in recognition of all the trouble it caused in the debate about Pluto's status as a planet.

Figure 10.16 Hubble Space Telescope photograph of Pluto (center) with its large moon Charon (lower right). In this photo are visible two newly discovered moons (towards the right): Nix and Hydra. The planet and its moons are named for deities associated with the underworld (NASA, ESA, H. Weaver (JHU/APL), A. Stern (SwRI), and the HST Pluto Companion Search Team).

Figure 10.17 Comet Hale Bopp passed through the inner Solar System in 1997. It will not get this close to the Sun again for almost 4,000 years (T. Nordgren).

discovered as they rise up out of the inky depths like ghostly whales from dark ocean waters? Is this region populated with its own collection of ice-worlds, and if so, how did they get here? No one yet knows.

At 10,000 AU (10^{15} m) one quadrillion yous, we come to the inner edge of the Oort Cloud, a great spherical shell containing as many as a trillion comet nuclei. While these iceballs are too small, faint, and far apart to ever allow us to see them directly, every few years a gravitational encounter or collision alters one's path just enough that it falls in towards the Sun on a 500,000 year trajectory. When, at last it's only a couple AU away, light from the Sun warms its icy surface for the very first time and treats us to the awe inspiring spectacle of a comet whose tail spans interplanetary distances. It is from tracing back the orbits of these long-period comets that we know there must be a vast reservoir of nuclei out there, a tenth of the way to the nearest stars.

At 100,000 AU, 10^{16} m, our view reaches the extreme outer edge of the Oort Cloud and we leave the Solar System behind. We are now so far from Earth, that if a spot light should suddenly appear here and be pointed Earthward, it would take a year for the light beam to reach the folks back home. We therefore devise a new unit of distance to mark the distance light travels in a year: the light-year.

Ten light-years away and we once more set our sights upon those objects that emit their own light: we have reached the stars. The closest star that's easily visible from the northern hemisphere is Sirius.[10] It's the brightest star in the night sky and visible to the lower left of Orion in winter. Southwest of Orion is Epsilon Eridani; it's an average looking star but the closest one yet known to have a planet of its own. Both of these stars are only visible in winter whereas in summer skies Vega is visible high over the canyon's walls. John Wesley Powell saw a similar sky on his first expedition down the canyons of the Colorado River when he passed close by to where I now stand.

> I do not sleep for some time, as the excitement of the day has not worn off. Soon I see a bright star that appears to rest on the very verge of the cliff overhead to the east. Slowly it seems to float from its resting place on the rock over the canyon. At first it appears like a jewel set on the brink of the cliff, but as it moves out from the rock I wonder that it does not fall. In fact, it does seem to descend in a gentle curve, as though the bright sky in which the stars are set were spread across the canyon, resting on either wall, and swayed down by its own weight. The stars appear to be in the canyon. I soon discover that it is the bright star Vega.
>
> John Wesley Powell, *The Exploration of the Colorado River and its Canyons*

We know the distance to stars like Sirius and Vega because their position in the sky, relative to more distant background stars, changes by a microscopic amount over the course of six months as the Earth moves from one side of the Sun to the other. This phenomenon is called parallax and it is the same reason that if you hold your thumb in front of your face and look at it through one eye and then the other, it moves back and forth relative to more distant objects in your field of view. The farther away the thumb (or star), the smaller the parallax shift.

One hundred light-years out, 10^{18} units of you, and the stars whose light we see tonight first left them while Powell continued his way down river. Many of these are well known stars because of their brightness in the evening sky. Stars like winter's orange Aldebaran in Taurus, or summer's blue Spica in Virgo are as bright as they are because they emit 350 and 23,000 times the light of the Sun respectively. These are big stars.

One thousand light-years away, 10^{19}m, and we soar amidst the nebulae. These are the gas clouds that span the distances between the stars and out of which new stars are continually born under gravity's incessant pull. We can see the biggest

[10] The closest star like our own is alpha Centauri at 4.4 light-years, but it is only visible from southern latitudes, and hardly ever makes it above the horizon from anywhere in the U.S. but the island of Hawai'i.

Figure 10.18 The Milky Way and bright stars set over Grand Canyon towards the west as viewed from Lipan Point along the South Rim. The bluish star Vega shines brightly at lower right just above the dark clouds along the horizon. City lights from Grand Canyon Village, Arizona; Las Vegas, Nevada; and St. George, Utah, are visible over the horizon reflecting off the undersides of the clouds. This viewpoint along the eastern portion of the canyon, near Desert View, has some of the darkest skies along the South Rim of the park (T. Nordgren).

Figure 10.19 Orion above the southern cliffs of the inner canyon as viewed from Phantom Ranch. Betelgeuse is the bright orange star at upper left. Beneath the three equally bright stars of Orion's Belt are four fainter stars that make up his sword. The faint pinkish 'star' (second from the bottom) of the sword is actually the Orion Nebula (also called M42) and is visible as a fuzzy 'star' to the naked eye (T. Nordgren).

Figure 10.20 The Lagoon Nebula (also called M8) is the bright red cloud of gas at bottom. Within it, new stars are being born that light up the rest of the cloud. Above it is the red and blue glow of the Trifid Nebula, another nursery of new stars (Steve Mazlin and Jim Misti).

and closest ones with our own eyes. In summer look to the Lagoon Nebula as a fuzzy cloud off the spout in Sagittarius' teapot. In winter the Orion Nebula is a fuzzy 'star' hanging in the sword from Orion's Belt. Both clouds glow by the light of hydrogen gas, lit up like neon road signs by the ultraviolet energy of hot young stars within.

Ten times farther away, at 10^{20} yous, ten thousand light-years distant, we come to the Milky Way that arcs overhead. In reality, the myriad stars that form that luminous band across the sky aren't really at any one distance. The Galaxy itself is on the order of 100,000 light-years across, but at the center of that vast spinning pinwheel of stars is a supermassive black hole 25,000 light-years away towards the bright constellation of Sagittarius.

At one hundred thousand light-years, 10^{21} meters distant from our eyes, we pass through the outer rim of our Galaxy. For those under southern skies, the Magellanic Clouds are small satellite galaxies of our own Milky Way. They and the Milky Way all reside inside a great spherical cloud of something we cannot see and so have come to call dark matter. The nature of dark matter is still a mystery, one of the many that makes science so exciting for those who have ever looked up at the stars and wanted to know more. It's exactly why I became an astronomer and when as a doctoral candidate in astronomy at Cornell University

Figure 10.21 The bright lights of the summer Milky Way rise above the canyon rim as seen from just north of the Grand Canyon. The brightest portion is towards the direction of the Milky Way's galactic center, twenty-five thousand light-years away (T. Nordgren).

I was able to conclude that dark matter halos extend for hundreds of thousands of light-years around galaxies like our own, I was just answering for myself the questions I had asked when I had first looked at the sky as a child.

Before we look further, let's take a moment to pause and think about how far we've come. We are now 100,000 light-years away at the edge of our Galaxy: 10^{21} yous. With only one or two tiny exceptions, everything we see with our own un-aided eye in the sky (by night or day) is within this one single solitary galaxy.

Beyond is the vast emptiness of intergalactic space, far emptier than the space we've seen between planets or stars. Within our Galaxy we see its contents as they looked less than one hundred thousand years ago. While this is long by the measure of the human race, even the shortest lived stars last for a million years or so. So while there are certainly stars we can see in the sky tonight that have died and disappeared while their light has been en route to us, by and large these are rare and what we see is what is really there.

Light, our time machine into the past, has so far failed to look far enough back into the Universe to see it sufficiently different than it looks today. Even here on Earth, a hundred thousand years only takes us back to the realm of the ice ages. One hundred thousand years is too short for changes in mountains or continents (besides the scouring action of glaciers and the changes in sea level caused by the ice ages). It is not too short for people, however, as we are back before any human being you ever heard of, before any history that was ever written, and before *homo sapiens* even walked out of Africa to eventually spread around the world.

Ten times farther away, at one million light-years we begin to bridge the vast emptiness between the galaxies. The Andromeda Galaxy is on the order of a million light-years away and for all but those few people with the most acute vision (and the very darkest skies), this is the most distant object most of us will ever see with our own eyes. In its fuzzy oval shape, visible at some time during the night for all seasons but spring, we see the combined light of a hundred billion stars in a spiral disk like our own Milky Way. The outer parts of the pinwheel (about where we are located in our own Galaxy) are too faint to be seen with the naked eye and all we see is the brightest part near its galactic center. If we could see its entire extent, the Andromeda Galaxy would appear in our sky as big as six full moons laid side by side. The Andromeda Galaxy (also called M31) and a second smaller pinwheel called M33 (plus all of their attendant satellite galaxies), make up our Local Group of galaxies of which the Milky Way is a part.

We know the distance to these galaxies by looking at the variation in light coming from some of the brightest stars orbiting within them. Cepheid variables are a type of star that change in brightness, growing alternately bright and dim, over a regular period like the rise and fall of a breathing chest. In 1912, the astronomer Henrietta Swan Leavitt discovered that the longer the pulsation took to repeat, the greater the star's average luminosity. As a result of this period-luminosity relation, Cepheids have become a standard candle that astronomers use for surveying the Universe.

Standard candles – a type of object with a known luminosity – are crucial for

Figure 10.22 Our nearest large galactic neighbor is the Andromeda Galaxy, a near twin to our own Milky Way. Visible just above and to the lower left of this spiral mass of a hundred billion stars are two small satellite galaxies, much like the Milky Way's satellites: the Large and Small Magellanic Clouds. The Andromeda Galaxy (also called M31) is typically the most distant object you can see with the naked eye (Jim Misti, Misti Mountain Observatory).

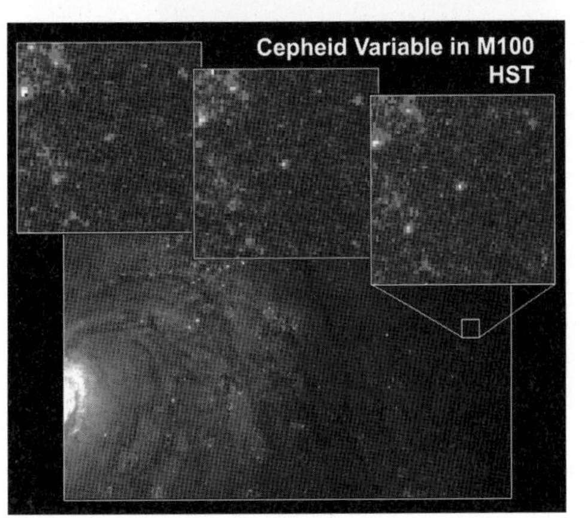

Figure 10.23 A sequence of Hubble Space Telescope images showing the outskirts of the galaxy M100. At center is a Cepheid variable star changing in brightness over the course of several days (Dr. Wendy L. Freedman, Observatories of the Carnegie Institution of Washington, and NASA).

astronomers because they allow us to accurately measure the distance to any such object. A 60 watt light bulb has a known brightness; it's a standard candle in your home. Move it twice as far away from you and it always gets four times dimmer. Move it three times farther away and it always gets nine times dimmer. If you know the neighbor down the street uses the same light bulb outside his

Figure 10.24 The spiral galaxy M81 is sometimes visible under the darkest of conditions to those with particularly acute vision. Like our Milky Way, it is a collection of over a hundred billion stars (NASA, ESA, and The Hubble Heritage Team (STScI/AURA)).

Figure 10.25 A typical small cluster of galaxies with a large elliptical galaxy at its center (NASA, ESA, and The Hubble Heritage Team (STScI/AURA), J. Blakeslee (Washington State University)).

front door, then comparing how bright the bulb looks from your house to how bright you know it should be, reveals how far away your neighbor lives.

We do the same thing with Cepheids. Find a Cepheid variable star in another galaxy and you immediately know its apparent brightness. Measure its pulsation period and you learn its absolute brightness. Since you know both how bright it looks and how bright it should look, the distance is simply how far away it must be to look as dim as it does. The astronomer Edwin Hubble found the first Cepheid variable star in the Andromeda Galaxy in 1924 and upon calculating its distance he once and for all determined that the spiral 'nebulae' we'd seen in the sky were islands of stars like our own Milky Way Galaxy, but vastly farther away.

Ten million light-years out, 10^{23} m, we pass another galaxy cluster like the Local Group; this one is centered around the beautiful spiral of M81. There are those who say they've been able to see this galaxy under unusually dark skies, but I'm honest enough to admit that they must have better eyes than I. Nevertheless, the light from this galaxy marks the furthest starlight you can see with the naked eye. This light comes predominantly from the very brightest stars within the galaxy which are exactly those stars that burn through their hydrogen fuel the quickest. In the time it takes their light to reach us, many of these stars will have already died. We begin to see a past which no longer exists. The light you see from M81 tonight (probably through a telescope eyepiece where it makes a lovely sight) left the stars of that galaxy when the Colorado Plateau was first raised and the river beside me first began to carve its way down to the sea. The entire canyon around me was carved while this light was en route to our eyes.

As we step ten times farther away our gaze passes the Virgo Cluster of Galaxies, a collection of 2000 galaxies (of which our Local Group is just one small part), and which itself is the heart of the Coma-Virgo Supercluster of galaxies. Here we cross the threshold of one hundred million light-years where galaxies and clusters of galaxies are as plentiful as the stars and star clusters were back home under pristine skies. Because of their great distance from us, these galaxies are utterly invisible to the naked eye, but through the eyes of sensitive cameras and telescopes they become works of art. These are the most distant galaxies in which individual Cepheids have been found and finding them was the primary motivator for building a space telescope named after Hubble. In the time it took their photons of light to make the journey here and be captured by the robotic

Figure 10.26 A photograph of part of the Coma-Virgo supercluster shows innumerable galaxies. Virtually every fuzzy spot in this image is an entire galaxy of stars (Jim Misti, Misti Mountain Observatory).

Ansel Adams of astronomy, every single layer of sediment and stone around me was laid here in place, one grain at a time.

One billion (one thousand million) light-years away and our vista point spans the largest structures the Universe has ever known. At the pace we have reached, we can no longer be bothered to pay attention to individual stars, or their clusters; galaxies themselves are just the barest bits of fluff. We are in the realm of galactic superclusters.

We know their distance because we know their velocity. The reason this is true has its origin only a couple of hours' drive from here at Lowell Observatory in Flagstaff, Arizona. In 1912 astronomer Vesto Slipher found that nearly every galaxy he observed through his telescope was moving away from us.[12] Seventeen years later, Edwin Hubble put these velocities together with his distances and found a simple relation: the farther away a galaxy is, the faster it is moving away from us. One of the key missions of the Hubble Space Telescope in the 1990s was to find as many Cepheids as possible in nearby galaxies so astronomer could precisely measure the value of Hubble's Constant: the ratio between velocity and distance for galaxies. The result is that for every 156 miles per second (250,000 meters per second) that a galaxy's velocity increases, it is ten million light-years farther away. Now all we need do is measure the recessional velocity (also called

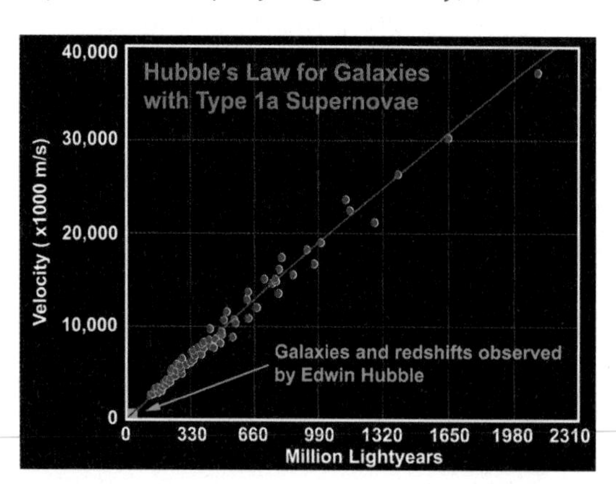

Figure 10.27 Hubble's Law showing that as galaxies get more distant, the velocity with which they are moving away from us increases. In this figure, the distance to galaxies is found using a special type of supernova (i.e., stellar explosion). If all Type Ia supernovae explosions are equally bright in reality, then we can tell how far away they and their host galaxies must be by comparing their standard brightness with how bright they look in our sky. The small blue square at lower left shows the range of distances and velocities over which Edwin Hubble originally used Cepheid variable stars to discover the expansion of the Universe (Modified from Robert P. Kirshner, 'Hubble's Diagram and Cosmic Expansion,' *Publications of the National Academy of Sciences*, vol. 101, no. 1, pp 8–13, Copyright (2004) National Academy of Sciences, USA).

[12] He determined this from passing each galaxy's light through a prism that split the light into its component wavelengths creating a spectrum. Elements within the stars that created the light produce well known absorption gaps at specific wavelengths (or colors). For galaxies moving away from us, these absorption bands are shifted to longer wavelengths (redder colors) as the light waves are spread out behind the galaxy: the faster the galaxy, the greater the redshift. Only a very few galaxies (including the Andromeda Galaxy) show a shift to blue wavelengths as the galaxy moves towards us bunching up the light waves in front of it.

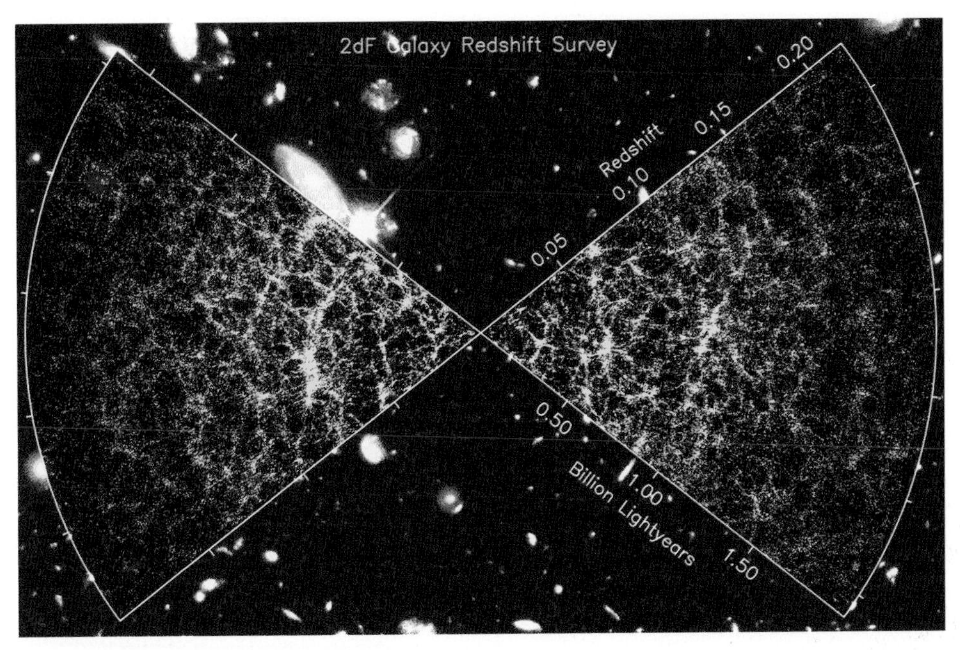

Figure 10.28 A slice through a three-dimensional map of the Universe made by applying Hubble's Law to the observed redshifts of 62,559 individual galaxies (yellow). On this scale great walls of galaxies are visible, along with numerous loops and voids. These are the largest structures ever observed in the Universe. The background image is from the Hubble Deep Field photo of a portion of the northern sky ((redshifts) The 2dF Galaxy Redshift Survey; (background) Robert Williams and the Hubble Deep Field Team (STScI) and NASA).

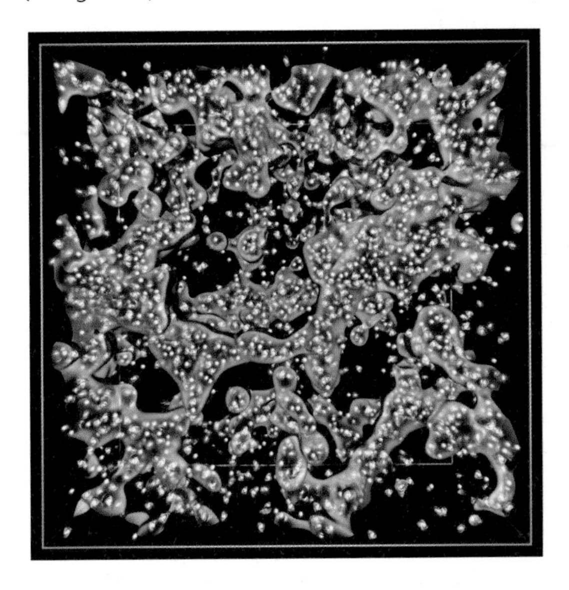

Figure 10.29 Illustration of the structure of galaxy clustering as revealed by the 2dF Galaxy Redshift Survey. Rather than being uniformly spread throughout the universe, galaxies (the little red and green blobs) are organized into structures and voids much like the cheese and holes in a block of Swiss cheese (The 2dF Galaxy Redshift Survey).

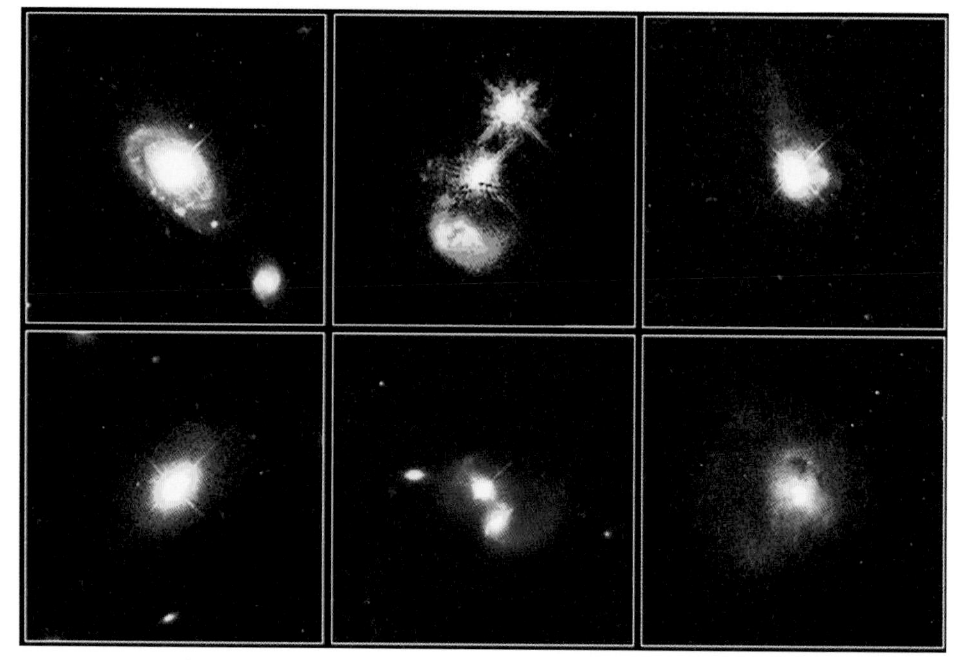

Figure 10.30 Hubble Space Telescope image of quasars and their faint host galaxies. Based on these images we now know that quasars are at the hearts of galaxies, including many near the edge of the visible Universe and thus far back in time when the Universe was younger. This image shows relatively nearby quasars at about 1.5 billion light-years distant (J. Bahcall (Institute for Advanced Study), M. Disney (University of Wales) and NASA).

the redshift) of any galaxy (no matter how small or far away) and with this relation we know its distance and can begin to map the Universe.

With the tools of Hubble's Law, we today survey vast groupings of galaxy clusters that trace out giant gossamer webs, loops, bridges, bubbles and voids. Here, on the very largest scale we find nature at work stringing together massive walls of galaxies nearly a billion light-years across. In the time it takes the light from the brightest stars in galaxies at one end to reach any potential alien astronomer on a planet around a star in a galaxy at the other end, the original stars have all long ago disappeared. To see these galaxies and the space they inhabit is to see the Universe as it looked almost ten percent of its age back in time. To sense how different these conditions could be, consider that a billion years ago on Earth, ancient blue-green algae stromatolites (of which there are fossils here within the canyon) were only just converting our planet's air into an oxygen atmosphere that later land animals could breathe. Were we able to travel back in time to the age of the universal structures we see in our telescopes, we would need a spacesuit to survive on our own planet.

Step again ten times farther away and we see a time before the Earth itself even formed. At ten billion light-years, 10^{26} m away we come to the realm of the most distant observable objects in our Universe. Here, at the centers of new-formed galaxies, supermassive black holes suck down enormous quantities of interstellar gas. The matter that spirals in towards the black hole's event horizon is heated to such high temperatures that it releases ten trillion times the energy of our Sun, rendering each ravenous monster a hundred times more luminous than an entire galaxy. For decades, these beacons were only visible across the intervening light-years as bright mysterious star-like points until the Hubble Space Telescope first detected the faint fuzzy outline of the surrounding galaxy. When we look at the inner workings of these quasars (a named derived from quasi-stellar object) we see how our own Galaxy looked soon after its birth nearly thirteen billion years ago.[13]

Towards these galaxies we aim our newest, most powerful telescopes in hopes of seeing the first generation of stars to ever turn on, and in time, produce the chemical elements we see around us today. The silicon in the rocks, the carbon in my bones, the oxygen in the air I breathe and in the water that carved this canyon were all made in the hearts of stars long before the Earth was even here. Only when those stars exploded and died, spreading their interior constituents throughout the Galaxy could later stars form from gasses enriched in these elements.

Here beyond the realm of the quasars our tour approaches an end. We are 10 billion light-years from home, 10^{15} astronomical units, 10^{26} meters, 10^{26} units of you: only 26 steps farther out into the Universe than the Earth at your feet. Beyond the last quasar is the blackness of space, and so perhaps we should say that on a moonless night in the Grand Canyon, or wherever you may be, the darkness itself is the most distant thing you can see for yourself.

This very blackness of space tells us something about our Universe. Imagine space extended infinitely far in all directions and had existed in its present form for infinite amounts of time with no creation, no change, and no end. This is, more or less, what astronomers thought the Universe to be before the early part of the twentieth century. This view, however, gives rise to an interesting paradox as made famous by the nineteenth century astronomer Heinrich Olbers (although he was not the first to pose it). If the Universe and all the stars and galaxies in it extend infinitely far in every direction – and (because light travels at a finite speed) if the Universe and all that's in it have been around for infinitely long – then no matter what direction we look in space, starlight should reach our eyes. In other words, every direction you look in such a Universe your eyesight should eventually come to rest on the surface of a star, and therefore the entire sky should appear as bright as the surface of the Sun.

To see how this is so, consider standing inside a forest. If the forest extends far enough in every direction then everywhere you look around, your line of sight eventually comes to rest on a tree. Only if the forest is small will you be able to see the empty meadow beyond your grove.

13 The supermassive black hole at the center of our own Galaxy has quieted down in its middle-age as all the available gas was long ago cleared away.

Figure 10.31 This Hubble Space Telescope Ultra-Deep Field image captures the light from 10,000 galaxies and is the deepest visible-light image ever made of our Universe. While a few stars are visible in this image (the bright objects possessing cross-shaped diffraction spikes) nearly every dot and fuzzy spot of light is a galaxy, the very faintest seen as little as 800 million years after the Big Bang. The entire picture spans an area one tenth that of the full Moon and was made from 800 exposures, over the course of 400 Hubble orbits around Earth, totaling 11.3 days of exposure time (NASA, ESA, S. Beckwith (STScI) and the HUDF Team).

Figure 10.32 Overlapping aspen trees in a forest near the South Rim of the Grand Canyon. If a forest is large enough, soon every direction you look will end on a tree. If the Universe is infinitely large and unchanging with time, then every sight-line into the sky should similarly end on a star. In such a Universe the sky at night would be as bright as the Sun. The fact that it is not means one of those two assumptions must be wrong (T. Nordgren).

How then can the sky be dark at night? There are only two possibilities. One possibility is that space is infinite but for some reason stars and galaxies inhabit only a small part of it. The problem is that since we see equal numbers of galaxies in every direction (on average) this option requires that we be located at the center of this small collection of galaxies in a vastly larger and empty Universe. Such a solution sounds ominously similar to our now long gone conviction that the Earth was the center of the Universe. Perhaps this is true, but the odds of it being so are vanishingly small.

The second solution is that the Universe is not static or unchanging. The reason every line of sight does not fall on the surface of a star is that there are many lines of sight that extend to a distance (meaning a time) when stars have not yet formed. In essence, they are directions for which starlight has not yet reached our eyes. It is this second solution that provides the reason we now understand for why the sky is dark at night. The evidence for this is provided by what Slipher saw, namely that nearly every galaxy is rushing away from us.

Did the Milky Way create a celestial faux pas? Are we at the exact center of a massive intergalactic explosion? The answer is neither, but rather that our Universe is expanding and as it does so it carries all the galaxies along with it. Imagine our Universe is the rubber fabric of a balloon. Every galaxy we know of, including our own, is a dot placed on the surface of this balloon. As we inflate the balloon every dot is pulled away from every other dot as the surface of the balloon expands. For any inhabitant of any one dot, it appears that every other dot is moving away. The farther away the two dots initially, the more fabric between them to expand and the faster the dots seem to diverge. The natural consequence of Slipher's redshifts and Hubble's velocity and distance relation is that we inhabit a Universe that must be expanding. Nor are we at its center, either; we are simply one amongst many billions of galaxies that are all being carried away from one another as space expands.[14]

The natural consequence of a Universe that's getting bigger is that in the past the Universe was smaller. For the first time scientific evidence pointed to a moment of creation. By its very motion, the Universe reveals that it was not eternal; while every night it is verified when the Sun goes down and we see the stars come out to shine in a dark night sky.

If we were to attempt to step another factor of ten farther away, we would come to a time before there were stars, gas, galaxies or quasars. In fact it would also be a time before the Universe itself, including the very idea of space and time. From the speed with which the galaxies are speeding away from us we

[14] For a universe that only exists on the surface of a balloon, imagine an inhabitant of one dot asking where the center of the Universe is found, or which galactic dot is closest to the center. The answer is no dot, or alternatively every one. The center of the balloon universe is at the center of the balloon itself and not on the surface at all. The same situation holds true for our universe as well.

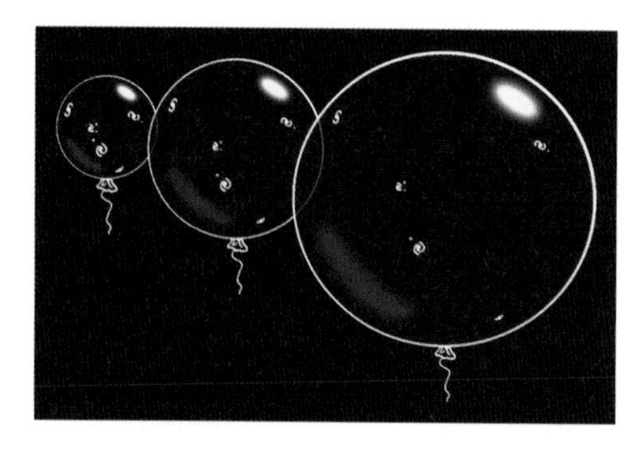

Figure 10.33 Galaxies on the surface of an expanding balloon. As the balloon gets bigger the distance between every galaxy gets larger. The farther apart two galaxies, the faster they are carried away from one another. This is precisely what we see in the Universe (T. Nordgren).

calculate that in order to be as distant as they now are, everything in the Universe, including the Universe itself, began to expand from a single point in space and time 13.7 billion years ago.[15]

Everywhere we look we see the darkness before the first stars let there be light. But all is not invisible. In the first moments of creation, everything in the Universe that would one day become matter and energy filled all available space with the heat of its creation. For 13.7 billion years the fires of the Big Bang have been left to cool as the space it permeates expands with the years. Today its faint glow is everywhere visible beyond the galaxies, and is our window back into the deepest of time.

This cosmic background radiation was first found by accident, by two scientists working at Bell Laboratories in New Jersey. In 1963, Arno Penzias and Robert Wilson were experimenting with technology that would eventually facilitate the world-wide use of cell-phones, but everywhere they pointed their receiver there was a background signal, a microwave hiss, for which they could find no known cause. At the same time, Robert Dicke, an astrophysicist at nearby Princeton University, had theorized that the still controversial Big Bang Theory would produce just such a background microwave hiss. He and his team of physicist were in the process of building a receiver to detect it when he received a phone call from the scientists at Bell Labs asking for his advice on what they had found. After sharing his hypothesis, he hung up the phone, and told his team, "Gentleman, we've been scooped." Without knowing it, Arno Penzias and Robert Wilson had discovered the final conclusive evidence for the Universe's

[15] Which raises an interesting question: If the space between galaxies is constantly expanding then are those quasars whose light we see really about 10 billion light-years away right now? The short answer is no. The light we see has been traveling for about 10 billion years, but in that time the quasars have continued to expand away from us. The most distant quasar currently known has a redshift of $z = 5.8$ (z is a number proportional to the shift in wavelength caused by its motion away from us), implying its light has been traveling to us for 12.7 billion years. Instead of being 12.7 billion light-years away, Hubble's Law implies that the universe's expansion has carried it to a current distance of 27 billion light-years. Rounded to the nearest power of ten though, it is still at our current step in our order of magnitude walk.

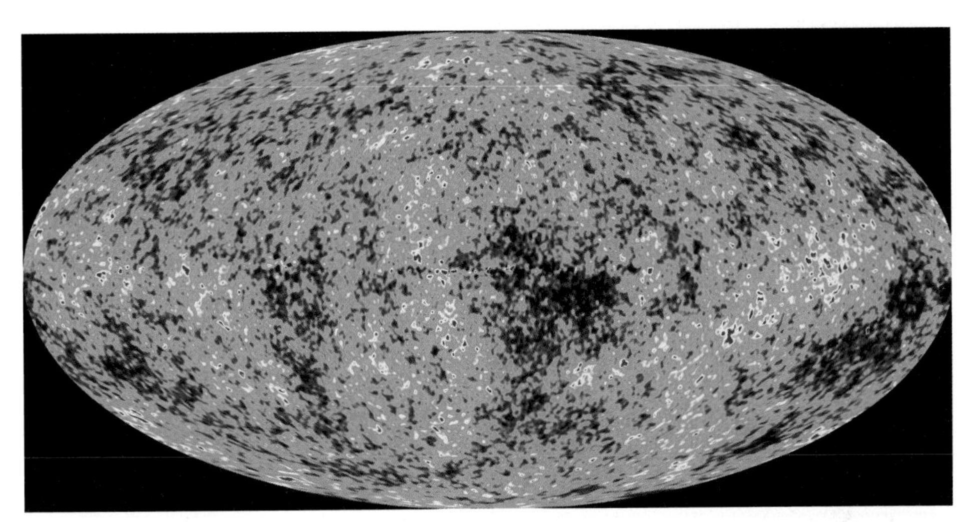

Figure 10.34 This is a map of the microwave (or radio) radiation emitted by the Universe as a faint glow left over from the explosive event of the Big Bang 13.7 billion years ago. If this map of the Universe (think of it like a flattened map of the sphere of the sky) were completely uniform, then no one region of the universe would be any different from any other and astrophysicists would have a hard time explaining why we live in a lumpy universe where there are stars, galaxies, superclusters, voids, and everything else you see around you. However, this map made by the WMAP spacecraft in 2003 shows tiny temperature fluctuations on the order of only plus or minus 200 microKelvin, roughly one ten thousandth of a degree (NASA/WMAP Science Team).

birth. In 1978 Penzias and Wilson were awarded the Nobel Prize in Physics for the paper they wrote about their serendipitous discovery; it was only two pages long.

Twenty five years later, NASA launched the Wilkinson Microwave Anisotropy Probe (WMAP) to map the miniscule temperature fluctuations within this background radiation. If there were no temperature fluctuations, meaning the Big Bang was utterly uniform, then there would be no way to explain the structure we see in the Universe today. But where temperatures fluctuated to just a fraction of a degree cooler in the primordial fireball, matter would one day move a little more slowly and gravity would take hold to begin the long, slow process of building the first galaxies, clusters, loops, and walls seen in our large-scale maps of the Universe. In WMAP's kaleidoscopic structure is the faint shadow of everything the Universe would become and everything we could subsequently see in the sky at night in this grandest of canyons.

Nearly fourteen billion years later (100 years ago) President Teddy Roosevelt dedicated Grand Canyon as a national monument. His actions set in motion a tradition of conservation that protects our view of the night sky above just as surely as it protects our views of the canyon below. While this possibility almost certainly never occurred to him at the time, he would have viewed it with pleasure had he known as he had a profound love of nature and knowledge. During his life, Roosevelt and his good friend, the naturalist William Beebe, had a "salutary ceremony" they'd take part in on crystal-clear evenings like the one I

Figure 10.35 President Theodore Roosevelt (lower left in dark suit and hat), 26th President of the United States, descends along the Bright Angel Trail into the depths of the Grand Canyon. In 1908 Roosevelt established by proclamation that the canyon be preserved as a national monument. In 1919, just three years after the formation of the National Park Service, Grand Canyon was declared a national park (Grand Canyon National Park Museum Collection, Image No. 05556).

see tonight. In his foreword to Roosevelt's book, *Book-Lover's Holiday in the Open*, Beebe wrote:

> After an evening of talk, perhaps about the fringes of knowledge, or some new possibility of climbing inside the minds and senses of animals, we would go out on the lawn, where we took turns at an amusing little astronomical rite. We searched until we found, with or without glasses [i.e., with or without binoculars or telescope], the faint, heavenly spot of light-mist beyond the lower left-hand corner of the Great Square of Pegasus, when one or the other of us would then recite:
>
> That is the spiral galaxy in Andromeda.
> It is as large as our Milky Way.
> It is one of a hundred million galaxies.
> It is 750,000 light-years away.[16]
> It consists of one hundred billion suns,
> each larger than our own sun.

After an interval Colonel Roosevelt would grin at me and say: "Now I think we are all small enough! Let's go to bed."

[16] Since Roosevelt's time the Cepheid period-luminosity relation has better refined Andromeda's distance to the currently accepted 2.5 million light-years.

Nearly a hundred years later, after a trip of 14 billion light-years that began here in the heart of the canyon he did so much to protect, so did we.

The next morning as I make my way back up the long trail out of the canyon I develop a whole new appreciation for the slow progress of centuries. While my feet tell me there is no way I can possibly make it, my brain knows that as long as I keep placing one foot in front of the next, I will get there eventually. But the effort is worth it for seeing the magnificence of sunrise from along the Tonto Platform at the rim of Grand Canyon's inner gorge. I remember the very first time I came this way, nearly a decade ago; afterwards I felt as if I had just taken the most important walk of my life and that everywhere else I would go would pale in significance. It's then that I'm reminded of the words of a professor of mine in graduate school who said when studying the clustering of some of the most distant galaxies in the Universe, "Humanity has no higher calling than science. To seek answers to such fundamental questions such as 'Where do we come from?' and 'How did we get here?' is the most important of pursuits." Such ends are always worth the effort.

Figure 10.36 The fuzzy light of the Andromeda Galaxy (lower left) shines down into the Grand Canyon on a beautifully starry sky. The light we see took 2.5 million years to reach us (T. Nordgren).

It's then that I am caught in my reverie on the trail by a group of fellow hikers who were in attendance at last night's talk. We're all of us eager for a reason to pause and rest (and perhaps prolong our time in this beautiful place) and so we stop and chat for a moment. We talk about what we saw last night (the Milky Way never fails to astound) and each has a question, like "Why is there matter?" or "What created the Big Bang?"

As a result of our conversation and my previous thoughts, I confess that with all due respect to President Roosevelt, the sky at night makes me feel far from small. Of all those stars and galaxies spread across 13.7 billion years of time (and even more light-years of space) there is not a single star that has the ability to appreciate its place in the cosmos. No spiral galaxy marvels at the amazing variety of astrophysical phenomena that produce it. No cloud of gas is ever left speechless by the thought of what it will someday form. No single atom of

Figure 10.37 A celestial landscape is captured in Hubble's cameras as powerful ultraviolet light from hot stars above slowly erodes away the dark clouds of cool gas and dust below. Within the protective dark layers, new stars and their potential planetary systems are born over slow spans of time. Will there someday be other forms of life here, able to look up at the stars and wonder about their own place in an ever expanding Universe? (NASA, ESA, and The Hubble Heritage Team (STScI/AURA), N. Smith (University of California, Berkeley)).

hydrogen has the capacity to appreciate that given nothing but the laws of physics and nearly 14 billion years of time it will someday find its way into the brain of a being able to contemplate the precious beauty of its own existence.

Stars create no art; nebulae are driven by no curiosity. In the words of the Hopi creation tale: we alone have the ability to seek the meaning of life. Astrophysics tells us how it is that we have come to be here, but it is only everything else that makes us more precious than stars that lets us draw meaning behind why we are.

See for yourself: powers of 10 through the Universe

Starting with the planet at your feet, there is an astronomical object visible at nearly every step of a "Power of 10" tour through the Universe. Rounding off to the nearest whole power of 10 (e.g., 1, 10, 100, etc) we will use the meter as our unit of measure (which has the benefit of being person-sized to the nearest power of 10). Many of these objects are located on the following monthly star maps. However, since many of these objects have already been described in earlier chapters, where appropriate, I've directed you to read the 'See for yourself' (SfY) section of the specific chapter.

10^0 m = 1m (you): See the Earth at your feet; it's the third planet from our Sun (no star map provided). Actual average height of person: 65 inches (1.6 m).

10^1 m = 10 m: The Phantom Ranch amphitheater, or the width of a scenic overlook for those on the canyon rim, (or the width of your own back yard).

10^2 m = 100 m: The width of the Colorado River, the thickness of the Kaibab formation, (or the length of your own street back home).

10^3 m = 1,000 m: The height of the Grand Canyon (or the length of your neighborhood or town). Average canyon height: 5,000 feet (1,500 m).

10^4 m = 10,000 m: The width of the Grand Canyon (or the width of your city). Typical canyon width: 4 – 18 miles (7 – 30 km).

10^5 m = 100,000 m: The altitude of meteors and the internationally defined edge of space: 60 miles (100 km).

10^6 m = 1,000,000 m: Satellites in Earth orbit, the International Space Station, Space Shuttle or Hubble Space Telescope. Consult Heavens-Above.com for locations and times to see them pass overhead. Average altitude of the Hubble Space Telescope: 360 miles (570 km).

10^7 m = 10,000,000 m: Geosynchronous satellites. While you can't see these directly, you *can* see the satellite dishes that some homes have pointed at these satellites that always stay over one spot on Earth. Actual distance: 22,500 miles (36,000 km).

10^8 m = 100,000,000 m: The Earth's natural satellite: the Moon. See the SfY section of Chapter 7 for identifying features on the Moon. Distance: 225,000 miles (360,000 km).

10^9 m = 1 billion meters: Nothing.

10^{10} m = 10 billion meters: Nothing; although if you were to make a tower of every person on Earth standing on the shoulders of every other, this is as far as it would reach: 7 million miles (12 million kilometers).

10^{11} m = 1 Astronomical Unit (AU): The nearest star: our Sun. Also, the inner solar system planets: Mercury, Venus, Mars. See SfY Chapter 5 for times and locations to see Mars at its closest. Distance to the Sun (and thus distance of 1 AU): 93 million miles (150 million kilometers).

10^{12} m = 10 AU: The outer Solar System giant planets such as Jupiter and Saturn. Saturn is the most distant object you can see in our Solar System with the naked eye. See SfY Chapters 3 and 4 for sky maps showing Jupiter and Saturn's position relative to the background stars each year until 2030. Saturn's actual distance at closest approach: 9 AU.

10^{13} m = 100AU: Pluto and the Kuiper Belt: While Pluto is only faintly visible through amateur telescopes in dark locations, we can see remnants of the icy worlds that orbit within the Kuiper Belt every year during meteor showers. Meteor showers occur when the Earth passes through the dusty remains of comets. Many of these comets (Halley's Comet most famously) have orbits that take them out into the Kuiper Belt. See SfY Chapter 7 for a table of bright annual meteor showers. In arriving at 100 AU we pass fully through the Kuiper Belt that extends between 30 and 50 AU from the Sun.

10^{14} m = 1000 AU: Sedna is an ice world not visible to the naked eye or amateur

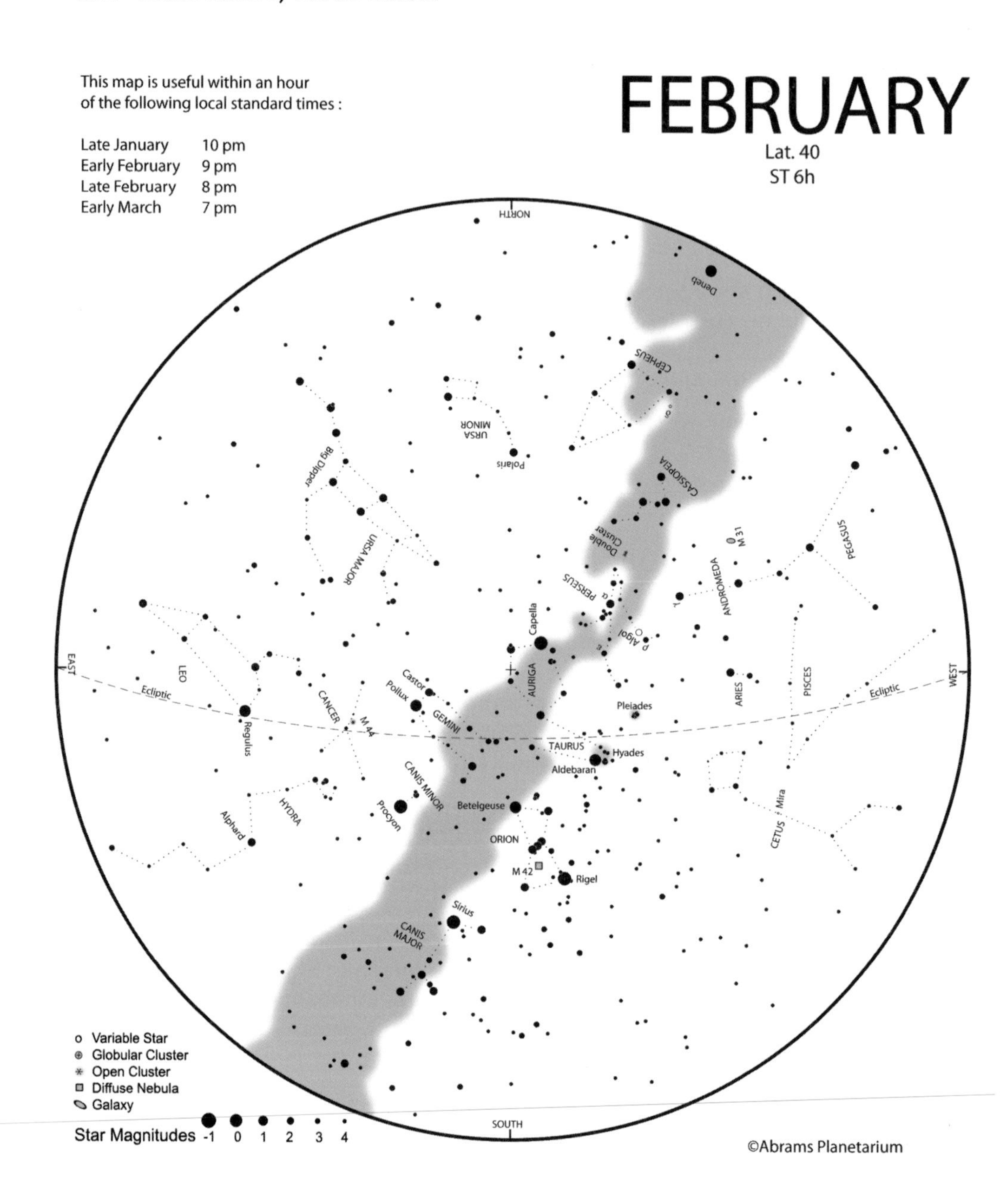

Hold the star map above your head with the top of the page pointing north. For those at mid-latitudes within the continental United States, the center of the map marked with a + will show the view directly overhead (the zenith) at the indicated times.

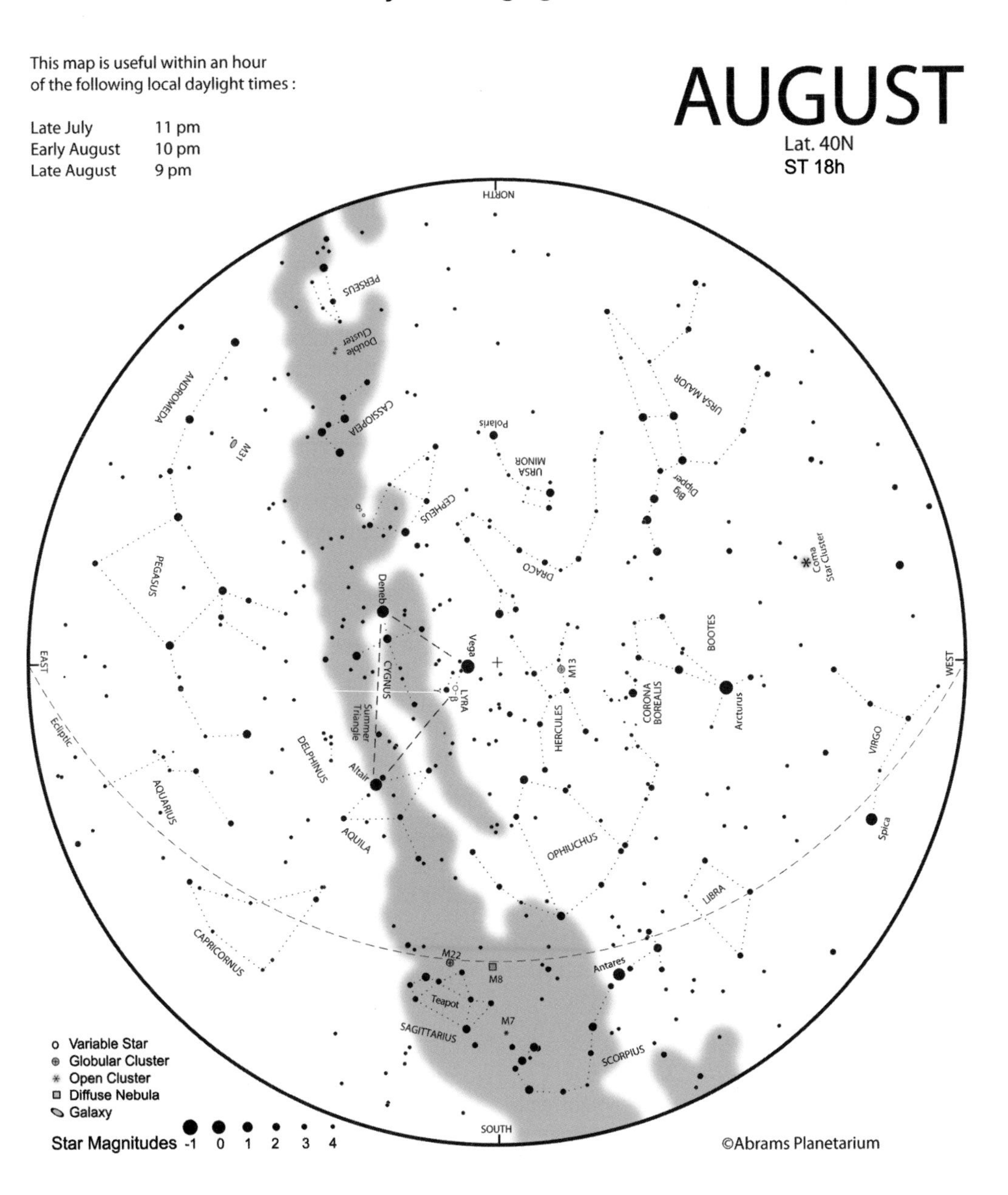
This map is useful within an hour
of the following local daylight times :
Late July 11 pm
Early August 10 pm
Late August 9 pm
AUGUST
Lat. 40N
ST 18h
NORTH
SOUTH
EAST
WEST
PERSEUS
Double Cluster
ANDROMEDA
CASSIOPEIA
M31
Polaris
URSA MINOR
URSA MAJOR
Big Dipper
CEPHEUS
DRACO
PEGASUS
Coma Star Cluster
BOOTES
Deneb
Vega
CYGNUS
LYRA
M13
HERCULES
CORONA BOREALIS
Arcturus
Summer Triangle
DELPHINUS
Altair
AQUILA
AQUARIUS
Ecliptic
OPHIUCHUS
VIRGO
Spica
LIBRA
CAPRICORNUS
M22
M8
Antares
Teapot
M7
SAGITTARIUS
SCORPIUS
Variable Star
Globular Cluster
Open Cluster
Diffuse Nebula
Galaxy
Star Magnitudes -1 0 1 2 3 4
©Abrams Planetarium

telescopes because of how faint the Sun is out at these distances. Its orbit carries it between 76 and 1000 AU.

10^{15} m = 10,000 AU: The inner edge of the Oort Cloud. Again we cannot see the icy snowballs that orbit in the Oort Cloud, but every so often when one of these frozen nuclei comes plunging into the inner solar system it creates a spectacular comet. Unlike Halley's Comet that comes around every 76 years, Oort Cloud comets are long period comets that are seen only once then never again over a human lifetime. Comet Hale Bopp in 1997 was such a visitor.

10^{16} m = 100,000 AU: This is the outer edge of the Oort Cloud and the boundary to our Solar System. It is also approximately the distance light travels in a year. One light-year (1 LY) = 63,000 AU.

10^{17} m = 10 LY: Sirius and Vega are two of the nearest bright stars easily visible in the northern hemisphere sky. In winter, Sirius is the brightest star in the sky (not including any planets that may be out) and is found by tracing the three stars of Orion's Belt down and to the east. In summer and fall, Vega is one of the three stars of the prominent Summer Triangle that will be high overhead. Actual distance to Sirius is 8.6 LY, actual distance to Vega is 25 LY.

10^{18} m = 100 LY: Aldebaran and Spica are two very bright and massive stars. Aldebaran is a red giant, noticeably orange in color and visible as one eye of Taurus the Bull in winter. Spica is a hot blue star much more massive than the Sun. To find it in summer, follow the handle of the Big Dipper and 'arc to Arcturus.' From Arcturus, continue on in the same direction and "spike to Spica." Aldeberan is actually 65 LY away, while Spica is 260 LY away.

10^{19} m = 1,000 LY: Gaseous nebulae like the Orion Nebula (also called M42) in winter (see SfY Chapter 8) or the Lagoon Nebula (called M8) in summer (see SfY Chapter 1). M42 is 1,300 LY away, M8 is 4,300 LW away.

10^{20} m = 10,000 LY: The band of the Milky Way. In summer we see the center of our Galaxy towards the constellation of Sagittarius (SfY Chapter 1). The supermassive black hole at the center of the Galaxy is 25,000 LY away.

10^{21} m = 100,000 LY: The Magellanic Clouds are satellite galaxies of our own Galaxy, the Milky Way. They appear as two clouds cut off from the band of the Milky Way, but they are individual galaxies in orbit around our own. Unfortunately they are only visible from the southern hemisphere. They are about 160,000 LY away.

10^{22} m = 1,000,000 LY = 10,000,000,000,000,000,000,000 you: The Andromeda Galaxy is a spiral galaxy that is a near twin to our own Milky Way. It is typically the most distant object in the Universe that you can see with your own eyes. It is visible all year long except for spring. For a photograph showing the position of this faint fuzzy oval in the sky see Figure 9.21 in Chapter 9. The Andromeda Galaxy is 2.5 million light-years away.

Further reading

The Fourth World of the Hopi by Harold Courlander (1971)
University of New Mexico Press, ISBN 0826310117

Roadside Geology of Arizona by Halka Chronic (1983)
Mountain Press, ISBN 0878421475

Reading the Rocks: The Autobiography of the Earth by Marcia Bjornerud (2005)
Westview Press, ISBN 081334249X

The Extravagant Universe: Exploding Stars, Dark Energy, And the Accelerating Cosmos by Robert P. Kershner (2005)
Princeton University Press, ISBN 0691058628

The Starry Room: Naked Eye Astronomy in the Infinite Universe by Fred Schaaf (1998)
Dover Publications, ISBN 0486425533

Hubble Heritage Site: The Hubble Space Telescope's photographic legacy
http://heritage.stsci.edu/

Starry sky national park (T. Nordgren)

11 Starry sky national park

The night sky is the world's largest national park with its beauty available to anyone who steps outside and looks up.

Geoff Chester, U.S. Naval Observatory

Where did we come from? How did we get here? What is out there? Are we alone in the Universe? These are the questions astronomy has sought to answer since the first person looked out into the night sky and wondered about the heavens. For most of human history the points of light in the sky, were just that: points of light. The only thing they had to tell us was contained entirely within their motions across the celestial sphere; astrology flourished.

Four hundred years ago, the telescope finally revealed that the planets, those special stars that wandered among the 'fixed' stars, were worlds potentially like our own. With Galileo's telescope we brought the heavens down to Earth. In the last fifty years we've achieved another milestone in our search for these cosmic answers by taking the Earth to the heavens in the form of human and robotic exploration of the moons and planets of our Solar System.

Yet with all of these advances, the answer to one of those fundamental questions is that everywhere we have

Figure 11.1 Here the Hubble Space Telescope peers into the heart of the Whirlpool Galaxy. Every dark dust lane along a spiral arm is where new stars are forming. Every red splotch is a glowing nebula like we see in Orion where baby stars are just turning on. Every blue cloud is a host of thousands of hot young stars potentially warming planetary systems of their own. In this one picture, how many millions of planets might right now be teeming with life looking back at us under a star filled sky? (NASA and The Hubble Heritage Team (STScI/AURA), N. Scoville (Caltech) and T. Rector (NOAO)).

Figure 11.2 The Galileo spacecraft captures a glimpse of South America and ice-covered Antarctica as it swings past the Earth on its way out to Jupiter (NASA/JPL).

looked we have found no sign of anyone else. But to paraphrase Shakespeare, is this a failure of the stars or of ourselves?

In December 1990 NASA's Galileo spacecraft flew by the Earth en route to explore the largest planet in our Solar System. The spacecraft had been launched from Earth a little over a year before, but without the power to reach Jupiter directly it had been sent to pick up additional speed by slingshotting first around Venus and then past the Earth. At Jupiter it would begin an ambitious program to learn about the king of planets and its host of strange, exotic moons. While previous spacecraft had merely flown by the planet on their way out of the Solar System (learning what they could in the few days it took to pass by) Galileo was a mission designed to finally answer questions about the Jovian system that earlier missions had only begun to ask.

One such question was the presence of life, if not in Jupiter's clouds, then perhaps on one of its moons. And while hardly any astronomers actually expected to find life floating amongst the Jovian clouds (although Carl Sagan and my old advisor, Ed Salpeter, had published a paper in 1976 describing what such life could be like) it was important to know if any failure to find it in the Jovian system, was due to Jupiter, or to our spacecraft. Fortunately, the Earth was one planet where Sagan was pretty sure life was plentiful. If Galileo failed to find life here, he reasoned, any failure to find it at Jupiter would be just as meaningless. So, as the Galileo spacecraft flew by the Earth, mission planners turned on Galileo's instruments, pointed them home, and asked the question, "Is there life on Earth?"

From 600 miles (960 km) above the day-lit Earth, Galileo showed our technology is all but invisible. Galileo's cameras saw no highways or cities. Everything we've ever built, every change we've made to our landscape, is utterly invisible from space. The rumor that the Great Wall of China is the only human engineering feet visible from the Moon is nothing but myth. What Sagan found was that to a passing planetary probe, we – as in all life on Earth – are revealed not by our engineering, but by our chemistry.

The Earth's infrared light was first aimed through a spectrometer that spread it apart into its component wavelengths. It showed the spectral signature of water, in one form or another, over nearly the entire planet's surface. Spectra of an apparently cloudless, featureless region of this water showed absorption of specific wavelengths of light due to extremely high amounts of oxygen and methane. Organic chemistry says that left alone, these two molecules will quickly react with one another to turn into water and carbon dioxide. The presence of methane in amounts 140 orders of magnitude greater than what chemistry says is possible means that some process must be occurring on Earth to continually replenish it. In fact, some process must also be replenishing the planet's oxygen, since nearly all of it should be removed from the atmosphere through reaction with the surface (for example, oxygen will react with iron in rocks to form rust). In short: Galileo scientists found the Earth has an atmosphere that shouldn't exist by the laws of inorganic chemistry.

Next, Galileo's camera captured images of Earth's day-lit side showing a range of light and dark regions not covered by water. Scientists called these regions 'land.' The bright regions of land reflect light all across the spectrum and so appear brown or reddish-tan. The dark regions, however, absorb relatively large amounts of red light and so would appear predominantly green. While many types of minerals are known that match the range of colors revealed in the bright regions, there are no known minerals that produce the range of colors seen in the green areas.

In their 1993 paper published in the prestigious journal *Nature*, Sagan and his collaborators concluded:

> The identification of molecules profoundly out of thermodynamic equilibrium, unexplained by any non-biological process; widespread pigments that cannot be understood by geochemical processes are together evidence of life on Earth without any *a priori* assumptions about its chemistry [i.e., without regards to "life as we know it"].

Galileo had found life on Earth, yet no such chemistry out of balance was found upon reaching Jupiter or any of its moons. Once again, we seemed to be alone; the discovery of any life there would evidently have to wait for a closer look, perhaps beneath the frozen oceans of one of its moons.

Seventeen years later another spacecraft passed by Earth, this time on its way to a comet. In 2014, the European Space Agency's Rosetta spacecraft will be the first to orbit a comet's nucleus and gently place a lander on its surface. If all goes as planned, scientists will be able to look closely at what is thought to be an example of the leftover remains of our Solar System's formation.[1]

And just as Galileo had done in 1990, Rosetta too turned on its instruments as it passed by the Earth and this time anyone asking if there was life here had a much easier time finding an answer. For Rosetta flew through the shadow of the Earth, and photographed our world at night.

[1] Four and a half billion years ago, comets were the building blocks of planets and possibly the origin of organic molecules (even nucleic acids) on Earth. Just as the Rosetta stone in Egypt allowed archeologists to piece together the language of a long vanished ancient civilization, the Rosetta spacecraft will allow astronomers to piece together the long vanished past of our own world and perhaps the ancient origin of life thereon.

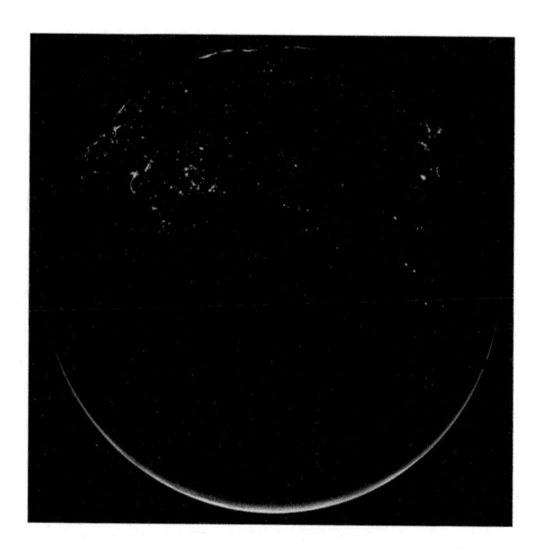

Figure 11.3 In 2007 the European Space Agency's Rosetta spacecraft passed by the Earth on its way to rendezvous with a comet. Visible in this photograph are the lights of our cities and civilization from post-sunset Europe in the upper left all the way to pre-dawn China and Japan in the upper right. In the very center, the lights of India outline the shape of the subcontinent while towards the west, the bright sinuous light of the Nile River snakes southward from the Mediterranean Sea (ESA ©2005 MPS for OSIRIS Team MPS/UPD/LAM/IAA/RSSD/INTA/UPM/DASP/IDA).

There in its cameras at a distance of 50,000 miles (80,000 km) above the Indian Ocean it caught sparkling lights from Europe to Africa and India to Asia. In Rosetta's photo of Earth, we see early evening's street lights turn on in England and along Italy's Adriatic coast. Along the life-giving waters of the Nile, night brings the lights of civilization huddled against the desert. To the east, midnight comes to the lights of Delhi and the outlines of the Indian subcontinent. Farther east, Rosetta's camera captures the early morning lights of Tokyo and Taiwan, still shining amid the oncoming glow of dawn.

The world at night shows the unmistakable sign of intelligent life on every hemisphere on Earth. No forest fire burns night after night with such regularity. No volcanic activity glows with such a constant light, never flowing or cooling. From coast to coast on every continent, the signs of our habitation announce themselves to the stars every night once the Sun goes down. From above, ours is a world of light where darkness is more the exception than the rule.

It's no wonder then that visitors from all over the world marvel at what can be seen in the night sky of America's national parks, oases of darkness, far from the bright lights of home. Those qualities that draw us to the parks by day – their unspoiled scenic vistas and backcountry wilderness – also make them especially beautiful at night. And while it may be the strange beauty of arches, canyons, and waterfalls that cause visitors to gaze in wonder by day, the bright shining stars and spectacular majesty of the Milky Way are equally strange and wonderful sights that captivate at night. To see all these sights together – the red rock arches illuminated by the pale milky glow of our Galaxy – places our planet and ourselves into the rich context of our Universe, and it is a sight no longer possible from much of our developed world.

Many parks, where televisions and movie theaters are rare, offer evening programs to educate us about these strange other-worldly sights. Over the last several years, Park Rangers have come to realize that their night sky programs are some of the most heavily attended of all the programs the park offers. At Grand

Figure 11.4 The Milky Way rises over Mesa Arch in Canyonlands National Park. The landscape is illuminated by nothing but starlight. In the distance the glow of several small cities are visible over the horizon (T. Nordgren).

Canyon National Park hundreds of visitors show up for each evening's ranger-led night sky walk along the canyon rim. To meet this demand, every ranger at Grand Canyon is now expected to be able to give an evening astronomy talk in addition to a more typical talk on the geological wonders of the canyon.

A number of these parks now partner with local astronomy clubs to set up telescopes on moonless summer nights when the sky is dark and the Milky Way stretches from horizon to horizon. In Joshua Tree National Park outside Los Angeles, you can find the local Andromeda Society with telescopes set up at the Hidden Valley picnic area one Saturday a month when there is no Moon in the sky to compete with the fainter light of our Galaxy. In Rocky Mountain National Park, the Estes Valley Astronomical Society mounts telescopes amid the park's aspens and elk of Upper Beaver Meadows on moonless Friday and Saturday nights during summer. Yosemite National Park in Northern California has so many astronomy clubs eager to set up telescopes on the dark heights of Glacier Point that local clubs from all over San Francisco and the Central Valley take turns pointing out the celestial sights to visitors on summer weekends.

Bryce Canyon National Park sits in the heart of southern Utah's starry sky and geologic wonderland and hosts the most organized night sky interpretation program of any national park. Permanent and seasonal rangers give multiple astronomy talks, often two a night on topics ranging from the beauty of a dark starry sky to black holes and general relativity. When the talks are done, and the sky is at its darkest, the audience is led over to a 'telescope field' behind the visitor center where volunteers operate multiple telescopes until the last family goes back to their campsite for the night.

Richard Blake is a mustachioed, grey-haired gentleman from Texarkana, Texas, who now travels the country from park to park volunteering his time and telescopes to show crowds the beauty of the night sky. Blake's deep bass, radio

Figure 11.5 National Park Service volunteer Amy Sayle reveals a close up view of our nearest star, the Sun, to visitors to Bryce Canyon National Park (T. Nordgren).

Figure 11.6 A field of telescopes set up by the Salt Lake Astronomical Society provides an unparalleled view of the cosmos to those attending Bryce Canyon's 8th Annual Astronomy Festival (T. Nordgren).

announcer's voice carries a hint of his Texas past as he describes star clusters and nebulae to every young child, European couple, or retiree on vacation. His knowledge of telescopes is encyclopedic and it is amazing how many conversations we have that eventually turn without notice to the wonders of this optical system or that expensive eyepiece. But while his technical knowledge is dizzying, his is a calm deep voice that has opened many a junior Park Ranger's eyes to the delights of Saturn and its rings.

This year I join Blake and Jim Closson, another old-time astronomy volunteer, in bringing our expertise to the public at Bryce. Together we talk about our favorite parks and night skies as we tell stories of the places we've been and the people we've known, marveling at how many places and friends we have in common as our paths have criss-crossed and barely missed one another over the years. This year we are all brought together to help the rangers with the 8th Annual Bryce Canyon Astronomy Festival.

What began in 2001 as a 'star party,' primarily for amateur astronomers looking to explore Bryce Canyon's dark skies with their telescopes, has over the last half dozen years been transformed into a four-day astronomy festival for the general public that draws visitors from all over the world. Hotel rooms in the neighboring towns now book up well in advance of the celebration in the park. Guest speakers range from NASA astronaut Story Musgrave, who was responsible for the first Space Shuttle mission to service the Hubble Space Telescope, to John Dobson, the master of 'sidewalk astronomy' whose cheap and simple Dobsonian telescope design has brought views of the Moon and planets to millions of people around the world.

During the day, we lead visitors on scale model Solar System walks and hold model rocket building workshops that complement the evening constellation shows and a vast field of telescopes arrayed after dark. Rather than being an event

Figure 11.7 A lone hiker descends by dim flashlight into Bryce Canyon's maze of bizarre rock formations. Overhead the Milky Way and Jupiter (the bright "star" just breaking through the clouds) produce a view above just as strange and spectacular as that below (T. Nordgren).

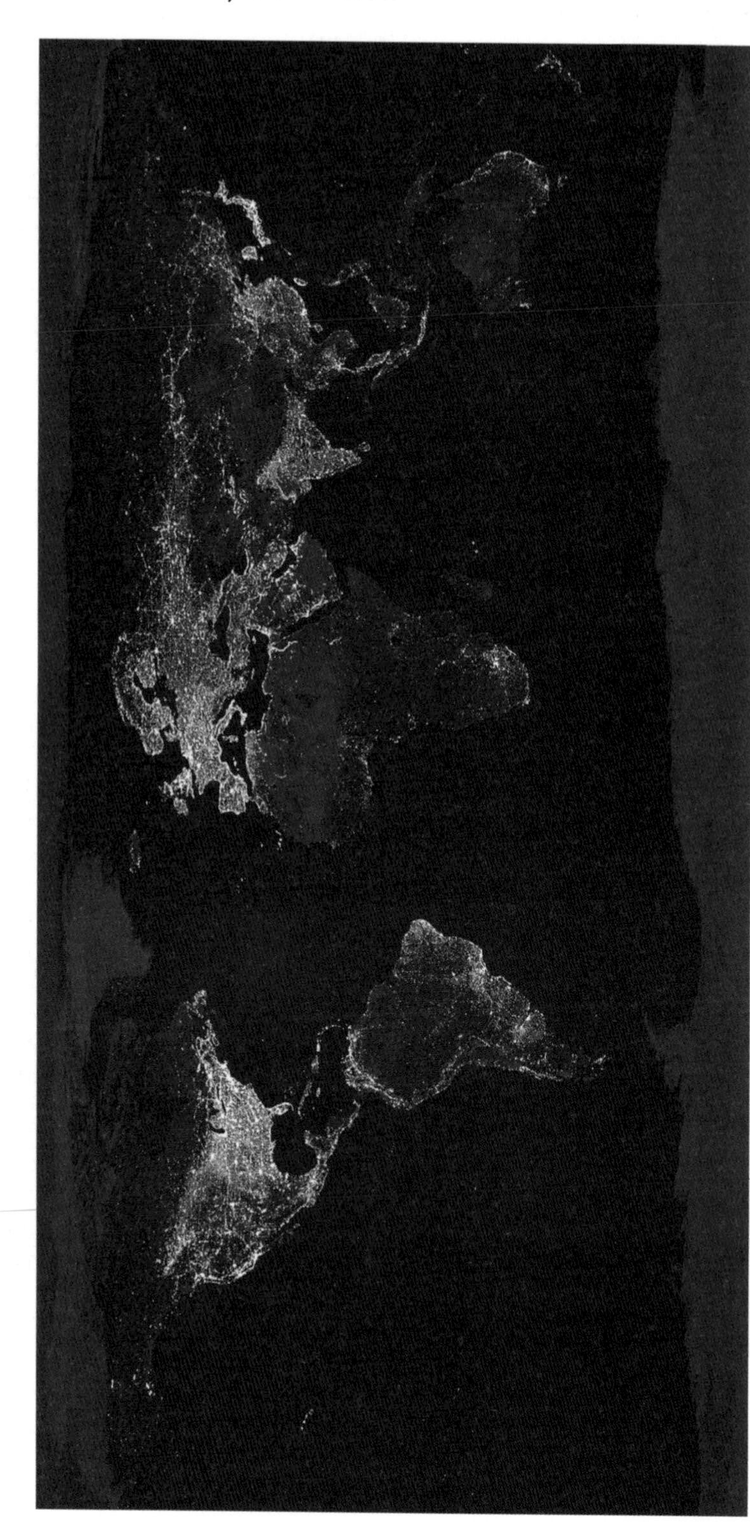

Figure 11.8 Mosaic of the Earth at night created using data from the Defense Meteorological Satellite Program (DMSP) Operational Linescan System (OLS). Designed to view clouds by moonlight, the OLS is used to map the location of permanent lights on Earth. In the U.S., cities along the interstate freeway system create a broad grid of illumination. In Russia, the Trans-Siberian railroad is a narrow tendril of light stretching from Moscow through the center of Asia to Vladivostok. As seen in the Rosetta fly-by image, the Nile River is ablaze with light. Can you find your city, state, or country? ((data) Marc Imhoff of NASA GSFC and Christopher Elvidge of NOAA NGDC; (image) Craig Mayhew and Robert Simmon, NASA/GSFC).

intended for just amateur astronomers, the festival's primary audience is now the public eager to learn more about this strange beautiful sky they see overhead. Only by revealing the pleasures of the night sky to everyone, young and old, astronomy enthusiast or casual stargazer, are we able to demonstrate that the wonders of Bryce Canyon National Park, and indeed, of the entire park system, don't end at sunset and that the beauty the park service is sworn to protect isn't found exclusively below the horizon.

This protection is key to the park service's mission. In its founding charter the National Park Service is mandated by the U.S. Congress "to conserve the scenery, the natural and historic objects and the wild life therein and to provide for the enjoyment of the same in such manner and by such means as will leave them unimpaired for the enjoyment of future generations...." While most people may think that preserving scenery is only a matter of protecting the physical presence of a forest, mountain, or waterfall, grassroots efforts are being made to expand

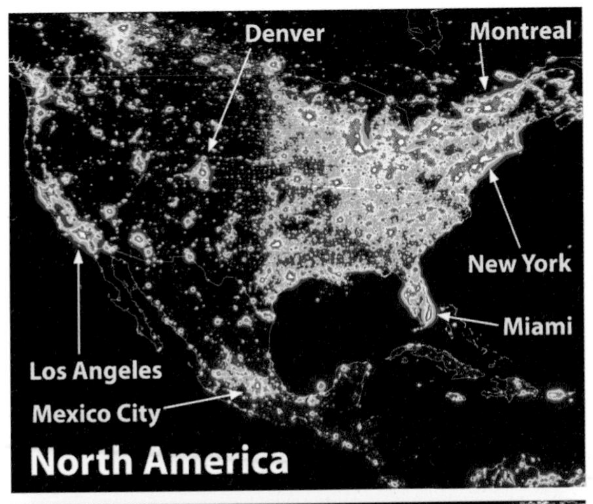

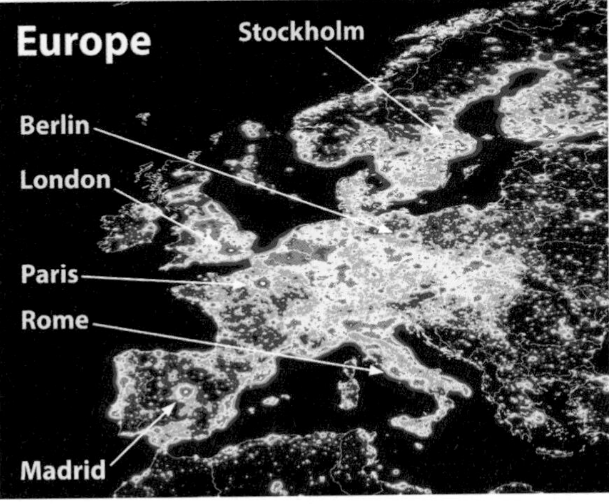

Figure 11.9 Map of the artificial night sky brightness seen at the zenith for locations in North America and Europe. Compare how much brighter skies are in the eastern half of North America compared to the west. Notice, there is no part of Europe as dark as the darkest parts of western North America where the majority of U.S. national parks are located (see the "See for yourself" section at the end of the chapter). France, however, does manage to be darker than neighboring countries (P. Cinzano, F. Falchi (University of Padova), C. D. Elvidge (NOAA National Geophysical Data Center, Boulder). Copyright Royal Astronomical Society. Reproduced from the *Monthly Notices of the Royal Astronomical Society* by permission of Blackwell Science).

this traditional resource definition to also include those qualities that affect how we experience the landscape.

These qualities include *viewscapes,* where atmospheric haze from industrial pollution renders distant mountains and canyon walls invisible on an otherwise clear day; natural *soundscapes,* which are the simple ability to hear the wind in the grass and birds in the trees as opposed to the throbbing roar of commercial airplanes or helicopter tours hovering overhead; and natural *lightscapes,* defined by the park service as, "a place or environment characterized by the natural rhythm of the Sun and Moon cycles, clean air, and of dark nights unperturbed by artificial light. Natural lightscapes, including dark night skies, are not only a resource unto themselves, but are an integral component of countless park experiences."

The man-made artificial light within this definition has come to be called light pollution. While six out of every ten Americans no longer live somewhere we can see the Milky Way, we are at least more fortunate than residents of other industrialized nations in that we have relatively nearby places where we still can. We have our national parks.

Look again at a picture of the planet at night. The light you see, the light that is at that moment illuminating the camera on the underside of a satellite in Earth orbit, is by and large the lights of businesses, homes, streets, and city squares. It's us. Every night when the Sun goes down and you turn on your porch light, and automatic timers turn on street lights, billboards, and flashing advertisements, roughly 50% of that light shines up above the horizon to light up birds, bats, airplanes, and spacecraft (and eventually the Moon, planets and someday distant stars).

If the Earth had no atmosphere, that would be the end of the story. Sadly (depending on how you look at it), the Earth does have an atmosphere. An atmosphere filled with the air we breathe, water droplets in the form of humidity and clouds, and particles of the dust, dirt and noxious compounds we call smog. All of these serve to create an obstacle course through which our upward streaming light reacts. With every encounter between a light ray and 'stuff' the light is reflected or scattered along a new trajectory. With enough stuff in the air, the light that leaves our simple street lamp is repeatedly scattered through the sky overhead until some of it is eventually reflected back down to us. The net result is that near urban areas scattered light reaches our eyes from every part of the sky. It is as if the sky itself is faintly glowing, and in reality it is. We have created an artificial day. This is light pollution.[2]

The more light that shines upwards, the brighter the glow becomes until

[2] Light pollution comes in three main forms, all detrimental. The type described above is called *sky glow*. A second form is *glare,* the light from a bulb that enters your eyes at night, ruining your dark adaptation, and your ability to see otherwise visible stars. The last form is *light trespass,* when other's lights shine onto land or property beyond their border. Each of these reduces the number of stars visible in the sky from even pristine, wilderness locations.

Figure 11.10 City lights from Moab, Utah (population 5,000) light up the sky, blotting out the Milky Way, and permanently lighting up the rock features within Arches National Park. Here the red sandstone of Park Avenue shines with reflected city lights through all hours of the night. Full darkness no longer comes to any of the plant and animal species that make this area their home (T. Nordgren).

eventually it is as bright as the faintest stars we can see and we lose the ability to distinguish them from the surrounding sky. After a thousand years' travel, the light from those distant stars, first generated in the nuclear fires deep within, is overwhelmed and lost in the last third of a millisecond amid the light of the far closer shopping mall that has probably already closed for the night.

The insidious aspect of light pollution is that the shopping mall need not be nearby, or even visible from your location, to render the night sky invisible. The light pollution of every grocery store, gas station and house light in town contributes to a light dome that expands outward from developed areas. The larger the community, or the more wasteful its outdoor lighting, the taller and wider the dome becomes until once dark locations hundreds of miles away begin to lose views of stars back in that direction. Stand outside Lowell Observatory on Mars Hill in Flagstaff, Arizona, and you can see the light dome of Phoenix, 125 miles (200 km) away competing with the glow of the galactic center in Sagittarius, twenty five thousand light-years distant.

The National Park Service Night Sky Team is a small group of Park Rangers and associated astronomers and researchers who for the first time are systematically looking to measure the amount of artificial light intruding into the night sky above the nation's parks. Based at Colorado State University, in Fort Collins, Colorado, they work with personnel at over 70 parks, measuring light pollution levels, helping reduce unintended light within the parks, and helping train Park Rangers in astronomy interpretation programs. While they now work with parks nationwide and have received glowing press from outlets as varied as the *New York Times* and the *Publications of the Astronomical Society of the Pacific*, the Night Sky Team had its origins in the efforts of a few individuals, working in their spare time, who were curious about what they saw happening overhead.

In 1997 Chad Moore was a Park Ranger and physical scientist stationed at Pinnacles National Monument in the mountains southeast of the San Francisco Bay. He'd loved astronomy as a kid, but like many people it had never been more than a casual hobby. His interest was renewed when early that spring he caught sight of Comet Hyakutake high in the cold dark skies of Bozeman, Montana.

Seeing a comet's tail extend over a third of the way across the sky he wondered, "Why did I ever let this go?"

Because of this renewed interest in the sky, he was able to tell that over the next two years (during a time that included a massive building boom in nearby Silicon Valley) the night sky above Pinnacles had gotten noticeably brighter. "As a scientist I realized that if I can see the difference, surely there must be some way to measure the change that was happening." Moore began searching around on the internet for any experts who could tell him what he could do. "What I heard back were variations on, 'Great question! If you find a way, let me know.' After about a dozen of these responses I thought, OK, I guess that's going to have to be me."

Three states away, Angie Richman was a ranger at Chaco Culture National Historical Park in northwestern New Mexico. Richman was finishing a degree in archeoastronomy at the University of New Mexico and was putting that training to use in a park where astronomy has been a part of the local culture for well over a thousand years. One night, she and her colleagues witnessed a rare display of the northern lights just visible over the northern rim of the canyon. "The aurorae were absolutely beautiful," she recalls, "but then we began to notice all the light domes around the horizon from distant towns. At first we even wondered if the aurorae were just lights from the city of Farmington. We were shocked at just how much stray light you could see within Chaco and we wondered if there might be some way to measure how much they were affecting the night sky above the park."

Over the next couple years Moore would connect and collaborate with Richman and other Park Rangers, professional astronomers, and dedicated night

Figure 11.11 Light from the distant cities of Crownpoint and Gallup, New Mexico, illuminate clouds along the horizon beside Fajada Butte in Chaco Culture National Historical Park. The inhabitants of this part of the Colorado Plateau have been observing the sky for thousands of years, but only in the last 100 has artificial illumination begun to alter that view (T. Nordgren).

Figure 11.12 Dan Duriscoe sets up the Night Sky Team's automated camera system on Hillman Peak within Crater Lake National Park during the summer of 2009. The camera is attached to a Celestron telescope mount and tripod and when commanded by a laptop computer automatically photographs the entire sky in a series of 45 exposures (D. Duriscoe).

sky enthusiasts all looking for the same answers. Perhaps the Park Ranger with the most astronomical experience was Dan Duriscoe, at Death Valley National Park. He'd watched the distant casinos in Las Vegas slowly light up the night in one of the most remote parts of the continental United States. Since the early 1990s he'd been looking into different ways to photograph the night sky to quantify the effects of all that light. After he and Moore joined forces in early 2000 Duriscoe came up with a computer controlled digital camera that could photograph the entire night sky in a series of quick exposures. Moore was impressed, "He's really an astronomical MacGyver when it comes to equipment."

Together, Duriscoe and Moore perfected their automated system for photographing the full hemisphere of sky over a particular location. The two developed software routines that from their data systematically calculate the different amounts and sources of light, both natural and artificial. With the help of astronomer Chris Luginbuhl at the U.S. Naval Observatory in Flagstaff, Arizona, they calibrated their data so that the results can be used by professional astronomers, while also allowing the general public to find out what can actually be seen overhead.

Today, members of the Night Sky Team travel the country making measurements above the nation's parks and monuments. The equipment they use consists of a digital charge-coupled device (CCD) camera used by many amateur astronomers which is similar to, but far more sensitive than, the digital cameras wielded by most park visitors. The CCD camera body is mounted to a regular Nikon 50mm f/1.8 camera lens and the entire unit is attached to a computer driven Celestron telescope mount. Using a laptop computer, the rangers run a computer routine that commands the mount to point the camera and photograph the entire night sky in a series of 45 separate 26 × 26 degree fields of view. Each exposure lasts 12 seconds, and in less than twenty five minutes they record the entire visible celestial hemisphere. Moore and Duriscoe

Figure 11.13 Chad Moore beside the Night Sky Team's camera on top of Mount Washington at Great Basin National Park in eastern Nevada (Kate Magargal).

then stitch together the 45 frames to produce a single 'fish-eye' mosaic of the entire sky showing what a typical park visitor would see on a clear night.

Unlike commercial digital cameras that make a full color image by recording light in three colors (red, green and blue), the sensitive CCD camera records only the combined intensity of all light falling on it, thus producing a black and white image (actually a grayscale) of the night sky. By restricting what light reaches the CCD using what astronomers call a Johnson V filter (a filter that only lets through light in the middle of the visible part of the spectrum between wavelengths of about 500 and 600 nm) the Night Sky Team restricts their measurements to only the light most visible to a person's dark-adapted eye while also insuring that the quantities extracted from their data are in a system in widespread use by professional astronomers.

The brightness of the sky's different features are found by first finding the average intensity of light in small regions one degree on a side and spaced regularly across the sky. This method produces 5069 individual measures of the sky's brightness in units of magnitudes per square arcsecond.[3] To better see variations in this sky brightness, they color code these brightness levels using a rainbow color-map where black and purple represent the darkest parts of the sky while red and white are the brightest. The resulting array of colors in each full-sky mosaic shows at a single glance how bright or dark the view would be for an average person and quickly allows comparison across the range of national parks and monuments.

To identify the brightness and extent of artificial light domes, their computer routines look for regions where the sky brightness along the horizon begins to exceed twice the average sky brightness levels in darker, unaffected parts of the mosaic. From repeated observations they can precisely measure and monitor these sources of artificial light noting changes over days, months and years.

[3] Astronomers use the magnitude scale to quantify the brightness of objects in space with increasing numbers counter-intuitively meaning increasing dimness. Arcseconds, on the other hand, are a measure of distance across the sky that is 1/3600th of a degree (for comparison, the full Moon is half a degree in diameter).

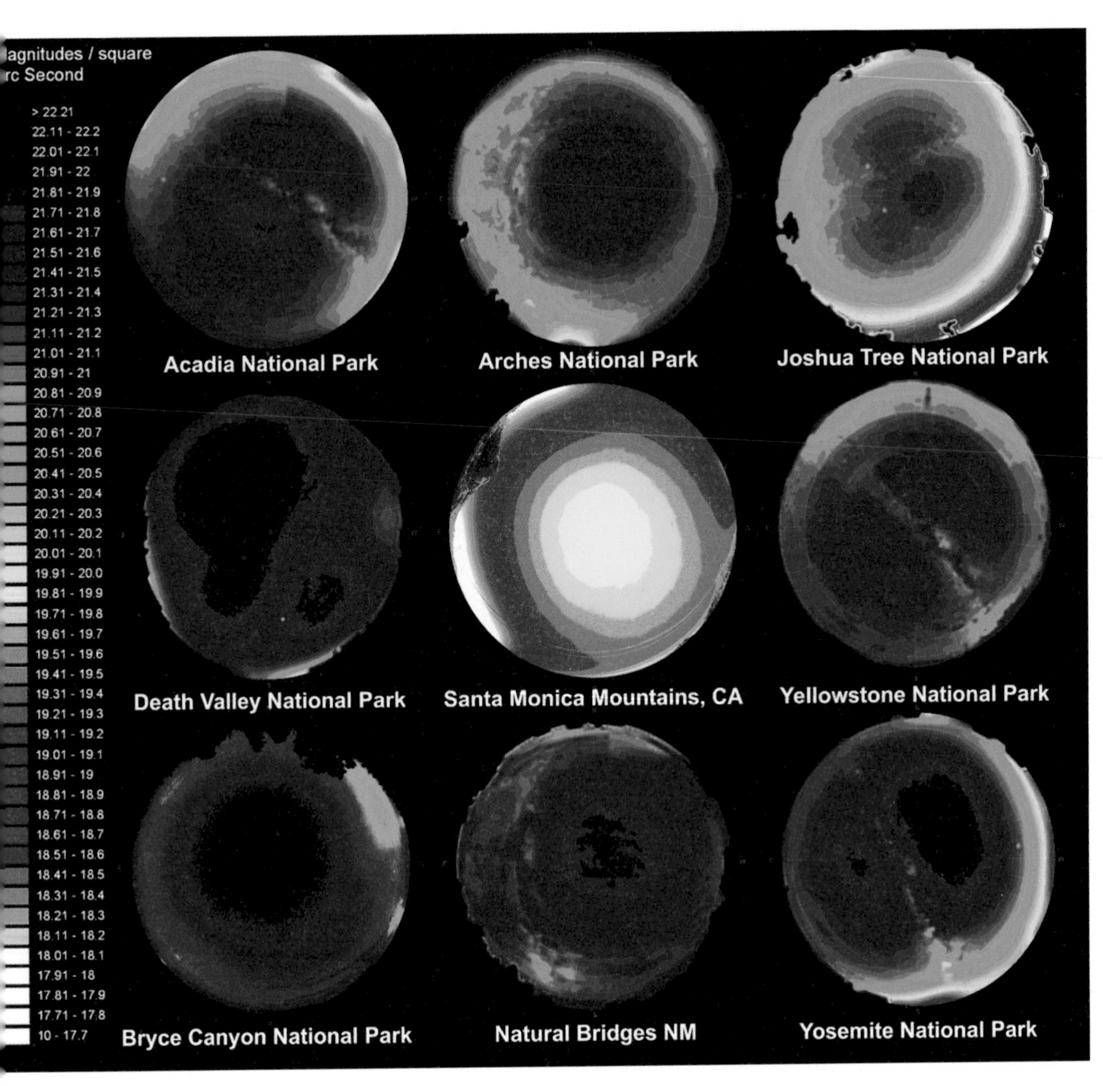

Figure 11.14 Nine all-sky brightness mosaics made by the NPS Night Sky Team (for each, North is up and East is left to match the view you have when looking up at the sky). The central image is of the sky over the Santa Monica Mountains National Recreation Area between Los Angeles and Malibu. This is a fairly typical representation of a suburban sky. The sky above eight other national parks and monuments are shown for comparison. Notice that while each is much darker, for all but Natural Bridges National Monument, artificial light domes from cities and towns are visible along the horizon. Moving clockwise from upper left, the most prominent sources of city lights in each are: Bar Harbor, ME, to north northwest (Acadia NP); Moab, UT, to south (Arches NP); Los Angeles, CA, along the western horizon (Joshua Tree NP); Old Faithful visitor area towards north (Yellowstone NP), Sacramento and San Francisco, CA, to west (Yosemite NP); nothing but natural air-glow (Natural Bridges); St. George, UT, to southwest (Bryce Canyon NP); Las Vegas, NV, to east and Los Angeles, CA, to south (Death Valley NP) (D. Duriscoe, C. Moore, NPS).

Figure 11.15 The Milky Way illuminates Racetrack Playa in Death Valley National Park. The sky and foreground were mosaicked using the NPS Night Sky Team camera (D. Duriscoe and C. Duriscoe).

Duriscoe and his wife, Cindy, have even taken some of the original full-sky mosaic photographs (before brightnesses are averaged and colors are added) to construct spectacular panoramic photographs of the Milky Way over the nation's parks. They are absolutely gorgeous works of art and dramatically illustrate the beauty that is there for anyone to see. In a single picture they capture the reason they have worked so hard to preserve our sky at night.

In addition to making sky measurements, members of the Night Sky Team travel the country working with rangers at local national parks and monuments to help reduce the sources of light pollution within the park. As a member of the team I have witnessed the build-up of anticipation as rangers at one park see the results from other parks and look with anticipation and some trepidation for when Moore and Duriscoe will be visiting them and measuring the quality of their skies.

Of the first several dozen parks the team visited, Natural Bridges National Monument showed the darkest, most remarkable skies. According to Duriscoe, Natural Bridges displayed the only 'Bortle Class 2' sky of any park he had yet seen.

The Bortle classification scheme is a method of rating how dark a location is for naked-eye and telescopic stargazing. Traditionally, most astronomers have spoken of a location's limiting magnitude: the faintest magnitude star that a person can see with the naked-eye. But this is a highly difficult quantity to determine as it depends on the viewer's visual acuity, and how much time and effort is expended to just barely tease out that very faintest star from the background sky. In addition, for those of us who do not normally carry around a table listing the magnitudes of faint stars, we have no way of communicating to others, or to ourselves days, weeks or months later just what the magnitude was of that faint no-name star we think we barely made out while we were camping.

For this reason and others, John E. Bortle at *Sky and Telescope* Magazine used his over 50 years of observing experience to devise a scale that was easier, more practical, and of wider utility to astronomer and novices alike. This is the scale as he described it in *Sky and Telescope* (February 2001). For definitions of specific terms, see the 'See for yourself' section at the end of this chapter.

Class 1: Excellent dark-sky site. The zodiacal light, gegenschein, and zodiacal band (*S&T*: October 2000, page 116) are all visible – the zodiacal light to a striking degree, and the zodiacal band spanning the entire sky. Even with direct vision, the galaxy M33 is an obvious naked-eye object. The Scorpius and Sagittarius region of the Milky Way casts obvious diffuse shadows on the ground. To the unaided eye the limiting magnitude is 7.6 to 8.0 (with effort); the presence of Jupiter or Venus in the sky seems to degrade dark adaptation. Airglow (a very faint, naturally occurring glow most evident within about 15° of the horizon) is readily apparent. With a 12.5-inch (32-centimeter) scope, stars to magnitude 17.5 can be detected with effort, while a 20-inch (50-cm) instrument used with moderate magnification will reach 19th magnitude. If you are observing on a grass-covered field bordered by trees, your telescope, companions, and vehicle are almost totally invisible. This is an observer's Nirvana!

Class 2: Typical truly dark site. Airglow may be weakly apparent along the horizon. M33 is rather easily seen with direct vision. The summer Milky Way is highly structured to the unaided eye, and its brightest parts look like veined marble when viewed with ordinary binoculars. The zodiacal light is still bright enough to cast weak shadows just before dawn and after dusk, and its color can be seen as distinctly yellowish when compared with the blue-white of the Milky Way. Any clouds in the sky are visible only as dark holes or voids in the starry background. You can see your telescope and surroundings only vaguely, except where they project against the sky. Many of the Messier globular clusters are distinct naked-eye objects. The limiting naked-eye magnitude is as faint as 7.1 to 7.5, while a 12.5-inch telescope reaches to magnitude 16 or 17.

Class 3: Rural sky. Some indication of light pollution is evident along the horizon. Clouds may appear faintly illuminated in the brightest parts of the sky near the horizon but are dark overhead. The Milky Way still appears complex, and globular clusters such as M4, M5, M15, and M22 are all distinct naked-eye objects. M33 is easy to see with averted vision. The zodiacal light is striking in spring and autumn (when it extends 60° above the horizon after dusk and before dawn) and its color is at least weakly indicated. Your telescope is vaguely apparent at a distance of 20 or 30 feet (10 meters). The naked-eye limiting magnitude is 6.6 to 7.0, and a 12.5-inch reflector will reach to 16th magnitude.

Class 4: Rural/suburban transition. Fairly obvious light-pollution domes are apparent over population centers in several directions.

Figure 11.16 The zodiacal light is visible beneath the glow of the Milky Way just to the right of Balanced Rock in Arches National Park. The Pleiades is the small bright patch within the yellowish triangular band of interplanetary dust scattering sunlight. If at first you don't see it, hold the picture at arm's length and it will stand out more (T. Nordgren).

The zodiacal light is clearly evident but doesn't even extend halfway to the zenith at the beginning or end of twilight. The Milky Way well above the horizon is still impressive but lacks all but the most obvious structure. M33 is a difficult averted-vision object and is detectable only when at an altitude higher than 50°. Clouds in the direction of light-pollution sources are illuminated but only slightly so, and are still dark overhead. You can make out your telescope rather clearly at a distance. The maximum naked-eye limiting magnitude is 6.1 to 6.5, and a 12.5-inch reflector used with moderate magnification will reveal stars of magnitude 15.5.

Class 5: Suburban sky. Only hints of the zodiacal light are seen on the best spring and autumn nights. The Milky Way is very weak or invisible near the horizon and looks rather washed out overhead. Light sources are evident in most if not all directions. Over most or all of the sky, clouds are quite noticeably brighter than the sky itself. The naked-eye limit is around 5.6 to 6.0, and a 12.5-inch reflector will reach about magnitude 14.5 to 15.

Class 6: Bright suburban sky. No trace of the zodiacal light can be seen, even on the best nights. Any indications of the Milky Way are apparent only toward the zenith. The sky within 35° of the horizon glows grayish white. Clouds anywhere in the sky appear fairly bright. You have no trouble seeing eyepieces and telescope accessories on an observing table. M33 is impossible to see without binoculars, and M31 is only modestly apparent to the unaided eye. The naked-eye limit is about 5.5, and a 12.5-inch telescope used at moderate powers will show stars at magnitude 14.0 to 14.5.

Class 7: Suburban/urban transition. The entire sky background has a vague, grayish white hue. Strong light sources are evident in all directions. The Milky Way is totally invisible or nearly so. M44 or M31 may be glimpsed with the unaided eye but are very indistinct. Clouds are brilliantly lit. Even in moderate-size telescopes, the brightest

Messier objects are pale ghosts of their true selves. The naked-eye limiting magnitude is 5.0 if you really try, and a 12.5-inch reflector will barely reach 14th magnitude.

Class 8: City sky. The sky glows whitish gray or orangish, and you can read newspaper headlines without difficulty. M31 and M44 may be barely glimpsed by an experienced observer on good nights, and only the bright Messier objects are detectable with a modest-size telescope. Some of the stars making up the familiar constellation patterns are difficult to see or are absent entirely. The naked eye can pick out stars down to magnitude 4.5 at best, if you know just where to look, and the stellar limit for a 12.5-inch reflector is little better than magnitude 13.

Class 9: Inner-city sky. The entire sky is brightly lit, even at the zenith. Many stars making up familiar constellation figures are invisible, and dim constellations such as Cancer and Pisces are not seen at all. Aside from perhaps the Pleiades, no Messier objects are visible to the unaided eye. The only celestial objects that really provide pleasing telescopic views are the Moon, the planets, and a few of the brightest star clusters (if you can find them). The naked-eye limiting magnitude is 4.0 or less.

Back home in Redlands, California, 60 miles (100 km) east of Los Angeles, I live under skies that are Bortle Class 8 at best. Just think about the magnitude of wasted light that implies. City lights from Southern California's suburban sprawl are so pervasive that to reach even Bortle class 4 or 5 requires me to drive an additional 90 miles (150 km) away from L.A.

Now consider the map of Utah's light pollution. From even a narrowly confined urban center like Salt Lake City (population 180,000), it's at least 100 miles (160 km) southwest as the crow flies across sparsely inhabited desert and salt flats before one has a hope of reaching unspoiled skies of even Bortle class 2. But to get to such a place requires leaving the interstates. So pervasive is our technology that our light fixtures have followed along behind us on these transportation arteries; billboards advertising fast food and gas stations now light up the night wherever we go. To get away from this glow altogether and see Bortle's "observer's Nirvana" I would need to find the road less traveled, and one way or another I was determined to do it.

Natural Bridges sits in the center of southern Utah's greatest dark, starry expanse.[4] Here, red rock bluffs contrast with pinyon's green, while both rise above meandering sandstone canyons of yellow, buff, and peach sandstone. Within the

[4] According to Duriscoe, the only reason Natural Bridges was Bortle class 2 instead of 1, was because of the presence of a thin, smoky haze from distant fires. This raises an important point I am all too familiar with in my smoggy southern California home: the quality of the night sky depends on both the darkness of your location and the clarity of your air. The more 'stuff' is in your air, the more it will scatter any upward pointing illumination: air pollution only makes light pollution worse.

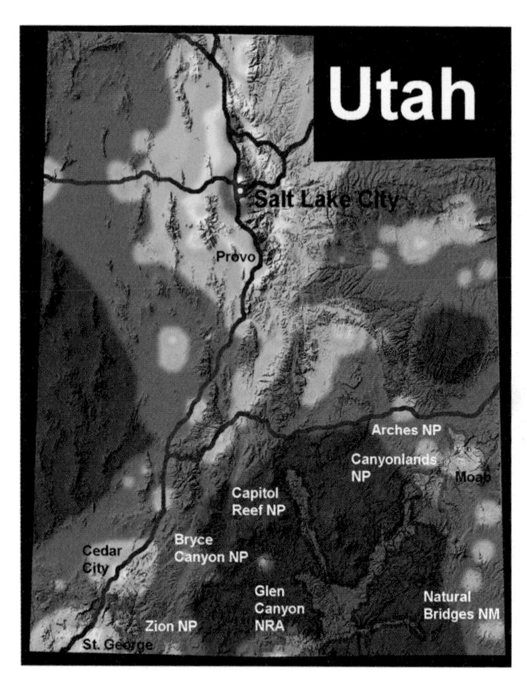

Figure 11.17 Map of Utah's light pollution showing national parks and monument boundaries. Several are located in the large dark region in south-central Utah (T. Nordgren, D. Duriscoe, NPS, after P. Cinzano, F. Falchi and C. D. Elvidge (2000) World Atlas of Artificial Sky Brightness, *Monthly Notices of the Royal Astronomical Society*, 318, 641-657).

canyon's overhanging walls, layers formed from alternating fossil sand dunes create cool alcoves and spectacular vistas. Pine trees grow tall within the canyon's shade and the meandering channel's floor is filled with the bright green of cottonwoods. Reflected in pools of early summer water, the sandstone bridges that give the monument its name stand bright against the pure blue sky.

In April 2007 Natural Bridges National Monument was declared the world's first International Dark-Sky Park based in part on the stunning night sky measurements that Moore, Duriscoe, and Richman had made during their survey of parks. Interstate freeways are absent here in Red Rock country and towns are infrequent. Much of the building boom that swept the rest of the U.S. left this stunningly beautiful area relatively unaffected. From overlooks along the monument's single loop road, a stargazer at night can look out and see hardly any sign of modern civilization. In the all-sky mosaic Duriscoe made from beside the road, the brightest object anywhere is not the light dome of a town along the horizon, but the center of our Milky Way Galaxy overhead.

For two nights I explored the monument's canyons and bridges by the light of a thin crescent Moon. My first night there clouds covered the sky just as the Moon was setting and the sky had begun to get dark. But on the second night, I was rewarded with a sight of my galaxy that I will never forget, as I saw with my own eyes all the dark dust lanes and features that until then I had seen only in photographs. I had never seen a sky so breathtaking and I was jealous of those who lived under such sights every night.

The park staff who call this visual wonderland home live in a small community of pre-fabricated buildings arranged out back of the visitor center. Here, kitchen windows look out onto terra-cotta hills and cool blue-green woodlands of pinyon and juniper. As I walk up to the superintendent's house to talk about her park's dark sky status, I cannot help but marvel at this cool, peaceful oasis compared with my home outside Los Angeles.

Figure 11.18 The center of our Galaxy rises beneath Owachomo Bridge in Natural Bridges National Monument. All of this detail in the Milky Way was visible to the naked eye after an hour's dark adaptation (T. Nordgren).

Sitting at her kitchen table with her dog Wookie, Superintendent Corky Hayes tells me that being designated the first International Dark-Sky Park is a reflection of more than what the park happens to have by virtue of its geography, it is also an acknowledgment of what they have chosen to do with it. At night Park Rangers offer astronomy programs to explain what is special about their dark sky and by day they demonstrate a commitment to preserving those skies by replacing old light fixtures with those that shine only downwards. "We also started to switch over to photovoltaic cells, solar panels, for electricity. These cells are so much quieter than what we had before that at night under the stars the silence is deafening. With the combination of natural nighttime viewscapes and soundscapes you can almost imagine yourself back in time when the first humans settled here nearly ten thousand years ago."

While at the time we spoke it was too early to tell if the new designation had had any appreciable effect on the level of visitation, Hayes did say there had been noticeable changes. "I receive phone calls from people interested about the dark sky designation and we get people now who have decided to stay longer than they would have. Those who might have just driven in and spent only an afternoon instead stop now and spend the night. The night sky is a feature they have added to their plans; I'm hearing a lot more enthusiasm."

The night sky has this effect on people. Two hundred miles (320 km) to the southwest, the town of Flagstaff, Arizona (population 60,000) was declared the first International Dark-Sky City as a result of local efforts to reduce upward shining lights that go back all the way to 1958. With both Lowell Observatory and the U.S. Naval Observatory within and just outside the city limits (as well as

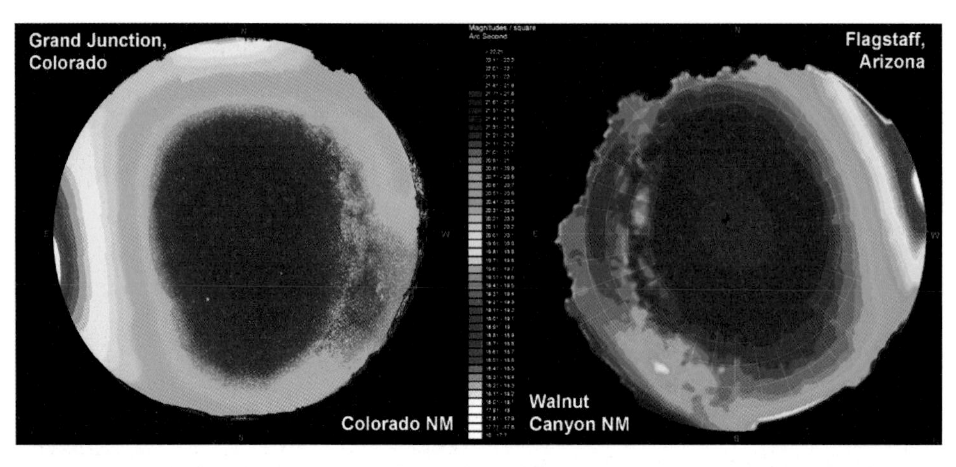

Figure 11.19 The effectiveness of Flagstaff, AZ lighting is revealed in these park service measurements. On the right is a night sky measurement made in Walnut Canyon National Monument immediately outside the Flagstaff city limits (population 60,000). On the left is a night sky measurement made in Colorado National Monument, immediately outside the similar sized city of Grand Junction, CO. Flagstaff has a vigorous lighting ordinance, Grand Junction does not. From Dan Duriscoe's observing notes for Walnut Canyon: ''Sky to zenith and east quite dark, lots of detail in Milky Way, faint stars easy to see. Light domes of Phoenix [125 miles to south] and Flagstaff [10 miles to west, right] separated from here, Flagstaff brighter but not much.... Considering the proximity to town, an excellent sky'' (C. Moore, D. Duriscoe).

the U.S. Geological Survey Astrogeology Branch and Northern Arizona University) there are probably more astronomers per capita here than any other city in the world. Thanks to Flagstaff's city ordinance requiring fully-shielded lighting and limits on the overall luminosity of remaining lights (the first such ordinance in the country) the city produces 23% less light visible in satellite images than that produced by other comparably sized cities. As a result, Lowell Observatory's evening programs that allow visitors to look at the sky through the telescope Percival Lowell first used to map the 'canals' of Mars are a major tourist attraction for the city. And even if you never make it up to the observatory, anyone standing on a downtown street corner of Flagstaff's neon-lit Route 66 can still look up and see the Milky Way arching overhead.

Other communities around the country, looking to preserve their view of the sky for professionals and the public, have followed Flagstaff's lead. Tucson, Arizona, home to the Kitt Peak National Observatory, has had strict lighting ordinances in place since 1972. In 2008 the town of Bar Harbor, Maine (on Mount Desert Island where Acadia National Park is located) passed a lighting ordinance to protect their rare dark skies along the brightly lit eastern seaboard. A year later, communities all over Mount Desert Island hosted their first annual Acadia Night Sky Festival to demonstrate to local businesses that the sky can be an effective tourist draw in addition to the more traditional sea and lobsters.

Perhaps most sweeping of all, the New Mexico Heritage Preservation Alliance

(NMHPA) declared the New Mexico Night Sky one of the state's most endangered cultural resources, prompting state legislators to pass a law protecting this resource for future generations. Peter Lipscomb in Santa Fe, New Mexico, is Director of the Night Sky Program for the NMHPA. He works with businesses and local communities helping educate them on what they can do to meet the requirements of New Mexico's law reducing wasted light and preserving the sky. In addition, he and Angie Richman worked on a state-wide initiative to introduce simple amateur observatories into state parks for the benefit of the local communities. Lipscomb explains, "Our hope is that in addition to the simple enjoyment these observatories bring, local communities will see them as an engine of economic growth by giving visitors to these parks a reason to stop and spend the night." As a predominantly rural state with at least a millennium-long tradition of astronomy among the local pueblo communities, Lipscomb says that "preserving the dark sky is about more than just astronomy, it's about preserving our historical, cultural, artistic, and spiritual identity."

Figure 11.20 Night on Isle Au Haut, part of Acadia National Park in Maine. These are perhaps the darkest skies anywhere on the east coast; distant city lights reflect off the undersides of clouds (T. Nordgren).

But while no state, city, or town with dark-sky lighting ordinances in place has ever reported an increase in crime or vandalism as a result, Lipscomb tells me that business owners with which he works often voice concern and reluctance, "As a result we risk losing our heritage overhead because the amusement park, the warehouse, or the car dealership that has already closed for the night has left its lights on for advertising and 'safety.'" And that is the trade-off we have been told we must make by electric companies, urban planners and architects: safety or the stars. An architect with whom I once worked, designed an astronomy deck for my university, and then surrounded it with thirty high-powered, poorly shielded lamps that made it the brightest spot on campus and blotted out half the stars that once were visible. When I pointed out the problem he asked, "Don't you want your students to be safe? Besides what do astronomers need to see the stars for any more, isn't that why you have the Hubble Space Telescope?"

We all want to be safe. From before the dawn of human history the dark night has been a scary place full of animals far bigger and nastier than we. Any small tribe of early humans who did not have a healthy fear of the dark, and did what they could to light up the darkness with torches and camp fires (the only form of urban lighting for the vast majority of human history) did not last long enough to create future generations of stargazers. Today, in addition to warding off wild things in the night, urban lighting helps us find the door with our keys, it tells us where curbs and pedestrians are when we are driving, and it helps us find that last motel that's open for the night. For all of these situations, however, the place we need illumination is on the ground. Pedestrians are walking by on the street; they are not swooping in from the sky. Potential villains are lurking behind bushes and down alleyways; they are not dropping in from roof tops and third story windows. And customers for businesses are, with very few exceptions, walking or driving by on the street, not flying overhead in personal airplanes. We need light at night, but we need it on the ground.

Consider a bare light bulb pointing upward on a lamp post. Turn it on and about 40% of the bulb's light shines downwards where it illuminates street corners and crosswalks. An additional 10% shines horizontally in all directions and into our eyes to create that other form of light pollution called glare. This light is totally counterproductive as it renders us partly blind to what may exist behind it or in shadows as our eyes are completely overwhelmed.[5] The remaining 50% of the bare light bulb shines upwards above the horizon. This is the light that does nothing but create sky glow and light up spacecraft. But you pay for every kilowatt of electricity needed to produce it, and the climate pays for every ton of coal burned to run the power plant needed to generate that electricity.

Estimates are that a single 100 watt incandescent bulb can consume 750 pounds of coal and emit 2000 pounds of carbon dioxide in one year. Imagine if half the light and half the energy could be redirected towards where it is needed thus requiring only half as many bulbs. Conservative calculations show the U.S. could save $2 billion a year by reducing this wasted lighting.

But even if energy were cheap there would still be reasons to improve how we use lighting. Sure, as an astronomer I'll always

Figure 11.21 Only 40% of the light from a typical decorative street lamp goes to illuminating the sidewalk and street where the public needs light for safety. It is estimated the U.S. wastes $2–5 billion a year through this type of unnecessary light and electricity (T. Nordgren).

[5] Driving at night through Grand Teton National Park, the only time I ever hit an elk was when a truck driving towards me with its high-beam lights on hid the silhouette of a small elk herd in its overwhelming glare. Absent the truck's overwhelming lights, my eyes would have easily picked out their shapes.

want darker skies, but if all that was affected were astronomers and astronomy enthusiasts, this would be a non-issue; there just aren't that many of us in a world with other problems. But light pollution affects more than just us. As just one example, consider how our lighting affects the non-human inhabitants around us. Many animals on this planet have evolved to take advantage of the darkness to migrate, breed, or forage. In the last few decades our city lights have created a sky as bright as the full Moon over much of the industrial world. Nocturnal birds have their daily life cycles disrupted, become visually distracted, and disoriented when migrating. The non-profit Fatal Light Awareness Program (FLAP) in Toronto, Canada estimates that of the 100 million birds killed over North America each year as a result of collisions with man-made structures, artificial lighting at night that blinds and disorients birds is a leading cause.

Outside Palm Beach, on Florida's Atlantic coast, sea turtles who hatch on sandy beaches by the light of a full Moon rising out to sea, possess hereditary training that says go to the light to reach the ocean's safety. Sadly, many are now finding that the brightest light is actually the local resort complex on the other side of the busy interstate. So for those of us who wouldn't dream of bulldozing a forest or draining a wetland we are destroying wildlife habitat just as surely as if we were paving it over with asphalt.

Look again at the map of the world at night. Each and every one of those lights burns for virtually every hour of every night of every year. And each year new lights are added so the maps are already out of date. Every single photon of light you see is wasted light, wasted electricity, wasted money, wasted power generation, wasted resources, and ultimately a wasted planet.

The International Dark-Sky Association (www.darksky.org) is a non-profit organization dedicated to increasing awareness of this problem and providing solutions for everyone from individual homeowners to entire communities looking for ways to address this waste. As it turns out, the solution is trivial: a simple light fixture, or a new shade on an existing light fixture that hangs just far enough below the light source (or any other reflecting surface within the light). The net result is that when looking down from above no part of the light source can be seen. Such a shade is called 'fully shielded,' or a 'full-cutoff' shade, and eliminates the source of light pollution in all of its forms.

In addition, you save more than just the night if you use a shade that's reflective or painted white inside. For this type of shade, nearly all the light that used to go to waste shining upwards, is instead reflected downward where it's actually needed. If, however, you don't need double the intensity downward, then you can replace the old light bulb with one that is half the wattage, saving energy and money. What's more, if you've been using a normal incandescent bulb that produces just as much heat as light, you can replace that bulb with an even lower wattage compact fluorescent bulb that produces almost no wasted heat and so is much more energy efficient to begin with.

Replacing a light bulb is also the perfect opportunity to ask yourself, "How bright do I really need this light to be?" The human eye doesn't work the same way in darkness as it does in light. Our eyes adjust to the brightest light source around. At night when our eyes adapt to a super-bright floodlight, everything outside its reach becomes even darker. Our efforts to create more light actually produce even inkier darkness. Conversely, lower wattage lights produce less glare, less contrast with the darkness, and allow our eyes to better adapt to the night and, counter-intuitively, actually see more.

Lastly, taking the time to install a new bulb or shade is also the time to ask yourself, "When do I need this light?" If safety is your concern, consider putting the light on a motion sensor so that light only appears when you, or a stranger, approach. A light that comes on at a prowler's movement, be it a thief on two legs or an animal on four, is far more effective than a single constant light casting dark shadows all night.

Through the use of fully shielded lights, a lower wattage light bulb, and a motion sensor trigger, you put exactly as much light as you need for safety, exactly where you want it, exactly when you need it, while also saving electricity, money, energy resources, and the night sky. With a simple fix, you achieve a win, win, win, win, win situation.

Such a solution, while simple is not always easy. Go into most big-chain home-stores and you are hard pressed to find any light fixtures that meet this standard (a simple on-line search of one retailer turns up 300 styles of outdoor wall lights, but only four that are fully shielded). But as energy prices continue to climb and people become more 'green' conscious in their purchases, this will likely change. For the time being the International Dark-Sky Association offers a buying guide and list of companies who manufacture these fixtures on their website.

But even if saving a few dollars while saving the environment still isn't reason enough, I am brought back to the architect I worked with who is out there as you

Figure 11.22 Examples of good and bad street lighting. On the left, a light bulb is encased in a typical globe lamp that throws light in all directions except straight down. Stars above are drowned out while strangers beneath are kept in dark shadows. On the right, the same light bulb (dashed circle) is surrounded by a full cut-off shade that allows light only out the bottom. The stars are revealed above, while friends are revealed below (T. Nordgren).

Figure 11.23 Bad and good house lamps illustrate the differences in lighting. On the left, we see harsh glare and upward-shining wasted light. On the right, light is placed exactly where it is needed (Chris Luginbuhl).

Figure 11.24 Light shines upward to illuminate the undersides of a roof on a typical college campus. Unless students are worried about attacks from large bats, this light and the electricity it uses is utter waste (T. Nordgren).

Figure 11.25 The National Park Service has begun retrofitting its light fixtures to be more night sky friendly and energy efficient. At Glacier National Park (left) this lamp shade outside Lake McDonald Lodge comes down below the level of the light bulb. The only light that escapes shines below the horizon. At Lake Yellowstone Hotel (right), park staff designed lamps that mimic historical features, but have the bulb placed at the top of the fixture so the only light they emit shines downward (T. Nordgren).

read this still designing more buildings and outdoor lighting. Why should he care if we can't see the stars? As it turns out, all life on Earth could depend on it. Sixty-five million years ago the dinosaurs and three quarters of all plant and animal species on Earth were wiped out when an asteroid hit the Yucatan Peninsula. Today most new asteroids and comets are found by amateurs and professional astronomers using small telescopes here on Earth. Many of these objects cross paths with the Earth's orbit, and with every year more are found that come dangerously close. While some of the known objects are worrisome, it's the currently unknown ones that are troubling.

When the day comes that one is found on a collision course, what we do about it depends on how far out the asteroid or comet is. Find it when it is far enough away to appear as nothing more than a faint dot or smudge against a dark starry sky, and the options are many. In this case, time and orbital dynamics work to our advantage: deliver a small nudge when there is plenty of time for trajectories to diverge and the crisis is averted. However, if we are unable to see a threatening asteroid until it's close enough to be unmistakable against a bright, light-filled sky, then the options may be none.

This may sound like a far-fetched worry, but the Earth has been hit many times in its past – just look at our next-door neighbor the Moon and the rest of our neighboring planets. In 1994 Jupiter was struck by a chain of small comets, the broken pieces of a cometary near-miss on a previous orbit. Over the course of a week, every telescope on Earth watched as one piece after another plunged into its atmosphere with such force that gargantuan fireballs blasted the cloud tops

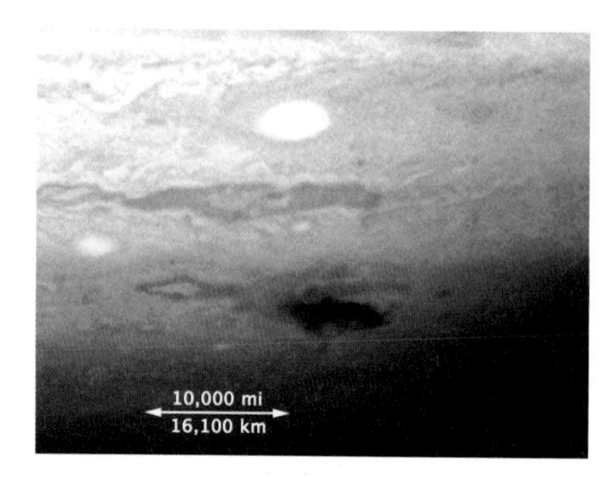

Figure 11.26 After Australian amateur astronomer Anthony Wesley discovered a dark spot on Jupiter's cloud tops, the Hubble Space Telescope was able to follow up with these high-resolution images of dark gasses streaming away from an unseen comet impact site. For comparison, the Earth is only 8,000 miles (12,800 km) in diameter (NASA, ESA, H. Hammel (Space Science Institute, Boulder, Co.), and the Jupiter Impact Team).

with great gaseous stains that lingered in the Jovian skies for months. Each impact's scar was larger than our entire planet. So the question is not a matter of if, but when.[6]

And yes, the Hubble Space Telescope is not affected by light pollution, but there are maybe a hundred professional telescopes on Earth, and many hundreds if not thousands of amateurs with telescopes as well. But there's only one Hubble, and that's because space telescopes are expensive. From design, through construction, to launch and all its servicing missions, the Hubble Space Telescope has cost about $10 billion as of 2009. This is 50 times more expensive than the twin Keck telescopes in Hawaii, the largest telescopes in the world. Space telescopes are an extraordinarily expensive undertaking, one that we astronomers cannot expect the public to fund with their tax dollars forever.

Yet we have been lucky. When a previous NASA administrator was willing to let Hubble die in space because he had decided it was too risky to service with new power packs and better cameras, the public spoke out in unison to the contrary. Because of its awe-inspiring images of everything from nearby Mars to galaxies on the edge of the Universe, Hubble has become the People's Space Telescope. Hubble has contributed something positive to the national psyche.

But how long will that enthusiasm continue, not just for this telescope but for all future astronomical exploration, when the last star winks out in the urban glow and the public no longer has even a passing personal connection to these heavenly wonders? Perhaps the images from space probes and space telescopes will become all the more precious to people as they recognize that only through

[6] A second comet appears to have hit Jupiter in the summer of 2009. This time there was no warning and no one saw the impact. But a lone amateur with a backyard telescope in Australia did happen to photograph the aftermath as a dark stain of gasses spread out across the planet's cloud-tops. Only once news of his discovery spread through the astronomical community could the world's big observatories turn their telescopes towards Jupiter to discover what had happened. The harder we make it for backyard astronomers like this one, the greater the risk to ourselves.

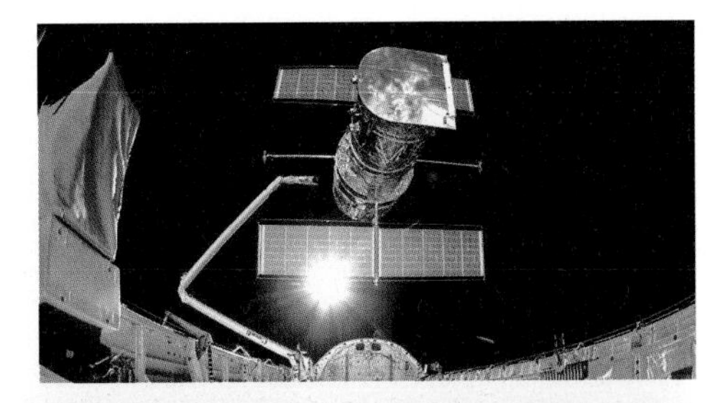

Figure 11.27 The Hubble Space Telescope is deployed from the cargo bay of the Space Shuttle Discovery on April 25, 1990. After nearly 20 years of operation Hubble has become the most famous telescope ever built and earned the name 'The People's Telescope.' (NASA/Smithsonian Institution/ Lockheed Corporation).

Figure 11.28 The Sombrero Galaxy, as revealed by Hubble, is a flat disk of gas and dust illuminated by the intense glow of starlight. Of all the countless Hubble images, this one is my favorite as I can actually imagine falling into this image and gliding across this galactic disk (NASA and The Hubble Heritage Team (STScI/AURA)).

these can they any longer see beyond the Earth's atmosphere. In that future, we astronomers will be their only window to the Universe beyond the Moon and Sun and the results of our research will be all the more precious.

Frankly, I doubt it. The reason is obvious to me every time someone looks at Saturn through my telescope. Everyone from grandchildren to grandparents gasps, shouts, and exclaims in delight when they see Saturn. It never fails, even through a small telescope or pair of binoculars. Every night when people see its rings and moons with their own eyes I hear the half-joking question, "Are you sure you aren't just holding a slide up there?" I know that every one of them has seen a picture of Saturn in far more detail and glory from spacecraft flying by or in orbit, but that in their hearts they recognize there is still something unspeakably special about seeing it with their own eyes. The very fact that people still go to see national parks in person when there are spectacularly beautiful photos available on the internet is how I know that when the last star is

Figure 11.29 The rings of Saturn are there for everyone to see through a simple telescope. I made this picture through my own 8-inch (20-cm) diameter telescope. No matter how many spacecraft may visit this spectacular planet and send back eye-popping pictures, there is no one who isn't thrilled by seeing this view with their very own eyes (T. Nordgren).

hidden behind a bright and featureless night sky, people will never be satisfied with a simple image on a computer screen.[7]

So instead, when the last star fades away behind the ugly orange glow of our own creation, I suspect that public interest in astronomy will simply fade away with it. After all, how do you convince someone to care about a forest wilderness who has never wandered in a meadow, climbed a mountain, or even seen a tree? When that day comes, when the public's interest dies, we take the first step on turning inward as a species. Our exploration of the Universe beyond our atmosphere stalls as we lose the will of those who pay our bills and for whom we therefore work. Astronomy is the single greatest gateway experience for new generations of scientists. Nearly every astronomer I know, can point to a transformative moment as a child, be it a first look through a telescope, a meteor shower, or the sight of the Milky Way on a night spent camping under the stars. With no night sky to fire the imagination of potential young Einsteins or Sagans, where do the new scientists come from? Science stalls. The creative engine that has motivated our understanding of the natural world from the time when our distant ancestors first developed agriculture and ultimately lent its energy to art, literature and music (everything, in fact, that we call the humanities) all of that energy over the last ten thousand years just sputters and stops.

There is a story I heard in graduate school about an astronomer testifying before Congress. Who the astronomer was, or what the occasion I have never learned. Suffice it to say, the congressman was curious why his constituents (you, me, all of us) should have to continue funding some form of astronomical research with our tax dollars. The astronomer answered, "Astronomy is like art; you can easily imagine a world without it, but would you really want to live there?"

With every faint galaxy lost to billboard lights our Universe gets smaller. The great Palomar 200-inch (5-meter) telescope was designed in the 1930s to look to

[7] I asked Story Musgrave, the question the architect asked me; why should we still care about seeing the stars when we have Hubble overhead? He replied, "Why should anyone care about going to Bryce Canyon when we can see it on TV?"

Figure 11.30 In 2008, the light from my 10 million neighbors illuminates the sky beneath Mt. Wilson Observatory in Southern California. For most evenings, the sky is a perpetual orange with little visible beyond our atmosphere save the Moon, and the brighter planets (in this case, Venus on the left and Jupiter on the right). A hundred years ago, Mt. Wilson was the site of the world's most powerful observatory above a tiny southern California hamlet called Los Angeles (Dave Jurasevich).

Figure 11.31 The lights of Southern California as photographed from the Mt. Wilson Observatory in 1908, a hundred years before the photograph in the previous figure. The bright knot of lights in the distance is downtown Los Angeles corresponding to the high-rise towers near the left edge of the 2008 photo. Edwin Hubble discovered the expansion of the Universe with the Mt. Wilson telescope (The Huntington Library).

the very edge of the visible universe. Nestled between the now brilliant cities of Los Angeles and San Diego (mere hamlets in the '30s) it can no longer do so. In the nearly 20 years since I first observed there, I witnessed a tsunami of commercial lights sweep past the mountain leaving it an island awash in a sea of illumination. Given the finite speed of light (and thus the farther away we look in space the longer ago we see it in time) a shrinking visible universe means we lose literal sight of where we all came from; the more light we shed, the less we see.

But light pollution is the one form of pollution that is 100% recoverable. Once the light's cut off, it leaves the sky at the speed of light. In relatively rural areas a single light can affect the view for a dozen miles around so that the actions of a single individual may be all that's necessary to hold the light at bay. Change a shade, change a bulb, change the way we use light and we preserve the stars in the national parks and the last remaining dark-sky oases. If other communities do the same, then we begin to push back the light to the cities where it belongs. The more we do, the more likely we are to get back the night sky of our mothers, and fathers and their parents before them.

But if we do nothing, then when the Milky Way disappears in even remote

Figure 11.32 Moonlight illuminates the high Sierras from Glacier Point in Yosemite National Park. Even as I stop to watch the setting Moon's light slowly rise up the face of Half Dome towards the left, I can see the summer Milky Way rise above me to the east. Towards my right the bright glow of Sagittarius marks the center of our Galaxy while the highest point of the milky band points down the local spiral arm and is the direction our Sun is heading as it orbits the galactic center once every 240 million years. Up until only a couple hundred years ago this is the view everyone had at night for all of human history. Compare this view with that from over Los Angeles to see exactly what we have robbed from ourselves and our children (T. Nordgren).

areas, and we are no longer able to see beyond our own atmosphere, we lose the ability to see that great sign in the sky that we are part of a larger Universe. We lose the glorious light of the galactic center and the cold pale light of the thin tenuous outskirts of our Galaxy; we lose our place in the Universe. As the fainter stars one by one wink out to shopping malls and parking lots we lose the constellations that every culture created to make sense of their world and the forces beyond their control. Up there in the stars are the morality plays that set civilization on its path back before there was writing. When we lose them we lose a direct visible connection to our ancestors and their hopes, dreams, fantasies, and fears. In short, we lose a tangible link to ourselves that gives life meaning beyond the here and now. And we pay for every kilowatt hour that we let it all just slowly slip away.

Are the stars worth saving? Come see for yourself. Come to the parks; bring your children. Come on a night when there is no Moon and the sky is clear and you can see the Universe as every generation of human beings from the earliest hominids nearly three million years ago to your great-grandfather's grandmother saw it just a few dozen decades ago. Come see what they saw; come see what

Figure 11.33 Poster by the author for the International Year of Astronomy 2009 advertising the wonderful views of the Milky Way that may be enjoyed in America's National Parks (T. Nordgren).

you've lost and your children are losing but could have back again exactly as it used to be. Come see the wonder of the cosmos with your own eyes. Come see the clues to who we are and where we come from. Come see the evidence for where we are and where we could go. Come see your home in the Galaxy. Come see the Milky Way.

See for yourself: light pollution

Sky Quality Meter: http://unihedron.com/projects/darksky/

Unihedron is a company that markets a small hand-held device for measuring the average brightness of the night sky (either directly overhead or in a specific direction). It is about the same size as a pack of cards and in only a few seconds will provide a reading of the average brightness overhead in units of magnitude per square arcseconds, just like the NPS Night Sky Team's sky brightness mosaics. Unihedron's SQM can be ordered online for about $100 and they maintain an online database of user's measurements from all over the world. See how your backyard or favorite national park compares.

The United States at night

The following images of the United States are close-up versions of those in Figure 11.9. They show the average sky brightness of the zenith (the point straight overhead) for locations around the U.S. and are made using satellite images of the world at night. They show light intensity levels as different colors. Black and purple represent the darkest locations with red and white the brightest. For ease of identification, interstate freeways are shown by dark blue lines, while national parks, monuments, and recreation areas shown in red. Find your town. How dark are the skies where you live? How far out into the surrounding countryside does your town's light dome extend? How far away from home do you have to go to get to a truly dark location? If the summer Milky Way can dimly be seen overhead in regions of light green or yellow, how far away are you from the nearest view of your own Galaxy? How dark is your nearest national park and what towns are affecting the night sky there?

Numbered on the maps and listed here are many of the national parks and monuments I visited for this book and the average SQM measurement I recorded (in units of magnitudes per square arcsecond). These numbers reflect what I happened to see on those nights I happen to be there. They are, by no means definitive. I have ordered them from the brightest to the darkest:

1. Great Smoky Mountains National Park: 20.60 magnitudes/square arcsecond
2. Yellowstone NP (Mammoth Hot Springs): 20.84
3. Joshua Tree NP (Jumbo Rocks): 20.90
4. Yosemite NP (Yosemite Valley): 21.32
5. Rocky Mountain NP (Bear Lake): 21.36

6. Acadia NP (Otter Point): 21.36
7. Arches NP (Courthouse Butte) : 21.44
8. Glacier NP (Many Glacier Hotel): 21.50
9. Chaco Culture National Historical Park: 21.59
10. Grand Canyon NP (Phantom Ranch): 21.85
11. Natural Bridges National Monument (Owachomo Bridge): 21.85
12. Bryce Canyon NP (Bryce Point): 21.87
13. Big Bend NP (Robbers Roost campsite): 21.90

Astronomical phenomena: zodiacal light

Zodiacal light is sunlight reflected off of tiny dust grains orbiting between the planets. Because the dust is confined to the plane of the Solar System we see it as a band after sunset and before sunrise extending upwards from where the Sun set (or will rise) along the ecliptic (the path the Sun Moon, and planets follow across the background constellations). The zodiac appears at its highest right after sunset and before sunrise during spring and fall. These are therefore the best time to see the zodiacal light as a triangular shaped glow, with a faint yellow, white, or bluish cast that differentiates it from city lights that appears as rounded orange-colored light domes along the horizon.It's normally only seen in Bortle Class 5 sites or better (see chapter for full discussion of the Bortle classification system).

Gegenschein

The gegenschein (or 'counterglow'), like zodiacal light, is sunlight reflected off of interplanetary dust. The gegenschein is different, however, by being a reflection from dust grains 180° away from the Sun as viewed from the Earth. This is incredibly difficult to see unless you are under a truly dark sky (Bortle class 1 or 2) with no Moon, haze or clouds. It is best seen in September-October or February-March when the plane of the ecliptic is high above the dimming effects of the horizon and the point opposite the Sun is free of the glow of the Milky Way. Observations are best around midnight (when the Sun is at its lowest) and the anti-solar point will be along the part of the ecliptic that is due south. The zodiacal band is the narrow glow that connects the more or less oval spot of the gegenschein to the triangular zodiacal light along the horizon.

M33

M33 is a faint spiral galaxy visible to the naked eye only under dark conditions (Bortle class 4 or better, very difficult by class 5). It is most visible in winter and is found near the Andromeda Galaxy (M31). Figure 9.21 in Chapter 9 shows a photograph of the Andromeda constellation. M33 is the faint fuzzy glow that appears just below the star beta (β) Andromedae. If instead of following the dashed line from β And to M31, you instead went the same distance in the opposite direction, you would land precisely on M33.

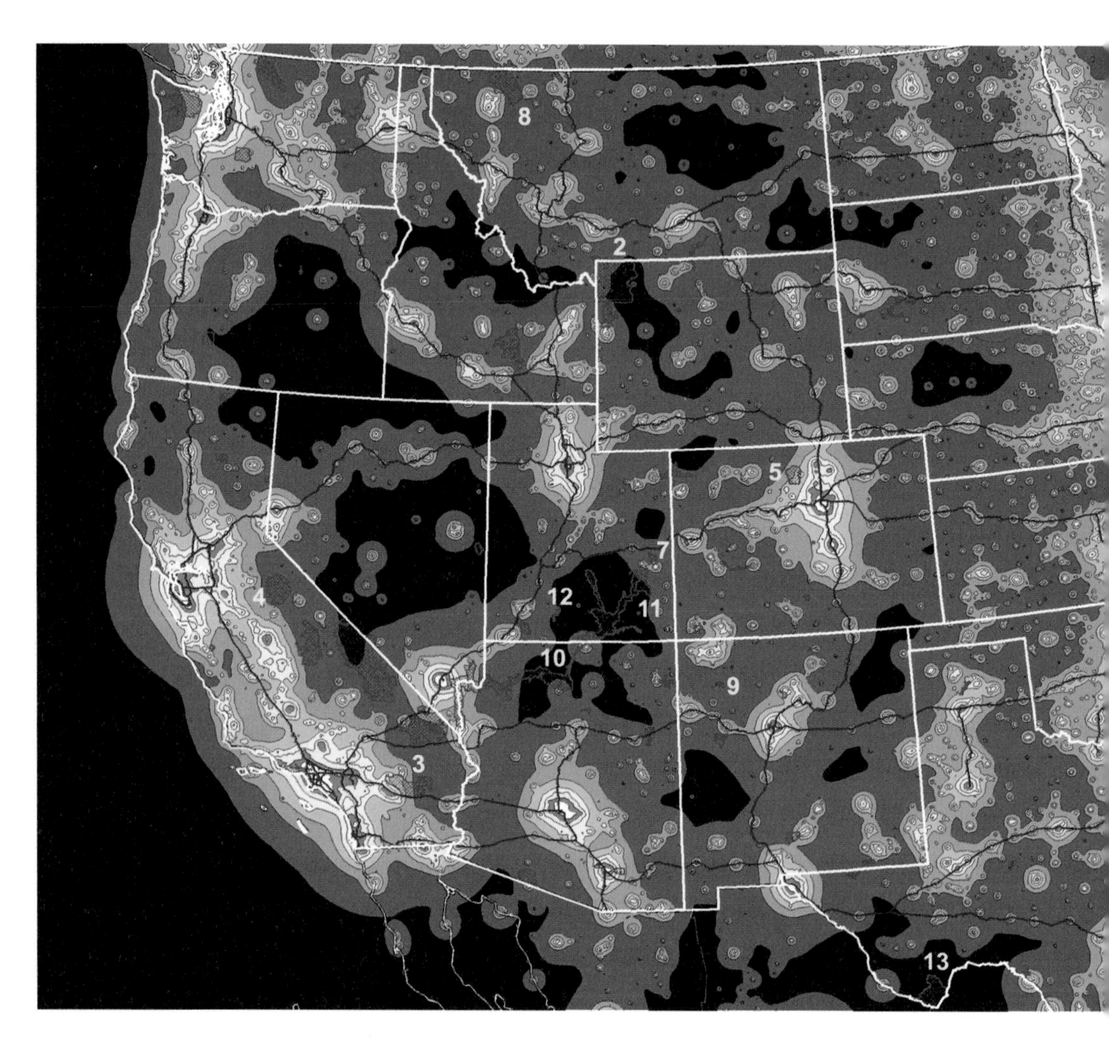

Maps of the artificial night sky brightness seen at the zenith for locations in the United States. National Parks, Monuments, Recreation Areas and other park units are shown in red. Interstate highways are shown in blue. The numbers correspond to parks with Sky Quality Measurements listed here (D. Duriscoe (NPS) after P. Cinzano, F. Falchi

(University of Padova), C. D. Elvidge (NOAA National Geophysical Data Center, Boulder). Copyright Royal Astronomical Society. Reproduced from the *Monthly Notices of the Royal Astronomical Society* by permission of Blackwell Science).

The Globe at Night: http://www.globe.gov/globeatnight/

The Globe at Night is an online opportunity for the public to monitor light pollution levels world-wide. Each year during moonless nights at the end of March, the Globe at Night asks people to go outside an hour after sunset and find the constellation of Orion. By comparing what you see with one of seven different charts (each showing Orion with a different number of stars visible) you can quickly identify the degree of light pollution in your area. By reporting your result online, you help build a world-wide database of current light pollution levels all made within the same period of several days.

Do for yourself

If you have enjoyed any of the 'See for yourself' activities under dark skies, then consider helping us protect those skies in some or all of the following ways:

Enjoy the night

If you are at a park, ask at the visitor center if they have an evening astronomy program. If so, go and take part and let the ranger know it is something you consider an important part of your visit to the park. If you are at home, go out tonight. At either place, let your eyes adjust to the dark and see what's visible. Try one of the 'See for yourself' activities in this book, perhaps one you haven't done before, or one you haven't done from your current location. Is it easier or harder to see some of the things compared to what you remember at a park, your home, or when you were younger? Ultimately we will only protect what we love and cherish.

Cover a flashlight

By covering your flashlight with red cellophane or a red filter, you can prevent it from disrupting your night vision when exploring the world after dark. With a bright flashlight all you see is what's illuminated by the light. With a dim red light, and full dark-adapted eyes, you will see everything else. Small flashlights work better than large ones. Let your eyes adjust to the dark and they'll work better than you think.

Shield your lights

Look around your house at the type of lighting you already have. Consider a new fixture with a full cut-off shield or install a shield on a pre-existing light. Some lights need only be tilted down. A simple plastic flower pot with a hole in the bottom works on an existing floodlight where looks aren't important. The International Dark-Sky Association (IDA) has links on their website for manufacturers that produce full cut-off shields and shielded lights. Look at the lights in your neighborhood and town; what fixtures do they use? How could they be improved?

Use light only when needed

Use motion sensors to turn outside lights on and off when needed. This costs less money, improves security, and reduces light pollution. Use timers for lights that are only needed in the evening or early morning.

Use less light

An efficient, shielded light fixture can use a smaller wattage light bulb and still be effective. Even a 25 or 40 watt incandescent bulb, or a 9 watt compact fluorescent, is enough to light a porch or driveway.

Talk to your neighbors

Share your appreciation of the night sky with your family, friends, neighbors, and community. Share something you've seen thanks to this book. Encourage them to make the night a better place for their communities and nearby parks. Those living near state and national parks are the ones who have the power to decide how much longer parks will be oases of starry-skies amid the growing desert of man-made light.

Build support for a community light ordinance

The IDA has examples on their website of ordinances that have proven effective in other communities like yours. Together we can protect our starry-sky heritage.

Further reading

National Park Service Night Sky Team
http://www.nature.nps.gov/air/lightscapes/

International Dark-Sky Association
http://www.darksky.org/

Atlas of World Light Pollution
http://www.lightpollution.it/dmsp/

The Fatal Light Awareness Program (FLAP) for protecting birds at night
http://www.flap.org/

The Globe at Night
http://www.globe.gov/globeatnight/

Flagstaff Dark Skies Coalition
http://www.flagstaffdarkskies.org/idsc.htm

Sky Quality Meter
http://unihedron.com/projects/darksky/

Bryce Canyon National Park Astronomy Program
http://www.nps.gov/brca/planyourvisit/astronomyprograms.htm

Island Astronomy Institute of Down East Maine (including Acadia National Park)
http://www.islandastro.org/

Astronomy Picture of the Day (APOD): 365 reasons a year for protecting our night sky
http://apod.nasa.gov/apod/

Index

Abenaki, 71, 230, 233 (Fig.)
Abenaki, Great Bear song, 233 (Fig.)
Acadia National Park, 69 (Fig.), 70–3, 88 (Fig.), 89 (Fig.), 98, 230 (Fig.)
  Cadillac Mountain, 89
  map, 77 (Fig.)
  Otter Cliffs, 89 (Fig.)
  Park Loop Road, 88 (Fig.)
  Sand Beach, 89
  sky brightness, 411 (Fig.)
Airglow, 413
Akins, Watie, 233 (Fig.)
Alcor, 232, 233 (Fig.)
Aldebaran, 303, 314, 373
Alluvial fans, 171, 172 (Fig.)
Alpha Centuari, 8
Altair, 10–12, 15, 16, 29, 301 (Fig.)
Anasazi, *see* Chacoans
Andromeda Galaxy (M31), 341 (Fig), 376, 377, 379, 388, 389 (Fig.)
Antares, 5, 28, 301, 314, 315
Apollo 16, 240, 241, 254 (Fig.), 266, 267 (Fig.)
Appalachian Mountains, 229–31, 240, 264, 265, 323
Aquila, the Eagle, 10, 29
Arches National Park, 161, 164–6, 168
  Balanced Rock, 165, 166 (Fig.)
  Courthouse Butte, 162 (Fig.)
  Delicate Arch, 165 (Fig.), 168
  Devils Garden, 166, 167 (Fig.), 171
  Fiery Furnace, 166, 167 (Fig.)
  map, 167 (Fig.)
  Park Avenue, 161, 407 (Fig.)
  sky brightness, 411 (Fig.)
  Salt Valley, 166, 167 (Fig.)
Arcturus, 314
Association of Amateur Variable Star Observers (AAVSO), 90 (Fig.), 102, 104
Asterism, 5
Asteroid, 242, 244
  belt, 242, 244, 256, 260
  Chicxulub Crater, 260
  dinosaur extinction, 259–61
  impact, 252–9
  Iridium, 259
  Near-Earth (NEAs), 244, 260
Astrology, 291
Astrometry, 333
Astronomical Unit (AU), 333
Astronomy clubs, 401
Atmospheres, 197
Atomic spectra, 295
Atomic structure of elements, 293 (Fig.), 294
Atoms, 293 (Fig.), 294
Averted-vision, 8
Aztec Ruins National Monument, 281

Bar Harbor, Maine, 72, 73 (Fig.), 75–7 (Fig.), 88, 89 (Fig.)
Barringer Crater, 255–8 (Fig.)
Bay of Fundy National Park, Canada, 75
Beehive Cluster (M44), 61 (Fig.), 356
Belt of Venus, 2
Betelgeuse, 301, 303, 304 (Fig.), 314
Big Bang, 386, 387 (Fig.)
Big Bend National Park, 23, 24
  map, 24 (Fig.)
Big Dipper, 5, 32, 96, 230, 232, 233 (Fig.)
Binary stars, 91, 93
Binoculars, 28, 99, 102, 144, 145, 184, 222, 266, 267 (Fig.)
Black hole 55, 56, 62
  event horizon, 61
  supermassive, 55, 56, 375, 383
Black smokers, 86 (Fig.), 87
Blackfeet, 190, 191 (Fig.)
Bortle Classification, *see* Light Pollution
Bryce Canyon National Park, 401–3 (Fig.), 405

Bryce Canyon National Park *cont.*
Astronomy Festival, 401 (Fig.), 403, 436
sky brightness, 411 (Fig.)

Callisto, 80, 99, 122 (Fig.), 124
Cancer, the Crab, 356
Canyonlands National Park, 169 (Fig.), 401 (Fig.)
Canyons of the Colorado by Powell, J. W., 156, 186, 373
Capitol Reef National Park, 179
Carbon cycle, 209–11, 218
Carbon dioxide thermostat, 209, 210, 218
Carbon dioxide, 166, 197–200, 202–4, 209–11, 212 (Fig.), 213 (Fig.), 217
Carbonate rocks, 204, 205, 217
Cassini radar, 132, 134 (Fig.)
Cassini spacecraft, 132, 133, 136, 137
Cassiopeia, 99, 15, 32
Celestial equator, 278
Celestial navigation, 325, 326
Cepheid variables, 377, 379, 380
Cetus, the Whale, 90, 102
Chaco Culture National Historical Park, 51, 271, 272, 286, 408
astronomy programs, 306, 307 (Fig.)
Casa Rinconada, 280–2, 286, 301 (Fig.)
diagonal doors, 279, 280 (Fig.)
Fajada Butte, 272 (Fig.), 273, 287, 288, 290
Great Houses, 272, 273, 275, 281
map, 273 (Fig.)
Peñasco Blanco, 291, 308, 311 (Fig.)
observatory, 306, 307 (Fig.)
Piedra del Sol petroglyph, 51 (Fig.), 52
Pueblo Bonito, 272 (Fig.), 274 (Fig.), 276 (Fig.), 279, 280, 287
Sun Dagger petroglyph, 287
Supernova pictograph, 291–3, 306
Chacoans, 271, 273, 274, 289
Chan, Marjorie, 178, 179
Cherokee, 230, 237
Cherokee, Moon tale, 236
Chimney Rock, 285, 286
Chumash Indians, 13–16
Chumash Milk Way story, 14–17,
Climate change models, 218
Climate change, 188, 190, 196, 202, 203, 208, 213, 215, 217, 218
Cold Faithful, 137 (Fig.), 139
Colorado Plateau, 156, 273
Colorado River, 156, 171, 179, 360, 362 (Fig.), 363
Coma Berenices, 23, 26, 32
Coma Star Cluster (Melotte 111), 26, 32
Comet, 242, 244, 248, 338, 372, 399
coma, 249
Hale Bopp, 248 (Fig.), 249 (Fig.)
Halley, 249 (Fig.), 251, 292
nucleus, 249
Shoemaker-Levy 9, 263 (Fig.)
tail, 249
Tempel-Tuttle, 249
Corona Borealis, 32, 231, 232
Cosmic microwave background radiation, 386
Cosmic rays, 305, 307, 309
Crab Nebula (M1), 306, 307 (Fig.), 315
Crater Lake National Park, 239 (Fig.), 409 (Fig.)
Craters, 239, 240, 254–6, 258
Cygnus Void, 12, 15, 16, 29
Cygnus, 10, 16, 26 (Fig.), 27, 29, 32

Dark matter, 28, 62, 63, (Fig.), 375
Dead Horse State Park, UT, 169 (Fig.)
Death Valley National Park, 163 (Fig.), 172 (Fig.), 173, 409, 412 (Fig.)
Eureka Dunes, 173
sky brightness, 411 (Fig.)
Declination, 278
Deep time, 359, 360, 363
Deneb, 10–12, 26 (Fig.), 29
Deuterium, 205, 206
Differentiation, 197, 262
Doppler Effect, 334 (Fig.)
Doppler Shift, 334–6, 380
Drake Equation, 349, 350
Duriscoe, Dan, 409, 412
Dust clouds, interstellar, 12, 13, 18, 19

Earth, 111 (Fig.), 134, 204, 208, 220, 221 (Fig.)
age 243, 244
atmosphere, 210, 211, 220, 382, 399
axial tilt, 208, 228, 229 (Fig.), 232, 234, 263, 277
climate, 234

Earth, *cont.*
  life, 398, 399
  map of lights, 404 (Fig.)
  supercontinents, 264, 265
  temperature variations, 212 (Fig.), 217–19
Eclipse, lunar, 38, 40–4 (Fig.), 63 (Fig.), 67 (Table), 68 (Table), 326, 327 (Fig.)
  Penumbra and umbra, 40, 42
Eclipse, solar, 35, 38, 51, 52, 57, 64, 65 (Fig.), 66 (Fig.), 346
  annular, 51, 64–6 (Fig.)
  of 1919, 59, 60 (Fig.)
  of 2017, 35, 37–9 (Fig.), 42, 51, 62
  Piedra del Sol petroglyph, 51 (Fig.), 52
Einstein, Albert 58, 60, 62
Enceladus, 132 (Fig.), 134–9
  geysers, 136 (Fig.), 137, 139
  Tiger Stripes, 135–8 (Fig.)
Entrada sandstone, 164–6 (Fig.)
Epicycles, 47
Equinox, spring and fall, 277, 287, 288, 311
Eris, 371
Erosion, 164, 173, 255, 361, 362, 363
Escape velocity, 54, 55
Europa, 80–6 (Fig.), 88, 121, 122 (Fig.), 124
  cycloidal cracks, 82 (Fig.), 83
  double ridge cracks, 84, 85 (Fig.)
  ice crust, 82–4
European Enlightenment, 113, 119, 322
Evening star, 222, 223 (Fig.)
Extrasolar planets (Exoplanets), 337
  51 Pegasi, 338, 339 (Fig.), 341 (Fig.), 353
  55 Cancri, 343, 344 (Fig.), 353, 354
  atmospheres, 346, 347 (Fig.)
  Cochran and Hatzes search, 339, 341, 343
  Doppler search method, 335–8
  Epsilon Eridani, 343, 353
  Fomalhaut, 344, 345 (Fig.), 352
  Gamma Cephei, 352
  hot Jupiters, 340–2 (Fig.)
  Marcy and Butler search, 337, 340
  Mayor and Queloz search, 338, 339, 340
  Pollux, 343, 352
  Rho Coronae Borealis, 342, 343 (Fig.), 353
  transits, 346
  Upsilon Andromedae, 341 (Fig.), 344, 353
Extraterrestrial life, 348–50

Fagre, Daniel, 213–17
Fall foliage, 227–9 (Fig.), 231 (Fig.), 232
Flagstaff, Arizona, 148, 149 (Fig.), 151, 152, 156 (Fig.)
  International Dark-Sky City, 417, 418 (Fig.)
Four Corners Region, 160, 168

Galactic center, 26 (Fig.), 27, 55, 56 (Fig.), 62
Galaxies, 95, 96 (Fig.)
Galilean satellites, 79 (Fig.), 80, 99, 113, 120, 144, 145
Galileo (the astronomer), 8, 46, 50, 54, 71, 72, 79, 80, 102, 112, 237, 238 (Fig.), 266, 322
Galileo (the spacecraft), 82–6 (Fig.), 126 (Fig.), 127, 129, 263 (Fig.), 398
Ganymede, 80, 122 (Fig.), 234
Gegenschein, 413, 431
Gemini, the Twins, 7, 352
General Relativity, 58, 59, 60, 61
Geysers, 106, 108, 109, 118
Glacier National Park, 188, 189 (Fig.), 190, 193, 199 (Fig.), 208, 212, 213, 216, 217, 225
  Going-to-the-Sun Road, 191 (Fig.), 194
  Grinnell Formation, 190
  Grinnell Glacier Trail, 190–2 (Fig.), 210, 213, 215 (Fig.), 225, 226 (Fig.)
  Grinnell Glacier, 188, 191–3, 211 (Fig.), 213–215 (Fig.), 225, 226 (Fig.)
  Helena Formation, 204, 210
  Jackson Glacier, 203 (Fig.)
  map, 191 (Fig.)
  Salamander Glacier, 215 (Fig.), 225, 226 (Fig.)
Glaciers, 187–90, 194 (Fig.), 208, 213, 217
  arête, 188, 189 (Fig.), 190, 195 (Fig.)
  cirque, 188–90, 193, 195 (Fig.)
  esker, 195 (Fig.)
  horn, 188, 190
  moraine, 187, 189 (Fig.), 194, 195, 214, 215
  retreat, 212–14, 216, 217, 219
  rock glacier, 195, 196 (Fig.)
  U-shaped valley, 187–9 (Fig.), 191 (Fig.), 193 (Fig.), 195
Globular clusters, 19

Grand Canyon National Park, 156, 253, 258 (Fig.), 358, 359, 362 (Fig.), 367, 387
astronomy programs, 401
cell phone audio tour, 367
evening sky programs, 370 (Fig.)
Lipan Point, 374 (Fig.)
map, 364
Phantom Ranch, 363, 366 (Fig.)
South Kaibab Trail, 362 (Fig.), 363, 365
Grand Staircase Escalante National Monument, 180
Grand Teton National Park, 35, 38, 62, 63 (Fig.), 114
map, 39 (Fig.)
Gravitational lens, 59 (Fig.), 62, 63 (Fig.)
Gravity, 52, 54, 55, 58, 60, 61, 72, 294, 295, 302
Great Smoky Mountains National Park, 227, 229, 234 (Fig.), 240, 252, 258, 264, 265
Abrams Falls, 243
Cades Cove, 241, 243, 265
map, 230 (Fig.)
Greenberg, Rick, 83, 84, 87, 104
Greenhouse effect, 119, 198, 200, 202, 205
Greenhouse gasses, 163, 199, 200, 205, 210, 213 (Fig.), 220

Hamelin Pool Marine Nature Reserve, Western Australia, 211 (Fig.)
Harriot, Thomas, 237, 270
Hawai'i Volcanoes National Park, 107 (Fig.), 129, 130 (Fig.), 207
Hayden Yellowstone Expedition, 114–17 (Fig.), 125, 141
Hematite, 173, 177–9, 180 (Fig.), 190
Hercules, constellation, 19, 32
Herschel, William, 17, 18 (Fig.), 19, 21, 57, 182
Hopi creation story, 357, 358
Hopi, 13, 273, 279, 289
Horoscopes, 4
Hubble Space Telescope, 344, 379, 380, 383, 384 (Fig.), 395, 424
Hubble's Constant, 380
Hubble's Law, 380–2, 386
Hydrogen, 294, 295

Ice-ages, 193, 208, 211, 234
Internal heat, 107, 110
International Dark-Sky Association, 421, 422, 434, 435
International Year of Astronomy (IYA), 237, 429 (Fig.)
Interstellar dust, 295 (Fig.)
Interstellar gas, 294–6, 373
Interstellar medium, 18, 20, 21, 55
Io, 80, 81 (Fig.), 120–2, 124, 125, 129
Loki Patera, 125, 126 (Fig.), 127, 128 (Fig.), 129, 130 (Fig.), 131
volcanoes, 121 (Fig.), 122, 124 (Fig.), 125–8
Iron-oxide, 163, 179
Island Astronomy Institute, 89 (Fig.)

Jackson, William Henry, 116–18, 141
Jefferson, Thomas, 318–20, 322
Joshua Tree National Park, 401
sky brightness, 411 (Fig.)
Jupiter, 12 (Fig.), 79, 81 (Fig.), 99, 100 (Fig.), 101 (Fig.), 111–13, 120, 121, 144, 145

Kargel, Jeff, 170, 194–6, 226
Kelvin temperature scale, 295
Kepler mission, 320, 346, 347 (Fig.)
Kepler, Johannes, 45–50, 322
Kepler's Laws, 49, 50 (Fig.), 51, 53–5, 57, 62, 80, 93, 131, 336
Krupp, E. C., 6, 33
Kuiper Belt, 113, 242, 244, 371

La Tonadora (Scorpion Woman), 14, 16 (Fig.)
Lagoon Nebula (M8), 5 (Fig.), 8, 10 (Fig.), 29, 298, 301 (Fig.), 315, 375 (Fig.)
Latitude, 326
Lava, 105, 107 (Fig.), 129, 240
Lewis and Clark, Corps of Discovery, 319–22, 324, 326, 327
Libration, 80
Light pollution, 2, 32, 89 (Fig.), 329, 404–7, 411 (Fig.), 415, 416 (Fig.), 427 (Fig.), 433–5
Bortle Classification, 412–15
ecological impacts, 407 (Fig.), 421
effective lighting, 422 (Fig.), 423 (Fig.)
Globe at Night, 434, 436
light ordinances, 417–19, 435

light pollution, *cont.*
safety or stars, 419, 420
sky glow, 406, 420 (Fig.)
solutions, 420–2, 427, 434, 435
Light-year, 8, 372
Little Ice-Age, 217
Local Group, 377
Longitude, 326, 327
Louisiana Purchase, 319, 323 (Fig.)
Lowell Observatory, 128, 148, 149 (Fig.), 153 (Fig.), 155, 159, 160, 380
Lowell, Percival, 147, 148, 151, 152, 154, 182
Luiseño, 6
Luiseño Milky Way story, 6
Luminosity, 12
Luna, 234
Lunatic, 234, 235
Lunation, 77
Lyra, the Harp, 10, 16, 29, 93, 102

M33, 377, 413, 431
M81, 378 (Fig.), 379
Magellanic Clouds, 375
Magnetic declination (or variation), 276, 277
Magnetic fields, 83, 84, 202
Magnitude scale, 3, 299
Malville, Kim, 52, 287, 316
Maple trees, 231 (Fig.), 232
Mariner 10, 119
Mars Exploration Rovers, 160, 161 (Fig.), 173, 180
Opportunity, 54 (Fig.), 173–7, 179, 180 (Fig.), 181
Spirit, 54 (Fig.), 163 (Fig.), 173, 174 (Fig.), 175 (Fig.)
Mars, 111 (Fig.), 112, 120, 147–54, 157, 169–75, 180–4, 188, 190 (Fig.), 196, 220
Argyre Basin, 194, 195 (Fig.), 199
atmosphere, 148, 152, 162, 163, 182, 195–7, 200, 202, 232
axial tilt, 232
blueberries, 177 (Fig.), 178, 180 (Fig.), 181
canals, 148, 151, 152 (Fig.), 153 (Fig.), 157, 182
Candor Chasma, 169 (Fig.)
civilization, 148, 156, 182
climate, 188, 232
Columbia Hills, 176 (Fig.)
core, 202
Eagle Crater, 177 (Fig.)
glaciers, 188, 189 (Fig.), 195 (Fig.), 196 (Fig.), 199, 201, 202
Gusev Crater, 173, 174 (Fig.), 176 (Fig.)
Hellas basin, 184, 185 (Fig.), 200 (Fig.), 201 (Fig.)
hoax, 151
iron-oxide, 202
map, 158 (Fig.), 185 (Fig.), 200 (Fig.)
Mariner 4, 159 (Fig.), 160 (Fig.)
Mariner 9, 159, 160 (Fig.), 168
Marsdials, 161, 163 (Fig.), 175 (Fig.), 176 (Fig.), 183
Meridiani Planum, 173, 174 (Fig.), 178 (Fig.), 180 (Fig.), 181
Olympus Mons, 197, 198 (Fig.), 199, 200 (Fig.)
opposition, 149, 150 (Fig.), 160, 183, 184 (Table)
polar ice cap, 147, 184, 185 (Fig.)
seasons, 232
Solis Lacus,˷Lake of the Sun¤, 152 (Fig.), 153 (Fig.)
subsurface ice, 195, 196, 197 (Fig.), 201 (Fig.)
Syrtis Major, 184, 185 (Fig.)
Tharsis Bulge, 160 (Fig.), 185 (Fig.), 200 (Fig.)
Valles Marineris, 160 (Fig.), 169 (Fig.), 189 (Fig.), 199, 200 (Fig.)
volcanoes, 197
water, 163, 168, 170–3, 18–2, 188, 190 (Fig.), 199
Mauna Loa, 107 (Fig.)
Mercury, 57, 61 (Fig.), 111 (Fig.), 112, 119 (fig.)
Mercury's orbital precession, 57, 58
Mesa Verde National Park, 287
MESSENGER spacecraft, 119 (Fig.)
M13, 19, 32
M51, 96, 103
Meteor Crater, 255, 256–8 (Fig.)
Meteor shower, 245–50, 252 (Table), 266
Gemininds, 250, 251 (Fig.)
Great Plains Indians, 246
Leonids, 245 (Fig.), 246, 247 (Fig.), 248, 249, 251

Meteor shower, *cont.*
  Orionids, 251
  Perseids, 247, 250
  quilt, 246, 247 (Fig.), 270
  radiant, 250, 251 (Fig.), 268 (Fig.), 269 (Fig.)
Meteor, 244–6, 247 (Fig.), 248, 250, 369
Meteoroid, 244
Meterorite, 244, 256, 258 (Fig.), 259
Micmac Great Bear star tale, 231, 232
Micmac, 230, 231
Milankovitch cycles, 208, 234
Milky Way, 1, 4–13, 16–20 (Fig.), 21–23, 26 (Fig.), 28, 32, 40, 41, 55, 96, 375, 376, 413
Miwok, 17
Mizar, 232, 233 (Fig.)
Moab, Utah, 161, 164, 182 (Fig.), 407 (Fig.)
Moersch, Jeff, 171
Monument Valley Navajo Tribal Park, 161 (Fig.), 162, 168
Moon, 35–8, 72, 78, 102, 110, 111 (Fig.), 234–9, 241, 252–4, 261, 369
  18.6 year cycle, 283, 284 (Fig.), 287, 288
  craters, 240, 241, 252, 253
  formation, 261–5
  full moon hikes, 235
  major and minor standstill, 284–6, 288
  Man in the Moon, 235, 236 (Fig.), 253
  map, 267 (Fig.)
  maria (orˆseas¤), 102, 239, 240, 241, 253, 254 (Fig.), 266, 267 (Fig.)
  markings, 235, 236 (Fig.), 239, 266
  mountains, 237–9 (Fig.), 240, 266
  orbit, 38, 77, 283, 286, 326
  phases, 36, 38, 235, 266
  recession, 264
  rise and set positions, 283–6
  Smoky Mountains, 240 (Fig.), 241 (Fig.)
  volcanoes, 240, 241
Moore, Chad, 407, 408, 409
Moqui marbles, 178–80
Moran, Thomas, 116, 117 (Fig.), 118, 123 (Fig.), 124, 141
Morning star, 222, 223 (Fig.)
Mt. Saint Helens, 105, 106 (Fig.), 210
Mt. Wilson Observatory, 19, 95 (Fig.), 427 (Fig.)

National Park Service Night Sky Team, 3, 33, 407, 409 (Fig.), 410, 412, 435
National parks (*see individual park entries*), 348, 400, 405, 406
  astronomy programs, 400, 401, 434
  darkest skies, 411 (Fig.), 430
Natural Bridges National Monument, 412, 415–17 (Fig.)
  astronomy programs, 417
  International Dark-Sky Park, 416, 417
  sky brightness, 411 (Fig.)
Navajo greeting, 310
Navajo sandstone, 164–6, 178–80
Neptune, 57, 111–13
Newton, Sir Isaac, 52, 53 (Fig.), 54, 60, 72, 98, 318
North Galactic Pole, 23, 26, 32
North Star, 228, 230, 325, 326
Northern Coal Sack Nebula, 13
Northwest Passage, 321, 328
Nuclear fusion, 296–8, 304, 305

Olbers' Paradox, 383, 384
Oort Cloud, 242, 372
Ophiuchus, 4
Orbital resonances, 121, 136
Orion Nebula (M42), 29, 297 (Fig.), 298, 314, 375 (Fig.)
Orion, the Hunter, 29, 32, 251, 296 (Fig.), 302 (Fig.), 311, 314, 434
Outgassing, 204

Pangaea, 264, 265
Pappalardo, Bob, 83–5
Perihelion, 150
Period of Heavy Bombardment, 253, 341
Periodic table of elements, 293 (Fig.)
Perseus, constellation, 90, 92 (Fig.), 99, 250
Petrified Forest National Park, 153, 154
Phobos, 54 (Fig.)
Phoenix Mars Lander, 196, 197 (Fig.)
Planetary nebula, 303, 304 (Fig.)
Planetary Society, 348, 356
Planetesimals, 331
Pleiades, 298, 302 (Fig.), 314
Pluto, 57, 371, 372 (Fig.)
Polaris (*see also* North Star), 23, 228, 230, 250, 276 (Fig.), 277, 325, 326
Portage Glacier, Alaska, 212

Powell Expedition of the Colorado, 155 (Fig.), 156, 360 (Fig.), 373
Powell, John Wesley, 147, 155, 156, 160, 179, 360
Pulsar planets, 337

Quasars, 382 (Fig.), 383, 386

Radioactive elements, 107, 110, 120, 199, 201
Rathbun, Julie, 128, 129, 131
Raven, 70
Retrograde motion, 47
Richman, Angie, 408
Rigel, 301, 314
Rocky Mountain National Park, 320, 324 (Fig.), 327, 328, 330 (Fig.)
  Alpine Visitor Center, 328
  astronomy programs, 329, 401
  map, 321 (Fig.)
  Trail Ridge Road, 325 (Fig.), 328 (Fig.)
Rocky Mountains, 323, 324, 327, 328
Roosevelt, Theodore, 360, 387, 388 (Fig.)
Rosetta Spacecraft, 399, 400 (Fig.)
Route 66, 255, 257, 259 (Fig.)

Sagan, Carl 69, 294, 370, 398, 399
Sagittarius A* (Sgr A*), 55, 56 (Fig.)
Sagittarius, 4, 5, 7, 8, 15, 19, 26 (Fig.), 27, 29, 32, 55, 301 (Fig.)
Salpeter, Ed, 398
Sapping, 165 (Fig.), 170 (Fig.), 171
Satellites, artificial, 64
Saturn, 111 (Fig.), 113, 131, 136, 137, 139, 145, 370
Science, 36, 46, 62, 82, 83, 87, 218, 291, 363, 389
Scientific method, 50, 53, 58, 195, 232, 322
Scorpius, 4, 5 (Fig.), 28, 301 (Fig.)
Seasons, 227, 228, 229 (Fig.), 232, 261, 263, 277
Sedna, 371
Seeing, 149, 151, 153 (Fig.), 184
Seiching, 75
Shimilaqsha, 14, 15, 16
Shoemaker, Gene, 257
Shooting star, 244, 245
Sirius, 18, 314, 373
Sky Quality Meters (SQM), 430
Sofaer, Anna, 287, 288
Solar corona, 35, 37 (Fig.), 61 (Fig.)
Solar System, 111, 141, 144, 241, 244, 332
  formation, 241, 242, 243, 331
Solstice markers (alignments), 279, 280 (Fig.), 281, 282 (Fig.), 287–91, 311
Solstice, summer and winter, 278 (Fig.), 279, 287, 288, 291, 311
Space-time, 58, 61, 62
Speed of light, 55
Spica, 373
Spitzer Space Telescope, 20 (Fig.), 21, 33
Squyres, Steve, 160, 161, 174, 181, 186
Standard candle, 377, 378
Stars, 297, 298, 299
  age, 299
  brightness (or luminosity), 299, 300 (Fig.)
  color, 301, 311, 314
  distance between, 22
  formation, 295–7
  mass, 299, 302, 303
  parallax, 373
  red giant, 303
  red supergiant, 303
  size, 300 (Fig.)
  spectrum, 335 (Fig.), 336
  temperature, 300 (Fig.), 301
  white dwarf, 303
Stellar evolution, 314, 315
Stellar wobble, 333, 335 (Fig.)
Stromatolites, 210, 211 (Fig.), 382
Summer Triangle, 10, 11 (Fig.), 12, 16 (Fig.), 29, 93, 346, 347 (Fig.)
Sun, 277, 278, 370
Sunrise and set positions, 277, 278, 279
Sunwatching shrine, 279
Superclusters of Galaxies, 380
Supernova, 292, 305, 306, 309, 380 (Fig.)
Supervolcano, 115
Sutcliffe, Ron, 285, 286, 287, 316
Symmetrical geography, 323, 324, 328

Taurus, the Bull, 26 (Fig.), 28, 32, 302 (Fig.), 314
Tectonics, 164, 166, 207, 210
Terraforming, 220, 221
Tidal forces, 72, 79, 80, 84, 91 (Fig.), 94–6, 98, 121, 136, 137, 234, 263, 264
Tidal heating, 82, 83, 87, 121, 122, 138 (Fig.)

Tidal locking, 78 (Fig.), 79, 92 (fig.), 263
Tides, 70–7, 234
  neap, 77
  perigean (astronomical), 77
  pools, 71 (Fig.), 98
  spring, 77
Titan, 132, 133 (Fig.), 134 (Fig.), 145
  lakes, 134, 135 (Fig.)
Tsimshian, 70
Tsimshian tide tale, 70, 71
Twain, Mark, 105, 129, 251
Tycho Crater, 236 (Fig.), 252, 253 (Fig.), 257 (Fig.), 260, 267 (Fig.)

Universe expansion, 385, 386 (Fig.)
Uranus, 57, 111 (Fig.), 113
Ursa Major, the Great Bear, 230, 231, 233 (Fig.)

Variable stars,
  Algol, 90, 92 (Fig.), 99
  Beta Lyrae, 93–5 (Fig.), 102
  Cepheids, *see* Cepheid Variables
  Mira (Omicron Ceti), 89, 90 (Fig.), 91 (Fig.), 102, 103 (Table)
Vega, 10, 11 (Fig.), 29, 93, 102, 373, 374 (Fig.)
Venus, 111 (Fig.), 112, 119, 203, 204 (Fig.), 205 (Fig.), 218, 222
  atmosphere, 203, 206, 208
  phases, 112, 222, 223 (Fig.)
  surface, 205 (Fig.), 206
  transit of 2012, 224 (Fig.), 225 (Table)
  volcanism, 206 (Fig.), 207 (Fig.)
Vikings 1 and 2, 120, 160 (Fig.)
Virgo Cluster of Galaxies, 379
Volcano, 105, 106, 114, 125, 239, 240
  caldera, 239 (Fig.)
  shield, 106, 120, 197
  stratovolcano, 105
  volcanism, 87, 107, 197, 210
Voyager spacecraft, 53, 81 (Fig.), 82
  Voyager 1, 120, 121 (Fig.), 124, 128, 131
  Voyager 2, 131, 135
Vulcan, 57

Walnut Canyon National Monument, 156
War of the Worlds, 157
Whirlpool Galaxy (M51), 96, 103
Wupatki National Monument, 156 (Fig.)

Yellowstone hotspot, 108 (Fig.), 115, 125
Yellowstone National Park, 106–8 (Fig.), 114, 115, 119, 125, 127, 137, 141
  caldera, 114–16 (Fig.), 125, 127, 136
  Castle Geyser, 107, 117 (Fig.)
  Grand Canyon of the Yellowstone, 118, 123 (Fig.), 124
  map, 116 (Fig.)
  Mt. Washburn, 115, 125
  Old Faithful Geyser, 108, 109 (Fig.), 110, 115, 135, 137, 139 (Fig.)
  sky brightness, 411 (Fig.)
  Upper Geyser Basin, 115, 117 (Fig.), 134
Yosemite National Park, 1, 4, 11 (Fig.), 15(Fig.), 17
  astronomy programs, 401
  Glacier Point, 3, 4 (Fig.), 7 (Fig.), 428 (Fig.)
  map, 4 (Fig.)
  sky brightness, 411 (Fig.)

Zodiacal band, 413
Zodiacal light, 413, 414 (Fig.), 431

Printing: Mercedes-Druck, Berlin
Binding: Stein+Lehmann, Berlin